Errata

Schott Series on Glass and Glass Ceramics

H. Bach, F. G. K. Baucke, D. Krause

Electrochemistry of Glasses and Glass Melts, Including Glass Electrodes

ISBN 3-540-58608-3

(1) The page numbers for the index entries should be ofset by +2. Thus, for example, the "accredited laboratories" index entry should refer to p. 228 not p. 226 etc.
The exceptions are index entries listing p. 1 or p. 2.

(2) On p. 141, the curve in part (c) of Fig. 2.35 is missing, and is given below:

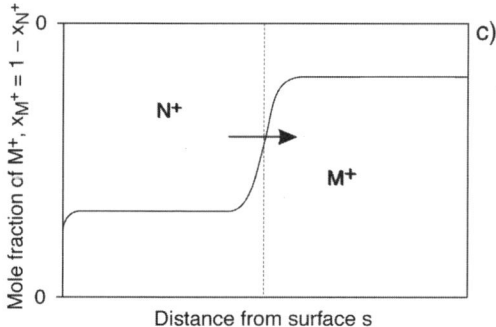

Fig. 2.35. Schematic diagrams of the three types of stable moving boundaries in glass membranes; see text for explicit explanation

Electrochemistry of Glasses and Glass Melts, Including Glass Electrodes

Springer

Berlin
Heidelberg
New York
Barcelona
Hong Kong
London
Milan
Paris
Singapore
Tokyo

Schott Series on Glass and Glass Ceramics
Science, Technology, and Applications

Low Thermal Expansion Glass Ceramics
ISBN 3-540-58598-2

Fibre Optics and Glass Integrated Optics
ISBN 3-540-58595-8

The Properties of Optical Glass
ISBN 3-540-58357-2

Thin Films on Glass
ISBN 3-540-58597-4

Electrochemistry of Glasses and Glass Melts,
Including Glass Electrodes
ISBN 3-540-58608-3

Surface Analysis of Glasses and Glass
Ceramics, and Coatings
ISBN 3-540-58609-1

Analysis of the Composition and Structure
of Glass and Glass Ceramics
ISBN 3-540-58610-5

Hans Bach
Friedrich G.K. Baucke
Dieter Krause

Editors

Electrochemistry of Glasses and Glass Melts, Including Glass Electrodes

With 146 Figures
and 22 Tables

Springer

Editors

Dr. Hans Bach
Dr. Friedrich G.K. Baucke, DSc.
Prof. Dr. Dieter Krause
Schott Glas
Hattenbergstr. 10
55122 Mainz, Germany

Die Deutsche Bibliothek - CIP-Einheitsaufnahme

Electrochemistry of glasses and glass melts, including glass
electrodes/Hans Bach... ed.. - Berlin; Heidelberg; New York;
Barcelona; Hong Kong; London; Milan; Paris; Singapore; Tokyo:
Springer, 2000
   (Schott series on glass and glass ceramics)
   ISBN 3-540-58608-3

ISBN 3-540-58608-3 Springer-Verlag Berlin Heidelberg New York

Springer-Verlag Berlin Heidelberg New York
a member of BertelsmannSpringer Science+Business Media GmbH

© Springer-Verlag Berlin Heidelberg 2001
Printed in Germany

Cover design: Meta Design, Berlin
Typesetting: Computer to film from editors data
Printed on acid-free paper   SPIN 10488292   56/3142PS 5 4 3 2 1 0

Foreword

This book, entitled *Electrochemistry of Glasses and Glass Melts – Including Glass Electrodes*, is one of a series reporting on research and development activities on products and processes conducted by the Schott Group.

The scientifically founded development of new products and technical processes has traditionally been of vital importance to Schott and has always been performed on a scale determined by the prospects for application of our special glasses. Since the reconstruction of the Schott Glaswerke in Mainz, the scale has increased enormously. The range of expert knowledge required could never have been supplied by Schott alone. It is also a tradition in our company to cultivate collaboration with customers, universities, and research institutes. Publications in numerous technical journals, which since 1969 we have edited to a regular schedule as *Forschungsberichte* – "research reports" – describe the results of these cooperations. They contain up-to-date information on various topics for the expert but are not suited as survey material for those whose standpoint is more remote.

This is the point where we would like to place our series, to stimulate the exchange of thoughts, so that we can consider from different points of view the possibilities offered by those incredibly versatile materials, glass and glass ceramics. We would like to share the knowledge won through our research and development at Schott in cooperation with the users of our materials with scientists and engineers, interested customers and friends, and with the employees of our firm.

Though the results documented in the volumes of the Schott Series are of course oriented to the tasks and targets of a company, we believe that readers can nevertheless – or just for that very reason – find demanding challenges for the development of process engineering, the characterization of measurement practice, and for applied research. Besides realizability, the profitability of solutions to customers' problems always plays a decisive role.

The first comprehensive presentation of research findings after the reconstruction of the factory in Mainz was edited by Prof. Dr. Dr. h.c. Erich Schott in 1959. It was entitled *Beiträge zur angewandten Glasforschung* – "contributions to applied glass research" (Wissenschaftliche Verlagsgesellschaft m.b.H., Stuttgart 1959). Since then, there has been an extraordinary worldwide increase in the application of glass and glass ceramic materials. Glass fibres and components manufactured from them for use in lighting and traffic engineering or in telecommunications, high-purity and highly homogeneous glasses for masks and projection lenses in electronics, or glass ceramics with zero expansion in astronomy and in household appliance technology are only some examples. In many of these fields Schott has made essential contributions.

Due to the breadth and complexity of the Schott activities, it takes several volumes to describe the company's research and development results. Otherwise it would be impossible to do full justice to the fundamental research work and technological innovation that is indispensable for product development, and to give an appropriate description of the methods of measurement and analysis needed for the development and manufacture of new products.

Four volumes, entitled *The Properties of Optical Glass, Low Thermal Expansion Glass Ceramics, Thin Films on Glass,* and *Analysis of the Composition and Structure of Glass and Glass Ceramics,* have already been published. Another two volumes, entitled *Surface Analysis of Glasses and Glass Ceramics, and Coatings* and *Fibre Optics and Glass Integrated Optics,* are in preparation and will be published in the next few years. Glasses for various applications in industry and science and their properties are being considered, and melting and processing technologies described.

With the presentation – in part detailed – of the work required for the development of successful products, Schott employees are giving all their interested colleagues working in the field of science and technology an insight into the special experiences and successes in material science, material development, and the application of materials at Schott. Contributions from scientists and engineers who work at universities and other research institutes and who played an essential role in Schott developments complete the survey of what has been achieved and prove the usefulness of the collaborations mentioned above.

In all the volumes of the series the fundamental issues from chemistry, physics, and engineering are dealt with, or at least studies are cited that enable or assist the reader to work his or her own way into the topics treated. Thus, the series may serve to fill gaps between the basic knowledge imparted by textbooks on material science and the product descriptions published by Schott. We see this as the best way to enable all our potential business partners who are not already familiar with glass and glass ceramics to compare these materials with alternatives on a thoroughly scientific basis. We hope that this will lead to intense technical discussions and collaborations on new fields of applications of our materials and products, to our mutual advantage.

Every volume of the Schott Series will begin with a chapter providing a general idea of the current problems, results, and trends relating to the subjects treated. These introductory chapters and the reviews of the basic principles are intended for readers dealing for the first time with the special properties of glass and glass ceramic materials and their surface treatment in engineering, science, and education.

Many of our German clients are accustomed to reading scientific and technical publications in English, and most of our foreign customers are better conversant with English than with German. We therefore decided to publish the Schott Series in English.

The publication of the Schott Series has been substantially supported by Springer-Verlag. We would like to express our special thanks to Dr. H. K. V. Lotsch and Dr. H. J. Kölsch for advice and assistance in this project.

The investment of resources by Schott and its employees to produce the Schott Series is, as already stated, necessary for the interdisciplinary dialogue and collaboration that are traditional at Schott. A model we still find exemplary today of a fruitful dialogue between fundamental research, glass research, and glass manufacture was achieved in the collaboration between Ernst Abbe, Otto Schott, and Carl Zeiss. It resulted in the manufacturing of optical microscopes that realized in practice the maximum theoretically achievable resolution. It was especially such experiences that shaped the formulation of the founding statute of the Carl Zeiss Foundation, and the initiative for the Schott Series is in accord with the commitment expressed in the founding statute "to promote methodical scientific studies".

Mainz, September 2000   Dieter Krause
                                           Vice President R & D (retd.)

Preface

All producers of glass and glass articles must deal with electrochemistry. Its importance for glass melts is quite obvious – they are good electrolytes, and their quality can be controlled by electrochemical means to a large extent. With glass articles, electrochemistry usually does not come to mind immediately. However, it is always involved, and some articles are fabricated because of the electrochemical properties of their glass parts, which make them highly suitable for application as electrochemical tools and sensors. Schott Glas, a special glass producer whose product range also comprises glass electrodes, related electrodes, and connected instrumentation, is involved in the electrochemistry of both glass melts and glasses and was one of the first of a small number of glass companies to maintain a laboratory exclusively devoted to electrochemical research and development. Founded in 1968, this "Electrochemical Laboratory" has launched investigations on some fundamental aspects of the electrochemistry in glass melts and on the functioning of glass electrode membranes, an item where nearly all electrochemical processes occurring at and in solid glasses can be studied.

The main aim of *Electrochemistry of Glasses and Glass Melts – Including Glass Electrodes* is to describe these research efforts and their results in detail. Bearing in mind that electrochemistry is generally perceived as unwieldy, difficult subject matter, we strove to ease access to our presentation by providing explicit, sound background knowledge.

Chapter 1 gives an overview of the electrochemistry as presently applied to glasses and glass melts. We focus on ten particularly interesting fields. One section, for instance, deals with active thin-layer systems, which are increasingly applied in large-scale production. Another topic to be discussed is pH, whose definition, standardization, and measurement are presently undergoing redefinition by IUPAC; the principles of the new *IUPAC Recommendations* are outlined.

Chapter 2 is concerned with solid glasses and their interaction with contacting phases, and in particular with the significance of interfacial equilibria between glass surface groups and ions dissolved in contacting solutions. These equilibria enable the functioning of glass electrodes, but they are also responsible for glass corrosion and for the field-driven transfer of ions from solutions into glasses. The latter mechanism is treated in a special section on glass electrolyses, which, used in combination with other methods (e.g., IBSCA and spectroscopy), yielded valuable results. It led, *inter alia*, to the measurement of concentration-dependent alkali ion and proton mobilities in protonated (proton for alkali ion-exchanged) glasses, and to the discovery of proton traps in these materials. The finding that the thermodynamically correct response of the glass electrode potential on pH (and pM), *in principle*, is subideal (below $2.303RT/F$), may come as a real surprise even to readers familiar with the application of glass electrodes. Further sections deal with the new dissociation mechanism of glass electrodes, with modern pH measurement and, finally, with the glass electrode mechanism in heavy water, which led to a mechanistic explanation of an empirical determination of pD values in D_2O.

Chapter 3 covers the electrochemistry of oxidic glass-forming melts, with special emphasis on their redox state, behaviour under non-isothermal conditions, and conductivities. Oxidic glass melts are characterized by their redox state via the intrinsic redox couple oxygen/"oxide", which is involved in all redox equilibria. We first treat the reduced or base component, "oxide", independent of how it is defined. The concept of optical basicity, although no thermodynamic concept, turns out to have clear advantages over all other basicity concepts developed. *Inter alia*, it allows one to characterize glass melts containing the same polyvalent ion but different alkalis by one number, the optical basicity, to establish electrochemical series of polyvalent ions and electrode potentials, and to estimate redox states of polyvalent ions in glass melts at melt temperature from their standard potentials in aqueous solution at $25\,°C$.

The oxidized component, oxygen, is of direct interest for industrial glass melting as its measurement suffices to characterize the redox state of production melts as long as the reduced component, or basicity, and thus the melt composition, is constant. This knowledge initiated the development of oxygen sensors. We describe their functioning and applicability under various conditions. Subsequently, fining of glass melts, the main process based on defined changes of the temperature-dependent redox state, is discussed in some detail. In addition, sulphate fining is briefly treated because of its worldwide application in the container and plate glass industry. Analytical methods for establishing thermodynamic standard data of redox equilibria (in particular *in situ* voltammetric techniques) are reported, followed by a description of the first accurate measurement of thermoelectric potentials of glass melts ever accomplished and a discussion of non-isothermal glass melts, which are more important for industrial glass melting than generally assumed. Another section explains the long-known bubble formation at the interface oxidic glass melts–zirconium silicate refractory. The chapter concludes with the description of a new cell concept for measuring absolute conductances of glass melts and its application to mixed-alkali glass melts.

In summary, all the information given in this book shows how much successful development and production depends on the sound understanding of physical and chemical processes, careful and consequent interpretation of experimental results, and on the availability of sophisticated tools. Often strong feedback and the willingness to reconsider established concepts are required to enable progress to be made. The literature cited should help the interested reader or teacher to find access to more detailed presentations and explicit discussions.

The book is conceived as a monograph. Its author, Dr. Friedrich G.K. Baucke, DSc., founded and headed the Electrochemical Laboratory until his retirement in 1996 and has held an Honorary Research Visiting Fellowship at the University of Aberdeen since 1982. Two other experts were kind enough to contribute sections on their own specialized work. Professor Dr. John A. Duffy, University of Aberdeen, explains his concept of optical basicity, and Dr. Detlef Köpsel, Schott Glas, offers detailed knowledge about sulphate fining, which is the most frequently applied fining procedure for mass-produced container and plate glasses.

We wish, above all, to thank the authors for their great care and perseverance in writing the book, and for their steady and pleasing cooperation. We also gratefully acknowledge the valuable help received from our colleagues at Schott and from employees of other institutions and industries. Special thanks go to Dr. Hans Bach, Schott Glas, for providing one of the first IBSCA instruments to the Electrochemical Laboratory and for many discussions on the subject of ion ablation, and to Dr. Renate Naumann, Max-Planck-Institut für Polymerforschung, Mainz, for extended, deep discussions about pH measurement. Last but not least, we thank the main coworkers of the Electrochemical Laboratory, Jutta Braun, Brigitte Mohr, Jutta van Harten, Gernot Röth, and Ralf-Dieter Werner, who over the years carried out the experiments whose results are now compiled in this book.

We are indebted to several employees at Springer Verlag, in particular to Christine Tsorpatzidis for copy-editing this volume, and to Peter Straßer as the responsible production editor. We are also grateful to Dr. Hans J. Kölsch for helpful discussions about this volume and for planning the Schott Series in general.

Special thanks, finally, are due to Wiltrud Witan, M.A., Schott Glas, for many suggestions, changes, and corrections regarding the English language and the clarity and fluency of the text and for preparing the computerized final version of the manuscript, and to Karin Langner-Bahmann, Schott Glas, for transforming the large number of often very raw line drawings of the manuscript into print-ready figures with given formats.

September 2000   Hans Bach, Friedrich G.K. Baucke, Dieter Krause

Contents

0. About Terminology

Dieter Krause

Electrochemistry as a special branch of physical chemistry handles the interactions of electrical energy with chemical reactions in an ionic conductor. The main fields are electrical conductivity, electromotive forces in galvanic chains containing solid or liquid electrolytes, polarization phenomena and potentials at electrodes, surfaces, and phase boundaries, or within membranes. Transport properties under external forces such as electric fields, concentration gradients, or external pressure, i.e., the transient behaviour in non-equilibrium situations, are of particular interest.

Like all natural sciences, electrochemistry describes its – on an atomic scale always very complex – processes by introducing suitable physical and/or chemical quantities that allow the formulation of simple expressions or formulae. Scientists working with results from electrochemistry are often educated in "pure" physics, "pure" chemistry, or engineering. Backed by their own tradition and language, their spontaneous understanding of terms such as "oxidation", "reduction", "redox couple" or "redox reaction", "partial pressure", etc., is likely to differ from the definitions given to those terms by glass electrochemists, and this may lead to misconceptions. To make the statements and results easily comprehensible to members of all disciplines, we would like explain the meaning the respective terms take in their specific scientific context.

In this book, electrochemistry is applied to (predominantly oxide) solid or liquid glasses. Ignoring all covalent parts of the bonding, these materials can simply be viewed as consisting of a variety of positive cations and only one type of anion, the doubly charged negative oxygen O^{2-}, which is defined as the intrinsic ground or reference state and briefly called "oxygen".

Because the optical polarizability in the absence of large polarizable cations (e.g. Pb) mainly depends on the (large) oxygen O^{2-} ions, the concept of "optical basicity" allows general statements to be made without specifying the cations (and thus the specific composition) in detail. The possible interaction of an oxide glass with neutral oxygen O_2 (physically dissolved or from the surrounding atmosphere) then results in the formation of (chemically bound) non-bridging singly charged oxygen O^-, which in electrochemical terminology is called the "oxidized form of oxygen", just following the definition of the term "oxidation" as transfer of an electron to a partner ion: $2O^{2-} + O_2 = 4O^-$. The same applies to the complementary process of "reduction of a cation".

Chemists usually describe redox couples by naming the cations, for example Fe^{3+}/Fe^{2+} as an abbreviation of a reaction equation of the type $2(2Fe^{3+}3O^{2-}) = 4(Fe^{2+}O^{2-}) + O_2$, whereas electrochemists name the anions and use the term "oxygen/oxide", meaning O_2/O^{2-}, where "oxide" denotes oxide ions instead of metal oxides.

The formation of free oxygen through redox reactions plays a major role in the fining process of glass manufacturing. The free oxygen has to escape from the melt for high-quality materials either by out-diffusion or by the formation of bubbles. There is always a danger of unintentional bubble formation if the "partial pressure of oxygen" in the melt is equal to or higher than the external pressure (i.e., the sum of atmospheric pressure – usually 1 bar – and the local hydrostatic pressure of the glass melt). In reality, bubble formation is not exclusively governed by oxygen but also by other gases with partial pressures of their own. The concept of "partial pressures" follows the physics of ideal gas mixtures (in an otherwise empty space), ignoring the inevitable nucleation phase of the bubbles. Conventional nucleation theory uses the surface energy (a continuums quantity) as a thermodynamic barrier, resulting in a singularity of the pressure in the limit of zero bubble diameter. Of course, thermodynamic fluctuations of the local composition and density cut this singularity to finite pressure values. But: this illustrative term "partial pressure" is misleading. Gases in a glass melt are not in empty space. The thermodynamic behaviour of components (ions) is governed by "activities" instead of "concentrations", where both of these quantities are proportional to the number of particles per volume, but may differ by orders of magnitude (and depend on concentration). The exchange of matter between two phases is controlled by Nernst's distribution law and the corresponding equilibrium constant of the mass action law. If one of these phases is a gas, in our case the atmosphere above a glass melt, the activity in the gas phase is identical with the concentration which then can be expressed by the pressure of this component. Nernst's distribution law in this special case then transforms into Henry's law. The distribution coefficient contains a factor for the number of oxygen (more general gas) molecules per volume of the melt, which for convenience is (unfortunately) called "partial pressure" also inside the melt. Thus the "concentration" or "partial pressure" (of oxygen or a gas) in the melt is not measured directly; it is always calculated from the comparison with a gaseous medium as reference in an electrochemical probe or sensor. This term describes the tendency for bubble formation relative to the external pressure. The experimental determinations of "partial gas pressures" for bubble formation give values ranging from well above 1 bar to approximately 3 bars.

Measurements of that type always determine "potential differences" between corresponding electrodes of a probe as the primary quantities. Speaking of "potential" in that context without naming the reference of zero potential is another usual but imprecise term. The "standard" zero potential for aqueous solutions is that of a hydrogen electrode at any temperature; for oxidic glass melts, ZrO_2, which is O^{2-} ion-conducting at high temperatures, is used in combination with a $Pt,O_2(1\,bar)$ electrode as standard reference because it can be directly immersed into the melt. Thus, a ZrO_2 electrode has the scheme $Pt,O_2(1\,bar)|ZrO_2|\dots$ or, used with air, $Pt,O_2(0.21\,bar)|ZrO_2|\dots$. For chalcogenide or other glass types other references would be convenient.

These preliminary remarks hopefully contribute to a pleasant and profitable reading of this unique book.

1. Overview

Friedrich G.K. Baucke

Electrochemistry is concerned with chemical and physico-chemical processes involving electrically charged particles, ions, and ionic groups. Examples for such processes are transport by diffusion and field-driven migration and homogeneous and heterogeneous ionic reactions causing an exchange of electric charges between different phases. Electrochemistry is mainly an interdisciplinary science and is involved in many fields of fundamental and applied research, for instance in energy conversion and storage, corrosion research, biochemistry, and medicine.

The electrochemistry of glasses and glass melts is best introduced by an electrochemical characterization of glasses. They consist of an irregular, rigid, atomic network of, for example, silicon or silicon and aluminium and oxygen. Part of the oxygens, the so-called non-bridging oxygens, bear a formal charge of minus one as in $O_{3/2}SiO^-$ (briefly $\equiv SiO^-$) or share a negative charge with other oxygens as in $(O_{3/2}AlO)^-$ (briefly $\equiv AlO^-$), where it is then divided between the four oxygens of the AlO_4 tetrahedron because the charge (plus three) and coordination number (four) of the aluminium are not equal. Due to this structure, the network contains locally fixed anionic groups whose negative charge is compensated by the positive charge of cations present in the network, of which mainly alkali ions have a small albeit finite mobility. Glasses are thus solid electrolytes with positive charge carriers. Glass melts are additionally subject to thermal motion of the network at high temperatures, which causes breaking and making of network bonds and, in principle, mobilizes also parts of the network. In addition, the high temperatures cause a vast increase of all chemical and electrochemical reaction rates.

Despite this simple structure, glasses undergo various electrochemical reactions. They arise from the different driving forces which can be applied, and from the "answers" of the cations and the anionic sites. Thus, the *mobile cations* follow concentration gradients, which cause intra- and interphasic ionic interdiffusion, react to potential gradients, which set up ionic migration, and respond to temperature gradients by establishing thermoelectric voltages (Seebeck effect) and thermoelectric diffusion (Soret effect). The *non-bridging oxygens* are distinguished by a more or less basic character because of their tendency to donate negative charge to neighbouring cations and cause a composition-dependent glass basicity. The tendency to donate charge actually being a tendency to donate electrons, the non-bridging oxygens represent also the reduced form of the redox couple oxygen/"oxide". It is inseparably connected to the glass and is involved in all redox reactions, for instance with polyvalent ion couples, and exerts a reducing (or oxidizing) power of glass melts on contacting metals and refractories. In addition, anionic groups located at the interface between glass and a solution are involved in heterogeneous equilibria with protons and/or alkali ions, which result in interfacial potentials (glass electrode effect) and subsurface interdiffusion (glass corrosion).

Over time, certain fields of electrochemical glass research attained special interest. Some fields are still the focus of attention, whereas others lost their attractiveness after a while, and this makes an overview of the electrochemistry of glasses and glass melts a snapshot characterizing the particular needs and interests at the time of its draft. In addition, its contents inevitably reflect the viewpoints and interests of the author and is also limited by the available space, as is the number of references given. They must be restricted to a few essential publications, which, however, do not represent a personal preference of the author. Despite these limitations, it is the author's aim to give an unbiased overview of the electrochemistry of glasses and glass melts by presenting those fields of electrochemical glass research and application which, according to the literature, are most frequently treated at present. Specific work done at the Electrochemical Laboratory of Schott Glas and reported in detail in Chaps. 2 and 3 will be appropriately referred to in this overview.

1.1 Solid Glasses

1.1.1 Glass Electrodes

The glass electrode demonstrates ideally the application of the electrochemistry of glasses, as its functioning is based on electrochemical processes between glass and contacting solutions. It was the first of the many sensors developed and represents a large business today. Estimated worldwide sales were \$1.55 billion in 1995 (after \$0.89 billion in 1988), of which the European market shared \$0.65 billion (after \$0.38 billion in 1988), both still with an upward trend [1.1]. These numbers apply to glass, reference, and single rod electrodes, excluding instrumentation such as pH meters, titrators, and processors. Among the many fields of application are the chemical industry, environmental and pollution control, medicine, biology, and life sciences. Detailed information about sensors is given in a comprehensive handbook [1.2] and a series of monographs [1.3].

A Unique Development

The glass electrode has experienced a unique development. Soon after the detection of the "glass electrode effect" (acidity-dependent potential) [1.4], the development of practically useful glass electrodes and of the understanding of their functioning have gone different ways. For constructing practical sensors and finding useful membrane glasses it sufficed to know *how glass electrodes operate*. This knowledge soon resulted from experimental work, which was fortuitously supported by the simultaneous development of electronics supplying instruments able to measure emfs of high-resistance electrochemical cells. Quite differently, the understanding of *why membrane glasses function* has been delayed for nearly seven decades by the lack of sufficiently sensitive and depth-resolving surface-analytical methods and thus of information about the ionic processes at the membrane glass surface.

Practical Design

The very first form of glass electrode, consisting of a large, thin-walled membrane glass sphere containing an internal solution and a "connecting" platinum wire and melted to a thick-walled, highly resistant shaft of the same Thuringian glass [1.5], has been maintained although all components have experienced drastic improvements. Special membrane glasses allow the use of small-size, shock-insensitive membranes with various shapes, which can be applied at pH values between 0 and 14 and temperatures above 100 °C and, in special cases, allow sterilization at 135 °C. Electrode shafts are made from chemically inert, high-resistance glasses and consist of two concentric tubes enclosing a grounded metal foil or, with combination electrodes, the reference electrode solution, both screening the internal buffer and reference electrode electrically. The internal solution has been replaced by a buffer with high buffer value and the connecting Pt wire by an internal reference electrode, usually of the second kind, with highly stable and reproducible potential. The arrangement, which sometimes contains a resistance thermometer, has high mechanical stability due to a firmly fixed electrode head that either carries an attached screened connecting cable or is equipped with a screw-on connecting plug. Similarly, reference electrodes have been improved. Combination electrodes combining glass and reference electrode have long been applied in laboratories and are increasingly introduced to industrial purposes.

Modifications and Alternatives

Interesting modifications are a redox-sensitive glass electrode with an electron-conducting membrane glass [1.6] and a replacement pH electrode containing a (tube-shaped) zirconia instead of a glass membrane for extreme pH values [1.7] and up to supercritical temperatures above 374 °C. The ion-sensitive field effect transistor (ISFET) [1.8] and its modifications were developed as alternatives to glass electrodes for special applications. They use the interfacial potential between the solution and a membrane, which controls a current across a semiconductor gate as the indicating means. ISFETs are attractive because of their robustness, small size, and possible multi-sensing application. The original claim, however, that they operate without a reference electrode, turned out to be incorrect.

Frequent attempts to replace the internal buffer solution of glass electrodes by a solid contact [1.9–11] generally failed, the main reason being the extreme isothermal intersection points of cells containing such glass electrodes. This is also valid for the electrochemically soundest proposal of intercallation compounds such as Prussian blue [1.12]. Micro- and ultramicro-glass electrodes with diameters down to 1 μm are mainly applied for biological, physiological, and medical purposes. They are usually fabricated by the researcher, who also operates them by means of micro-manipulators.

Origin of Electrode Response

The origin of the glass membrane potential has interested scientists since its detection, and many physical phenomena have been proposed as its cause. Theories, however, were doomed to remain hypotheses as long as the ionic processes at the membrane surface remained unknown due to the lack of appropriate experimental methods. This concerns also the ion exchange theory proposed in 1937 [1.13], which is based on the thermodynamics of an assumed exchange of alkali ions of the glass for hydrogen ions of the solution. It can, *in principle*, neither confirm the underlying reaction nor deduce any other mechanism of potential formation. Nevertheless, it remained the most long-lived theory of the glass electrode because scientists concluded from the well-behaved thermodynamic Nicolsky equation that the assumed ion exchange was the underlying physical reality. Surprisingly, it lived on after the first contradictory experimental results had been obtained. Thus, Boksay's first subsurface concentration profile [1.14] should already have cast doubts on ion exchange as the origin of the membrane potential, but this contradiction was not seen. Meanwhile, more experimental information and especially the systematic application of IBSCA in combination with other methods have elucidated the reaction that is responsible for the pH- and pM-dependent potential of the membrane glass [1.15], see Chap. 2, particularly Sect. 2.9.

Present knowledge is that the membrane potential is an effect of the interfacial equilibrium of glass surface groups with ions in the solution. It involves negatively charged, anionic groups, which represent a charge density at the glass surface and constitute the potential of the membrane glass, whose dependence on the ion activity is thus obvious. The interfacial equilibrium satisfies the basic physical requirement for any potential formation, which is a charge separation between the phases, here glass and solution, as recently also stated for other ISEs [1.16]. It is also responsible for a number of secondary reactions, for instance subsurface interdiffusion, which is the onset of the various kinds of glass corrosion.

The complex of all interdependent reactions, termed *dissociation mechanism*, was verified experimentally in light and heavy water, whereas the ion exchange theory was experimentally disproved, see Sect. 2.9. The thermodynamic treatment of the interfacial equilibrium revealed a basically subideal response of pH (and pM) glass electrodes, which explains their well-known sub-Nernstian response. It also gives detailed information about the equilibrium, yielding the dependence of the membrane potential on the activity of the charged surface groups ("internal slope") and the dependence of their activity on pH (and pM). The dissociation mechanism is also responsible for the hydrogen gas-dependent potential of the Pt,H_2 electrode on protonated glasses, glass(H)|Pt,H_2, see Sect. 2.8.

1.1.2 Glass Electrolysis

Glass electrolysis was carried out rather early [1.17] but has not been widely applied although it is of considerable industrial and scientific interest. Two kinds of electrolyses must be distinguished. Either the glass network acts as a stable matrix, in which the mobile cations of the glass are shifted and replaced by other cations, or it is unstable and undergoes additional structural and/or chemical changes. Which of these processes is observed depends on the stability of the network relative to the strength of the electric field that must be applied to transfer and shift the cations, and thus also on the applied electrical contact. Silicate glasses generally show field-driven cation drifts within a stable network, whereas borate glasses frequently exhibit network changes in addition to ion drifts.

Electrodes and Electrical Contacts

Mainly the anodic contact determines the course of the electrolysis. It causes not only the transfer of cations into and their migration in the network, but can also extract and discharge anions and thus change the glass structure if this is energetically possible. Cathodic contacts, in contrast, merely accept and/or discharge cations leaving the glass and only rarely, if at all, change the network. Table 1.1 gives a rough overview. Non-blocking electrolytic and oxidizable metallic contacts cause field-driven transfer of sterically unhindered cations into and their migration within a stable glass network. Blocking electrodes are inert metals, which neither offer nor accept cations from the glass although, according to *Ernsberger*, true ion blocking can be realized only with difficulty, if at all [1.18]. Pseudo-blocking anodic contacts are electrolytes or oxidizable metals offering cations to the glass, which are not transferred into

Table 1.1. Anodic electrical contacts and electrodes for glass electrolyses

Application to stable glass networks	Application to unstable glass networks

• Non-blocking contacts
  – oxidizable metals – sterically unhindered
    transfer of generated ions
  – electrolytes (salts, salt melts, solutions)
    sterically unhindered transfer of ions
• Blocking electrodes
  – inert metals (platinum)
• Pseudo-blocking contacts
  – oxidizable metals – transfer of generated
    ions is sterically hindered
  – electrolytes (salt melts, solutions)
    transfer of ions is sterically hindered
	• (Pseudo-)blocking electrodes
	  – inert metals (platinum)
	  – oxidizable metals

the network because of their large radius (sterical hindrance). Blocking or oxidizable pseudo-blocking metallic electrodes applied to unstable glasses transfer and shift cations *and* cause structural and compositional changes of the glass network. However, this is only a rough classification omitting various additional effects and overlapping electrolytic reactions.

Technical Applications

Glasses are mainly applied as solid electrolytes (see Table 1.1, "Non-blocking contacts") as indicated by the following examples.

Sodium-sulphur batteries make use of the formation energy of sodium polysulphide, which is stored during the electrolytical decomposition of the compound (battery charging) and released during its formation (discharging) [1.19]. The half-cells are separated either by β''-alumina or by a sodium-containing glass, which serve as sodium-conducting solid electrolytes. Special sodium silicate, aluminosilicate, and borate glasses have been developed for this application [1.20]. They must resist liquid sodium and sulphur under electrolytic working conditions at temperatures above 350 °C, the minimum operating temperature of the cells. The high resistance of the glass can be overcome by large cell constants, which, however, require the application of large numbers of thin-walled glass tubes, whose fabrication presents a technical problem. Sodium-sulphur batteries are of interest for automobile traction and stationary energy storing because of the abundance and low price of the reactive elements. The declining number of publications, however, suggests a decrease of research in this field despite the high significance of energy storage.

Lithium-containing glasses are developed for application in rechargeable solid-state, preferably micro-batteries, which are frequently constructed in thin-film technology [1.21]. Different from sodium-sulphur batteries, the electrode reactions involve a transfer of lithium ions from solid electrodes (lithium metal, lithium compounds) into the glass electrolyte (if liquid electrolytes are absent), which, however, poses serious problems, even at low temperatures.

Planar waveguides for coupling and branching light conducted by light pipes are frequently fabricated by electrolysis [1.22]. For this purpose, for instance a sodium-containing substrate glass with a stripe of a modified glass with a larger refractive index at its surface is subjected to an electrolysis employing an anodic molten sodium salt. The ionic migration during this electrolysis shifts the glass stripe into deeper glass regions so that a completely "buried" light-conducting path is generated, whose distance from the surface is determined by the electrolysis time.

Glass Electrolyses for Scientific Purposes

The understanding of most electrochemical processes described in Chap. 2 has been possible by the application of glass electrolysis. A special technique is the combination of electrolysis with the non-electrochemical surface-analytical method IBSCA, see Sects. 2.4 and 2.6.6, which constitutes a modification of the well-known moving boundary method as applied in solutions. This "modified moving boundary method" (m.m.b.m) gives even more detailed information about ionic migration than the original technique because it yields the concentration dependence of the moving boundary instead of indicating only its position as in solutions. Thus the transfer number of cations across the glass–solution interface was determined, sterical effects of ions were detected, concentration-dependent mobilities of lithium ions and electrolytically introduced protons were measured, see Sect. 2.6, the mechanism of the glass electrode response was elucidated, see Sect. 2.9, and diffusion potentials in leached subsurface layers were determined, see Sect. 2.7.4. In addition, glass electrodes with electrolytically protonated membranes disproved the ion exchange theory of glass electrodes, Sect. 2.9. The absence of a mixed-alkali effect and the existence of proton traps (see Sect. 2.6.7) in protonated glasses, both detected by glass electrolyses, are of interest for the further development of the "dynamic structure model" for ion migration in glasses [1.23].

Electrolysis of borate glasses with blocking electrodes causes a change in the anodic glass network in addition to the migration of alkali ions, see Sect. 2.6.8. The most surprising result of this work was the generation of an anodic glass layer completely freed of its sodium oxide content (35 mol%) and the observation that the "empty" glass could be "refilled" with Na_2O by a reverse electrolysis. This effect awaits further investigation.

1.1.3 Ionic Transport: Conductivity

Although it has been known for 125 years that glasses are ionic conductors, the process of ionic conduction is still not fully understood. This holds true for a.c. conductivity, which is tightly connected with the study of relaxation phenomena, as well as for d.c. conductivity, and it is also evident from the discussion of the mixed-alkali effect in Sect. 1.1.4. Such a gap in knowledge has no parallel in other fields of glass science and is particularly surprising in view of the great number of publications and conferences on the subject. It has been explained by (a) the extremely fast-growing amount of experimental data, (b) the difficult treatment of irreversible processes in tremendously complex non-crystalline materials, and (c) the different and often diverse educational and research backgrounds of the scientists involved [1.24]. This publication was written as a possible common basis for future experimental and theoretical work.

Empirical results

A.C. Conduction. The most recent and perhaps the most comprehensive experimental result in a.c. conduction is the measurement of complete conductivity spectra from near-d.c. conductivity to the teraohm range of multicomponent glassy materials and oxidic melts at various temperatures [1.25]. The existence of a high-frequency plateau of the hopping conductivity was verified by subtracting the vibrational far-infrared part (which is proportional to the square of the frequency) from the experimental spectrum. The spectra of all glasses and melts measured showed low-frequency scaling. In addition, scaling properties of conductivity spectra were thoroughly treated recently by *Roling* [1.26].

C.C. Conduction. Electrolysis, which represents long-time d.c. conduction of glasses with constant composition, has been treated in some detail in Sect. 1.1.2., where electrolyte solutions served as well-defined, ideal nonblocking contacts. Now we focus on the dependence of the d.c. conductivity on the composition of single glasses or, more exactly, the modifier (alkali) content of glasses. The state of understanding this dependence is obvious by considering the following well-known experimental facts [1.27]. (a) The d.c. conductance σ of single modified glasses exhibits a power law dependence on the modifier content c_+,

$$\sigma \propto c_+^{\gamma} . \tag{1.1}$$

(b) The activation energy of d.c. conductivity depends logarithmically on the glass modifier content,

$$E_{\mathrm{a}} \propto -\ln c_+ . \tag{1.2}$$

(c) The exponent γ in (1.1) is proportional to the negative inverse temperature at constant modifier content c_+,

$$\gamma \propto -\frac{E_a}{k_B \ln c_+} \frac{1}{T} \ , \tag{1.3}$$

because it is also known that $\sigma \propto \exp(-E_a/k_B T)$. It is very surprising to somebody not active in this field that none of these three experimentally well-established, apparently simple, relationships can be thoroughly explained on the basis of glass composition and glass structure [1.27].

Models Describing the Empirical Results

Several models have been developed to explain d.c. and a.c. conductivity phenomena in glasses. After the "classical" Anderson–Stuart model [1.28] had been found to show serious drawbacks, two more advanced models were proposed, the *jump relaxation model* (JRM) by *Funke* [1.29] and the *dynamic structure model* (DSM) by *Bunde* [1.30], which were subsequently combined to give the *unified site relaxation model* (USRM) [1.31]. As a further development of these approaches, *Funke* introduced the *concept of mismatch and relaxation* (CMR) [1.32], which provides a general basis for understanding conductivity spectra on the basis of structural dynamics. However, this concept has also been criticized recently [1.33].

According to the CMR, the glass network contains two different sites into which an alkali ion A^+ can jump, "good" sites $\bar{A}$ that "fit" and comfort the ion after the jump, and "bad" sites $\bar{B}$ that can only be occupied if the alkali ion "spends" extra energy, the "mismatch" energy. The sites are not rigid. An ion, after a jump from an $\bar{A}$ site into a $\bar{B}$ site, where it has a higher energy, has the chance either to fall back into its original $\bar{A}$ site or to rearrange the potential distribution ("coulombic cage") of the $\bar{B}$ site ("to dig its own hole"), which is determined by the surrounding ions, so that it becomes the new equilibrium position of the ion. In the first case, the jump does not, in the second case, it does contribute to the a.c. conductivity of the glass. The probability of the jump to be "successful" or "unsuccessful" depends in a different way on the time the ion remains at the new $\bar{B}$ site. The CMR thus explains the dependence of the a.c. conductance of the glass on the measuring frequency. Application of appropriate jump and probability data has yielded quantitative agreement with experimental results. The mutual effect of neighbouring ions on their coulombic potential distributions shows that the CMR actually takes into account that ion migration (and diffusion) is a complicated concerted event involving all alkali ions and sites of the glassy material.

1.1.4 Mixed-Alkali Effect

The mixed-alkali effect (MAE), also called the mixed mobile ion effect, the poly-alkali effect and, in Russian papers, the neutralization effect, has caused a great number of experimental investigations and publications in glass science. It is exhibited by glasses containing a constant sum and various ratios of molar concentrations of two alkali ions. If any property of such glasses, as a function of the mole fraction x of the alkali ions, is distinguished by an extreme value, the glass series is said to show the MAE, at least according to its present definition. The extreme can deviate from the "ideal" value of the assumed linear function by several orders of magnitude.

The MAE is found with various glass properties, for instance with density, refractive index, molar volume, hardness, thermal expansion, elastic modulus, dielectric constant, and gas permeability, and is largest with properties connected with ion movement, for example electric conduction and activation energy, which are of prime interest here, and diffusion coefficient, internal friction, dielectric loss, and viscosity. Its occurrence is independent of the network composition; it is also found with alkaline-earth ions, in oxidic glass and salt melts, with porcelains and crystalline materials.

Understanding the MAE

Although already known in 1883 [1.34], detected for conductivity in 1925 [1.35], and a continuous subject of research ever since, the MAE is still not fully understood. Of the eight theories already put forward by 1976 [1.36] none explains the MAE for all properties concerned. The continuous lack of understanding is evident from four papers published at rather different times, two reviews, in 1969 [1.37] and 1976 [1.36], and two research papers, in 1989 [1.38] and 1998 [1.27], which emphasize unanimously the necessity of obtaining more experimental information despite the large amount of research already done on this phenomenon.

The recently proposed "unified site relaxation model" (USRM) is promising [1.31]. It yields, *inter alia*, the conclusion that the MAE is not a direct consequence of mixing different cations, for instance A^+ and B^+ in a glass, as has always been assumed, but rather results from mixing A^+ ions with $\bar{B}$ sites and B^+ ions with $\bar{A}$ sites. Glasses containing only one site are thus not expected to exhibit the MAE. This conclusion is verified by several ion-exchanged glasses [1.39–41] and by protonated lithium silicate glasses [1.42]. They contain only sites created by lithium ions during cooling of the glass melt and can be occupied by both lithium ions and protons without sterical hindrance. Protonated glasses do not show the two typical features of the MAE, a conductivity minimum and a cross-over of proton and lithium ion mobility curves, see Fig. 2.31.

The Basic Problem

Despite the various theories and approaches to explain the MAE, the basic problem is obviously not that the MAE, but that the properties causing the MAE, are not sufficiently understood, and this deficiency results from a lack of detailed experimental information, see Sect. 1.1.3. An example is the conductivity of glasses. The only information about the ion transport in mixed-alkali glass gained from conductance measurements is the mole fraction-dependent *effective* or *overall mobility* of the mobile ions,

$$u_{\text{overall}}(x_i) = \frac{\sigma(x_i)}{z \, F \, c^0} \, , \tag{1.4}$$

where $c^0 = (c_i + c_j) = (x_i + x_j)c^0$ is the total molar concentration of alkali ions I and J, x is the mole fraction and z is the ionic charge, here equal to unity. It consists of the contributions of the mole fraction-dependent *average mobilities* $\bar{u}$ of the ions,

$$u_{\text{overall}}(x_i) = x_i \bar{u}_i(x_i) + x_j \bar{u}_i(x_j) \, , \tag{1.5}$$

which are unknown unless they are obtained by a special technique, for instance the modified moving boundary method (m.m.b.m.) [1.42], see Sects. 2.4 and 2.6. But even if the average mobilities of the ions are known, the mole fraction-dependent *individual mobilities* u forming them,

$$\bar{u}_i(x_i) = \frac{1}{x_i} \int\limits_{x=0}^{x=x_i} u_i(x)\mathrm{d}x \, , \tag{1.6}$$

remain obscured. The most direct information about the ionic transport in glasses is thus not obtained from conductance measurements and additional techniques.

Nevertheless, the average mobility can yield significant information about the individual mobility. This was proved by proton traps in protonated lithium silicate glasses [1.42], see Sect. 2.6.7. They are a consequence of a special structural unit of the glass network, the mixed twin siloxy group $(\equiv \text{SiO}^- \cdots \text{HOSi} \equiv)^- \text{Li}^+$, which exchanges its lithium ion for migrating lithiums with much higher probability than the proton. Trapped protons thus have a much smaller *individual mobility* than protons of other structural configurations, see Sect. 2.6.7.

The lack of detailed experimental knowledge about the ionic transport is also obvious from the existence of the MAE of conductance and activation energy in glass melts [1.38, 43, 44], see Sect. 3.9.3 and in particular Figs. 3.65 and 3.66. It can only be understood if glasses retain structural entities that are responsible for the MAE below T_g, as also concluded by a theoretical treatment of the experimental data by *Pfeiffer* [1.45]. Theories of the MAE based on the structure of solid glasses, consequently, must include this close relationship between glasses and melts.

It should also be rediscussed whether extreme values, by which the MAE is presently *defined*, really characterize the underlying physico-chemical phenomena. "Non-idealities" [1.38], as defined in Sect. 3.9.3, seem to support an alternative treatment.

Future attempts at explaining the MAE must thus strive for better understanding of the glass properties causing it, instead of developing new models, and the obvious way to reach this goal is to gather more detailed information by applying more sophisticated and perhaps new experimental methods.

1.1.5 Glass Corrosion, Glass Durability

Glasses are usually regarded as chemically inert materials, as indicated by the frequently used term "glass durability". But glasses do corrode, and their corrosion, as that of metals, is an electrochemical process. While metal corrosion is an electrode reaction involving an electron exchange that leads to an oxidation of the metal, corrosion of glasses, which already consist of oxides, is an electrolytic reaction between the glass and a gas, a liquid, or a solid that results in subsurface layers of "transformed glass" as corrosion products.

Mechanism

The mechanism of glass corrosion is treated in detail in Sect. 2.7 and in [1.46]. Briefly, the basic process between glasses and solutions leading to corrosion is the formation of an interfacial equilibrium between glass surface groups and dissolved cations. It results in a coverage of the glass surface with protons (for instance as silanol groups) and/or alkali ions (as siloxy alkali groups), besides a minute concentration of negatively charged groups. Contrary to the generally assumed inert glass surface, through which ions exchange simply according to their relative free energies in the two phases, the "electrochemically structured" glass surface, not being in equilibrium with the underlying glass, sets the quantitative conditions for subsurface interdiffusion. Depending on the kind of ion attached to the surface groups, the alkali ions of the glass either exchange for protons ("glass leaching") or alkali ions from the surface ("alkali/alkali exchange") or for both. In addition, the solution attacks the glass network, which results in its dissolution, so that, after some time, a steady state of the transformed surface layer develops ("incongruent dissolution"). The formation period and thickness of this steady-state subsurface layer depend strongly on the solution and the glass composition. The formation period of the layer lasts from days to months so that the steady state is not always observed; the layer thickness is from the nanometre to the hundreds of micrometre range and frequently even leads to peeling of the surface layer.

Unlike the glass–solution interaction, glass corrosion in humid atmospheres principally modifies the appearance because interfacial equilibria, removal of exchanged material by dissolution, and a steady state of subsurface layers are excluded. Instead, layers of exchanged material form on the surface, the exchange rate being determined by various types of interdiffusion between atmosphere, layer, and glass.

The corrosion mechanism has been investigated by chemical surface profiling, see Sect. 2.7 and [1.46]. Further work particularly concerns the action of water in, and structure changes of, transformed surface layers, for which a variety of modern surface-analytical techniques are applied; see the thorough discussions of advanced surface analysis by *Bach* [1.47, 48].

Weathering

Special conditions can override the described corrosion mechanism in solutions. Window panes and container glasses are exposed to repeated or continued removal of the alkali compounds formed (carbonates, hydroxides), whereas condensing water vapour under stagnant conditions results in high and locally differing corrosion rates. The ion exchange increases the pH, and the pH increase is the faster the smaller the ratio of droplet volume to glass surface area. This leads to an autocatalytic effect: the pH increase increases the dissolution rate of the network and vice versa so that the glass surface is damaged at an accelerated rate.

A pH *decrease* by glass corrosion is basically excluded, and a reduced pH, if nevertheless observed, must be traced to a different cause. An example is the pH decrease of solutions enclosed in ampoules, which was asserted to be an "acid error" due to glass corrosion [1.49] but proved to be caused by carbon dioxide faultily admitted during flame sealing the containers by the authors [1.50].

Significance of Glass Corrosion: Examples

Despite the high chemical resistance of modern glasses, glass corrosion is a serious problem in several areas. Ancient glasses and stained windows are particularly endangered by the worsening environmental conditions [1.51]. A fundamental solution to this problem has not been found, as organic and inorganic coatings, as well as their combinations (ORMOCERs) isolate glasses from the environment only for limited periods.

Perhaps the most critical process with respect to glass corrosion is glass coating because traces of exchanged alkali ions from substrates can prevent the adhesion of coatings and impair their properties [1.52]. Means to avoid this effect are chemically resistant substrate glasses (silica glass, Duran®), surface modifications of the substrate glass (sulphurizing [1.53]), and intermediate protective layers (SiO_2, PbO [1.54]).

Glass fibres for reinforcement of cement are subject to the most extreme corroding conditions because of the alkalinity (pH $>$ 12) of the cement, in particular during curing, the mechanically damaging effect of its crystals, and varying, extreme temperatures. Even AR (alkali-resistant) glass with large zirconia content undergoes heavy degradation but, surprisingly, nevertheless improves the mechanical stability of the cement [1.55].

1.1.6 Active Thin-Layer Systems

Coating with thin layers and layer systems provides glasses with physical and chemical properties not found in the uncovered bulk materials, for instance with electrical conductivity, inertness to corrosion, and optical constants. Coated glasses are a major industrial product, and the output is increasing. Details are discussed in a recent monograph [1.56]; an overview of the different types of coatings is presented in Fig. 1.1.

Most coatings are "static" coatings with constant physical properties. However, variable layer systems have been developed as well. Their optical constants can be changed by applying an external electrical voltage. They are called active thin-layer systems or electrically switchable coatings [1.57]. Two main types of such layer stacks have been constructed.

(a) In "field-effect devices" the optical properties of one of the layers are changed by an external electrical voltage applied to two sandwiching transparent (for instance ITO) electrodes. The optically active materials are either liquid crystals or small needle-like particles dispersed in a liquid [1.57], both of which are oriented when an electric field is applied. Field effect devices can thus be switched discontinuously between an optically transparent and an absorbing or diffusely reflecting state, where the transparent state, however, must be maintained by a constant voltage. A great number of liquid crystal devices have been developed. The major application is in PC monitors.

(b) "Electrochromic" or "chromogenic" devices contain a layer of a reversibly variable, "electrochromic" redox substance whose reduced and oxidized states are characterized by different colours [1.58, 59]. Application of an external electric voltage causes a reversible change of the redox state of the layer and a connected change of the optical constants of the device. Because of electroneutrality, the chromogenic electrode reaction necessitates a complimentary reaction of a counter electrode. Electrochromic devices are thus reversible thin-layer batteries with optically indicated energy content [1.60].

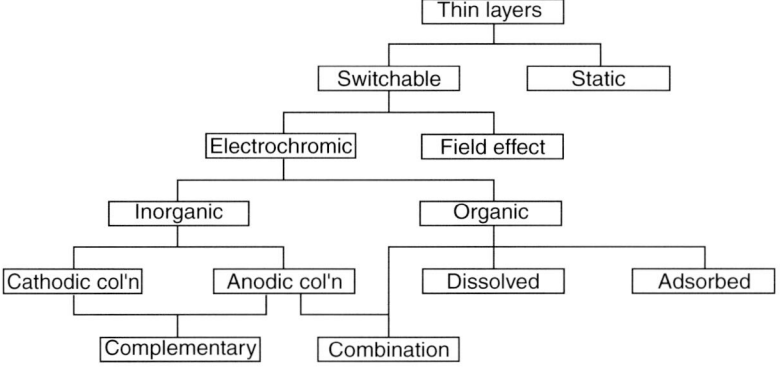

Fig. 1.1. Overview of the various types of glass coatings

They are an interesting subject of the electrochemistry of glass products although the glass involved acts merely as the substrate and cover glass for the active layer system.

Electrochromic Reactions

Electrochromic devices with two different types of redox reactions have been developed.

(a) *Redox reactions of inorganic oxides* are solid-state electrochromic reactions, independent of whether the reduced state of the oxide is coloured, as the bronze of colourless tungsten trioxide, which has a deep blue colour due to an intervalence transfer absorption mechanism [1.61],

$$WO_3 + (M^+ + e^-) \rightleftharpoons W^{6+}_{1-x}W^{5+}_x O_{3-x}(OM)_x \; , \tag{1.7}$$

or whether the oxidized state is the coloured form of the chromogenic material, as the brownish black oxy-hydroxide of colourless nickel hydroxide [1.62],

$$Ni(OH)_2 \rightleftharpoons Ni^{3+}_x Ni^{2+}_{1-x} O_x(OH)_{2-x} + x(H^+ + e^-) \; . \tag{1.8}$$

Both the cathodic (1.7) and the anodic colouration reaction (1.8) are non-stoichiometric reactions with continuously varying x, which make the oxides and devices continuously colourable. Besides, any colouration depth (and x) reached is stable after the voltage has been switched off, provided appropriate counter-electrodes are applied. Thus, both reactions exhibit memory. These properties make inorganic electrochromic oxides suited for special applications, for instance in automotive rear-view mirrors [1.63], where a continuously changeable reflection is required, and for automobile sunroofs, sun glasses and large-area displays, where continuous currents for maintaining certain device states must be avoided. Electrochromic devices involving both a cathodically and an anodically colouring chromogenic layer are particularly effective because of their simultaneous colouration (and bleaching).

(b) *Redox reactions of organic chromophores* are reactions of single molecules [1.64]. Examples are bipyridilium derivatives or viologens,

$$[R\text{–}\overset{(+)}{N}C_5H_4\text{–}H_4C_5\overset{(+)}{N}\text{–}R']^{2+} + e^- \rightleftharpoons [R\text{–}\overset{(+)}{N}C_5H_4\text{–}H_4\overset{\bullet}{C}_5N\text{–}R']^+ \; , \tag{1.9}$$

where R and R′ are different or equal substituents, the dot is a single electron, and the positive charges (shown localized on N in the equation) and the single electron are delocalized over the appropriate pyridine rings. The dication is colourless, whereas the radical cation is deeply coloured due to an intra-molecular charge transfer absorption, a well-known mechanism in organic chemistry. The chromophore is dissolved in an appropriate solvent, which also contains the counter substance, for instance ferrocene, which is simultaneously oxidized when the chromophore is reduced in an electrochromic cell. Counter-diffusion of the electrolysis products leads to spontaneous back reaction, which can only be retarded, but not eliminated, by high solvent viscosity or by adsorbing the chromophore on an appropriately modified, for instance TiO_2-covered, electrode [1.65]. Organic electrochromic systems thus have a limited memory. Although the redox state of the chromophore molecules changes discontinuously at the electrodes according to (1.9), organic chromogenic devices colour and bleach continuously because of the time-dependent concentration changes during the electrolyses. As indicated in Fig. 1.1, organic and anodically colouring inorganic chromogenic systems can be combined to give optically very effective chromogenic devices.

Present State and Future Developments

It is surprising that chromogenic devices did not come onto the market until about thirty years after *Deb* published the effect [1.66] and about seventeen years after the preliminary sample of an electrochromic automotive rearview mirror was presented to the public [1.67]. The constant increase of the number of patents shown in Fig. 1.2 (160 patents per year between 1978 and 1998) indicates that the delay is not caused by a declining interest in chromogenic systems but by technical difficulties. The continuing interest is confirmed by the large number of companies and institutions presently developing electrochromic devices (29 in Europe, 15 in North America, three in South America, and 14 in Asia, preferably Japan) [1.57].

Serious difficulties arise from the high requirements on practically applied chromogenic devices and from the extremely small thickness of electrochromic layer systems, which is in the micrometre range. Depending on the type of device, they consist of three to six layers (electrodes, electrolytes, reflectors).

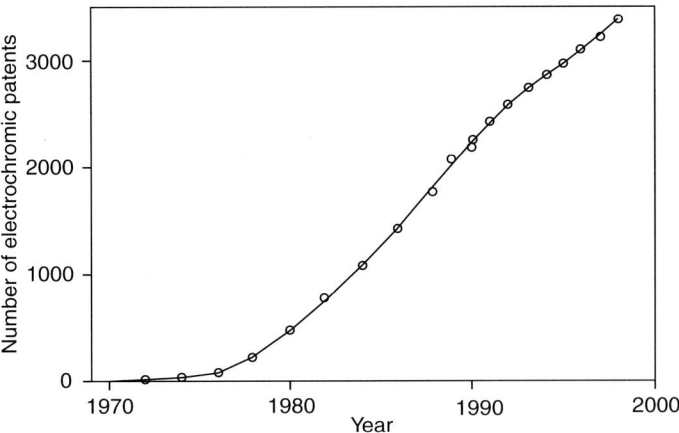

Fig. 1.2. Number of patents granted for electrochromic devices worldwide as a function of time between 1969, the year of *Deb's* first publications [1.66], and 1998

Both anodic and cathodic reactions must be nearly 100% reversible with negligible side reactions, and the reaction rate must be maintained for several years. All layers must be free of measurable failures and, for instance in automotive rear-view mirrors, must withstand temperatures up to 120 °C over a cycle life of at least 10 000, and often 100 000 cycles. These requirements explain the long delay and the great effort in research and development that was necessary to make chromogenic devices ready for industrial production.

1.1.7 Electrometric pH Measurement

pH was initially defined in terms of hydrogen ion concentration, $pH_c = -\log c_H$, by *Sørensen* in 1909 [1.68]. After the significance of the ion activity had become general knowledge, he redefined it in terms of hydrogen ion activity, $pH = -\log a_H$, in 1924 [1.69] and thus laid the foundations for modern pH measurement. Although pH has meanwhile become the most frequently determined quantity in fundamental and applied chemistry, the subject is under active discussion at present because of its unique character. Involving a single ion activity, pH, *in principle*, cannot be measured. This seems surprising because the hydrogen ion activity (and thus pH) determines all acidity-controlled reactions and equilibria in aqueous solutions and is indispensable for describing these reactions quantitatively. Because the glass electrode is the most frequently used pH sensor, the present development in electrometric pH measurement is of immediate interest and will be briefly described in the following.

Several so-called pH scales (an unfortunate term coined in the 1940s [1.70]) have been established. They are defined by standard buffer solutions with assigned standard pH values used for calibrating pH-measuring cells. Two of them, the multi-standard scale of the National Institute of Standards and Technology (NIST) [1.71] and the single standard scale of the British Standard Institution (BSI) [1.72], existed when the latest and still valid *IUPAC Recommendations* on pH were issued in 1985 [1.73]. Because "pH experts were divided and IUPAC was unable to recommend, ..., one or other of the two approaches to pH scale definition", the document was designed as a "compromise recommendation" allowing the use of both scales in parallel despite the metrologically intolerable consequence that two pH values are obtainable for each unknown solution measured. But it was wisely issued as an interim document with the intention to replace it by amended recommendations when better knowledge would allow one to "recommend a thermodynamically significant and metrologically sound pH scale".

This better insight, in particular into the metrology of pH measurement, had been gained in 1996 when four more metrological shortcomings were uncovered by a critical analysis of the 1985 document [1.74]. They emphasized the necessity for speedy replacement, and so the German representative of IUPAC Commission V.5 initiated an Interdivisionary Working Party on pH in 1996. It was formally established in 1997 [1.75] with the aim of drafting

new recommendations, for which the basic principles had been proposed in
the author's above-mentioned critical paper [1.74]. The work has meanwhile
advanced to a state that makes a brief outline of its contents worthwhile.

Principles of pH Measurement for New IUPAC Recommendations

The new recommendations will be based on one definition of pH, namely that
proposed by *Sørensen* in 1924 [1.69], in terms of molality,

$$\text{pH} = -\log a_\text{H} = -\log(m_\text{H}/m^0) , \tag{1.10}$$

where m^0 is standard molality ($1\,\text{mol}\,\text{kg}^{-1}$). The link between definition and
assigned and measured pH values will be given by traceable procedures,
traceability being defined by an international metrological standard [1.76].
The concept of traceability, consequently, will eliminate any arbitrary "scale"
defining pH, and the term "scale" will not appear in the new document.

The practical link between pH definition and measured pH will be accomplished by pH standards with two different metrological qualities, primary
and secondary standards, and by two methods for calibrating practical pH
meter-electrode assemblies, the two- and the multi-point calibration, each of
these assignments and procedures having stated uncertainties.

Primary pH Standards – Determination of pH(PS)

Only buffer solutions that meet stringent requirements are suited as primary
standards. Their pH(PS) are determined by a "primary method of measurement" [1.76], which is a direct method yielding pH(PS) without reference
to other standards of the same quantity. The only potentiometric cell that
qualifies for these measurements is the Harned cell,

$$\text{Pt}|\text{H}_2(p_{\text{H}_2} = 101.352\,\text{kPa})|\text{buffer PS, Cl}^-(m_\text{Cl})|\text{AgCl}|\text{Ag}|\text{Pt} , \tag{1.I}$$

because (a) the potential of the Pt,H_2 electrode indicates the hydrogen ion
activity in a thermodynamically correct way, (b) the Ag|AgCl electrode is a
correctly working electrode of the second kind, and (c) the cell is without liquid junction. The effect of chloride, which is added to the buffer solution for
the correct functioning of the Ag|AgCl electrode, is eliminated by extrapola-
tion to zero chloride concentration. The remaining trace activity coefficient of
chloride due to dissolved AgCl is calculated by an extra-thermodynamic convention, for instance the Bates–Guggenheim convention [1.77], and applied
to the evaluation of pH(PS). Details are given in Sect. 2.10.3.

pH(PS) is a conventional number with a typical uncertainty of 0.004.
Traceability to the SI system requires one to include the uncertainty associated with the Bates–Guggenheim convention, which, for instance, is esti-
mated as 0.007 at an ionic strength of $0.05\,\text{mol}\,\text{kg}^{-1}$ so that pH(PS) traced to

SI units has a total uncertainty of 0.011 under these conditions. SI-traceable pH values, however, will only rarely be needed.

Because of unavoidable small batch-to-batch differences of pH(PS) values, the new document will merely contain a table with pH(PS) (between $5\,^\circ$C and $50\,^\circ$C) as guiding numbers, and it will be recommended that national and international metrological institutes issuing primary standard materials provide certificates – together with the substances – that inform about their actual pH(PS) values to be applied.

Secondary pH Standards – Determination of pH(SS)

Certain buffer solutions that do not qualify for measurements in the Harned cell or are not subjected to it for time-saving or economic reasons will be termed secondary standards in the new document. Their pH(SS) are "assigned by comparison with a primary standard of the same quantity" [1.76]. Because, with one exception, this comparison is carried out in cells with liquid junction, pH(SS) have larger uncertainties than the pH(PS) from which they are derived.

The new document will recommend four cells with liquid junction, which are to be applied according to the specific properties of the secondary buffers to be assigned, the primary buffers used as standards and their relative pH values. The so-called *differential pH cell* [1.78] comparing buffer solutions with the same nominal composition and yielding pH(SS) with the smallest uncertainties is discussed in detail in Sect. 2.10.4. The other cells employ a concentrated KCl solution as reference or electrolyte bridge solutions and also the Pt,H_2 electrode, except for one cell, which uses a glass electrode. These cells are preferably recommended for assigning pH(SS) that differ from the pH(PS) of the applied standard. An uncertainty budget will be provided for each cell, and the accredited laboratories carrying out the measurements will be asked to supply a certificate with each certified material that informs about its actual pH(SS) values.

Calibration of pH Meter-Electrode Assemblies

Practical pH measurements are necessarily carried out in cells with transference employing a single or combination glass electrode,

$$\text{ref. electr.}|\text{KCl}(c \geq 3.5\,\text{mol dm}^{-3})\|\text{sol'n pH(S) or pH(X)}|\text{glass electr. ,}$$
$$(1.\text{II})$$

whose potential difference includes a liquid junction potential. Because of various systematic and random effects, these cells must be calibrated with primary or secondary pH standards, for which two calibration procedures will be recommended. The *two-point or bracketing procedure* [1.79] uses two standard buffers for establishing an interpolation line, whereas the calibration function of the *multi-point calibration procedure with linear regression* [1.80] is obtained from five standard pH values, thus yielding pH values with smaller uncertainties than two-point calibration and also identifying safely offsets and outliers. A third method, *one-point calibration*, employs one standard buffer and yields mainly pH estimates. But this orientational method is economical and often quite valuable for practical purposes.

1.2 Isothermal Glass-Forming Melts

1.2.1 Redox State and Basicity

The reduction-oxidation (redox) state of glass melts is of high interest for several reasons. For the glass industry the correct choice of the redox state means economical and ecological glass melting and optimum control of the glass quality because it helps avoid unnecessary dissolution of platinum metal parts, precipitation of easily reducible metals, extensive evaporation of melt components, and unwanted colouration of melts due to certain poly-valent elements present in certain redox states. Most significant, temperature-dependent equilibria of redox ions are the basis of redox fining, the removal of gaseous impurities by means of reversibly generated oxygen bubbles, see Sect. 3.5.1. The fundamental interest in the redox state, on the other hand, is concerned with the electron donor power of the oxygens and with melt basicity and the connected redox state of dissolved polyvalent ions.

Oxygen Partial Pressure

The first step in investigating these processes and equilibria was the development of potentiometric oxygen sensors [1.81]. They consist of a platinum measuring and a zirconia reference electrode and are constructed in various ways, for scientific work in laboratories and for melt control in glass melting tanks with normal and long lifetimes. Besides, the measured emfs can be evaluated for isothermal and non-isothermal glass melts, see Sect. 3.4. Meanwhile, technical oxygen sensors for glass melts are also commercially available [1.82].

However, it was soon realized that the mere knowledge of the oxygen partial pressure of production glass melts suffices only as long as the concentrations of the other redox components of the produced melt are constant. What was thus needed was either a technical sensor for the on-line analysis of all polyvalent ions present in the melt and/or the determination of temperature-dependent thermodynamic data of the redox couples. The latter is presently carried out by means of voltammetric methods, in particular square-wave voltammetry (SWV), by which the polyvalent ions in the melt are analysed *in situ* at the temperatures of interest [1.83], see Sect. 3.5.2.

SWV is a convenient and fast technique, but it must be kept in mind that results are discussed in terms of ionic concentrations and not activities, which, however, may not cause serious errors as long as the charges of the polyvalent ions involved are not too different. A number of electrochemical series of redox equilibria and elements have been determined on this basis. Oxygen partial pressures measured at defined temperatures thus yield the complete redox conditions of the melt when they are evaluated on the basis of these thermodynamic data. However, the polyvalent element present in the melt must be known. If this is not the case or if appreciable concentrations of more than one polyvalent element are dissolved, on-line technical sensors must be applied to determine the concentrations of the polyvalent ions present in the melt, and this makes the evaluation more cumbersome.

Melt Basicity

Several basicity concepts have been developed in the past. All of them are more or less based on the activity of "oxide", $a_{O^{2-}}$, or oxidic groups, a_{SiO^-}, which, being single ions or ionic groups, cannot be measured *in principle*. Besides, such measurements involve a liquid junction potential whose magnitude is principally unknown. Optical basicity, which is based on optical measurements and thus excludes these difficulties, is a measure of the electron donor power of the oxygens of glass melts and thus indicates "correct" basicities, although it is not a thermodynamic concept [1.84, 85]. In many melts, it does not have to be measured but can be calculated from experimentally determined increments if the melt composition is known, see Sect. 3.3.

Optical basicity has been applied to a large number of redox couples. That the concept rests on a sound basis is evidenced by the fact that it combines the individual curves log of redox ratio as a function of alkali oxide content into one general curve according to the relationship

$$\log[\mathrm{M}^+]/[\mathrm{M}^{(m+n)+}] = a\Lambda + b , \tag{1.11}$$

where Λ is the optical basicity, and a, with few exceptions, is a negative constant. Constants a and b of many polyvalent ion couples, electrode potentials, and equilibria between two different ion couples have been tabulated, see Sect. 3.3. As optical basicity is a general concept and, for instance, characterizes also aqueous solutions, it was possible to predict approximate redox equilibria as functions of the optical basicity of melts at $1400\,^{\circ}\mathrm{C}$ and air atmosphere by using standard redox potentials in aqueous solution at $25\,^{\circ}\mathrm{C}$.

Thermodynamic Paradox

According to the general equation for the redox equilibrium,

$$4\mathrm{Fe}^{3+} + 2\mathrm{O}^{2-} \rightleftarrows 4\mathrm{Fe}^{2+} + \mathrm{O}_2 , \tag{1.12}$$

where the basicity is symbolized by the oxide ion O^{2-}, it should be expected corresponding to Le Chatelier's principle that high basicity favours the lower oxidation state of the polyvalent element. However, except for a few cases, for instance copper, the contrary is observed. Despite many attempts to explain the effect, this so-called thermodynamic paradox is not completely understood. It was stated that it must be considered for an explanation that equilibrium (1.12), with respect to basicity, is not an equilibrium within a single phase but involves melts with different compositions and, consequently, different free standard energies and equilibrium constants [1.86].

Redox Reaction between Glass Melt and Refractory

Redox reactions are not restricted to glass melts but are also observed in refractories used as melting tank material. An example is bubble formation at the interface melt-zirconium silicate (ZS), Fig. 1.3. This long-known reaction

Fig. 1.3. Redox reaction between ZS refractory and an alkali-containing oxidic glass melt. During temperature increase, oxygen bubble formation starts at $1230\,^\circ$C at the surface containing the most highly oxidized redox impurity, here at the upper surface

has at last been elucidated [1.87], see Sect. 3.8. It is caused by frozen-in redox impurities of ZS, which, at melt temperatures, oxidize oxide of the melt to give oxygen. The internal reduction of the impurity is enabled by finite electronic and alkali ion conductivities of ZS, which link the half-reactions at the different locations. The solid-state reaction is an example of serious technical disturbances caused by traces of impurities that do not even appear in the product. It is surprising that it is an electrochromic reaction, see Sect. 1.1.6, and it is certainly the electrochromic reaction at the highest temperature ever reported.

1.2.2 Transport Properties: Electrical Conductivity

Glass melting tanks are increasingly heated by electricity for economical and ecological reasons and because the direct energy supply to the melt allows much better temperature control during the melting process and thus guarantees a higher quality than heating by combustion. Due to the large conductivity of glass melts, the electrical energy is transferred via the Joule effect, the melts being the heating resistors. Inert metals, for instance platinum plates or, more frequently, molybdenum rods, are used as electrodes. Because of high melt conductances and small average cell constants of the conductivity cells represented by the melting tanks, currents up to hundreds of kiloamperes are supplied by applying relatively small voltages in the range of several volts. The distribution of current densities within the melts is extremely non-uniform, the current densities increase with decreasing distance from the electrodes and have maximum values at the electrode surfaces, in particular with rod-shaped electrodes. Accordingly, the temperature has a maximum near the heating electrodes, which can cause intolerable chemical and electrochemical reactions, for instance deposition of reducible metals. In addition, the conductance of glass melts has a positive temperature coefficient and tends to cause auto-deregulation of the current-density distribution.

These conditions necessitate an exact knowledge of the temperature-dependent conductance of glass melts. However, a literature search resulted in a number of conductivity cells, none of which satisfied the requirements with respect to accuracy, reproducibility, and (relatively) easy handling [1.88], see Sect. 3.9.1. Thus, a cell with a new concept was developed and further improved at Schott Glas [1.88, 89, 90]. The cell, consisting of an alumina (or silica) cell tube and an upper and a lower platinum electrode, is immersed in the melt, which is contained in a platinum crucible. The lower electrode is formed by a flat-ended cylinder protruding from its bottom. The upper electrode is designed such that bubbles sticking to its lower surface can be removed by rotation. The arrangement is heated in a special electrically heated tube furnace that causes a constant temperature within the cell tube. Unlike conventional cells, which are operated on the basis of standardization with standard liquids such as electrolyte solutions and molten salts, this cell yields absolute conductances. The measured data are independent of temperature-dependent changes of density and surface tension, evaporation of melt components, which would cause composition changes, and electrical conduction through the atmosphere above the melt, and can be checked for reproducibility by varying the cell length during the measurements. However, its successful application depends on the corrosion resistance and electrical conductivity of the alumina tubes applied, which necessitates checking each charge of tubes purchased.

The conductance cell is permanently applied for research and technical purposes at Schott Glas. The mixed-alkali effect of conductivity in oxidic glass melts, which was measured by means of the cell [1.38], is discussed in Sect. 3.9.3. Meanwhile, the cell has also been applied at several research institutes and glass-producing companies, see for instance [1.91, 92].

1.3 Non-Isothermal Oxidic Glass-Forming Melts

In industrial glass melting, the situation is more complicated than that described in the preceding section because technical glass melts usually are not in an isothermal state but subject to often large temperature gradients. This applies to discontinuous as well as continuous melting units. In discontinuous melting, the size of containers, crucibles, and moulds has constantly increased in past decades so that isothermal melting and pouring has become more and more difficult and, in many cases, even impossible. An example is the 8.2 m diameter telescope mirrors produced by Schott [1.93], the pouring of which necessarily causes rather large temperature gradients across the mould. Contrary to discontinuous melting, where the temperature of the batch changes as a function of time, in continuous glass melting the melt flows through the tank, which has a constant local temperature distribution. Thus, the temperature of each volume element of the melt changes as a function of location and time. Consequently, all discontinuous melting units *in principle* produce glass melts under non-isothermal conditions.

Temperature gradients *per se* are of no concern during glass melting; this also applies for discontinuous melters. The contact of melts with metals applied, for instance, as heating electrodes or protecting tubes of thermocouples does not present any danger either as long as they are well insulated from each other and their extension allows them to be within isothermal parts of the melt. However, electronic conductors in melting tanks are often incidentally short-circuited, for instance by stray currents via ground or construction elements of the tanks. They thus represent frequently short-circuited non-isothermal electrochemical cells, which are sources of cell reactions that may produce intolerable results, for instance oxygen bubble formation or deposition of molten metals from the melt.

Despite these difficulties and in view of a worldwide gross production of glass valuing approximately US$165 billion p.a. [1.94], literature on non-isothermal glass melts is surprisingly scarce. There is one paper reporting on qualitative observations [1.95] and three reports on (faulty) attempts to measure thermoelectric potentials of glass melts [1.96–98]. Quantitative data, although available for salt melts, ceramics (e.g., zirconia, see Sect. 3.4.3), and solid glasses, are totally lacking for non-isothermal glass melts.

To gain better insight into the non-isothermal character of the production tanks, Schott Glas developed a method for determining thermoelectric voltages in glass melts (see Sect. 3.6.). Measurements are conducted by means of two zirconia electrodes such as those developed as reference electrodes for measuring oxygen partial pressures. They are ideally suited to these measurements: the measuring platinum electrodes, which are sensitive to redox components, are electronically insulated from, but ionically connected to the melt. (This condition was not met in earlier attempts to measure thermoelectric voltages [1.96–98].) Moreover, a "standard Seebeck coefficient" was defined by referring thermoelectric voltages to 1 bar oxygen partial pressure in both zirconia electrodes in order to replace the misleading and often contradictory terminology generally used in this field. The Soret effect, which describes thermodiffusion, is not of interest for technical glass melting because the small diffusion rates [1.99, 100] and the agitated glass melts exclude the formation of detectable concentration gradients.

Thermoelectric voltages between metals, the main practical problem of continuous melters reported here, cannot be directly measured with sufficient accuracy for several reasons. But they are obtained by combining non-isothermal voltages measured by means of zirconia electrodes with temperature-dependent isothermal emfs of cells for measuring oxygen partial pressures as described in Sect. 3.4.1. The resulting data yield information about whether the short circuit of metals in certain non-isothermal glass melts and at certain temperatures and temperature differences can result in the formation of oxygen bubbles and, in addition, give maximum possible quantitative shifts of oxygen reboil temperatures to higher or lower values. Whether oxygen bubbles are actually formed depends also on non-thermodynamic conditions in the melting tanks, for instance on the hydrodynamics of the melt and the relative size of the short-circuited metal parts. The measurements allow one to identify the thermodynamic possibility of oxygen bubble formation and, in positive cases, to choose dynamic conditions that exclude this complication. Discussing other non-isothermal reactions on the basis of standard Seebeck coefficients is possible, but requires additional data. Thermoelectric potentials are also required for evaluating emfs of oxygen sensors applied in non-isothermal glass melts, see Sect. 3.6.

References

1.1 B. Elvers, S. Hawkins, W. Russey (Eds.): *Ullmann's Encyclopedia of Industrial Chemistry*, 5th ed., Vol. B 6 (VCH, Weinheim 1995) p. 488
1.2 H.-R. Tränkler, E. Obermeier (Eds.): *Sensortechnik, Handbuch für Praxis und Wissenschaft* (Springer, Berlin, Heidelberg 1998)

1.3 W. Göpel, J. Hesse, J.N. Zemel (Eds.): *Sensors. A Comprehensive Survey* (VCH, Weinheim 1989) and subsequent volumes

1.4 M. Cremer: "Über die Ursachen der elektromotorischen Eigenschaften der Gewebe, zugleich ein Beitrag zur Lehre von den polyphasischen Elektrolytketten", Z. Biol. **47**, 562–608 (1906)

1.5 F. Haber, Z. Klemensiewicz: "Über elektrische Phasengrenzkräfte", Z. Phys. Chem. **67**, 385–431 (1909)

1.6 A.A. Belyustin, A.M. Pisarevsky, G.P. Lepnev, A.S. Sergeyev, M.M. Shultz: "Glass electrodes: a new generation", Sens. Actuators B **10**, 61–66 (1992)

1.7 L.W. Niedrach: "Elektroden für Potentialmessungen in wässrigen Systemen bei hohen Temperaturen und Drücken", Angew. Chem. **99**, 183–191 (1987)

1.8 P. Bergfeld: "Development of an ion-sensitive solid-state device for neurophysiological measurements", IEEE Trans. Biomed. Eng. **17**, 70–71 (1970)

1.9 K.-D. Kreuer: "Solid potentiometric pH electrode", Sens. Actuators B **1**, 286–292 (1990)

1.10 A.M. Pisarevskii, A.S. Sergeev, P.M. Tolstikov: "Temperature coefficient of glass electrodes with alloy inner contact for pH measurement", Russ. J. Electrochem. **30**, 841–845 (1994)

1.11 W. Vonau, H. Kaden: "Glass membrane electrodes with solid state internal contacts for pH and pNa measurements", Glass Techn. Ber. Glass Sci. Technol. **70**, 155–160 (1997)

1.12 A. Noll, V. Rudolf, E.W. Grabner: "A glass electrode with solid internal contact based on Prussian blue", Electrochim. Acta **44**, 415–419 (1998)

1.13 B.P. Nicolsky: "Theory of the glass electrode I", Acta Physicochim. USSR **VII**, 597–610 (1937)

1.14 G. Bouquet, S. Dobos, Z. Boksay: "Untersuchung der Oberflächenschicht des Glases", Ann. Univ. Sci. Budapest (Rolando Eötvös Nominatae), Sect. Chim. **6**, 5–13 (1964)

1.15 F.G.K. Baucke: "Glass electrodes: why and how they function", Ber. Bunsenges. Phys. Chem. **100**, 1466–1474 (1996)

1.16 E. Pungor: "Ion-selective electrodes – analogies and conclusions", Electroanalysis **8**, 348–252 (1996)

1.17 F. Quittner: "Einwanderung von Ionen aus wässriger Lösung in Glas", Ann. Phys. IV. Folge **85**, 745–769 (1928)

1.18 F.M. Ernsberger: "Ion conduction in oxide glasses: blocking electrodes and space charge", Phys. Chem. Glasses **36**, 152–153 (1995)

1.19 C.A. Levine, R.G. Heitz, W.E. Brown: "The Dow sodium sulfur battery", in *Proc. 7th Intersociety Energy Conversion Engineering Conference* (Am. Chem. Soc., Washington, DC 1972) pp. 50–55

1.20 G. Gutmann, H. Kistrup, F.G.K. Baucke, G. Müller: "Sodium ion-conducting glass electrolyte for sodium/sulphur batteries", US Patent 42 37 196 (1980)

1.21 S.D. Jones, J.R. Akridge, F.K. Shokoohi: "Thin film rechargeable Li batteries", Solid State Ionics **69**, 357–368 (1994)

1.22 S.I. Najafi (Ed.): *Introduction to Glass Integrated Optics* (Artec House, Boston 1992)

1.23 M.D. Ingram: "Optical basicities and structural dynamics in glassy materials", J. Non-Cryst. Solids **222**, 42–49 (1997)

1.24 K.L. Ngai: "A review of critical experimental facts in electrical relaxation and ionic diffusion in ionically conducting glasses and melts", J. Non-Cryst. Solids **203**, 232–245 (1996)

1.25 K. Funke, B. Roling, M. Lange: "Dynamics of mobile ions in crystals, glasses and melts", Solid State Ionics **105**, 195–208 (1998)

1.26 B. Roling: "Scaling properties of the conductivity spectra of glasses and supercooled melts", Solid State Ionics 105, 185–193 (1998)

1.27 A. Bunde, K. Funke, M.D. Ingram: "Ionic glasses: history and challenges", Solid State Ionics 105, 1–13 (1998)

1.28 O.L. Anderson, D.A. Stuart: "Calculation of activation energy of ionic conductivity in silica glasses by classical methods", J. Am. Ceram. Soc. 37, 573–580 (1954)

1.29 K. Funke: "Jump relaxation in solid electrolytes", Prog. Solid State Chem. 22, 111–195 (1993)

1.30 A. Bunde, M.D. Ingram, P. Maass: "The dynamic structure model for ion transport in glasses", J. Non-Cryst. Solids 172–174, 1222–1236 (1994)

1.31 J.E. Davidson, M.D. Ingram, A. Bunde, K. Funke: "Ion hopping processes and structural relaxation in glassy materials", J. Non-Cryst. Solids 203, 246–251 (1996)

1.32 K. Funke: "A modified jump relaxation model for fragile glass-forming ionic melts", Z. Phys. Chem. 206, 101–116 (1998)

1.33 J.R. MacDonald: "Critical examination of the mismatch-and-relaxation frequency-response model for dispersive materials", Solid State Ionics 124, 1–19 (1999)

1.34 R. Weber: "Über den Einfluß der Zusammensetzung des Glases auf die Depressions-Erscheinungen der Thermometer", Berliner Akad. Wiss. II, 1233–1238 (1883)

1.35 G. Gelhoff, M. Thomas: "Die physikalischen Eigenschaften der Gläser in Abhängigkeit von der Zusammensetzung. I. Über das elektrische Leitvermögen von Gläsern", Z. Techn. Phys. 6, 544–554 (1925)

1.36 D.E. Day: "Mixed alkali glasses – their properties and uses", J. Non-Cryst. Solids 21, 343–372 (1976)

1.37 J.O. Isard: "The mixed alkali effect in glass", J. Non-Cryst. Solids 1, 235–261 (1969)

1.38 F.G.K. Baucke, R.-D. Werner: "Mixed alkali effect of electrical conductivity in glass-forming silicate melts", Glastechn. Ber. 62, 182–186 (1989)

1.39 R.H. Doremus: "Exchange and diffusion of ions in glass", J. Phys. Chem. 68, 2212–2218 (1964)

1.40 R.H. Doremus: "Mixed alkali effect and interdiffusion of Na and K ions in glass", J. Am. Ceram. Soc. 57, 478–480 (1974)

1.41 G. Tomandl, H.A. Schaeffer: "Relation between the mixed alkali effect and the electrical conductivity of ion-exchanged glasses", in The Physics of Non-Crystalline Solids, ed. by G.H. Frischat (Trans Tech Publ., Aedermannsdorf 1977) pp. 480–485

1.42 F.G.K. Baucke: "Electrochemistry and glass structure", J. Non-Cryst. Solids 129, 233–239 (1991)

1.43 K.A. Kostanyan,: "Investigation of the conductivity neutralization effect in fused borate glasses", in The Structure of Glass, Proc. of the 3rd All-Union Conference on the Glassy State, Vol. 2, Leningrad 1959 (Consultants Bureau, New York 1960) pp. 234–236

1.44 R.E. Tickle: "The electrical conductance of molten alkali silicates. Part I, Experiments and results", Phys. Chem. Glasses 8, 101–112 (1967)

1.45 Th. Pfeiffer, R. Müller, R.-D. Werner: "Transport phenomena in oxidic glass-forming melts", Ber. Bunsenges. Phys. Chem. 100, 1503–1507 (1996)

1.46 F.G.K. Baucke: "Ionic processes between glasses and solutions", in Analysis of the Composition and Structure of Glass and Glass Ceramics, Schott Series on Glass and Glass Ceramics, ed. by H. Bach, D. Krause (Springer, Berlin, Heidelberg 1999) pp. 405–421, 447–449

1.47 H. Bach: "Advanced surface analysis of silicate glasses, oxides and other insulating materials: a review", J. Non-Cryst. Solids **209**, 1–18 (1997)

1.48 H. Bach: "Surface analysis and the control of product properties", in *Surface Analysis of Glasses and Glass Ceramics, and Coatings*, Schott Series on Glass and Glass Ceramics, ed. by H. Bach, D. Krause (Springer, Berlin, Heidelberg) Sect. 1.3, to be published

1.49 J.P. Surmann, I. Bosse: "Abfall des pH-Wertes wäßriger Salzlösungen in Ampullen", Pharm. Ind. **54**, 66–68 (1992)

1.50 F.G.K. Baucke: "Konstante pH-Werte wäßriger Salzlösungen in Ampullen aus Neutralglas", Pharm. Ind. **54**, 886–889 (1992)

1.51 H. Römich: "Historic glass and its interaction with the environment", in *Current Topics in Glass and Ceramics Conservation* (James & James, London 1999) pp. 5–14

1.52 F.G.K. Baucke: "Corrosion of glasses and its significance for glass coating", Electrochim. Acta **39**, 1223–1228 (1994)

1.53 R.W. Douglas, J.O. Isard: "The action of water and of sulphur dioxide on glass surfaces", J. Soc. Glass Technol. **33**, 289–335 (1949)

1.54 H. Bach: "Eine Methode zur Bestimmung von Mikrokonzentrationsgradienten an Glasoberflächen mit Ionenätzung", in *Proc. Int. Congr. Glass* (ICG, Versailles 1971) pp. 155–170

1.55 V.T. Yilmaz, E.E. Lachowski, F.P. Glasser: "Chemical and microstructural changes at alkali-resistant glass fiber–cement interfaces", J. Am. Ceram. Soc. **74**, 3054–3060 (1991)

1.56 H. Bach, D. Krause (Eds.): *Thin Films on Glass*, Schott Series on Glass and Glass Ceramics (Springer, Berlin, Heidelberg 1997)

1.57 C.M. Lampert: "Smart switchable glazing for solar energy and daylight control", Solar Energy Mat. Solar Cells **52**, 207–221 (1998)

1.58 C.M. Lampert, C.G. Granqvist (Eds.): *Large-Area Chromogenics: Materials and Devices for Transmittance Control*, SPIE Institutes for Advanced Optical Technologies, Vol. IS 4 (SPIE, Bellingham 1990)

1.59 C.G. Granqvist: "Electrochromic tungsten oxide films. Review of progress 1993–1998", Solar Energy Mat. Solar Cells **60**, 201–262 (2000)

1.60 F.G.K. Baucke: "Electrochromic mirrors with variable reflectance", Solar Energy Mat. **16**, 67–77 (1987)

1.61 F.G.K. Baucke, J.A. Duffy: "Darkening glass by electricity", Chemistry in Britain **21**, 643–646 (1986)

1.62 K. Bange, F.G.K. Baucke, B. Metz: "Properties of electrochromic nickel oxide coatings produced by reactive evaporation", SPIE **1016**, 170–175 (1988)

1.63 F.G.K Baucke: "Reflectance control of automotive mirrors", SPIE Institute Series IS 4, 518–538 (1990)

1.64 P.M.S. Monk, R.J. Mortimer, D.R. Rosseinsky: *Electrochromism: Fundamentals and Applications* (VCH, Weinheim 1995) Part II B, Organic Systems, pp. 124–182 – see also reviews given on p. 139

1.65 R. Cinnsealach, G. Boschloo, S.N. Rao, D. Fitzmaurice: "Electrochromic windows based on viologen-modified nanostructured TiO_2 films", Solar Energy Mat. Solar cells **55**, 215–223 (1998)

1.66 S.K. Deb: "A novel electrophotographic system", Appl. Opt. Suppl. **3**, 192–195 (1969)

1.67 F.G.K. Baucke: "Beat the dazzlers", Schott Information 1e, 11–17 (1983)

1.68 S.P.L. Sørensen: "Enzymstudien. II. Über die Messung und die Bedeutung der Wasserstoffionenkonzentration bei enzymatischen Prozessen", Biochem. Z. **21**, 131–200 (1909), continuation: 201–304 (1909); Compt. Rend. Trav. Lab. Carlsberg **8**, 1 (1909)

1.69 S.P.L. Sørensen, K. Linderstrøm-Lang: "On the determination and value of π_0 in electrometric measurements of hydrogen ion concentrations", Compt. Rend. Lab. Carlsberg **15**, 1–40 (1924)

1.70 R.G. Bates, G.D. Pinching, E.R. Smith: "pH standards of high acidity and high alkalinity and the practical scale of pH", J. Res. NBS **45**(5), 418–429 (1950)

1.71 R.G. Bates: "The modern meaning of pH", Crit. Rev. Anal. Chem. **10**, 247–278 (1981)

1.72 A.K. Covington: "Recent developments in pH standardisation and measurement for dilute aqueous solutions", Anal. Chim. Acta **127**, 1–21 (1981)

1.73 A.K. Covington, R.G. Bates, R.A. Durst: "Definition of pH scales, standard reference values, measurement of pH and related terminology", Pure Appl. Chem. **57**, 531–542 (1985)

1.74 F.G.K. Baucke: "The definition of pH. Proposal of improved IUPAC recommendations", in *Traceability of pH Measurement*, Lectures delivered at the 126th PTB Seminar, PTB-Bericht W-68, ed. by P. Spitzer (Physikalisch-Technische Bundesanstalt, Braunschweig 1997) pp. 10–20

1.75 F.G.K. Baucke, P. Spitzer, R. Naumann: "pH controversy revisited" (Letter to the Editor), Anal. Chem. News & Features **70**(7), 226 A (April 1, 1998)

1.76 BIPM, IEC, ISO, OIML: *International Vocabulary of Basic and General Terms in Metrology*, 2nd ed. (ISO, Geneva 1994) pp. 45–47

1.77 R.G. Bates, E.A. Guggenheim: "Report on the standardization of pH and related terminology", Pure Appl. Chem. **1**, 163–168 (1960)

1.78 F.G.K. Baucke: "Differential-potentiometric cell for the restandardization of pH reference materials", J. Electroanal. Chem. **368**, 67–75 (1994)

1.79 R.G. Bates: "Electrometric pH determination", Chimia **14**, 111–126 (1960)

1.80 F.G.K. Baucke, R. Naumann, Ch. Alexander-Weber: "Multiple-point calibration with linear regression as a proposed standardization procedure for high-precision pH measurements", Anal. Chem. **349**, 3244–3251 (1993)

1.81 F.G.K. Baucke: "High-temperature sensors for oxidic glass-forming melts", in *Chemical and Biochemical Sensors* Vol. 3, *Sensors. A Comprehensive Survey*, ed. by W. Göpel, J. Hesse, J.N. Zemel (VCH, Weinheim 1992) pp. 1155–1180

1.82 Kühnreich & Meixner: "Redox control in glass melts", Glastechn. Ber. Glass Sci. Technol., for example: **71**, XVI (1998)

1.83 C. Montel, C. Rüssel, E. Freude: "Square-wave voltammetry as a method for the quantitative in-situ determination of polyvalent elements in molten glass", Glastechn. Ber. **61**, 59–63 (1988)

1.84 J.A. Duffy: *Bonding, Energy Levels and Bands in Inorganic Solids* (Longman, Harlow, Essex, England 1990) Chaps. 6 and 8

1.85 J.A. Duffy, M.D. Ingram: "An interpretation of glass chemistry in terms of the optical basicity concept", J. Non-Cryst. Solids **21**, 373–410 (1976)

1.86 F.G.K. Baucke, J.A. Duffy: "The effect of basicity on redox equilibria in molten glasses", Phys. Chem. Glasses **32**, 211–218 (1991)

1.87 F.G.K. Baucke, G. Röth: "Electrochemical mechanism of the oxygen bubble formation at the interface between oxidic melts and zirconium silicate refractories", Glastechn. Ber. **61**, 109–118 (1988)

1.88 F.G.K. Baucke, W.A. Frank: "Conductivity cell for molten glasses and salts", Glastechn. Ber. **49**, 157–161 (1976)

1.89 F.G.K. Baucke, J. Braun, G. Röth, R.-D. Werner: "Accurate conductivity cell for molten glasses and salts", Glastechn. Ber. **62**, 122–126 (1989)

1.90 F.G.K. Baucke, R.-D. Werner: "Temperature-dependent mixed alkali effect in silicate melts", in *Glass 89, Proc. XV. Int. Congress on Glass 1989*, Vol. 2a,

Properties of Glass. New Methods of Glass Formation. Techn. Sessions, ed. by O.V. Mazurin (NAUKA, Leningrad 1989) pp. 242–246

1.91 O. Svensson: "Electrical conductivity of glasses in the composition range of 24% PbO lead glass", Glasteknisk Tidskrift **35**(1), 5–11 (1980)

1.92 O. Swensson: "Electrical conductivity of glasses in the composition range of 24% PbO lead crystal. – Complementary measurements, Part II", Glasteknisk Tidskrift **35**(2), 37–40 (1980)

1.93 H. Höness, A. Jacobsen, K. Knapp, T. Marx, H. Morian, R. Müller, N. Reisert, A. Thomas: "Production of Zerodur® in special shapes", in *Low Thermal Expansion Glass Ceramics*, Schott Series on Glass and Glass Ceramics, ed. by H. Bach (Springer, Berlin, Heidelberg 1995) pp. 143–169

1.94 Schott Glas: "Schott group, facts and figures", brochure (Schott Glas, Mainz 1999)

1.95 E. Plumat, F. Toussaint, M. Boffe: "Formation of bubbles by electrochemical processes in glass", J. Am. Ceram. Soc. **49**, 551–559 (1966)

1.96 W. Oldekop: "Über thermoelektrische Erscheinungen an Gläsern", Glastechn. Ber. **29**, 73–78 (1956)

1.97 D.E. Carlson, C.E. Trzeciak: "Thermoelectric effects in ion conducting glasses", Phys. Chem. Glasses **14**, 10–15 (1973)

1.98 Y. Ukyo, K.S. Goto: "Absolute thermoelectric power of ZrO_2–CaO–MgO–Y_2O_3 solid electrolytes and liquid PbO–SiO_2 systems", Tetsu-to-hagane **69**, 67–72 (1983)

1.99 W. Jost, K. Hauffe: *Fortschritte der physikalischen Chemie*, Vol. 1, *Diffusion, Methoden der Messung und Auswertung*, ed. by W. Jost (Steinkopff, Darmstadt 1972) pp. 249–254

1.100 K.S. Goto: *Solid State Electrochemistry and Its Application to Sensors and Electronic Devices*, Materials Science Monographs, Vol. 45 (Elsevier, Amsterdam 1988) pp. 105–110

2. Electrochemistry of Solid Glasses

Friedrich G.K. Baucke

2.1 Introduction

Glass and electrochemistry? The obviously "inert" character of most solid glasses seems to exclude the application of electrochemistry to these materials. In fact, one-component as well as multi-component glasses appear to be more suited as electrically insulating materials than as substances on which electrical or electrochemical experiments can be conducted. Nevertheless, a chapter covering more than one half of this book is devoted to the electrochemistry of these electrically "dull" materials. Is there a misunderstanding, or do reactions take place which one does not normally realize without specific instrumental support?

Indeed, glasses would not corrode, energy-storing secondary sodium-sulphur batteries with glass electrolyte could not be constructed, and glass electrodes – the analytical tool probably most often and most widely used in research and industrial laboratories, for process control, in medicine, geographic and deep-sea investigations, environmental research and control – would not be available without electrochemical processes within and at these materials. What prevents the electrochemical reactions from being easily observed is the extremely small electric conductance, that is, the small velocity with which the charged atoms, the ions, respond to the driving forces of electric field, concentration gradient and temperature gradient and, consequently, the slow rates with which glasses react with contacting solid, liquid, and gaseous phases. These reactions are complicated processes due to the glass structure, which consists of a network of so-called network-forming elements and oxygen and network-modifying cations, which are the slow charge carriers mentioned above. In addition to the cations, also the network can react with components of contacting phases, for instance solutions, which complicates the situation considerably.

Glass electrodes are ideally suited for investigating the electrochemistry of glasses because nearly all imaginable electrochemical reactions take place in and at their membrane. Because, in addition, the functioning of the glass electrode was far from clear and its elucidation promised the development of new and better membrane glasses, Schott Glas in the early 1970s decided to let the Electrochemical Laboratory do research in this field. The project was particularly promising because a surface-analytical technique with a depth resolution of only a few nanometres (ion bombardment for spectro-chemical analysis, IBSCA) had just been developed at Schott. It allowed the investigation of extremely thin subsurface regions of glasses before, during, and after defined reaction periods and thus literally offered a view into the glass and informed about the time dependence of depth-dependent reactions between glasses and contacting phases, and of equilibria eventually established. This work has been carried on for more than two decades, although not with top priority because of other responsibilities of an industrial electrochemical laboratory, and resulted in the so-called *dissociation mechanism* of the glass electrode.

This mechanism is based on physical processes at and below the membrane glass surface and differs fundamentally from the numerous attempts in the past to explain the functioning of glass electrodes. Due to the lack of sufficiently sensitive analytical tools, the earlier explanations were based on the experimental information available at that time, which was potentiometric and thermodynamic data (in the broadest sense of the word). They thus necessarily represented theories because thermodynamics, in principle, gives no information about the course of a reaction. The electrochemistry of the functioning of glass electrodes, consequently, remained obscured from the detection of the "glass electrode effect" by Cremer in 1906 and the first attempt at its explanation by Haber and Klemensiewicz in 1909 until information about the processes at glass membranes was obtained by means of appropriate surface-analytical techniques, for instance IBSCA.

In parallel to the application for studying diffusional processes at the glass surface, IBSCA was also applied to investigate field-driven ionic migration into and within membrane glasses. These measurements secured the results obtained and the conclusions drawn and yielded new information about ionic transport processes in glasses, several of which are not directly connected with their functioning as electrode membranes. The work has thus developed into a rather complete picture of the electrochemical processes in and at ionically conducting oxidic membrane glasses in contact with solid, liquid, and gaseous phases and under the driving forces of concentration and electric potential gradients. The main aspects and results will constitute the second part of this book. The relevant literature will be cited in the respective sections.

Chapter 2 starts with a section on the phenomenological, operational basis of glass electrodes. It is intended for those readers whose daily work does not concern, and who are thus less familiar with, glass electrodes and their application but who need this basic knowledge for understanding the material presented in the following sections. In particular, it is made clear that the glass electrode is not an electrode in the strict sense of this term but an arrangement that allows one to measure the pH or pM dependence of the electric potential difference between two electrolytes, a solid electrolyte, the ion-sensitive glass membrane, and a liquid electrolyte, the solution of interest. The origin of this membrane potential and the numerous consequences of the electrochemical reaction (and equilibrium) which it reflects are shown to be the centre point of the entire work reported.

Section 2.3 gives an overview of past theories of glass electrode functioning, which had been developed when the processes at the membrane surface were not yet known. They were based on a number of ingeniously invented ionic reactions at and in the membrane surface, all of which agreed with the thermodynamics of glass electrodes, which, however, does not mean that the imagined processes are the reactions actually taking place. What is amazing is the great number of invented reactions, which range from an equilibrium between a subsurface water phase in the glass surface with the solution over an exchange of alkali ions of subsurface glass ranges versus hydrogen ions of the solution to the transfer of fast protons (which do not exist in silicate glasses) with mobilities many orders of magnitude larger than those of alkali ions through the glass membrane. The main ideas of these theories will be briefly described.

The experimental methods applied to the investigation of the reactions at the membrane are briefly described in Sect. 2.4. In particular, IBSCA and nuclear radiation analysis (NRA), a depth profiling method for protons in glasses, will be mentioned. Besides, the "modified moving boundary method", an application of a technique for measuring transport numbers in electrolyte solutions to the determination of time-dependent concentration profiles caused by interdiffusion and field-assisted migration, will be briefly reported in this experimental section and discussed in more detail in Sect. 2.6.6.

After these more introductory sections, the following three Sects. 2.5–7 report on the processes whose common origin is the interface between the glass surface and the solution. This phase boundary neither is nor behaves as an inert plane, as often tacitly assumed in the past, but bears functional groups of the glass which form equilibria with ions (hydrogen, alkali) in contacting solutions. They are heterogeneous equilibria, which involve charged particles of both phases. They represent the central process of the dissociation mechanism and are responsible for the three most significant electrochemical reactions of electrode membrane glasses.

(a) Involving negatively charged anionic groups bound to the glass surface, as they do, they are the cause of a potential of the glass relative to that of the solution and thus of the pH- and pM-dependent glass electrode response.

(b) Involving a solution-dependent coverage of the glass surface with alkali ions and/or protons, which are attached to the surface groups at equilibrium, they cause an interdiffusion of these ions at the surface versus ions of the glass interior. It is the first step of glass corrosion, whose course is thus determined by the composition of the solution (pH, pM) contacting the glass.

(c) The cations attached to the glass surface in equilibrium with those in the solution are those individual ions that are transferred anodically into the glass and replace the ions of the bulk glass drifting towards the cathodic membrane surface when an electric field is applied to the membrane. Due to large exchange current densities of the surface equilibria, the ions leaving the surface are immediately replaced by ions of the same kind from the solution so that the equilibrium is practically not polarized at moderate current densities. The relative individual ion concentrations transferred into the glass are thus equal to the relative concentrations at the anodic surface, whereas their subsequent migration rates within the glass are determined by their concentration-dependent mobilities. Membrane electrolyses, consequently, reveal the solution-determined quantitative coverage of the membrane surface with protons and/or alkali ions, and the combination of the surface concentrations thus determined with the respective potentials of the membrane glass, see (a), confirm quantitatively the dissociation mechanism.

In Sect. 2.5, equations for the interfacial potential of the glass membrane are derived for protons and alkali ions on the equilibrium- and on an electrode-kinetic basis. An unexpected result of these derivations is that the thermodynamically correct pH and pM response is subideal, $k' < k = 2.303\,RT/F$, independent of glass composition and solvent (H_2O, D_2O). Although surprising for most researchers at first glance, this finding explains thermodynamically the sub-Nernstian response, which has occasionally been reported in the literature and was also confirmed by our own measurements.

Section 2.6 reports on field-assisted ion migration in glasses. The solution-determined field-driven transfer of ions from their position at the surface groups into the glass was applied as a means to generate protonated layers with defined depth-independent proton concentrations and to determine mole-fraction-dependent mobilities of both the lithium ions of the glass and the replacing protons. It will be seen that, within the entire concentration range, protons are less mobile than alkali ions, which accounts for the distinct boundary between protonated layers and bulk glasses and which is explained by a new migration mechanism. At small proton concentrations, certain arrangements of anionic groups act as proton traps. It is also shown that the transfer of ions into glasses depends on the relative steric conditions of ion and membrane.

Different from solution-contacted lithium silicate glasses, electrolyses on sodium borate glasses employing metal electrodes (platinum, lead, amalgams) cause the well-known migration of alkali ions towards the cathodic surface and a simultaneous, equivalent, anodic removal of oxide ions from the glass network, as reported in Sect. 2.6.8. Thus, the total alkali oxide content of $35\,mol\%$ is removed from the glass, and it is extremely surprising that it is reintroduced when the electric field is reversed.

In Sect. 2.7, ion exchange and subsequent corrosion of membrane glasses are reported in some detail. The exchange of alkali ions for protons ("leaching") differs basically from an exchange of different alkali ions, in which protons are not involved. The (inter-)diffusion voltage within leached surface layers was determined experimentally, yielding a magnitude of approximately −70 mV (glass negative). Also reported is the interaction of membrane glasses with humid atmospheres (Sect. 2.7.2), which, due to the different conditions at the glass surface, exhibits temperature-dependent mechanisms fundamentally different from the mechanism observed with contacting solutions.

Section 2.8 reports on a hitherto unpublished investigation of the potential formation at the protonated glass/platinum interface in hydrogen-containing atmospheres. The electrode glass(H$^+$)|Pt,H$_2$ exhibits the theoretically correct hydrogen (H$_2$) function, as does the Pt,H$_2$ electrode in aqueous solution. It is shown that functional surface groups of the glass are also involved in this equilibrium and, in principle, result in a subideal response also of the hydrogen slope. However, the effect is obviously too small to be detectable experimentally.

After Sect. 2.9, which summarizes all results relevant to the functioning of glass electrodes and describes the dissociation mechanism in detail, Section 2.10 reports on the application of glass electrodes, especially for pH measurements and shows by the present activities of IUPAC that this field is not an established analytical science but that the definition of pH, the standardization of pH measurements, and the calibration of pH-measuring glass electrode-containing cells being are redefined at present and will be newly recommended on the modern basis of metrology.

Chapter 2 finally concludes with Sect. 2.11 on the functioning and application of glass electrodes in heavy water. It is shown that the glass electrode mechanism in D$_2$O is the same as in H$_2$O and that the potential is also subject to a subideal response. The knowledge of the mechanism explains a well-known empirical method of pD determination by the different dissociation constants of the glass surface groups in H$_2$O and D$_2$O solutions and by the different subsurface diffusion potentials developed in these solvents (deuteron effect). Also, a related empirical method of pNa determination is explained by the different thermodynamic properties of the membrane glass in the two solvents (deuterium oxide effect). In addition, it is found that the sodium error of (pH, pD) glass electrodes is significantly smaller in heavy than in light water solutions.

2.2 Operational Basis and Response of Glass Electrodes

This section gives a short survey of the effect on which the glass electrode is based, the construction of these sensors, and their application in electrochemical cells for the measurement of ion activities in solutions. It is the basis of the following treatment of the electrochemistry of solid glasses and was written particularly for those readers whose daily work is not concerned with glass electrodes and their applications. It is thus kept as brief as possible and at the same time sufficiently informative to ensure that the effects of electrochemistry shown by glass electrodes and their explanation can be understood. In addition, numerous monographs and overview articles on the operational basis and application of glass electrodes are available and should be consulted for further information, see, for instance, [2.1–7]. However, they also provide only the basis for the material treated, although often in a more detailed form.

In principle, all oxidic glasses exhibit an electrical potential difference or Galvani voltage ε_m with respect to contacting aqueous and most non-aqueous solutions, whose magnitude depends on the activities of certain cations, i.e., hydrogen and alkali ions, in the liquid. The glass electrode functioning is based on this specific property. Different from metal–metal ion and redox electrodes, however, the glass–solution system is not an electrode in the electrochemical sense as it is not characterized by an electron exchange between the phases but consists of a solid and a liquid electrolyte, and an ionic process at their interface is responsible for the potential difference. For the purpose of measuring the dependence of this Galvani voltage on the ion activities in the contacting solution (in principle, an absolute Galvani voltage is not measurable), the ionic reaction at the interface must be coupled electrochemically to the electrode reactions of two terminal metal electrodes, which can then be connected to a voltmeter. This is done by extending the glass–solution system by appropriate additional phases, which results in a glass electrode cell as, for instance, represented by the basic cell scheme

$$\left| \begin{array}{c} \text{(ext.) ref.} \\ \text{electrode} \end{array} \right| \begin{array}{c} \text{ref. elec-} \\ \text{trolyte} \end{array} \left\| \begin{array}{c} \text{meas'g} \\ \text{solution} \end{array} \right| \begin{array}{c} \text{glass} \\ \text{membrane} \end{array} \left| \begin{array}{c} \text{int. ref} \\ \text{electrolyte} \end{array} \right| \begin{array}{c} \text{int. ref.} \\ \text{electrode} \end{array} \right| .$$

$$\qquad \varepsilon_r \qquad\qquad \varepsilon_j \qquad\qquad \varepsilon_m \qquad\qquad \varepsilon_{m,i} \qquad\qquad \varepsilon_{r,i} \qquad\qquad (2.\mathrm{I})$$

Cell scheme (2.I) shows the added phases and demonstrates that an additional Galvani voltage is introduced at each new interface so that the measurable electromotive force, or briefly emf, E', of cell (2.I) is the sum of all Galvani voltages of the arrangement.

$$E' = \varepsilon_{r,i} + \varepsilon_{m,i} + \varepsilon_m + \varepsilon_j + \varepsilon_r . \qquad (2.1)$$

The conditions are chosen such that the Galvani voltages introduced by the extension of the cell and thus their sum are constant,

$$\sum \varepsilon_n = \varepsilon_{r,i} + \varepsilon_{m,i} + \varepsilon_j + \varepsilon_r = \text{const} . \qquad (2.2)$$

The Galvani voltages ε_r and $\varepsilon_{r,i}$ of the reference and the internal reference electrode, respectively, which are usually electrodes of the second kind, as silver-silver chloride, calomel (mercury-mercurous chloride) or Thalamid® (thallium amalgam-thallous chloride) electrodes, are fixed by the composition of the reference and internal reference electrolyte, respectively, and the Galvani voltage $\varepsilon_{m,i}$ at the interface glass-internal reference electrolyte is kept constant by the constant pH due to the buffer action of this solution. The liquid junction potential ε_j between the reference and the measuring solutions depends slightly on the composition of the measuring solution (at constant composition of the reference electrolyte) and thus introduces an unavoidable small uncertainty into the emf but can be regarded as sufficiently constant for most applications. The emf E' of cell (2.I), consequently, is a unique function of the Galvani voltage ε_m between the glass and the measuring solution and thus of the potential-determining ionic activity of interest,

$$E' = \sum \varepsilon_n + \varepsilon_m = \text{const} + \varepsilon_m(\text{pH}, \text{pM}) ,\qquad (2.3)$$

where the activities of hydrogen and alkali ions have been expressed by their negative logarithms, $\text{pH} \equiv -\log a_H$ and $\text{pM} \equiv -\log a_M$, respectively. It is mentioned that both ε_m and $\varepsilon_{m,i}$ consist actually of two parts, the phase boundary potential difference between the glass and the solution and a diffusion voltage below the glass surface, which is determined by the composition of the glass surface. These details, however, are of no concern for the operation of glass electrodes as described here and will be treated in detail later.

Cell (2.I) contains the Galvani voltages between the membrane glass and both the internal reference electrolyte and the measuring solution. It should thus be expected that the difference of these voltages is zero if these solutions are identical. This, however, is not exactly the case; a small difference ($\varepsilon_{m,i} - \varepsilon_m$) is always observed, which is called the asymmetrical Galvani voltage or, briefly, asymmetrical potential of the glass membrane. It is larger with sodium- than with modern lithium-containing glasses and can slowly change with time. Its cause is not exactly known although the effect has been amply treated both experimentally and theoretically.

In practice, the electrode glass is employed as a several tenths of a millimetre thick membrane melted to a tube or shaft of a glass with high electrical resistance, which contains the internal reference electrode and internal reference electrolyte as shown in Fig. 2.1. Besides, the (external) reference electrode and reference electrolyte are combined in a glass tube containing a liquid junction device, which allows contact of the measuring and reference solution but excludes their convective mixing. The glass electrode cell is thus divided into two practical units, the glass electrode and the reference electrode, which are immersed in the measuring solution as shown by the practical glass electrode cell scheme

$$|\text{reference electrode}\|\text{measuring solution}|\text{glass electrode}| ,\qquad (2.II)$$
$$\qquad\qquad \varepsilon_j \qquad\qquad\qquad \varepsilon_m$$

which shows the operational cell parts but hides their construction and mode of response and can even be misleading because, due to the functioning of the glass membrane, the glass electrode is not strictly an electrode in the electrochemical sense. Glass and reference electrode are often combined in one practical unit, which is called a combination or single-rod electrode, Fig. 2.2.

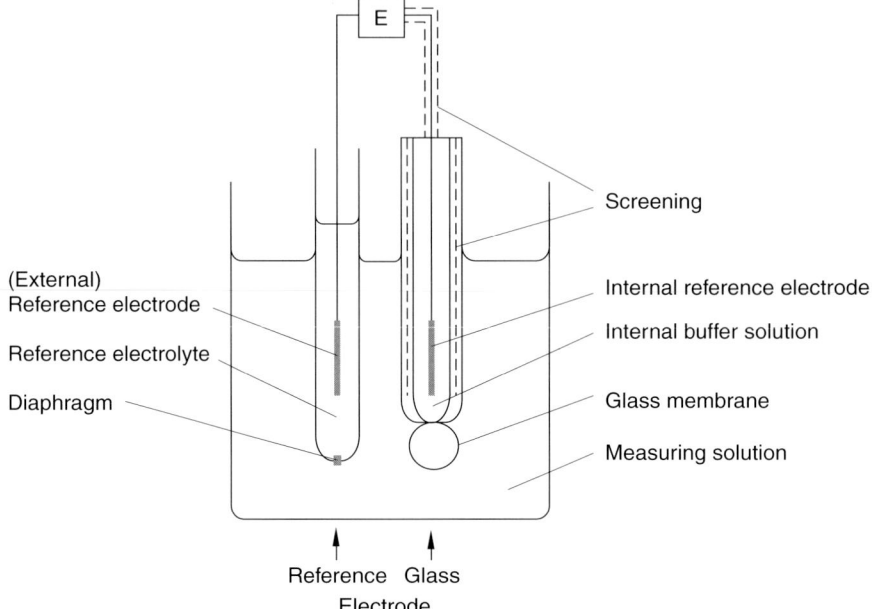

Fig. 2.1. Schematic diagram of an electrochemical cell consisting of a glass and a reference electrode

Due to the high resistivity of the membrane glass, which, nevertheless, completes the electrical circuit, glass electrode cells are distinguished by high resistances, which are usually of the order of several $10^8 \, \Omega$. This necessitates the use of measuring instruments with an input resistance larger by a factor of 10^3 and good screening of glass electrodes and glass electrode leads. Modern pH meters have input resistances of more than $10^{14} \, \Omega$ and thus easily meet this requirement. They are economically available and widely used, and it may be interesting to realize the electrical power given off by a cell connected to such an instrument. For example, a cell with an emf of 1 V connected to a pH meter with an input resistance of $10^{14} \, \Omega$ supplies $10^{-14} \, W$, which, for instance, would heat $1 \, mm^3$ water by 1 K within the unbelievably long time of 1.34×10^4 years. The example demonstrates the tremendous development achieved also in the field of electronics, which has supported the development of practical glass electrode cells and their application.

The response of the Galvani voltage ε_m of the electrode membrane and thus of the emf E' of glass electrode cells to hydrogen and alkali ions depends in a characteristic way on the membrane glass composition. Silicate glasses exhibit a linear dependence of their Galvani voltage on pH within a pH range of $\Delta pH \geq 9$. Figure 2.3 presents the pH response of a pH glass electrode cell at $25\,°C$ and $50\,°C$. The cell is symmetrical, i.e., the reference and internal reference electrodes are of the same kind, for instance consisting of silver-

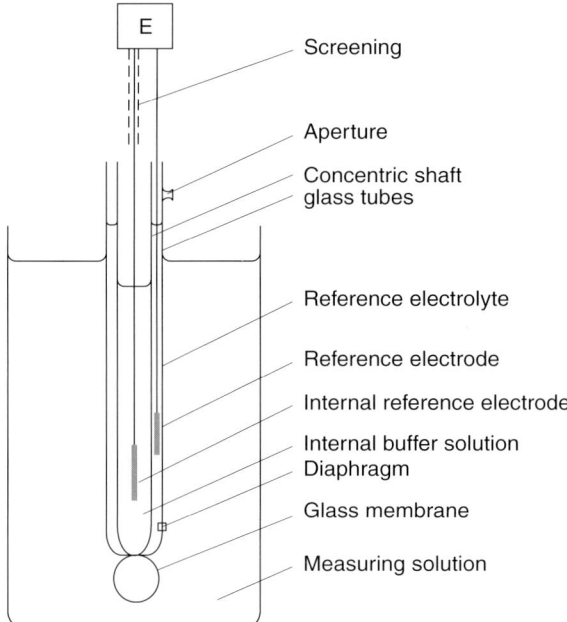

Fig. 2.2. Schematic of a single-rod or combination electrode, containing both glass and reference electrode and the diaphragm connecting reference and measuring solutions in one unit. The reference electrolyte also screens the internal reference electrode and the internal buffer solution electrically. The aperture allows refilling and exchange of reference electrolyte

silver chloride electrodes. The pH of the internal buffer solution is 6.8, and the zero-pH emf, i.e., the pH value of the measuring solution at which the cell exhibits zero electromotive force, thus has a similar value. The temperature-dependent slope is slightly below the Nernstian slope, which is $dE/d\,pH = -2.303\,RT/F$, for instance $-59.16\,mV$ per $\Delta\,pH = 1$ at $25\,°C$, see Sect. 2.5.3. As indicated in Fig. 2.3, at high pH values and large alkali concentrations, deviations from the straight line towards more positive emfs are observed. They are caused by a sensitivity of the membrane glass also to alkali ions, especially to sodium ions, and this deficiency is thus called the alkali or sodium error of the pH glass membrane.

The sodium sensitivity of pH electrode membranes depends on the composition of the silicate glass and is strongly increased, for instance, by the aluminium or boron content of the glass. The influence of these elements is so powerful that only a few mole percent of Al_2O_3 or B_2O_3 make the glass predominantly sodium-sensitive. Figure 2.4a shows the pH response of a symmetrical pNa glass electrode cell at various pNa values at $25\,°C$ and demonstrates the much wider range of sodium ion sensitivity compared to that of the pH glass electrode cell as given in Fig. 2.3. In a certain way, the

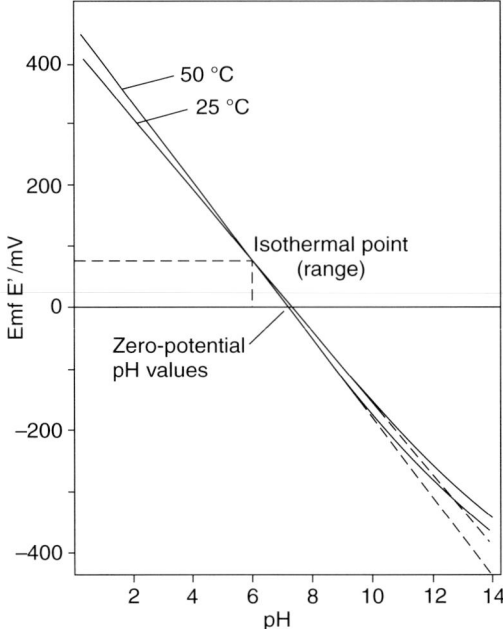

Fig. 2.3. pH response of a symmetrical pH glass electrode cell at 25 °C and 50 °C. The cross-over of the isotherms, called the isothermal point, differs slightly from $E' = 0$ because of the individually temperature-dependent Galvani voltages of the cell

situation is reversed in that such glass electrodes are mainly sodium-sensitive and exhibit a "hydrogen error" at small pH values, which is caused by a "rest sensitivity" of the glass to pH. Figure 2.4b, presenting the pNa response of the same glass electrode cell, here at various pH values, demonstrates this and, for the particular membrane glass, indicates the respective minimum

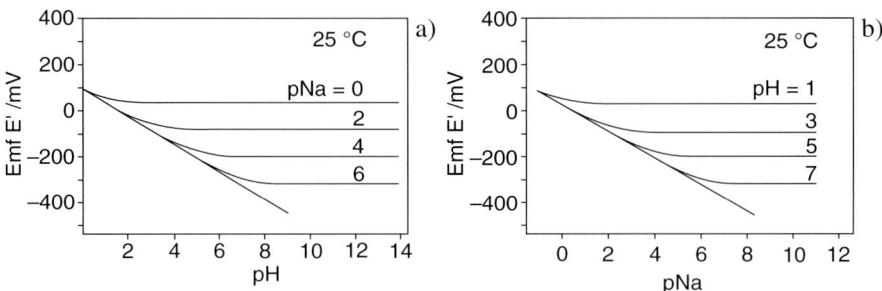

Fig. 2.4. Response of a pNa glass electrode-containing cell at 25 °C. (**a**) emf as a function of pH at various, constant pNa values, (**b**) emf as a function of pNa at various pH values

pH values to be chosen if the glass electrode cell is to be applied for sodium activity measurements. It also shows that the slope of the Galvani voltage with respect to the sodium ion activity of the solution is slightly below the Nernstian slope as in the case of pH-sensitive glasses. The internal buffer solution of the cell yielding the data given in Fig. 2.4, incidentally, had a pNa of ≈ 0.5 at any pH ≥ 3.5.

At very low pH values, i.e., at pH < 0.5, deviations of the Galvani voltage of pH membrane glasses towards more negative values have been reported. The effect is called the acid error and has been observed mostly with sodium-silicate-based glasses. However, because modern electrode glasses are lithium instead of sodium silicate glasses, which show no detectable or only small acid errors at even extremely low pH, this deficiency is of minor significance for practical pH measurements and has received little or no attention in the past two decades.

Because the emf of glass electrode cells contains several constant Galvani voltages, each of which has an individual reproducibility and temperature dependence, see (2.2), and because of the solution dependence of the liquid junction potential and the slightly inconstant asymmetry potential of the glass membrane, the characteristic of the glass electrode cell, i.e., the emf as a function of pH or pM, which is also called an isotherm, Figs. 2.3 and 2.4, must be determined by calibration at each measuring temperature. For pH this is usually done by means of two or more standard buffer solutions, whose standard pH values have been assigned by national standard organizations. Currently, two such standard pH scales with slightly different standard pH values due to different philosophies are recommended by IUPAC, and two different modes of calibration, i.e., with two and with more than two standard buffers, have been proposed. If only limited precision and accuracy of the measurements are required, both standard scales and calibration methods can be applied without complication. However, pH standardization has become an internationally widely discussed, complex field, especially for measurements with high requirements. This subject will be treated in detail in Sect. 2.10.

It may also be mentioned that glass electrodes work in non-aqueous solutions in much the same way as in water. Because of the different autoprotolysis constants of the numerous solvents of interest, a standard pH scale must be established in each of them but only a limited number has been set up at present due to the large amount of work connected with standardization. A special case is deuterium oxide where glass electrodes respond to pD, which is defined in a similar way as pH, pD $= -\log a_D$, as they do to pH and where a standard pD scale has been established. A detailed discussion of the relative emfs of a glass electrode cell in light and heavy water is given in terms of the *dissociation mechanism* of the glass electrode in Sect. 2.11.

2.3 Historical Notes

Perhaps no other physico-chemical method has experienced as many and as different theories and explanations of their functioning as the glass electrode. They differ in their main points but overlap in other respects and were sometimes published at the same time, and it is rather difficult to divide them into characteristic groups. Still in 1964, *Schwabe* and *Suschke* refereed on the subject and described four simultaneously existing theories [2.8]. Without doubt, the reason for this parallel development of explanations was the lack of information about the ionic processes taking place at and below the glass surface causing the ion-activity-dependent Galvani voltage between the membrane glass and the solution. Scientists involved in the subject could measure emfs of glass electrode-containing cells, i.e., thermodynamic data, and their dependence on ionic activities, temperature, and glass composition and could electrolyse the membrane glass, i.e., introduce and replace ions in the glass by means of electric fields, which, surprisingly, was not carried out very often, and could measure resistance changes during the electrolyses. These experiments, however, could result, at best, in information about gross changes in the glasses and their influence upon the electrode response and on speculations about the possible functioning of the glass electrode, which were called theories. Even the advent of radioactive elements in the 1950s, which were used as labelled ions in solution and membrane glass [2.9, 10], yielded only gross amounts of the ions exchanged between the phases but not their depth-dependent concentrations, which would have been needed for elucidating the reactions and equilibria at the glass–solution interface. The evolving theories were thus, although often ingenious, combinations of these experimental results with thermodynamic first principles but did not inform about the mechanism of glass electrode functioning.

These historical notes can neither give a complete list of publications on the functioning of the glass electrode nor present a complete survey of past theories and their mutual interdependence. Rather, this section offers an overview of the development of the theories of the past to give an impression of what the situation was like in the early 1970s, when we started the investigations reported in Chap. 2. In particular, an analysis is presented of the origin, the many modifications, and the reasons for the longevity of the ion exchange theory. Ion exchange was first suggested in 1923/1924, and the ion exchange theory is still cited as the basis of glass electrode functioning, although it is a thermodynamically based theory like most other theories of that time. For more details of the theories developed in the past, refer to the numerous monographs and review articles available [2.4, 5, 11–21].

Giese, in 1880, was probably the first to ever construct and use a glass electrode, albeit without realizing the significance of his experiments. Thus, *Helmholtz*, during his Faraday Memorial Lecture in 1881, presented Giese's Daniell cell [2.22] involving a glass membrane between the $ZnSO_4$ and $CuSO_4$ solutions,

$$|Zn|ZnSO_4|glass|CuSO_4|Cu| , \qquad\qquad (2.III)$$

which exhibited the same emf as Daniell cells containing a liquid junction device, usually called a semipermeable membrane at that time, instead of the glass although the construction of the cell excluded any surface conduction due to humidity, Fig. 2.5. What was to be shown was the possibility of constructing electrochemical cells with extremely high internal resistance; what was not clear to either scientist was the way the semipermeable membrane was coupled electrochemically to the two adjacent solutions; and what was actually achieved but remained obscured was a pH glass electrode cell at zero-potential pH because the two neutral solutions of cupric sulphate and zinc sulphate obviously had approximately the same pH. The scientists concentrated mainly on the ionic conduction of the glass and missed the practical significance of the experiment although a different cell constructed by Giese,

$$|Hg|glass|solutions(HNO_3, H_2O, NH_3, and others)|Pt| , \qquad (2.IV)$$

had even clearly shown a strong dependence of the emf on the kind of solution contacting the glass, especially between nitric acid and ammonia, despite the ill-defined contacts of glass and solution [2.23].

Neither was the electrochemical coupling of the solutions to the glass clear to *Cremer*, who took up the idea of glass being a semipermeable membrane for his studies of physiological systems in 1906 [2.24]. However, to his

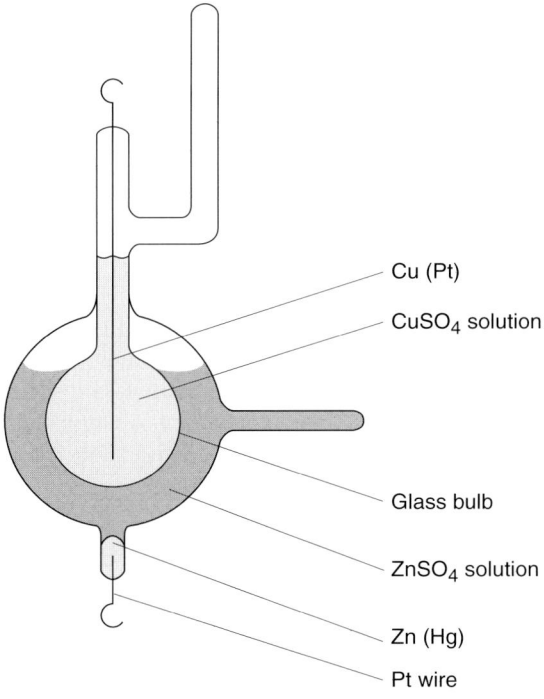

Cu (Pt)

CuSO$_4$ solution

Glass bulb

ZnSO$_4$ solution

Zn (Hg)

Pt wire

Fig. 2.5. Giese's Daniell cell with a "semipermeable glass membrane", the first glass electrode cell ever constructed, as presented by, and reproduced after, *Helmholtz* [2.22]

surprise, the experiments showed a distinct character of the glass–solution interface as, fortuitously, he applied solutions of acids and bases with various concentrations in such cells with a glass membrane. The emfs he measured depended strongly and systematically on the hydrogen ion concentration of the solutions (the term pH had not been coined), and he verified conclusively the glass membrane as the cause of his observation by comparing a cell having the cell scheme

$$\left|Zn\left|ZnSO_4\right\|NaCl\right\|H_2SO_4\left|glass\right|NaCl\|ZnSO_4\left|Zn\right| \tag{2.V}$$

with the same cell without an intermediate glass membrane,

$$\left|Zn\left|ZnSO_4\right\|NaCl\right\|H_2SO_4\|NaCl\left|ZnSO_4\right|Zn\right| , \tag{2.VI}$$

which, being completely symmetrical, exhibits zero emf in contrast to the hydrogen ion concentration-dependent emfs of cell (2.V). He saw the meaning of his measurements but still did not realize their practical significance despite many systematic measurements, although he created glass electrodes with a bulb-shaped membrane, which, in principle, are still applied today, see for instance Figs. 2.1 and 2.2 of Sect. 2.2.

In 1909, the year *Sørensen* proposed the pH value as the negative logarithm of the hydrogen concentration, pH $= -\log c_{H^+}$ [2.25], *Haber* and *Klemensiewicz* reported their systematic application of Cremer's bulb-type glass electrodes in combination with reference electrodes to the titration of acids and bases [2.26]. Their titration curves were nearly identical with those obtained with Pt,H$_2$ electrodes, but they realized that their glass electrode cells, unlike the latter, could be used in the presence of reducing and oxidizing compounds in the solutions. By applying glass electrodes, they thus performed the first acidimetric titrations in redox component-containing solutions.

In addition to their practically significant experiments, *Haber* and *Klemensiewicz* suggested the first theory about the functioning of glass electrodes [2.26]. By modifying *Nernst's* thermodynamic derivation of a potential difference between mixed salt crystals and their saturated solutions [2.27], and knowing *Schott's* [2.28] and *Förster's* [2.29–31] publications on the interaction of glasses with water, they assumed that the water-containing surface layers formed on the glass during contact with water, which they called the "water phase" of the glass, contained constant concentrations of hydrogen and hydroxyl ions. They obtained an equation which represented their titration curves rather well but which does not reflect the sodium error of the glass, whose existence was, of course, not known at that time.

Haber's water phase theory, which, incidentally, was later termed "phase boundary theory" by *Dole* [2.32], was taken up by *Hughes* as late as in 1928, who modified it by assuming a constant pH of the water-containing glass surface layer due to the buffer action of silicic acid and sodium silicate, which, according to his assumption, were formed during the glass–water interaction

[2.33]. In addition, Hughes explained the sodium error, which had meanwhile been realized, by the break-down of the glass network at pH ≈ 9, which is the approximate magnitude of the first dissociation constant of silicic acid, and rejected the sodium sensitivity of the membrane glass as its cause. It is certainly of interest that even in 1955 *Britton*, in his book, preferred Hughes' buffer action theory of pH glass electrodes [2.14].

A modification of Haber's water phase theory was suggested by *Gross* and *Halpern* [2.34], who assumed that electrolytes, for instance acids and bases, distributed between the water-containing glass surface range and the solution until equilibrium was attained. In that way they arrived at a distribution equilibrium which represents the functioning of the glass electrode. However, as *Dole* pointed out [2.32], the time necessary for the formation of the distribution equilibrium is much too long to be in accordance with the relatively fast response exhibited by glass electrodes, particularly during titrations.

In 1931, *Lengyel* put forward a theory which assumed adsorption as the basis of glass electrode functioning [2.35, 36], although in 1920 *Freundlich* and *Rona* had already rejected adsorption as a possible basis of glass electrode functioning [2.37]. Lengyel tried to verify this idea by means of electrodes with quartz membranes, which do not contain any alkali ions for an exchange against ions from the solution and which, according to his conviction, do not form surface layers by uptake of water. All reactions, besides adsorption, were thus believed to be excluded. Indeed, the quartz membranes responded to hydrogen and alkali ion activities but exhibited remarkably strong sub-theoretical slopes. Nevertheless, these experiments are of some fundamental interest, and there will be more to say about them in Sect. 2.9.

As recently as in 1990, *Cheng* tried to revive the idea of electrode functioning being based on ion adsorption [2.38, 39]. He rejects the fact that glass electrode cells are galvanic cells and thus a kind of battery and proposes that the emf of such cells, in reality, is made up of two "capacitors" located at the interfaces between the membrane and the contacting solutions and that the capacity of these condensers is caused by adsorbed ions. Cheng claims that the Nernstian equation has been "misused" whenever it was applied to glass electrodes, but he suggests no alternative except the very basic condenser equation, which, however, offers no possibility of calculating measured voltages from ionic activities in the solutions. In addition, any interface between two electrolytes or between an electrolyte and an electrode metal represents a condenser with a solution-dependent capacity and has been widely treated as such, see for instance [2.40]. Because some of the experiments reported by *Cheng* are fundamentally questionable [2.39], the future development of his "capacitor theory" must be awaited.

Another electrochemical phenomenon proposed as the basis of glass electrode functioning is ionic diffusion. Thus, *Cremer*, in 1924, assumed permeability of electrode glasses to hydrogen ions and obtained a relationship for the pH-dependent emf of glass electrode cells [2.41] by applying *Donnan's* equation of semipermeable membranes [2.42] as well as *Henderson's* equation for liquid junction potentials [2.43, 44]. The idea of a permeability of glass was extended to sodium ions by *Michaelis*, which resulted in an expression yielding also the sodium error [2.45]. A further extension of the theory of the permeability of glass to water reported by *Dole* resulted in a relationship for the acid error of sodium silicate glasses [2.46]. However, these theories were experimentally disproved by *Schwabe* and *Dahms* in 1959 [2.47] and again by *Hammond* in 1962 [2.48], who showed by means of electrolyses with tritium-labelled anodic solutions that "glass is not permeable to hydrogen ions". This conclusion, which has recently been questioned [2.49], see, however, [2.50], will be modified in Sect. 2.6.3.

Recently, *Abe* and coworkers [2.51] tried to revive an old theory created by *Dole* in 1931, according to which pH glass electrodes function due to the diffusion of protons and alkali ions through the "boundary layer" between glass and solution [2.32]. Abe and coworkers base the revival on two grounds. (1) They claim the observation that hydrogen-bonded protons are highly mobile in several alkaline-earth phosphates and some other glasses [2.52, 53], where they exhibit mobilities of up to 10^8 times as high as those of added alkali and silver ions. (2) From the fact that the characteristic infrared absorption band at $2900\,\mathrm{cm}^{-1}$ caused by these protons is also observed weakly with silicate pH membrane glasses, they conclude that highly mobile hydrogen-bonded protons are contained also in alkali, for instance lithium silicate glasses as used for pH electrodes. The authors thus assume that these intrinsic (impurity) protons connect the opposite surfaces of the glass membrane ionically. However, they have never verified experimentally the high mobility of protons in silicate glasses. In general, protons are much less mobile than alkali ions in silicate glasses [2.54, 55]. In particular, protons with an infrared absorption band near $2900\,\mathrm{cm}^{-1}$ were found to be trapped in mixed lithium siloxy-silanol groups, $[\mathrm{SiOH}\cdots\mathrm{OSi}]^-\,\mathrm{Li}^+$, [2.56], see also Sect. 2.6.7, and have extremely low, if not negligible, mobilities, which obviously renders the proposed fast proton theory of pH glass electrodes obsolete.

The ion exchange theory has existed for the longest time and is still often cited today although it belongs to the thermodynamic theories. What are the reasons for its longevity? In 1937, *Nicolsky* took up the idea [2.57], already suggested by *Schiller* in 1924 [2.58] and *Horovitz* in 1925 [2.59], that an exchange of alkali and hydrogen ions between glass and solution might be the cause of pH glass electrode functioning. He assumed that this ion exchange, which, for sodium ions, is represented by

$$\mathrm{Na}^+(\text{glass}) + \mathrm{H}^+(\text{solution}) = \mathrm{Na}^+(\text{solution}) + \mathrm{H}^+(\text{glass}) \qquad (2.4)$$

led to a reversible equilibrium in reasonably short times, and he called the equilibrium constant the ion exchange constant,

$$K_{\mathrm{H\,Na}} = \frac{a_{\mathrm{Na}}a'_{\mathrm{H}}}{a_{\mathrm{H}}a'_{\mathrm{Na}}}, \qquad (2.5)$$

where a and a' are the activities in the solution and in the glass, respectively, of the ions indicated. He treated this assumed interfacial reaction thermodynamically and obtained the following well-known equation for the emf E of the glass electrode cell,

$$E = E^0 + \frac{2.303\,RT}{F}\log(a_H + K_{H\,Na}a_{Na})\,,\tag{2.6}$$

which was later given his name. It was certainly a great success that the Nicolsky equation described the experimental results rather well, although less exactly in the sodium error range. Nevertheless, the ion exchange theory is entirely thermodynamic in character in that it is not based on definite bond energies of the ions to the glass but only on energy differences. It describes, consequently, any ion exchange, independent of the solid material and the locus of the solid phase to which the ions are attached, as also stated by *Eisenman*, who treated glass as an ideal ion exchanger [2.10], and by *Nicolsky* and coworkers [2.60]. According to this indeterminate character, various locations have been suggested in the literature, for example "in the glass" [2.60], "in the hydrated layer" [2.61], and "in the surface layer of the glass" [2.60], and several different sites have been proposed, for instance "ionogenic groups with different acidity" [2.60] and "ion exchange sites" [2.62]. In this situation, the glass electrode functioning cannot be interpreted by a mechanism, but can, at best, be described quantitatively by the Nicolsky equation.

However, the experimental success of the Nicolsky equation led to the far-reaching erroneous conclusion that the assumed ion exchange which it was based on was also the mechanism through which the glass electrode functioned. This conclusion triggered a tremendous amount of theoretical and experimental work by which the researchers tried to do the impossible, that is, to refine a non-existent mechanism by means of the assumption of more mechanistic details in order to fit the thermodynamic Nicolsky equation to the experiments. Thus, *Nicolsky* published the first variant of his theory in 1953 [2.60], *Nicolsky* and coworkers, in 1962, applied the mass action law to dissociating groups and activity coefficients to ions in the glass phase and thus developed the generalized ion exchange theory [2.60]; *Eisenman* in 1957, introduced a fitting parameter into the Nicolsky equation and, in 1962, developed his selectivity theory as an addition to the ion exchange theory [2.63]; and *Boksay*, in 1971, proposed the presence of cationic sites in the glass, which, however, were not further defined [2.55]. Finally, *Eisenman* derived his equation [2.62, 63]

$$E = E^0 + \frac{n2.303\,RT}{F}\log\left[a_H^{1/n} + (K_{HNa}^{pot}a_{Na})^{1/n}\right]\tag{2.7}$$

by combining his improved version of the Nicolsky equation with an expression for the diffusion voltage resulting from the treatment of ion exchangers. The modified ion exchange constant in (2.7) is given by

$$K_{\mathrm{H\,Na}}^{\mathrm{pot}} = \left(\frac{u_{\mathrm{Na}}}{u_{\mathrm{H}}} \right)^n K_{\mathrm{HNa}} \,. \tag{2.8}$$

It consists of Nicolsky's (reversible) ion exchange constant K_{HNa}, the ratio of the (irreversible) mobilities u_{Na} and u_{H} in the glass or glass surface layer, assumed to be concentration-independent, and Eisenman's fitting parameter n, which originates from the assumption that glass is a "perfect ion exchanger obeying n-type non-ideal behaviour, $n = \mathrm{d}\ln a/\mathrm{d}\ln c$" [2.10]. Today it is known from concentration profile measurements that, especially with pH membrane glasses, the ions causing the phase boundary potential difference are often not identical with the ions interdiffusing in the subsurface range of the glass because of sterical conditions, which hinder the penetration of ions into glasses [2.65]. This is also a basic difference between membrane glasses and most ion exchangers, see Sect. 2.9.2. Besides, ionic mobilities in glasses and subsurface layers are known to be concentration-dependent [2.54]. Phase boundary potential difference and diffusion voltage can thus not generally be combined in a simple way as in (2.7) but must be treated separately. The main value of (2.7) therefore seems to be the possibility of fitting membrane potentials to experimental results.

The long-lasting firm belief in ion exchange as the glass electrode mechanism is obvious from several papers which appeared in the mid-1960s, where Schwabe and Lengyel discussed at length the relative magnitude of sodium ion and hydrogen ion activity coefficients in the glass phase to be applied in the Nicolsky equation [2.8, 66–68]. In 1976, when detailed information about the ion distribution in subsurface layers of electrode glasses had been available for some time, *Isard* published a paper in which he drew the conclusion "that the presence of the hydrated layer on the surface does not affect the (ion exchange) equilibrium potential" [2.69]. Even *Wikby*, who provided much information about the ionic processes below the glass surface, paradoxically discussed his new results on the basis of the ion exchange theory [2.70], which his results actually contradicted. Obviously, the great success of Nicolsky's thermodynamic equation handicapped the development of new ideas on the glass electrode mechanism for at least five decades [2.71], a period which could therefore be termed the Nicolsky hiatus [2.72] in analogy to the Nernstian hiatus (*Bockris* [2.73]) in electrode kinetics some decades earlier.

The first information about processes below the glass surface in contact with aqueous solutions was provided by Boksay and coworkers, who measured concentration profiles of cations interdiffusing in the glass [2.74, 75]. His first, and thus historical, sodium concentration profile below a sodium silicate glass surface, measured by the ingenious as well as cumbersome method of fractional dissolution of the leached glass in fluoric acid and subsequent analysis of the resulting solutions [2.74], is depicted in Fig. 2.6. It shows that a subsurface layer develops in the glass due to an exchange of its sodium ions for hydrogen ions from the solution (whose presence was verified later). Additional concentration profiles given in the same paper demonstrate the time

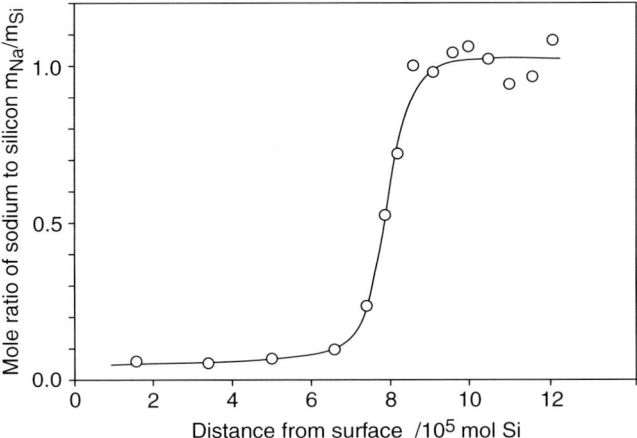

Fig. 2.6. The first concentration profile of an element (sodium) below a glass surface measured by, and reproduced after, *Boksay* and coworkers [2.74]

dependence of the growing layer thickness. Actually, this information already contains the first strong evidence that an exchange of ions as shown in Fig. 2.6 does not represent the ion exchange as assumed by Nicolsky for the potential formation, because the experimentally observed exchange periods are far too long compared to the usual response times of glass electrodes. For example, the profile shown in Fig. 2.6 developed within 72 h, whereas glass electrodes respond within the ten-millisecond range [2.76, 77]. Boksay's results should thus have already cast serious doubt on the ion exchange theory in 1964 but, as outlined above, no doubts have ever been voiced in the literature. On the contrary, the term "ion exchange" has remained mostly undefined, also among the advocates of the ion exchange theory, until today and has created much confusion among scientists involved in glass electrode research and even more so among those not concerned with the theoretical aspects of the subject.

After 1969, *Wikby* and coworkers provided much additional valuable information about the processes in subsurface glass layers by combining Boksay's method with a new pulse technique for measuring differential resistances, by which bulk glass and surface layer resistances could be separated [2.70, 78–82]. At the same time, *Bach* published a method for measuring continuous concentration profiles in subsurface glass layers by means of ion-ablating the glass with monoenergetic argon ions and simultaneously recording the intensity of characteristic spectral lines of the elements of interest [2.83, 84]. Due to its depth resolution of 3–5 nm, the technique is well suited to investigate processes at and below electrode glass surfaces, and it is also very practicable because of the short measuring periods, which are in the range of one hour per concentration profile. The technique was later called ion bombardment for spectrochemical analysis (IBSCA). Its combination with other electrochemical and non-electrochemical methods, such as coulometry, spectroscopy, electrolysis, and potentiometry, was applied to electrode membrane glasses, and the results were later confirmed in some cases by a new method of hydrogen profiling called nuclear reaction analysis (NRA) [2.85]. This work, together with information provided by other scientists, resulted in the dissociation mechanism of the glass electrode [2.72, 86, 87], which, *inter alia*, will cover Chap. 2 of this volume.

2.4 Experimental Methods for the Investigation of Glass–Solution Interfaces

Reactions between membrane glasses and solutions involve subsurface glass regions with extensions from several micrometres down to the nanometre and even subnanometre range. Their investigation, consequently, necessitates surface-analytical methods with extremely high depth resolution. One of these techniques is ion bombardment for spectrochemical analysis (IBSCA), which was developed [2.83, 84] and described in detail [2.88] by *Bach*, and applied to electrode membrane glasses in combination with other electrochemical and non-electrochemical methods. In some cases, the results thus obtained were confirmed by nuclear reaction analysis (NRA), a method that was developed several years later than IBSCA [2.85, 89, 90] and allows the direct measurement of hydrogen concentration profiles, which are not obtainable by the application of IBSCA. The principles of these methods and of their application to membrane glasses are described in this section.

A schematic diagram of an IBSCA apparatus is presented in Fig. 2.7. The glass specimen to be analysed is bombarded *in vacuo* with monoenergetic argon ions with 5.6 keV energy, in most cases under an angle of $75°$, which causes a slow ablation of the glass in even planes parallel to its original surface. The ablation rate is of the order of 2.5–5.0 nm min^{-1}. During the ablation, certain components of the glass are excited in the collision cascade below the glass surface, leave the glass in the excited state, and emit their surplus energy as characteristic spectral lines with intensities which are proportional to their concentration at the momentary surface caused by the ablation. Recording the intensity of an appropriate line of an element of interest as a function of time yields an intensity–time plot, $I = f(t)$, which can be transformed into a concentration–depth plot, $c = f(s)$, the concentration profile, perpendicular to the glass surface. This is possible because the intensities of the lines emitted are proportional to the immediate concentration sputtered,

$$c = \frac{c^0}{I^0} I \ , \tag{2.9}$$

where c^0 is the concentration of the element within the bulk glass, and I^0 is the intensity of the spectral line emitted when the bulk glass is ablated.

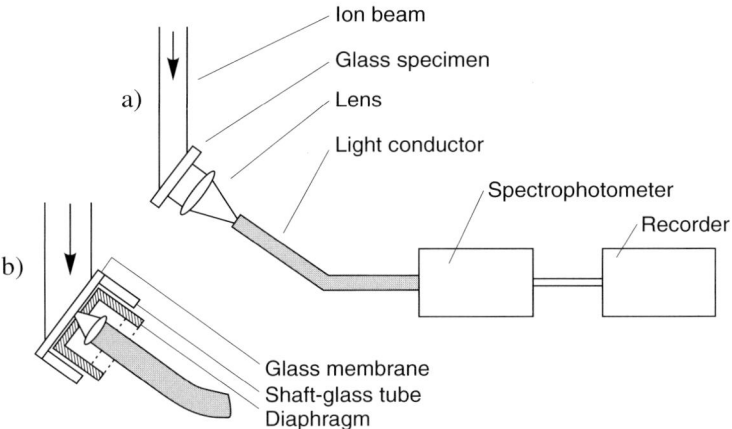

Fig. 2.7. Schematic presentation of the arrangement of ion bombardment for spectrochemical analysis (IBSCA) for measuring subsurface concentration profiles of elements in glasses

Besides, the sputter depth s is obtained from the ablation time t,

$$s = \frac{\mathrm{d}s}{\mathrm{d}t}t \, , \tag{2.10}$$

because the ablation rate $(\mathrm{d}s/\mathrm{d}t)$ is known from interferometry of etch pits generated by the same beam density during long sputtering periods of known duration. As an example, Fig. 2.8 shows an interferometric photograph of an etch groove with 270 nm per fringe that was developed by means of a 5.6 keV argon ion beam with $10\,\mu\mathrm{A\,cm^{-2}}$ ion density under an angle of incidence of

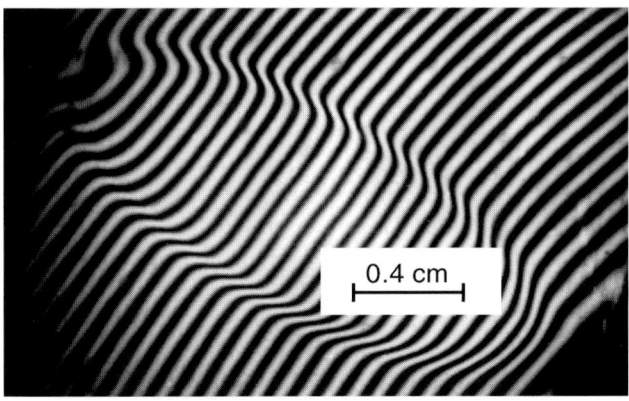

Fig. 2.8. Interference micrograph of an etch groove developed during 150 min of bombardment of a lithium silicate glass with 5.6 keV argon ions with an ion density of $10\,\mu\mathrm{A\,cm^{-2}}$. Angle of incidence: 75°, 270 nm per fringe

75°. The depth resolution of the technique is 3–5 nm and is thus well suited for the investigation of glass electrode membranes.

Hydrogen concentration profiles are not measurable directly by IBSCA but can be obtained indirectly from differential concentration profiles of other elements, for instance alkali, in combination with infrared spectra because hydrogen forms its own indicator in the form of OH groups, which exhibit strong IR absorption bands in the wavelength region 2.7–4.7 μm. Because, in addition, these bands indicate the character of the OH group [2.91], it is possible under certain conditions to characterize the environment of the oxygen atoms to which the hydrogens are bound.

An independent confirmation of hydrogen concentration profiles obtained by a combination of IBSCA and infrared spectroscopy is given by the application of nuclear reaction analysis (NRA). This technique, which yields hydrogen depth profiles directly, is based on the nuclear reaction $^{1}H(^{15}N,\alpha\gamma)^{12}C$, where protons react with ^{15}N under generation of ^{12}C and ^{4}He and emission of 4.43 MeV γ radiation [2.85, 89, 90, 92]. The characteristic resonance energy is 6.385 MeV with an energy width of 0.1 MeV. For the analysis, the glass specimen is subjected *in vacuo* to a beam of monoenergetic ^{15}N ions, Fig. 2.9. Starting with an energy slightly below the resonance energy E_{res}, the beam energy is increased stepwise , for instance E_1 and E_2. The ^{15}N ions penetrate the glass and lose energy by elastic collisions with the atoms of the glass until they have attained the resonance energy, which enables the nuclear reaction with hydrogen, for instance at $s_{res,1}$ and $s_{res,2}$. The yield of the γ radiation, which is proportional to the hydrogen concentration at the location of the reaction, is measured as a function of the beam energy and

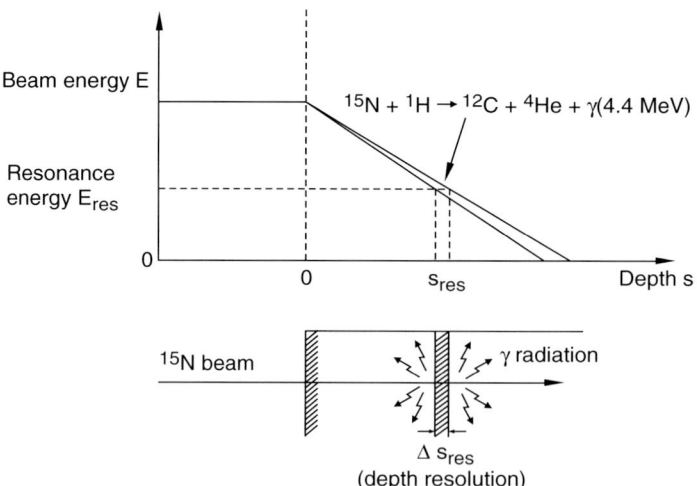

Fig. 2.9. Principle of the measurement of subsurface hydrogen concentration profiles in glasses by nuclear reaction analysis (NRA)

thus represents the hydrogen concentration depth profile because the average specific energy loss of the ^{15}N in the solid can be calculated from corresponding cross sections and stopping powers. The depth resolution of NRA is in the range 5–10 nm, which is slightly inferior to that of IBSCA, and decreases with depth because of energy straggling of the penetrating ^{15}N ions; for silicon it was reported to be 14 nm at a depth of 400 nm [2.92].

The concentration profile of one or several elements measured by IBSCA after a certain interaction period of a glass sample with a solution represents a snapshot of the state of the reaction between the phases, and several such snapshots taken in identical samples after different interaction periods characterize the time course of the reaction taking place. An example, given in Fig. 2.46 of Sect. 2.7.1, is the interdiffusion of lithium ions of a lithium silicate glass and hydrogen ions of a solution whose time dependence is demonstrated by several lithium concentration profiles below the glass surface measured after various periods of diffusion time. Similarly, the field-driven migration of ions in glasses can be followed by measuring concentration profiles after various migration periods. The experimental arrangement is sketched in Fig. 2.10. This application, of which two examples are given in Fig. 2.16 of Sect. 2.6.2, represents a modification of the well-known moving boundary method for measuring ionic migration velocities and transport numbers in electrolyte solutions [2.93] and will be treated in detail in Sect. 2.6.6. It is emphasized here, however, that the modification for glasses gives more information about migration processes than the original method in that IBSCA yields the concentration profiles at the moving boundary between the different migrating ions, in addition to its position, whereas the original method informs only

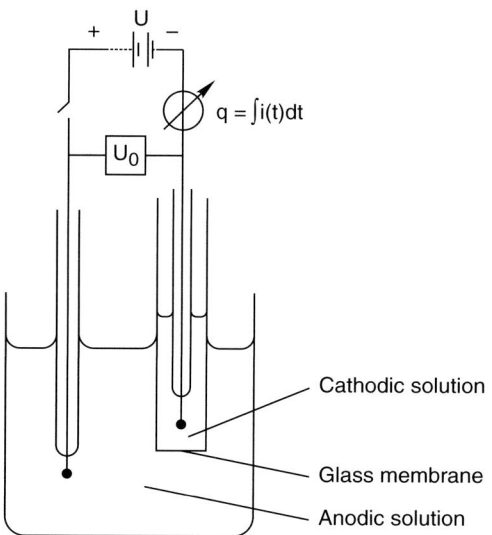

Fig. 2.10. Schematic of the electrolysis arrangement for the modified moving boundary method

about the position of the boundary between the different migrating ions in the solution.

2.5 Glass in Contact with Solutions – Interfacial Equilibria

Although contact of glasses with water or aqueous solutions is an event happening continuously all over the world, it is not generally known that it involves a series of chemical and electrochemical processes which are connected with, and dependent on, each other in a complicated way. These reactions take place also between membranes of glass electrodes and solutions subjected to measurements with these sensors and can be studied particularly efficiently by means of these arrangements because membrane glasses represent empirically developed model glasses for the investigation of electrochemical processes in and at solid glasses. In addition, the construction of glass electrodes allows one to obtain a very important quantity connected with the glass–solution interaction, i.e., the (relative) Galvani voltage or difference of the internal electrical potentials between the glass and the solution. This information, together with surface analyses of elements below the glass–solution interface during the time-dependent interaction, have yielded details and the time course of the reactions and allowed the separation and identification of the single reaction parts. It is thus now possible to describe the sequence of the intertwined, mutually dependent reaction steps taking place between electrode glasses and solutions.

The primary and fundamental reaction – concluded from and verified by several different experiments – and the electrochemical interfacial equilibrium it leads to are described quantitatively in this section. This description is preceded by an overview of the reactions which avoids any thermodynamic and kinetic equations except for some easily comprehensible homogeneous and heterogeneous equilibria, and thus gives a qualitative picture of why and how glass electrode membranes function and what, often disturbing, reactions are initiated by the fundamental electrochemical equilibrium between the glass and the solution. The description will start with the basic structure of oxidic glasses and their electrochemical characterization, which leads directly to the basic equilibrium between the glass and the solution, and to the subsequent reactions, such as ionic interdiffusion in the subsurface range of the glass.

Based on this qualitative understanding of the glass–solution interaction, it will be easy to follow the basically simple quantitative derivation of the ionic activity-dependent glass electrode potential in the pure pH and pM ranges as well as in the transition region between these ranges, which will be conducted on a thermodynamic and kinetic basis. Of particular interest will be a rigorous treatment of the thermodynamic expression for the electrode potential because it leads to the conclusion that the slope of the pH-dependent Galvani voltage, $\varepsilon = f(\text{pH})$, of the glass electrode can basically not be "ideal", if "ideal" is characterized by the factor $(RT \ln 10)/F$, as in the case of metal electrodes, but must be smaller because of the special character of the interfacial equilibrium. It is also shown that this effect has been known as the so-called sub-Nernstian response of glass electrodes for a long time.

2.5.1 Qualitative Description

Electrode membrane glasses are multi-component glasses consisting basically of oxides of so-called glass-forming elements, for instance silicon, aluminium, or boron, and oxides of glass-modifying elements such as alkali and alkaline earths [2.94]. The glass-forming oxide, for example SiO_2 in a silicate glass, forms SiO_4 tetrahedra with silicon as the centre atom, which are connected to each other via the oxygen atoms and thus establish an irregular spatial network in which each silicon is bound to four oxygens and each oxygen to two silicons. The glass-modifying oxides interrupt these chains of silicon and oxygen so that some of the oxide atoms are bound only to one silicon and thus bear one negative charge according to

$$O_3Si\text{–}O\text{–}SiO_3 + Na_2O \rightarrow O_3SiO^- + 2Na^+ + {}^-OSiO_3 , \qquad (2.11)$$

where sodium is taken as an example for the network-modifying element. As indicated by (2.11), the charges of the terminal oxides, or so-called siloxy groups, $\equiv SiO^-$ (the three bars denote bonds to tetrahedral oxygen atoms), are balanced by the positive charges of the network-modifying cations introduced, which can be positioned at several different locations near the negative groups in the more or less open glass network, the alkali ions being able to change their position relatively easily. Additional glass-forming elements that are threefold positively charged, such as aluminium or boron, are also built into the network as four-coordinated atoms. In this case, the negative charge is not attached to a terminal oxide but is distributed over the four oxygen atoms of the, for instance, AlO_4 tetrahedron connecting the aluminium to the silicon atoms of the surrounding tetrahedra. This is often indicated by formulations such as $\left[AlO_{4/2}\right]^-$ or $\left[\equiv SiOAl \equiv\right]^-$. The negative charge also of these groups is balanced by the positive charge of the network-modifying cations, which are basically their cause. In addition, many electrode membrane glasses contain so-called intermediate oxides such as BeO, La_2O_3, TiO_2, etc., which act partly as network-forming and partly as network-modifying oxides and influence the structure and thus the properties of the glasses or their melts.

According to this structure, electrode membrane glasses are characterized electrochemically as solid electrolytes consisting of an irregular, three-dimensional network of oxygen and one or more glass-forming elements with fixed negative anionic groups, such as $\equiv SiO^-$ (siloxy) or $\left[AlO_{4/2}\right]^-$, and containing charge-balancing cations including alkali ions with finite, although small, mobilities [2.71]. For the application as membrane glasses, one of the alkali ions must have unit transport number, that is, it must carry 100% of the electricity when the glass is subjected to an electric field [2.87].

The given structure implies two properties which are essential for the functioning of these materials as electrode membrane glasses [2.87].

- The mobile cations respond to concentration and potential gradients (temperature gradients are of no concern here) and can leave the network if charge neutrality of the glass is maintained, for example, by their replacement by other cations. The substitution, however, is limited by sterical conditions, i.e., the ionic radius of the replacing ions must be sufficiently small not to hinder their partial or complete accommodation by the vacancies offered by the leaving ions in the network. The mobile alkali ions couple the ionic processes at the opposite surfaces of electrode membranes electrically and enable their integration into electrochemical cells, such as cell (2.I) of Sect. 2.2, and the completion of the corresponding electrical circuits.

- Anionic groups that are located at the immediate glass surface are distinguished groups in that they belong to both the glass to whose network they are bound and the neighbouring phase they contact. If this is a solution, as in all cases of glass electrode application, these groups are functional surface groups which establish equilibria with ions in the liquid and are thus subject to the mass action law. Such heterogeneous, so-called interfacial, or phase boundary, equilibria which involve the charged form of the functional groups at the glass surface, however, differ fundamentally from equilibria of ions in homogeneous solutions because they are subject not only to chemical driving forces, as homogeneous equilibria, but also to an electrical driving force caused by the charge separation between the glass and the solution, which counteracts and, at equilibrium, balances the chemical driving force. A chemical equilibrium becomes an electrochemical equilibrium whenever charged particles at an interface are involved.

For example, the dissociation of silanol ($\equiv$SiOH) groups attached to a glass surface that is contacted by an aqueous solution,

$$\equiv SiOH(s) + H_2O(sol'n) \rightleftarrows \equiv SiO^-(s) + H_3O^+(sol'n) , \tag{2.12}$$

(where s and sol'n denote glass surface and solution, and the three bars in front of the silicons stand for bonds to oxygen atoms) generates siloxy groups at the glass surface and hydronium ions in the solution [2.86, 87] and, due to this charge separation between the phases, an electric field perpendicular to the interface, which counteracts the dissociation. The glass surface is electrified, and the electrochemical phase boundary equilibrium is reached when the chemical and the electrical driving forces balance. This is the case at extremely small siloxy concentrations [2.87, 95] because, as is well known, the potential difference between two phases depends strongly on the charge density at their interface. As a consequence, the dissociation of surface silanol groups is much weaker than that of an acid with the same chemical free energy of dissociation in a homogeneous solution, for instance of silicic acid, under otherwise identical conditions. Heterogeneous dissociation constants of surface groups are thus much smaller than homogeneous dissociation constants of corresponding compounds.

The dissociation of silanol groups at glass surfaces, incidentally, corresponds to the dissociation of metal atoms under formation of hydrated metal ions in the solution and electrons at the metal surface,

$$M(\text{solid}) + n H_2O(\text{sol'n}) \rightleftarrows e^-(M) + M(H_2O)_n^+(\text{sol'n}) . \tag{2.13}$$

At electrochemical equilibrium, the chemical driving force of the metal dissociation is balanced by the driving force of the electric field generated by the charge separation at the interface metal/solution. The main difference is the state of the electrons in the solids, which are delocalized electrons in the conduction band of the metal, (2.13), and localized electrons in the valence band of negatively charged oxygens, for example of siloxy groups, at the glass surface, (2.12) [2.71].

A silicate glass surface subject only to the dissociation equilibrium thus contains mainly silanol groups and a minute concentration of siloxy groups, which, however, can basically not be neglected because they are the participants in the phase boundary equilibrium that impose an electric charge on the glass phase and thus generate the hydrogen activity-dependent or pH-dependent Galvani voltage between the glass and the solution. Equation (2.12), consequently, describes the pH response of glass electrodes in a qualitative way [2.86, 87]. An increase of the hydrogen ion activity in the solution, corresponding to a pH decrease, for instance, shifts the equilibrium to the left and increases the silanol and decreases the siloxy group concentrations by equal small amounts. The connected reduction of the negative charge density at the glass surface corresponds to an increase of the electrical potential of the glass relative to that of the solution (which remains constant because of the large hydrogen ion reservoir) and thus presents the pH response of the membrane glass.

For small hydronium ion and relatively large alkali ion, M^+, activities in the solution, the surface groups are subject to a different phase boundary equilibrium. A silicate glass is again chosen as the example for consistency. Under these conditions, the siloxy groups at the glass surface establish an equilibrium with alkali ions as indicated by (2.14) [2.71, 87]

$$\equiv SiO^-(s) + M^+(\text{sol'n}) \rightleftarrows SiOM(s) . \tag{2.14}$$

The glass surface is covered mainly by siloxy salt groups and contains a minute concentration of siloxy groups as in the case of silanol dissociation [2.87, 95]. Equation (2.14) thus represents the alkali ion or pM response of glass electrodes in the same way as (2.12) reflects the pH response. An increase of the alkali ion activity in the solution, corresponding to a pM decrease, for example, shifts the equilibrium to the right side and increases the siloxy salt and lowers the siloxy group concentrations by minute equal amounts. The resulting reduction of the negative charge density at the glass surface causes a corresponding increase of the electrical potential of the glass relative to that of the solution and thus demonstrates qualitatively the pM response of the membrane glass.

The formation of substantial siloxy salt concentrations at the glass surface according to (2.14) is not expected at first sight because alkali salts of strong and weak acids are generally completely dissociated in homogeneous solutions. However, the negative charge at the glass surface counteracts the chemical driving force of dissociation of the siloxy salt groups, or, in other terms, favours the association of siloxy groups with alkali ions. Heterogeneous association constants of functional groups at the glass surface, consequently, are much larger than homogeneous association constants of corresponding compounds (which are generally much smaller than unity). This statement is tantamount to what was said above about the heterogeneous dissociation of functional surface groups.

In the intermediate pH range between 100% hydronium ion and 100% alkali ion response of the membrane glass, which is called the transition pH range, the functional surface groups are simultaneously subject to silanol dissociation and siloxy salt formation. The single equilibria, (2.12) and (2.14), can be combined to give the overall equilibrium, (2.15), in this pH region,

$$\equiv \mathrm{SiOH(s)} + \mathrm{H_2O(sol'n)} + \mathrm{M^+(sol'n)} \rightleftarrows \equiv \mathrm{SiOM(s)} + \mathrm{H_3O^+(sol'n)} \,, \tag{2.15}$$

which shows the pH and pM dependence of the relative surface concentrations of silanol and siloxy salt groups [2.71, 87]. For instance, the smaller the pH relative to the pM value of the solution, the larger is the concentration of the silanol relative to that of the siloxy salt groups at the glass surface. Although the siloxy groups cancel in (2.15), their surface concentration and thus the Galvani voltage of the membrane glass are determined by the pH and the pM of the solution also in the transition pH range. This is seen by a different combination of (2.12) and (2.14), yielding

$$\equiv \mathrm{SiOH(s)} + \mathrm{H_2O(sol'n)} \rightleftarrows \equiv \mathrm{SiO^-(s)} + \mathrm{H_3O^+(sol'n)}$$
$$+$$
$$\mathrm{M^+(sol'n)}$$
$$\downarrow\uparrow$$
$$\equiv \mathrm{SiOM(s)} \tag{2.16}$$

and is also demonstrated by the quantitative treatment in Sect. 2.5.4. According to (2.12), (2.14), and (2.15), glass surfaces always contain silanol or siloxy salt groups, or both, and siloxy groups (or their analogues, such as $[AlO_{4/2}]H$, $[AlO_{4/2}]M$, and $[AlO_{4/2}]^-$) whose relative concentrations are determined by the solution composition. A solution-contacted glass surface is thus not an inert surface, as has always been tacitly assumed for ionic interdiffusion and migration between solutions and glasses, but is character-ized by a defined composition and charge density and could be called an *electrochemically structured glass surface.*

While, however, the functional surface groups are in electrochemical equi-librium with the ions in the solution, they are often not with the ions of the glass, i.e., on the opposite side of the phase boundary, because the surface composition differs in most cases from the composition of the glass. Depend-ing on the time of glass/solution contact, this generates more or less large ionic concentration gradients below the glass surface and an interdiffusion of the ions of the glass versus the ions attached to the surface groups. The diffusional exchange of ions between glasses and contacting solutions is thus driven by the concentration gradients established by the composition of the glass interior and that of the *electrochemically structured glass surface* in equilibrium with the solution and not simply by the relative ionic activities of glass and solution [2.87]. If, for example, a lithium silicate-based glass is in contact with a lithium ion-containing solution which generates mainly $\equiv$SiOH groups at its surface, there will be an interdiffusion of lithium ions of the glass versus hydrogen ions from these silanol groups at the glass surface (and subsequent reactions which are not of interest here), and the formation of a hydrogen ion-containing subsurface glass layer will be observed. If, in contrast, a similar glass containing a few percent Al_2O_3 contacts the same solution, the surface will be covered mainly by $[AlO_{4/2}]Li$ groups, which do not lead to an interdiffusion of different ions, and, consequently, the compo-sition of the subsurface glass range will not change.

During the interdiffusion, the phase boundary equilibrium is perturbed by the removal of ions from the surface groups and by ions of the bulk glass trespassing on the equilibrium and leaving the glass for the solution. These processes can lead to a polarization, that is, a shift or change of the Galvani voltage between glass and solution. Whether the Galvani voltage is or is not polarized by the disturbance, and, if so, to what degree, is a matter of the kinetic conditions. Like homogeneous equilibria, also heterogeneous or phase boundary equilibria are not rigid but dynamic in character, which means that they proceed in both dissociating and associating direction with equal rates [2.71, 87], see (2.12) or (2.14) and Sect. 2.5.5. Because of the charges borne by the ions exchanging between surface groups and solution, the reaction rates represent current densities perpendicular to the glass surface, dissociation, for example of $\equiv$SiOH groups, representing an anodic and the (reverse) asso-ciation, for example of $\equiv$SiO$^-$ groups and H$^+$ ions, a cathodic current den-sity. The phase boundary equilibrium is thus characterized by equal anodic and (absolute) cathodic current densities, which are termed (equilibrium) exchange current densities, and it is plausible that the interfacial equilibrium is the more stable and, consequently, the corresponding Galvani voltage between glass and solution is the less polarizable, the larger the exchange current density at the glass surface. Phase boundary equilibria at membrane glass surfaces have surprisingly large exchange current densities, which are even comparable with such values of metal electrodes with medium polarizability [2.71]. The interfacial equilibria between glass surface and solution are thus practically not perturbed by the ionic interdiffusion between glass surface and glass interior, except at the beginning of the interdiffusion when the concentration gradients are extremely large and the interdiffusion rate is at its maximum. This means that also the Galvani voltage between glass and solution and thus the response of electrode membrane glasses according to (2.12), (2.14), and (2.15) are independent of the interdiffusion processes below the glass surface. Phase boundary equilibria thus offer unperturbed and, consequently, constant ionic concentration ratios during the interdiffusion with ions of the glass. In other words, the glass surface acts as a phase with constant composition [2.87, 96].

The different ions interdiffusing in the subsurface region of electrode membrane glasses have different and concentration-dependent mobilities and diffusion coefficients and thus cause a diffusion voltage between their origins, i.e., the *electrochemically structured glass surface* and the bulk glass. The diffusion voltages below both surfaces of electrode membranes must thus be added to the emf E' of glass electrode-containing cells as given by (2.1) in Sect. 2.2. Fortunately, however, they are constant as long as the boundary conditions determining their magnitude are constant, as predicted by *Eisenman* [2.10]. Thus, pH glass membranes exhibit constant diffusion voltages as long as the surface contains only silanol groups (besides the minute concentration of siloxy groups). At 100% pH response, the diffusion voltage is independent of pH. Within the alkali error pH range, i.e., if also siloxy salt groups and thus alkali ions are present at the glass surface, however, the diffusion voltage depends on the relative concentrations of the different groups and thus changes with pH at constant pM although the alkali ions may not participate in the interdiffusion [2.87, 96].

pM-selective membrane glasses exhibit a more complicated behaviour because, due to the different glass structure caused by their Al_2O_3 content, alkali (e.g. sodium) ions are not hindered sterically from diffusing into the glass as they are with pH glasses. If thus the alkali ion of a pM glass and the alkali ion attached to its surface are different, they interdiffuse in the subsurface range of the glass and cause a time-dependent diffusion potential, whose change rate decreases with time. pM glass electrodes must thus be "conditioned" in solutions containing the alkali ion M^+ to be measured and generating a constant $[AlO_{4/2}]M$ group concentration at their membrane surface until a nearly constant diffusion potential has developed and the measured emfs are no longer subject to intolerable changes during measurements [2.87, 96].

Such glass electrodes must be stored under the same conditions also between measurements.

Due to its stability, the phase boundary equilibrium also determines the relative concentrations of different ions which are transferred from a solution into the glass by means of an electric field. The ions attached to the surface groups leave the surface for the glass and are replaced by the same ions from the solution, i.e., the interfacial equilibrium is unperturbed as long as the exchange current density is much greater than the electrolyzing current density. Electrolyses have been carried out extensively in order to investigate interfacial equilibria and ionic transport properties of the bulk glass as well as to verify the dissociation mechanism of the glass electrode [2.54, 55, 97]; they will be treated in detail in Sect. 2.6. What is to be emphasized at this point, however, is the nature of the *electrochemically structured glass surface*, which plays a significant role in all reactions between solutions and glass surfaces and the glass interior. The large magnitude of the exchange current density suggests the conclusion that the functional surface groups are in direct contact with the solutions they are in equilibrium with because the participation of any groups below the glass surface would significantly reduce the exchange current densities because of the solid state diffusion involved. The interfacial equilibrium, consequently, must be an equilibrium between the solution and the very glass surface, which thus acts as a two-dimensional phase with constant composition during the transport of ions from the solution into the glass by diffusional as well as migrational processes. This is also confirmed by the response rate of electrode glass membranes, which is basically extremely large [2.76, 77, 98]. The representation of the electrode membrane as a three-layer membrane, |surface layer|bulk glass|surface layer|, suggested in the literature [2.99], must thus obviously be replaced by that of a five-layer membrane, scheme

$$|\text{glass surface}|\text{subsurf. layer}|\text{bulk glass}|\text{subsurf. layer}|\text{glass surface}| \, ,$$

$$(2.\text{VII})$$

(see also Fig. 2.11), where glass surface stands for *two-dimensional electrochemically structured glass surface* and the subsurface layers consist of glass modified to various degrees by ion exchange, hydration and so forth, which are secondary processes and not directly connected with the electrode response of the membrane glass. The entire complex of the primary phase boundary equilibrium involving surface coverage and potential formation and the subsequent secondary reactions including ionic interdiffusion, sterical hindrance, formation of diffusion voltages, hydration, and others must be understood if the functioning of glass electrodes and the large number of phenomena shown by these sensors are to be explained. All these processes are summarily called the dissociation mechanism.

2.5.2 Quantitative Treatment

As has been qualitatively explained above, anionic groups of glasses positioned at the immediate glass surface become functional groups when the glass is brought into contact with, for instance, aqueous solutions. Their negative charge is no longer completely balanced by the positive charge of the cations of the glass but they are involved in interfacial electrochemical equilibria with cations of the solution. This involvement leads to certain relative concentrations of their different forms and thus of the attached ions at the surface and to a potential of the glass relative to that of the solution (charge separation) because also the dissociated form of the functional groups is included in the equilibrium. According to the specific conditions, two different phase boundary equilibria are formed by the surface groups. Their acidic form dissociates, and the degree of dissociation is determined by the hydronium ion activity of the solution, or their dissociated form associates with alkali ions of the solution, and the degree of association, or surface salt formation, is given by the alkali ion activity of the solution. Both dissociation of the acidic form and surface salt formation are observed in the so-called transition range between these regions, where dissociation and association overlap. The states of the single and combined equilibria are given by the dissociation and association constants, that is, by the electrochemical standard free energies of dissociation and association, and by the ionic activities in the solution.

These equilibria will be treated quantitatively in the following [2.71, 2.72, 2.87], whereas any subsequent reactions initiated by the phase boundary equilibria, for instance interdiffusion and the connected formation of diffusion potentials in subsurface glass layers, which are part of the electrode potential, will be treated separately in Sect. 2.7 [2.72, 2.87]. The following treatment is thus solely devoted to the fundamental phase boundary or interfacial equilibria. In order to arrive at consistent equations that are as clear as possible, both dissociation and association of the same group, i.e., the silanol/siloxy group, will be treated although the surface salt formation of this group is restricted to the sodium error range and does not show up over an extended

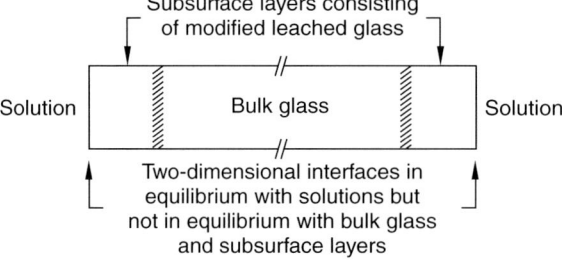

Fig. 2.11. Schematic presentation of the cross section of a glass electrode membrane showing the five-layer membrane according to the dissociation mechanism

pH range as in the case of alkali-selective membrane glasses. Introduction of corresponding thermodynamic data of other groups, however, can easily adjust the equations to other functional groups of interest.

Dissociation of Acidic Surface Groups – pH Dependence of Galvani Voltage

An electrochemical cell containing a glass electrode for the purpose of pH measurements is typically represented by the cell scheme

$$|\text{ref. el.}|\text{KCl(conc'd)}\|\text{sol'n, pH}|\text{glass membrane}|\text{int. buffer}|\text{int. ref. el.}|$$
$$\varepsilon_r \qquad\quad \varepsilon_j \qquad \varepsilon_m \qquad\qquad\qquad \varepsilon_{m,i} \qquad \varepsilon_{r,i} \,, \qquad (2.\text{VIII})$$

which is a variant of cell (2.I) of Sect. 2.2 for pH measurements. With modern glass electrodes, the membrane glass is a modified lithium silicate glass, which is the origin of the functional silanol/siloxy group at its surface. ε_m and $\varepsilon_{m,i}$ are the Galvani voltages between the opposite surfaces of the glass membrane and the measuring and internal reference solutions, respectively, which are of interest here. Because, however, the composition of the internal reference solution is constant, the following treatment is limited to the interface between the measuring solution and the outer glass surface and to the equilibrium that leads to the formation of the Galvani voltage ε_m and thus to the pH response of cell (2.VIII). The interfacial equilibria at the opposite surfaces of the glass membrane are connected electrically by the ionic conductivity of the glass.

The dissociation of the surface silanol groups is represented by

$$\equiv \text{SiOH(s)} + \text{H}_2\text{O(sol'n)} \rightleftarrows \text{SiO}^-(\text{s}) + \text{H}_3\text{O}^+(\text{sol'n}) \,. \qquad (2.17)$$

Equation (2.17), incidentally, was first concluded and proposed as the pH-determining equilibrium responsible for the glass electrode response from work based on concentration profile measurements in 1974 [2.86]. Equilibrium of reaction (2.17) is thermodynamically characterized by zero change of the electrochemical free energy of dissociation,

$$\Delta\overline{G}_{\text{D,H}} = G_{\text{SiO}^-} + G_{\text{H}_3\text{O}^+} - G_{\text{SiOH}} - G_{\text{H}_2\text{O}} - F\varepsilon_m = 0 \,, \qquad (2.18)$$

where G is the chemical free energy of the particles and groups indicated, $\varepsilon_m = \varphi_{\text{gl}} - \varphi_{\text{sol'n}}$ is the Galvani voltage between glass membrane and solution, and F is the Faraday constant. The minus sign in front of the potential term of (2.18) results from writing the silanol dissociation equilibrium, (2.17), opposite to the direction it should be written according to cell (2.VIII), where the membrane potential is ε_m according to the recommendations by IUPAC. Introduction of the chemical standard free energy of dissociation,

$$\Delta G^0_{\text{D,H}} = G^0_{\text{SiO}^-} + G^0_{\text{H}_3\text{O}^+} - G^0_{\text{SiOH}} - G^0_{\text{H}_2\text{O}} \,, \qquad (2.19)$$

and of activities, a, in solution and of surface activities, a', results in

$$\Delta\overline{G}_{D,H} = \Delta G^0_{D,H} + RT \ln \frac{a'_{SiO^-}\, a_{H_3O^+}}{a'_{SiOH}\, a_{H_2O}} - F\varepsilon_m = 0 \ , \tag{2.20}$$

and, after rearrangement and introduction of the dissociation constant according to

$$\Delta G^0_{D,H} = -RT \ln K_{D,H} \tag{2.21}$$

in

$$\varepsilon_m = -k \log K_{D,H} + k \log\frac{a'_{SiO^-}}{a'_{SiOH}\, a_{H_2O}} - k\,\mathrm{pH} \ . \tag{2.22}$$

In this equation, $k = 2.303\, RT/F$ is the so-called Nernstian slope, and pH $\equiv -\log a_{H_3O^+}$. Equation (2.22) represents the pH dependence of the Galvani voltage of the glass membrane. For example, a decrease of pH, corresponding to an increase of the hydronium activity, increases the silanol and decreases the siloxy concentration by equal absolute amounts, $\Delta c'_{SiOH} = -\Delta c'_{SiO^-}$, which means a reduction of the negative charge density at the glass surface and thus a more positive potential of the glass membrane relative to the potential of the solution, which, on the other hand, is constant due to the large hydronium ion reservoir. Because of the well-known strong dependence of the potential of a phase on its surface charge, however, the concentration of the siloxy groups and even more so their pH-induced change are much smaller than the concentration of the silanol groups,

$$\left|\Delta c'_{SiO^-}\right| < c'_{SiO^-} \ll c'_{SiOH} \approx c'^0 \ , \tag{2.23}$$

where c'^0 is the total concentration of functional surface groups, which, at pure pH function of the glass membrane, consists of nearly 100% silanol groups. The second term on the right side of (2.22) can thus be assumed constant with good approximation and can be combined with the first term to form the standard Galvani voltage of silanol dissociation at the glass membrane,

$$\varepsilon^{0'}_H = -k \log K_{D,H} + k \log\frac{a'_{SiO^-}}{a'_{SiOH}\, a_{H_2O}} \ , \tag{2.24}$$

so that, with this approximation, the Galvani voltage of the glass membrane as a function of pH is represented by

$$\varepsilon'_m = \varepsilon^{0'}_H - k\,\mathrm{pH} \ . \tag{2.25}$$

Introduction of (2.25) into the equation of the emf of a glass electrode-containing cell, e.g. (2.1) of Sect. 2.2, finally, yields

$$E' = \sum \varepsilon_n + \varepsilon_{\mathrm{H}}^{0'} - k\,\mathrm{pH} \ , \tag{2.26}$$

representing the cell emf as a function of pH, or

$$E' = E_{\mathrm{H}}^{0'} - k\,\mathrm{pH} \ , \tag{2.27}$$

if the first two terms on the right side of (2.26) are combined to form the standard emf of the glass electrode cell

$$E_{\mathrm{H}}^{0'} = \sum \varepsilon_n + \varepsilon_{\mathrm{H}}^{0'} \ . \tag{2.28}$$

The standard emf, $E_{\mathrm{H}}^{0'}$, (2.28), contains basically two approximations: (1) the approximation of a constant, i.e., pH-independent, liquid junction voltage between measuring solution and reference electrode solution and (2) the approximation of a negligible change of the siloxy surface concentration when the pH of the solution is changed. These quantities always appear together when measurements are performed on glass electrode cells with transference, where their sum is taken into account by standardization. A method of a separate measurement of the second effect will be treated in connection with the rigorous examination of the phase boundary equilibria in Sect. 2.5.3. It is again emphasized that contributions of diffusion voltages within subsurface ranges of the glass to the Galvani voltage of the glass membrane as well as to the emf of glass electrode cells will be treated separately in Sect. 2.7.4.

Salt Formation of Surface Groups – pM Dependence of Galvani Voltage

The derivation of a quantitative relationship of the Galvani voltage as a function of the alkali ion activity of the solution corresponds to that presented above for the pH-dependent Galvani voltage. Nevertheless, it is briefly given in the following for completeness' sake and because it shows some peculiarities that should be noted. The basis of the derivation is also cell (2.VIII), but with the difference that the measuring solution contains alkali (e.g. sodium) ions, to which the glass membrane is exclusively responsible. As stated above, the glass membrane consists also of a silicate-based glass characterized by siloxy-based functional surface groups despite the rather limited alkali-responsive pH range of these entities.

The association of the functional surface groups with alkali, M^+, ions is represented by

$$\equiv \mathrm{SiO}^-(s) + M^+(\mathrm{sol'n}) \rightleftarrows \equiv \mathrm{SiOM}(s) \ , \tag{2.29}$$

where the alkali ions are hydrated ions, which, however, is not noted in the equation. Equilibrium is characterized by zero electrochemical free energy of association

$$\Delta \overline{G}_{A,M} = G_{SiOM} - G_{SiO^-} - G_{M^+} + F\varepsilon_m \; , \tag{2.30}$$

where, in contrast to the corresponding equation (2.18), the Galvani voltage-containing term is positive because cell (2.VIII) and (2.30) are written in the same direction, as requested by IUPAC in this case. Introduction of the standard free energy of association or of the association constant, respectively,

$$\Delta G^0_{A,M} = G^0_{SiOM} - G^0_{SiO^-} - G^0_{M^+} = -RT \ln K_{A,M} \; , \tag{2.31}$$

and of activities a in solution and a' at the surface of the indicated ion and functional surface groups, respectively, results in

$$\Delta \overline{G}_{A,M} = \Delta G^0_{A,M} + RT \ln \frac{a'_{SiOM}}{a'_{SiO^-} \, a_{M^+}} + F\varepsilon_m = 0 \; , \tag{2.32}$$

or, after rearrangement,

$$\varepsilon_m = k \log K_{A,M} + k \log \frac{a'_{SiO^-}}{a'_{SiOM}} - k\,pM \; . \tag{2.33}$$

Because the siloxy concentration is very small also in the association equilibrium and the glass surface is mainly covered with siloxy salt groups, the second term on the right side of (2.33) can be assumed approximately constant and can be combined with the first term to give the standard Galvani voltage of interfacial association at the glass membrane,

$$\varepsilon^{0'}_M = k \log K_{A,M} - k \log \frac{a'_{SiO^-}}{a'_{SiOM}} \; , \tag{2.34}$$

so that, with this approximation, the pM-dependent Galvani voltage between the membrane glass and the solution is represented by

$$\varepsilon'_m = \varepsilon^{0'}_M - k\,pM \; . \tag{2.35}$$

Introduction of this expression into (2.3) of Sect. 2.2, finally, yields the emf of a glass electrode cell at 100% alkali ion selectivity,

$$E' = \sum \varepsilon_n + \varepsilon^{0'}_M - k\,pM \; , \tag{2.36}$$

and

$$E' = E^{0'}_M - k\,pM \; , \tag{2.37}$$

if the first two terms on the right side of (2.36) are combined to give the standard emf of the pM glass electrode cell. Equation (2.37) describes the pM dependence of the emf of a pM-measuring glass electrode cell on the basis of two approximations: that of a solution- and thus pM-independent liquid junction potential between measuring and reference solutions, and that of a constant siloxy group concentration at the glass surface, which always appear together when measurements on cells with transference are conducted, as in the case of pH glass electrode-containing cells, see above. Contributions of diffusion voltages to both the Galvani voltage of the glass membrane and the emf of glass electrode cells will be treated in Sect. 2.7.

2.5.3 Rigorous Examination of Interfacial Equilibria – the Sub-Nernstian Response

Equation (2.25), which was derived in Sect. 2.5.2, represents the Galvani voltage between glass electrode membrane and solution as a function of pH and describes the experimentally observed response of electrode glasses at 100% hydrogen ion selectivity, that is, outside the acid and sodium error ranges, rather well. It reflects the well-known linear dependence of the glass electrode potential on the pH value in the intermediate pH range and corresponds to the general assumption that, ideally, the glass electrode should exhibit the theoretical slope $k = (RT \ln 10)/F$,

$$\frac{\mathrm{d}\varepsilon'_m}{\mathrm{d\,pH}} = -k \;, \tag{2.38}$$

as does, for instance, the Pt,H_2 electrode. The approximation contained in the standard Galvani voltage $\varepsilon_H^{0'}$ of the interfacial equilibrium, see (2.24) of Sect. 2.5.2, is thus indeed justified for practical applications of (2.25) to pH measurements by means of glass electrode cells with transference.

That the assumption of pH-independent concentrations of the siloxy and silanol groups at the glass surface is not correct fundamentally, however, is demonstrated by

$$\varepsilon_m = -k \log K_{\mathrm{D,H}} + k \,\log \frac{a'_{\mathrm{SiO^-}}\, a_{\mathrm{H_3O^+}}}{a'_{\mathrm{SiOH}}\, a_{\mathrm{H_2O}}} \;, \tag{2.39}$$

which is the rigorous form of the membrane potential, (2.22), Sect. 2.5.2. It shows, for example, that an increase of the hydronium ion activity, for thermodynamic reasons, increases the silanol group activity (and concentration) and decreases the siloxy group concentration by an equal absolute amount because of the mass balance, $\Delta c'_{\mathrm{SiOH}} = -\Delta c'_{\mathrm{SiO^-}}$, and, in turn, increases the electrical potential of the glass membrane because of the reduced density of negative charges at the glass surface. However, this potential change could not be understood if the siloxy concentration was constant. Because of the strong dependence of the potential of a phase on the charge density at its surface [2.100], the siloxy concentration (and activity) is much smaller than the silanol concentration, and its change caused by a pH variation is even smaller [2.95],

$$\left|\Delta c'_{\mathrm{SiO^-}}\right| < c'_{\mathrm{SiO^-}} \ll c'_{\mathrm{SiOH}} \cong c'^0 \;, \tag{2.40}$$

(c'^0 is the total concentration of functional surface groups), so that, nevertheless, it could be concluded that siloxy concentration changes are not detectable experimentally.

However,

$$\varepsilon_{m,1} - \varepsilon_{m,2} = k \log \frac{a'_{\mathrm{SiO^-},1}\, a'_{\mathrm{SiOH},2}\, a_{\mathrm{H_2O},2}}{a'_{\mathrm{SiO^-},2}\, a'_{\mathrm{SiOH},1}\, a_{\mathrm{H_2O},1}} - k(\mathrm{pH}_1 - \mathrm{pH}_2) \;, \tag{2.41}$$

giving the practically relevant potential difference as a function of a pH change, demonstrates that this conclusion may not be correct. Because the siloxy concentration is much smaller than the silanol concentration, the absolute siloxy activity ratio is always larger than the absolute silanol activity ratio, which is close to unity, and also larger than the activity ratio of the water in dilute solutions,

$$\left| \frac{a'_{SiO^-,1}}{a'_{SiO^-,2}} \right| > \left| \frac{a'_{SiOH,2} \, a_{H_2O,2}}{a'_{SiOH,1} \, a_{H_2O,1}} \right| , \tag{2.42}$$

which means that the first (logarithmic) term on the right side of (2.41) reduces the "ideal" slope k. Consequently, the theoretical, thermodynamically correct, response of glass electrodes is represented by

$$\frac{d\varepsilon_m}{d\,pH} = -k \left(1 - \frac{d \log \frac{a'_{SiO^-}}{a'_{SiOH} \, a_{H_2O}}}{d\,pH} \right) , \tag{2.43}$$

and, in principle, is smaller than the generally assumed "ideal" slope according to (2.38) [2.95]. Thus, the difference is measurable.

Indeed, the subideal slope can be determined experimentally and has been known for a long time as the so-called sub-Nernstian response of pH glass electrodes . Thus, "too small" practical potential slopes are generally obtained when glass electrode potentials are directly compared with the potential of the platinum–hydrogen electrode by means of cell (2.IX) without transference,

$$\left| Pt, H_2 \right| buffers\ with\ pH_1\ and\ pH_2 \left| glass\ electrode \right| , \tag{2.IX}$$

or when they are indirectly referred to those of the platinum–hydrogen electrode by means of cells

$$\left| ref.\ electrode\ (KCl,\ m \geq 3.5\,mol\,kg^{-1}) \right.$$
$$\left\| buffers\ with\ pH_1\ and\ pH_2 \right| glass\ electrode \right| \tag{2.X}$$

and

$$\left| ref.\ electrode\ (KCl,\ m \geq 3.5\,mol\,kg^{-1}) \right.$$
$$\left\| buffers\ with\ pH_1\ and\ pH_2 \right| Pt, H_2 \right| \tag{2.XI}$$

with transference in which an identical reference electrode is employed. The practical glass electrode slope k_{gl} obtained by these measurements is usually referred to the theoretical slope k by the so-called electromotive efficiency, $\alpha = k_{gl}/k$ [2.20], or by the electromotive loss factor, $n = 1 - \alpha$ [2.95], which are used to adapt (2.25) to the experimental results by means of

$$\varepsilon_m = \varepsilon_H^{0'} - k_{gl}\,pH = \varepsilon_H^{0'} - \alpha k\,pH = \varepsilon_H^{0'} - (1-n)k\,pH \tag{2.44}$$

and

$$\frac{d\varepsilon_m}{d\,pH} = -k_{gl} = -\alpha k = -k + nk \;, \qquad (2.45)$$

where nk is the electromotive loss,

$$nk = \frac{d\varepsilon_m}{d\,pH} - \frac{d\varepsilon'_m}{d\,pH} \;. \qquad (2.46)$$

Some papers reporting on the sub-Nernstian response may be given as examples. *Haugaard* observed already in 1938 that the glass electrode slope, in his words, has "values close to, but even with the best electrodes always a little smaller than, the theoretical value" [2.101]. In 1950, *Kratz* cited several papers from which $1 > \alpha \geq 0.996$ can be estimated [2.102]. *Bates*, referring to specifications given by the British Standards Institution [2.103], reported that electromotive efficiencies could be as small as $\alpha = 0.995$ [2.20]. *Covington*, in his paper on high-precision measurements, gives emfs that yield $\alpha \geq 0.997$ [2.104], and from precise data on several commercial glass electrodes published by *Light* [2.105], electromotive efficiencies between $\alpha = 0.9987$ and 0.9992 are calculated. In addition, our recent measurements on cells (2.X) and (2.XI) employing multiple point calibration with linear regression yielded $\alpha = 0.997$–0.998 for a glass electrode with a lithium-silicate-based membrane glass [2.106], and, finally, our direct comparison of commercial glass electrodes with a platinum–hydrogen electrode in cell (2.IX) resulted in electromotive efficiencies $\alpha = 0.9967 \pm 0.0005$ at $25\,°C$ and $\alpha = 0.9972 \pm 0.0008$ at $50\,°C$ [2.95].

These and other publications suggest with high probability that the experimental sub-Nernstian response represents the theoretical, that is, the thermodynamically correct, slope of glass electrodes according to (2.43), and the electromotive efficiencies at $25\,°C$ and $50\,°C$ of the paper cited above [2.95] confirm this conclusion because the data were obtained under conditions that excluded safely any other cause of the subideal response:

- Liquid junction effects were excluded.
- Influences of acid and sodium errors were ruled out.
- Electric shunts (short circuits) of the glass electrodes, which would also have resulted in a pH-independent decrease of the ideal slope (see Sect. 2.12) were negligible.
- An effect of the diffusion potentials within subsurface layers of the membrane glasses was excluded because they are pH-independent at 100% hydronium ion selectivity, as predicted by *Eisenman* [2.107] and more recently confirmed experimentally [2.72, 87].

The sub-Nernstian response can thus be explained thermodynamically by the phase boundary equilibrium, that is, by the *dissociation mechanism*, the electromotive efficiency being represented by

$$\alpha = 1 - \frac{d \log \frac{a'_{\mathrm{SiO^-}}}{a'_{\mathrm{SiOH}} \, a_{\mathrm{H_2O}}}}{d \, \mathrm{pH}} = 1 - \frac{d \log Q}{d \, \mathrm{pH}} \,, \tag{2.47}$$

and the electromotive loss factor by

$$n = \frac{d \log \frac{a'_{\mathrm{SiO^-}}}{a'_{\mathrm{SiOH}} \, a_{\mathrm{H_2O}}}}{d \, \mathrm{pH}} = \frac{d \log Q}{d \, \mathrm{pH}} \,, \tag{2.48}$$

where the fraction of the activities in the logarithm of (2.47) and (2.48), which has been equated to Q for practical reasons, can be approximated by the siloxy activity, $Q \cong a'_{\mathrm{SiO^-}}$, because the change of both silanol and water activities, and their product, with pH are close to unity, see above.

The sub-Nernstian response, consequently, is not just an artifact of glass electrodes, as has been assumed in the past, but, obviously, has a thermodynamic origin and a fundamental significance. Due to this thermodynamic basis, it yields a quantitative insight into the phase boundary equilibrium between glass and solution [2.95], of which generally only the overall dependence of the Galvani voltage on the hydronium ion activity in the solution is given by the Nernst equation, see for example (2.25). Besides, it yields a quantitative difference between homogeneous and heterogeneous electrochemical equilibria. For example, a *homogeneous* dissociation of a weak monobasic acid AH in an aqueous solution,

$$\mathrm{AH(sol'n)} + \mathrm{H_2O(sol'n)} = \mathrm{A^-(sol'n)} + \mathrm{H_3O^+(sol'n)} \,, \tag{2.49}$$

with the dissociation constant

$$K_{\mathrm{a}} = \frac{a_{\mathrm{A^-}} \, a_{\mathrm{H_3O^+}}}{a_{\mathrm{AH}} \, a_{\mathrm{H_2O}}} \tag{2.50}$$

is characterized by the derivative of the activities of the participating particles with respect to the pH value of the solution that is unity,

$$\frac{d \log \frac{a_{\mathrm{A^-}}}{a_{\mathrm{AH}} \, a_{\mathrm{H_2O}}}}{d \, \mathrm{pH}} = 1 \,. \tag{2.51}$$

For the *heterogeneous* dissociation of an acidic functional group (e.g., silanol) attached to a glass surface, in contrast, the corresponding derivative is given by the electromotive loss factor n , (2.48), which is smaller than unity by two to three orders of magnitude. Electromotive loss factors for several values of α in the range reported in the literature are given in the second column of Table 2.1. They demonstrate quantitatively that the surface activity of the siloxy groups is determined not only by the chemical driving force of the dissociation, as is the activity of the anion $\mathrm{A^-}$ in homogeneous solution, (2.50), but also, and even to a high degree, by the counteracting potential of the glass, that is, actually by their own activity.

Table 2.1. Detailed data of the phase boundary equilibrium at the glass–solution interface of pH glass electrodes for various electromotive efficiencies α, 25 °C ($n =$ electromotive loss factor)

$\alpha = \dfrac{k_{gl}}{k}$	$\dfrac{d \log Q}{d\,pH} = n$ [a]	$\dfrac{a'_{SiO^-,1}}{a'_{SiO^-,2}}$	$\dfrac{\Delta c'_{SiO^-}}{c'_{SiO^-,2}}$ [b]	$\dfrac{d\varepsilon_m}{d \log Q} = -\dfrac{\alpha k}{n}$ [c]	$\dfrac{d\varepsilon_m}{d\,pH} = -\alpha k$ [d]
		$(\Delta pH = -1)$	% $(\Delta pH = -1)$	V	V
0.9990	0.0010	0.9977	−0.23	−59.10	−0.05910
0.9985	0.0015	0.9966	−0.34	−39.38	−0.05907
0.9975	0.0025	0.9943	−0.57	−23.61	−0.05901
0.9960	0.0040	0.9908	−0.92	−14.73	−0.05892
0.9950	0.0050	0.9886	−1.14	−11.77	−0.05886

a $\dfrac{d \log Q}{d\,pH} = \dfrac{d}{d\,pH}\left(\log \dfrac{a'_{SiO^-}}{a'_{SiOH}\,a_{H_2O}}\right) \simeq \dfrac{d \log a'_{SiO^-}}{d\,pH}$

b $\dfrac{\Delta c'_{SiO^-}}{c'_{SiO^-,2}} = \dfrac{c'_{SiO^-,1} - c'_{SiO^-,2}}{c'_{SiO^-,2}} \simeq \dfrac{\Delta a'_{SiO^-}}{a'_{SiO^-,2}}$

c $\dfrac{d\varepsilon_m}{d \log Q} \simeq \dfrac{d\varepsilon_m}{d \log a'_{SiO^-}}$ ("internal slope" of membrane potential)

d) $\dfrac{d\varepsilon_m}{d\,pH} = -\alpha k = -(1-n)k$ ("practical slope" of membrane potential)

Table 2.1 also lists ratios of siloxy activities, $a'_{SiO^-,1}/a'_{SiO^-,2}$ (third column) and approximate relative changes of the siloxy concentration as obtained by

$$\frac{\Delta c'_{SiO^-}}{c'_{SiO^-,2}} \simeq \frac{\Delta a'_{SiO^-}}{a'_{SiO^-,2}} = \left(10^{n(pH_1 - pH_2)} - 1\right) , \tag{2.52}$$

(fourth column) both for unit pH change, $\Delta pH = -1$. They show quantitatively the corresponding small changes of the siloxy activity (and concentration) with pH, which have been mentioned only qualitatively in the foregoing sections. For example, at $\alpha = 0.9975$, which is approximately the average of the electromotive efficiencies given in the literature, a tenfold increase of the hydronium ion activity, or a change of $\Delta pH = -1$, results in a siloxy concentration decrease of only -0.57%, the ratio of the siloxy activities being 0.9943.

Also, the strong dependence of the electric potential of the glass on the surface activity of the negatively charged siloxy groups is obtained quantitatively from the electromotive loss factor. Thus, writing the potential slope, $d\varepsilon_m/d\,pH$, according to the chain rule,

$$\frac{d\varepsilon_m}{d\,pH} = \frac{d\varepsilon_m}{d \log Q}\frac{d \log Q}{d\,pH} , \tag{2.53}$$

rearranging, and inserting the known quantities $d\varepsilon_m/d\,pH = -(1-n)k$ and $d \log Q/d\,pH = n$ yields the potential slope referred to the logarithm of the siloxy group activity at the glass surface,

$$\frac{\mathrm{d}\,\varepsilon_m}{\mathrm{d}\log a'_{\mathrm{SiO}^-}} \cong \frac{\mathrm{d}\,\varepsilon_m}{\mathrm{d}\log Q} = -\frac{1-n}{n}k \ , \tag{2.54}$$

which we have termed the *internal slope* of the membrane potential for obvious reasons [2.100]. As seen in the fifth column of Table 2.1, it amounts to tens of volts (!) per decade of siloxy activity. For example, at $\alpha = 0.9975$, it is $23.61\,\mathrm{V}$ per $\Delta\mathrm{pSiO} = 1$ (if pSiO, according to pH, is defined as the negative logarithm of the siloxy surface activity, $\mathrm{pSiO} \equiv -\log\left(a'_{\mathrm{SiO}^-}/\mathrm{mol\,dm}^{-2}\right)$). Table 2.1 also shows that the internal slope increases rapidly with increasing electromotive efficiency and approaches infinity as α approaches unity. This should be expected because a constant siloxy activity at changing pH is physically meaningless according to the dissociation mechanism, which attributes the glass membrane potential to the charge density established by the siloxy groups at the glass surface.

The situation can be summarized as follows. It has been known empirically in the past that pH changes cause defined changes of the glass membrane potential, (2.38), and that the potential slope is subject to the "artifact" of a subunit electromotive efficiency α, (2.45). According to the *dissociation mechanism*, however, this pH-caused potential change occurs in two steps. (1) The pH change causes primarily a minute change of the activity of the negative form of the functional groups at the glass surface, (2.48), and (2) these small changes generate large changes of the potential of the glass, (2.54). The combination of these internal functions represents the overall pH dependence of the potential,

$$\frac{\mathrm{d}\,\varepsilon_m}{\mathrm{d}\,\mathrm{pH}} = \frac{\mathrm{d}\,\varepsilon_m}{\mathrm{d}\,\mathrm{pSiO}}\frac{\mathrm{d}\,\mathrm{pSiO}}{\mathrm{d}\,\mathrm{pH}} = -n\frac{1-n}{n}k = -(1-n)k = -\alpha k \tag{2.55}$$

with its "artifact", the subunit electromotive efficiency α.

So far, the rigorous treatment of phase boundary equilibria has been limited to silanol groups, that is, to the pH function of silicate membrane glasses, and the question arises whether there are similar dependencies (a) for other electrode functions, for example for the pD and pM functions of the siloxy group, (b) for other functional groups, for instance for the $\left[\mathrm{AlO}_{4/2}\right]^-$ group, and (c) for different groups characterizing different glasses. As expected from the above derivations, this can be shown theoretically although the experimental proof remains to be given in the future.

Various Electrode Functions of Silicate Glasses

Because the siloxy concentration at the glass surface is much smaller than the silanol concentration, $c'_{\mathrm{SiO}^-} \ll c'_{\mathrm{SiOH}} \cong c'^0 = \mathrm{const}$, and the activity of water in dilute solutions is unity, the internal slope of the membrane glass with respect to the pH function of the silanol group, (2.54), is given by

$$\frac{\mathrm{d}\,\varepsilon_m}{\mathrm{d}\log\dfrac{a'_{\mathrm{SiO}^-}}{a'_{\mathrm{SiOH}}\,a_{\mathrm{H_2O}}}} \cong \frac{\mathrm{d}\,\varepsilon_m}{\mathrm{d}\log a'_{\mathrm{SiO}^-}} = -\frac{1-n_{\mathrm{SiOH}}}{n_{\mathrm{SiOH}}}k \ , \tag{2.56}$$

where n_{SiOH} is the electromotive loss factor of the corresponding pH function. For the same silicate membrane glass in deuterium oxide solutions and the connected pD function, a corresponding relationship can be written [2.108],

$$\frac{\mathrm{d}\,\varepsilon_m}{\mathrm{d}\log \frac{a'_{\mathrm{SiO}^-}}{a'_{\mathrm{SiOD}}\,a_{\mathrm{D_2O}}}} \cong \frac{\mathrm{d}\,\varepsilon_m}{\mathrm{d}\log a'_{\mathrm{SiO}^-}} = -\frac{1-n_{\mathrm{SiOD}}}{n_{\mathrm{SiOD}}}k\;,\tag{2.57}$$

where n_{SiOD} is the electromotive loss factor of the pD function, and, finally, a similar equation results also for the pM function of the particular silicate membrane glass [2.72, 95],

$$\frac{\mathrm{d}\,\varepsilon_m}{\mathrm{d}\log \frac{a'_{\mathrm{SiO}^-}}{a'_{\mathrm{SiOM}}}} \cong \frac{\mathrm{d}\,\varepsilon_m}{\mathrm{d}\log a'_{\mathrm{SiO}^-}} = -\frac{1-n_{\mathrm{SiOM}}}{n_{\mathrm{SiOM}}}k\;,\tag{2.58}$$

where n_{SiOM} is the electromotive loss factor of the pM response of the glass. It is obvious from these equations that, except for the influence of activity coefficients of the surface groups, the electromotive loss factors for the different functions of the silicate glass must be approximately equal,

$$n_{\mathrm{SiOH}} \cong n_{\mathrm{SiOD}} \cong n_{\mathrm{SiOM}}\;,\tag{2.59}$$

because the dependence of the membrane potential on the logarithm of the siloxy activity, the internal slope, is obviously independent of the cation involved in the phase boundary equilibrium and thus attached to the functional groups at the glass surface. This conclusion must be valid also for the transition range between two electrode functions, for instance between 100% pH and 100% pM response, where, consequently, a response-independent, constant electromotive efficiency of the overlapping different responses must be expected.

Electrode Functions of Glasses Other than Silicate Glasses

Because the same argument as given above for silicate glasses can be applied to any other functional surface group, for example to $\left[\mathrm{AlO_{4/2}}\right]^-$, it must be concluded that the magnitude of the sub-Nernstian response also of glasses other than silicate membrane glasses is independent of the cation to which it responds [2.72, 95]. Thus, for example, the electromotive loss factors of the pH, pD, and pM functions of a glass characterized by $\left[\mathrm{AlO_{4/2}}\right]^-$ groups are expected to be approximately equal,

$$n_{\mathrm{AlOH}} \cong n_{\mathrm{AlOD}} \cong n_{\mathrm{AlOM}}\;,\tag{2.60}$$

where the functional group is indicated by AlO for the sake of simplicity.

Different Electrode Functions of Different Glasses

Finally, it must be expected that, except for activity coefficients of the surface groups, the internal slope is independent also of the individual character of the functional surface groups that cause the potential [2.95],

$$\frac{d\varepsilon_m}{d\log a'_{\mathrm{SiO}^-}} \cong \frac{d\varepsilon_m}{d\log a'_{\mathrm{AlO}^-}} \cong \frac{d\varepsilon_m}{d\log a'_{\mathrm{R}^-}} , \tag{2.61}$$

where R^- stands for the negative form of any functional surface group. Consequently, the electromotive loss factor (and the electromotive efficiency) must also be expected to be approximately equal for all electrode membrane glasses,

$$n_{\mathrm{SiOH}} \cong n_{\mathrm{AlOH}} \cong n_{\mathrm{RH}} \cong n_{\mathrm{RD}} \cong n_{\mathrm{RM}} , \tag{2.62}$$

in particular because the derivatives (2.56)–(2.58) do not contain individual equilibrium constants.

It thus results that, except for influences of activity coefficients, the sub-Nernstian response must be basically independent of the membrane glass composition and the selectivity range chosen. For pH glasses, this seems to be confirmed by the relatively narrow range of the electromotive efficiencies reported and measured (see above), which are between 0.995 and 0.999, the majority of values being around 0.9975. Nevertheless, in order to substantiate this interesting conclusion, we have proposed that electromotive efficiencies of more glass electrodes with different membrane glass compositions should be measured [2.95]. The findings should also be verified for the alkali-selective range of glass electrodes, for instance, by directly comparing the response of pNa or pK glass electrodes with that of sodium or potassium amalgam electrodes, respectively, which do not exhibit sub-Nernstian response, in cells according to cells (2.IX) and (2.X) and (2.XI).

The thermodynamically caused sub-Nernstian response is basically not restricted to glass electrodes but must be expected also with other electrodes if the following three conditions are met.

(a) *Basic condition*: The functioning of the electrode must be the result of a phase boundary equilibrium involving a charged form of functional groups bound to the electrode surface whose concentration (activity), given by the ionic activities in the solution, determines the Galvani voltage between the electrode and the solution.

(b) *Quantitative condition*: The activity changes of the charged form of the functional surface groups caused by (opposite) activity changes of the ions in the contacting solution must be sufficiently large to reduce the Nernstian slope measurably.

(c) *Excluding condition*: Other causes which also yield pH-independent reductions of the "ideal" potential slope, for instance electrode shunts, must be excluded.

As shown above, glass electrodes meet these conditions. (a) The *dissociation mechanism*, by which the sub-Nernstian response is explained, was verified experimentally [2.72, 87]. (b) According to literature data and our own measurements, the effect is obviously large enough to be measurable, and (c) other causes, particularly electrode shunts, were excluded during measurements as far as possible. The ion exchange theory, however, fails to offer an explanation of the sub-Nernstian response because it is based on an exchange of different cations between glass and solution, which is not further specified and does not involve individual charged surface groups of the glass (see Sect. 2.3), and thus does not meet the *basic condition* given above. SiO_2 and Al_2O_3, which are extensively applied in ion-sensitive field effect transistors (ISFETs), also show sub-Nernstian response, which has been discussed in the literature, see for instance [2.109]. The site-binding theory of these sensors proposed by *Yates* in 1974 [2.110], the same year the dissociation mechanism was initially suggested [2.86], assumes differently charged surface groups of the oxides in equilibrium with the dissolved cations and thus meets the *basic condition*, and it may well be that the *quantitative condition* is also met, which, however, necessitates a reconsideration of the points made in [2.109] and probably further specific experiments. The mechanism of PVC-based ion-selective electrodes (ISEs), which also exhibit sub-Nernstian behaviour [2.111], is based on diffusion of complexed cations through the membrane and obviously does not meet the *basic condition* so that in all probability the non-ideality of these sensors has a different origin. Finally, sub-Nernstian response is neither expected nor observed with redox, metal–metal ion, and crystal membrane (e.g., LaF_3, $AgCl$) electrodes because the electrons and cations, respectively, responsible for the Galvani voltage between electrode and solution do not represent localized charges so that the *basic condition* for the thermodynamically caused sub-Nernstian response is not met by these electrodes and common sensors.

Appendix

Electromotive Efficiencies of Glass Electrodes as Obtained by Cell (2.IX) and Cells (2.X) and (2.XI). In order to measure electromotive efficiencies and electromotive loss factors, potentials of the respective glass electrodes and those of a Pt,H_2 electrode are compared directly in cell (2.IX) or in cells (2.X) and (2.XI). The equations derived for α and n of these arrangements differ slightly from the respective equations (2.47) and (2.48) [2.95] and will be compared in this appendix for the sake of completeness although the difference is negligible under normal conditions, i.e., in dilute solutions.

The emf of cell (2.IX) without transference is the sum of the Galvani voltages ε_{gl} and ε_{Pt,H_2} of the glass and the Pt,H_2 electrodes, respectively,

$$E_{Pt,gl} = \varepsilon_{gl} - \varepsilon_{Pt,H_2} = \varepsilon_{gl}^0 + \varepsilon_m - \varepsilon_{Pt,H_2} , \tag{2.63}$$

where ε_{gl}^0 is the (constant) sum of the Galvani voltages of the internal reference electrode and between the internal membrane glass surface and the internal solution of the glass electrode. The Galvani voltage between the outer membrane surface and the solution is given by (2.39) above, and that of the Pt,H_2 electrode is established by the equilibrium

$$1/2\,H_2(g) + H_2O(\text{sol'n}) \rightleftharpoons H_3O^+(\text{sol'n}) + e^-(Pt) \,, \tag{2.64}$$

where g denotes gas. Taking into account that the standard Galvani voltage of the Pt,H_2 electrode is defined to be zero at all temperatures, the emf of cell (2.IX) is given by

$$E_{Pt,gl} = \varepsilon_{gl}^0 - k\log K_{D,H} + k\log\frac{a'_{SiO^-}\,a_{H_3O^+}}{a'_{SiOH}\,a_{H_2O}} - k\log\frac{a_{H_3O^+}}{p_{H_2}^{1/2}\,a_{H_2O}} \tag{2.65}$$

or, after rearranging, by

$$E_{Pt,gl} = \varepsilon_{gl}^0 - k\log K_{D,H} + k\log\frac{a'_{SiO^-}}{a'_{SiOH}} + \frac{k}{2}\log p_{H_2} \,, \tag{2.66}$$

whose differentiation with respect to pH at constant hydrogen pressure and rearrangement yields

$$\frac{1}{k}\frac{dE_{Pt,gl}}{d\,pH} = \frac{d\log\frac{a'_{SiO^-}}{a'_{SiOH}}}{d\,pH} \,. \tag{2.67}$$

Equation (2.67) demonstrates that measurements on cell (2.IX) result in what was termed the "ideal" electromotive loss factor n' (and "ideal" electromotive efficiency α') [2.95],

$$n' = \frac{d\log\frac{a'_{SiO^-}}{a'_{SiOH}}}{d\,pH} \,, \tag{2.68}$$

which differs from the electromotive loss factor n, (2.48), by the lack of the water activity,

$$n' = n + \frac{d\log a_{H_2O}}{d\,pH} \,. \tag{2.69}$$

However, n can be equated to n' if the glass electrode is transferred between dilute solutions, for instance between standard buffer solutions, whose water activity is close to unity.

A corresponding derivation for cells (2.X) and (2.XI) results in

$$n' = 1 - \frac{dE_{gl}}{dE_{Pt,H_2}} \tag{2.70}$$

or, for two individual buffer solutions with pH_1 and pH_2, in

$$n' = 1 - \frac{[E_{gl}(pH_1) - E_{gl}(pH_2)]}{[E_{Pt,H_2}(pH_1) - E_{Pt,H_2}(pH_2)]} \quad , \tag{2.71}$$

where E_{gl} and E_{Pt,H_2} are the emfs of cells (2.X) and (2.XI) at the indicated pH values, respectively. A further derivation shows that n and n' obtained from these cells do not contain the activity of water either and are thus also "ideal" values. An equation for α' according to (2.71), incidentally, was given by *Bates* [2.20].

2.5.4 The Transition Range Between the pH and the pM Response of Glass Electrodes

Between 100% pH and 100% pM sensitivity, there is the so-called transition range in which the glass membrane potential is a function of both hydronium and alkali ion activities [2.72, 87]. The functional surface groups are simultaneously subject to the dissociation and association equilibrium, which are interconnected by the activity of the siloxy groups at the glass surface and the Galvani voltage of the glass membrane, which take part in both equilibria. The overall equilibrium is formally given by the combination of (2.17) and (2.29) of Sect. 2.5.2,

$$\equiv SiOH(s) + H_2O(\text{sol'n}) + M^+(\text{sol'n}) \rightleftharpoons \equiv SiOM(s) + H_3O^+(\text{sol'n}) \ , \tag{2.72}$$

which, however, does not show explicitly the participation of siloxy groups in the equilibrium and thus the simultaneous pH and pM dependence of the Galvani voltage between glass and solution because the siloxy groups cancel in (2.72). But a combination of the equilibria in the form given in (2.16) in Sect. 2.5.1 demonstrates that the siloxy groups not only participate in but connect both equilibria. For example, an increase of the hydronium activity at constant alkali ion activity of the solution increases the silanol activity at the expense of the siloxy salt activity at the glass surface, which is only made possible by the presence of siloxy groups, whose activity decreases slightly and thus makes the potential of the glass more positive.

For the derivation of the pH-dependent and pM-dependent potential of the glass membrane in the transition range, it is advantageous to replace thermodynamic by stoichiometric equilibrium constants, because they allow one to treat concentrations instead of activities, and to re-introduce activity coefficients and water activity into the final equations. Thus, the total concentration of functional surface groups is the sum of the concentrations of silanol, siloxy salt, and siloxy groups,

$$c'^0 = c'_{SiOH} + c'_{SiOM} + c'_{SiO^-} \ , \tag{2.73}$$

which is not necessarily the case with the corresponding activities. The stoichiometric dissociation, $K''_{D,H}$, and association, $K'_{A,M}$, constants are given by

$$K''_{D,H} = K_{D,H} \frac{\gamma'_{SiOH} a_{H_2O}}{\gamma'_{SiO^-}} \tag{2.74}$$

and

$$K'_{A,M} = K_{A,M} \frac{\gamma'_{SiO^-}}{\gamma'_{SiOM}} , \tag{2.75}$$

respectively, and the total thermodynamic equilibrium constant of the overall equilibrium, (2.72), is the product of the dissociation and association constants,

$$K_{D,H} K_{A,M} = \frac{a'_{SiOM} a_{H_3O^+}}{a'_{SiOH} a_{M^+} a_{H_2O}} , \tag{2.76}$$

whose transformed stoichiometric version to be applied here is given by

$$K''_{D,H} K'_{A,M} = \frac{c'_{SiOM} a_{H_3O^+}}{c'_{SiOH} a_{M^+}} . \tag{2.77}$$

It can readily be shown that a simple transformation of (2.77) yields the silanol concentration at the glass surface as a function of the hydronium and alkali ion activities,

$$c'_{SiOH} = \frac{a_{H_3O^+}(c'^0 - c'_{SiO^-})}{(a_{H_3O^+} + K''_{D,H} K'_{A,M} a_{M^+})} , \tag{2.78}$$

whose introduction into one of the equations derived for the Galvani voltage in Sect. 2.5.2, for instance into (2.22), results in the rigorous expression for the Galvani voltage between glass and solution as a function of the hydronium and alkali ion activities,

$$\varepsilon_m = -k \log K''_{D,H} + k \log \frac{c'_{SiO^-}}{c'^0 - c'_{SiO^-}} + k \log(a_{H_3O^+} + K''_{D,H} K'_{A,M} a_{M^+}) . \tag{2.79}$$

The first two terms on the right side of (2.79) can be combined with good approximation, at least for practical applications, to give the standard Galvani voltage of the overall equilibrium,

$$\varepsilon_H^{0''} = -k \log K''_{D,H} + k \log \frac{c'_{SiO^-}}{c'_{SiOH} + c'_{SiOM}} = -k \log K''_{D,H} + k \log \frac{c'_{SiO^-}}{c'^0} . \tag{2.80}$$

The Galvani voltage of the glass membrane as a function of both ionic activities in the transition range is thus represented by

$$\varepsilon'_m = \varepsilon_H^{0''} + k \log(a_{H_3O^+} + K''_{D,H} K'_{A,M} a_{M^+}) \tag{2.81}$$

or, if the electromotive efficiency (see Sect. 2.5.3) is introduced into (2.81), by

$$\varepsilon_m = \varepsilon_H^{0''} + \alpha k \log(a_{H_3O^+} + K''_{D,H} K'_{A,M} a_{M^+}) \ . \tag{2.82}$$

In addition, the emf of a glass-electrode-containing electrochemical cell is obtained by introducing the Galvani voltage into the general equation (2.1) of Sect. 2.2,

$$E' = E_H^{0'} + k \log(a_{H_3O^+} + K''_{D,H} K'_{A,M} a_{M^+}) \ , \tag{2.83}$$

where $E_H^{0'}$ is the standard emf of the cell,

$$E_H^{0'} = \sum \varepsilon_n + \varepsilon_H^{0''} \ . \tag{2.84}$$

The standard emf as defined by (2.84) contains two approximations. (a) The liquid junction voltage between the measuring and the reference electrode solution, which is contained in $\sum \varepsilon_n$, is assumed to be independent of the composition of the measuring solution. (b) The siloxy group concentration included in the standard Galvani voltage $\varepsilon_H^{0''}$ is assumed constant. Besides, the emf E', (2.83), contains the "selectivity product", i.e., the product of the stoichiometric equilibrium constants $K''_{D,H} K'_{A,M}$, (2.77), which is also taken to be constant. It must thus be expected that (2.83) yields emfs that deviate more or less from experimental data in the transition range, as is known for the original Nicolsky equation [2.63], (2.6) of Sect. 2.3, which has the same form.

In order to gain information about the possible extent of these deviations, the thermodynamic equilibrium constants and thus the activity coefficients and the water activity are reintroduced into the Galvani voltage, (2.79), yielding

$$\begin{aligned}
\varepsilon_m = &- k \log K_{D,H} + k \log \frac{a'_{SiO^-}}{a_{H_2O} \, \gamma'_{SiOH} \, (c'_{SiOH} + c'_{SiOM})} \\
&+ k \log(a_{H_3O^+} + K_{D,H} K_{A,M} \frac{\gamma'_{SiOH} \, a_{H_2O}}{\gamma'_{SiOM}} a_{M^+}) \ .
\end{aligned} \tag{2.85}$$

This expression shows that the main deviations seem to be caused by the factor in front of the alkali activity. It contains the ratio of the activity coefficients of the silanol and siloxy salt groups, whose magnitude changes with high probability when the silanol concentration changes from unity to zero (corresponding to a respective change of the siloxy salt concentration from zero to unity), i.e., in the transition range although the single activity coefficients of the uncharged groups are close to unity. Larger deviations, however, are caused by the diffusion voltage in the subsurface region of the glass membrane, which changes appreciably in the transition range, as will be shown in Sect. 2.7.4. Unfortunately, quantitative data of surface activity coefficients and their concentration dependence, which would be of great interest with respect to (2.85), are not available.

The selectivity product $K''_{D,H} K'_{A,M}$ characterizes the membrane glass with respect to its selectivity. The relative magnitudes of the hydronium ion activity $a_{H_3O^+}$ and the product $K''_{D,H} K'_{A,M} a_{M^+}$, (2.82), determine whether a glass electrode exhibits pH or pM response, or both, in a particular solution. In addition, the rearranged form

$$\frac{c'_{SiOH}}{c'_{SiOM}} = \frac{a_{H_3O^+}}{K''_{D,H} K'_{A,M} a_{M^+}} \tag{2.86}$$

of the overall equilibrium (2.77) shows that the ratio of the same quantities, right side of (2.86), also determines the relative concentrations of silanol and siloxy salt groups at the glass surface. The electrode response, (2.82), and relative surface concentrations (or relative surface coverage), (2.86), thus always appear in parallel. This important fact was verified experimentally by means of ionic electromigration in connection with potential formation measurements, see Sects. 2.6.3 and 2.11.

It is advantageous for the further discussion to define the negative logarithm of the denominator of the right side of (2.86) as the transition pH value,

$$\begin{aligned} pH_{tr} &= -\log(K''_{D,H} K'_{A,M} a_{M^+}) \\ &= p(K''_{D,H} K'_{A,M} a_{M^+}) \\ &= p(K''_{D,H} K'_{A,M}) + pM \ . \end{aligned} \tag{2.87}$$

It is thus easily seen from (2.82) and (2.86) that the magnitude of pH_{tr} relative to the solution pH informs us about both the membrane response, (2.82), and the relative surface coverage, (2.86), of the glass. As indicated in Fig. 2.12, which presents the emf of a pNa glass electrode-containing cell as a function of pH at three constant pNa values, the transition from 100% pH to 100% pNa selectivity extends over a certain pH region $pH_{tr} \pm b$, where b depends on the degree by which 100% selectivity is to be attained. For all practical cases, $b = 2$ will suffice, see Fig. 2.12. Thus, at pH $= (pH_{tr} - 2)$, the membrane potential obtained from (2.82) differs by only $+0.26\,mV$ from that for 100% pH response. With increasing difference b, the deviation decreases rapidly, so that the glass electrode exhibits practically pure hydronium ion selectivity and its membrane surface shows nearly 100% coverage with silanol. At pH $\geq (pH_{tr} + 2)$, on the other hand, practically 100% pNa response is observed, and the surface contains nearly 100% siloxy sodium groups. Within

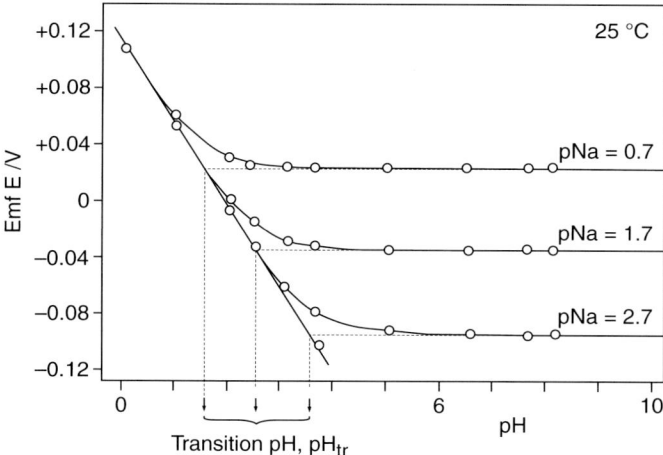

Fig. 2.12. Experimental emf of a pNa glass electrode cell as a function of pH at three constant sodium ion activities (pNa) demonstrating the transition pH value, pH_{tr}, (2.87), between the pH-selective and pNa-selective pH ranges. Membrane glass: rate-cooled lithium aluminosilicate glass

the transition range, at $(pH_{tr} - 2) \leq pH \leq (pH_{tr} + 2)$, the glass shows mixed pH and pNa response, and the glass surface is covered by both silanol and siloxy sodium groups. At exactly pH_{tr}, finally, pH and pNa have an equal effect on the electrode potential, and the membrane surface carries equal concentrations of SiOH and SiONa groups.

The selectivity product of practical glass electrodes is either obtained from plots as shown in Fig. 2.12 or, more accurately, by measuring two emfs, one, E'_1, at 100% pH response and the other, E'_2, at 100% pM response of the electrode employed. The selectivity product is then obtained by

$$p(K''_{D,H} K'_{A,M}) = (pH - pM) + \frac{E'_1 - E'_2}{k} \ , \tag{2.88}$$

as is easily shown. This procedure, however, is difficult or impossible to apply to pH glass electrodes because their 100% pM response range is difficult to reach or not attainable at all. The measurement of potentials at high pH, in addition to that at low pH, is then the only possibility to determine an approximate pH_{tr}.

Table 2.2 gives typical examples of the selectivity product (in the form of $p(K''_{D,H} K'_{A,M})$ values) of pH-selective silicate and pNa-selective aluminosilicate glasses. It shows, for instance, that the selectivity product for the hydronium–lithium ion couple differs by nine orders of magnitude for these different glass types. Also, the selectivity product of the aluminosilicate glass differs for the hydronium–lithium and hydronium–sodium ion couples by more than one order of magnitude. Its magnitude depends even on the cooling rate of the membrane glass causing, for instance, a "lithium error" of the

Table 2.2. Selectivity products of a lithium silicate pH and a lithium aluminosilicate pNa electrode membrane glass at different thermal states of the glass and at different temperatures

Glass electrode	Membrane glass	Thermal state	Ion couple	Temperature $/°C$	$p(K''_{D,H} K'_{A,M})$ [a]
pH	Li silicate	quenched	H^+, Li^+	25	12.1
				50	11.3
				75	10.6
		rate-cooled	H^+, Li^+	25	11.3
				50	10.6
				75	9.9
pNa	Li Al silicate	rate-cooled	H^+, Li^+	25	2.44
				50	2.24
			H^+, Na^+	25	0.94
				50	0.76

[a] $p(K''_{D,H} K'_{A,M}) = -\log(K''_{D,H} K'_{A,M})$

rate-cooled pH-selective silicate glass (cooling rate: $3.75\,\mathrm{K\,min^{-1}}$) larger than that of the quenched glass (cooling rate: $> 2000\,\mathrm{K\,min^{-1}}$) by $\Delta pH_{tr} = -0.8$ at the same lithium activity. As will be discussed in Sect. 2.6.3, this effect of the thermal history depends mainly on the entropy of the phase boundary equilibrium and thus on the degree of order at and around the surface groups.

In contrast to the selectivity product, which is easily attainable, the single values of the dissociation constants $K''_{D,H}$ and the association constants $K'_{A,M}$ are not available. Probable magnitudes, however, can be obtained by considering the charge-dependent potential of glasses [2.72]. As stated before, the potential of the glass causes the heterogeneous dissociation of both acidic and salt groups at its surface to be much weaker than the homogeneous dissociation of related molecules containing the same or similar (acidic and salt) groups in solution. This is especially obvious with siloxy salt groups, which exist in nearly completely associated form at a glass surface in its alkali-sensitive state, whereas alkali silicates in solution are dissociated by practically 100%, independent of pH. The $pK''_{D,H}$ value of an acidic surface group, consequently, can be assumed to be larger than the $pK_{D,sol'n}$ value of the corresponding monomolecular acid in homogeneous solution by a certain difference, say Δ_D,

$$pK''_{D,H} = pK_{D,sol'n} + \Delta_D . \tag{2.89}$$

The known selectivity product, formulated with this modified expression for the heterogeneous dissociation constant, consequently, yields a heterogeneous association constant for each assumed difference Δ_D if, in addition, the homogeneous dissociation constant of a related dissolved acid is known:

$$pK'_{A,M} = p(K''_{D,H} K'_{A,M}) - (pK_{D,sol'n} + \Delta_D) . \tag{2.90}$$

Because (2.89) yields also the appropriate heterogeneous dissociation constant, this simple procedure allows one to calculate pairs of heterogeneous dissociation and association constants of glass surface groups, whose products are equal to the selectivity product. Although the differences Δ_D are chosen completely arbitrarily, the process excludes certain combinations of constants that are evidently physically meaningless and yields others that can be selected as probable numbers because of evidence or due to additional information.

Table 2.3 shows combinations of heterogeneous dissociation and association constants obtained for silanol dissociation and siloxy lithium association, which are based on the homogeneous dissociation constant of monomeric silicic acid reported in the literature [2.112] and on $p(K''_{D,H}\,K'_{A,Li})$ values for the pH-selective silicate glass given in Table 2.2. It is obvious from these data that the heterogeneous dissociation constant must be smaller than the homogeneous dissociation constant by at least four orders of magnitude because the association constant must be assumed to be at least $3 \times 10^2\,\mathrm{kg\,mol^{-1}}$, which, however, still seems to be much too small. Rather, it appears that the difference is of the order of $\Delta_D = 6$, or even larger, which results in a combination of a surface dissociation constant of $1.6 \times 10^{-16}\,\mathrm{mol\,kg^{-1}}$ and an association constant of $3 \times 10^4\,\mathrm{kg\,mol^{-1}}$.

Corresponding combinations of surface equilibrium constants can be deduced by means of association constants $K_{A,\mathrm{sol'n}}$ of the monomolecular salts in homogeneous solution, which, however, have generally much greater uncertainties because of the nearly complete dissociation of alkali salts, if they are known at all. If, in this case, the heterogeneous association constant $K'_{A,M}$ of

Table 2.3. Possible combinations of heterogeneous dissociation constants $K''_{D,H}$ of silanol groups and association constants $K'_{A,Li}$ of siloxy lithium groups of a rate-cooled pH-selective silicate glass, based on the homogeneous dissociation constant $K_{D,\mathrm{sol'n}}$ of silicic acid and the selectivity product $K''_{D,H}\,K'_{A,Li}$ of the glass (see Table 2.2)

$pK_{D,\mathrm{sol'n}}$ [a]	$p(K''_{D,H}\,K'_{A,Li})$ [b]	Δ_D [c]	$pK''_{D,H}$ [d]	$pK'_{A,Li}$ [e]	$K''_{D,H}$ mol kg^{-1}	$K'_{A,Li}$ kg mol^{-1}
9.8 [2.112]	11.3	2	11.8	−0.5	1.6×10^{-12}	3×10^{0}
		4	13.8	−2.5	1.6×10^{-14}	3×10^{2}
		6	15.8	−4.5	1.6×10^{-16}	3×10^{4}
		8	17.8	−6.5	1.6×10^{-18}	3×10^{6}
		10	19.8	−8.5	1.6×10^{-20}	3×10^{8}

[a] $pK_{D,\mathrm{sol'n}} = -\log K_{D,\mathrm{sol'n}}$
[b] $p(K''_{D,H}\,K'_{A,Li}) = -\log(K''_{D,H}\,K'_{A,Li})$
[c] $\Delta_D = pK''_{D,H} - pK_{D,\mathrm{sol'n}}$
[d] $pK''_{D,H} = -\log(K''_{D,H}/\mathrm{mol\,kg^{-1}})$
[e] $pK'_{A,Li} = -\log(K'_{A,Li}/\mathrm{kg\,mol^{-1}})$

the surface groups is assumed to be larger than the homogeneous association constant by a certain difference Δ_A,

$$pK'_{A,M} = pK_{A,\text{sol'n}} - \Delta_A , \tag{2.91}$$

the corresponding heterogeneous dissociation constant of the acidic form of the surface groups is obtained by means of

$$pK''_{D,H} = p(K''_{D,H} K'_{A,M}) - (pK_{A,\text{sol'n}} - \Delta_A) . \tag{2.92}$$

Table 2.4 contains possible combinations of heterogeneous dissociation and association constants as obtained for the rate-cooled aluminosilicate pNa glass given in Table 2.2. The homogeneous association constant was assumed to be $K_{A,\text{sol'n}} = 10^{-4}$, which corresponds to 99% dissociation of the salt in water. It seems reasonable to exclude from the resulting pairs of data association constants below $K'_{A,\text{Li}} = 10^6 \text{ kg mol}^{-1}$ because the corresponding heterogeneous dissociation constant of the [$\equiv$ SiOAl $\equiv$]H groups is with high probability not larger than $K''_{D,H} = 3 \times 10^{-9} \text{ mol kg}^{-1}$. The pNa glass thus seems to have constants $K'_{A,\text{Li}} \geq 10^6 \text{ kg mol}^{-1}$ and $K''_{D,H} \leq 10^{-9} \text{ mol kg}^{-1}$.

As shown by the data in Tables 2.3 and 2.4, the proposed procedure gives at least ranges of orders of magnitude for the equilibrium constants, which had not been available in the past. In addition, they definitely confirm the large change that dissociation and association constants of monomolecular acids and salts, respectively, undergo when they or an equivalent form are attached to a glass surface and act as functional surface groups.

Table 2.4. Possible combinations of heterogeneous dissociation constants $K''_{D,H}$ of [$\equiv$ SiOAl $\equiv$]H groups and association constants of $K'_{A,\text{Li}}$ of [$\equiv$ SiOAl $\equiv$]Li groups of a pNa-selective aluminosilicate pNa-selective glass based on an assumed homogeneous association constant $K_{A,\text{sol'n}}$ and the selectivity product $K''_{D,H} K'_{A,\text{Li}}$ of the glass (see Table 2.2)

$pK_{A,\text{sol'n}}$ [a]	$p(K''_{A,H} K'_{A,\text{Li}})$ [b]	Δ_A [c]	$pK'_{A,\text{Li}}$ [d]	$pK''_{D,H}$ [e]	$K'_{A,\text{Li}}$ kg mol^{-1}	$K''_{D,H}$ mol kg^{-1}
4	2.44	2	+2	0.44	10^{-2}	3.6×10^{-1}
		4	± 0	2.44	10^0	3.6×10^{-3}
		6	-2	4.44	10^2	3.6×10^{-5}
		8	-4	6.44	10^4	3.6×10^{-7}
		10	-6	8.44	10^6	3.6×10^{-9}
		12	-8	10.44	10^8	3.6×10^{-11}

[a] $pK_{A,\text{sol'n}} = -\log(K_{A,\text{sol'n}}/\text{kg mol}^{-1})$
[b] $p(K''_{D,H} K'_{A,\text{Li}}) = -\log(K''_{D,H} K'_{A,\text{Li}})$
[c] $\Delta_A = pK_{A,\text{sol'n}} - pK'_{A,\text{Li}}$
[d] $pK'_{A,\text{Li}} = -\log(K'_{A,\text{Li}}/\text{kg mol}^{-1})$
[e] $pK''_{D,H} = -\log(K''_{D,H}/\text{mol kg}^{-1})$

2.5.5 Kinetics of Interfacial Equilibria

Chemical equilibria are never static in character but are dynamic states, the opposite reactions proceeding with equal rates in both directions. The same applies to electrochemical equilibria of metal and sparingly soluble salt electrodes in equilibrium with solutions, a subject that has grown to the tremendously wide and important field of electrode kinetics, which covers most of the electrochemistry of today. For an introduction into electrode kinetics, please refer to the excellent text books available, for instance to [2.40, 113]. It has been found that the equilibria at solution-contacted glass electrode membranes can also be treated on an electrode-kinetic basis and that this results in the equations for the membrane potential as a function of pH and pM which have been obtained from thermodynamics [2.71, 72, 87]. Because this derivation is of basic significance for the dissociation mechanism, it is briefly presented in the following.

Homogeneous Equilibria

Let us start the discussion with a brief repetition of the kinetics of equilibria in homogeneous solution. An example is the dissociation equilibrium of a monobasic acid AH in aqueous solution,

$$AH(\text{sol'n}) + H_2O(\text{sol'n}) \rightleftarrows A^-(\text{sol'n}) + H_3O^+(\text{sol'n}) , \qquad (2.93)$$

whose second-order dissociation and (reverse) association reactions proceed with the respective equal reaction rates,

$$v_d = k_d\, c_{AH}\, c_{H_2O} \qquad (2.94)$$

and

$$v_a = -k_a\, c_{A^-}\, c_{H_3O^+} . \qquad (2.95)$$

The proportionality constants called rate constants are given by the respective standard free energies of activation according to

$$k_d = \frac{kT}{h} e^{-(\Delta G_d^{0\neq}/RT)} \qquad (2.96)$$

and

$$k_a = \frac{kT}{h} e^{-(\Delta G_a^{0\neq}/RT)} , \qquad (2.97)$$

whose meaning is explained in Fig. 2.13, where the energy of the proton is plotted as a function of the non-directional distance between the reacting particles. In order to surmount the energy summit at distance x_s in the

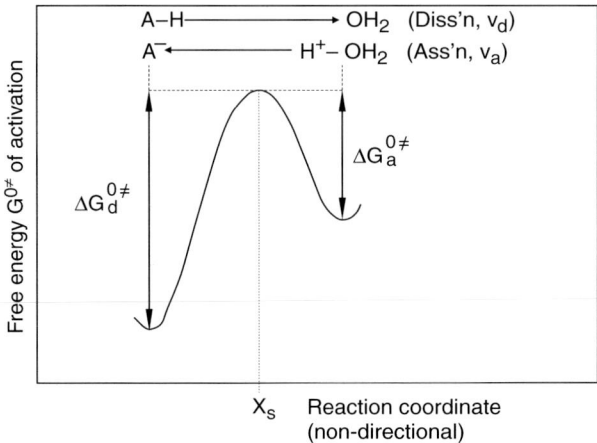

Fig. 2.13. Schematic energy–distance plot of the proton for a homogeneous disso-ciation of a monobasic weak acid and the reverse (association) reaction in aqueous solution. The reaction coordinate is non-directional. $\Delta G_d^{0\neq}$ and $\Delta G_a^{0\neq}$ are standard free energies of activation of dissociation, respectively association

dissociation or association direction, the proton, as long as it does not tun-nel the energy barrier, must attain the activation energy of dissociation or association, respectively. They consist of a chemical and an electrical part, which, however, are not separable. Equal dissociation and association rates at equilibrium yield

$$v_0 = v_d = v_a = k_d\, c_{AH}\, c_{H_2O} = -k_a\, c_{A^-}\, c_{H_3O^+} \ , \tag{2.98}$$

whose rearrangement gives (2.99), which demonstrates the kinetic meaning of the stoichiometric dissociation constant,

$$\frac{k_d}{k_a} = K_D' = \frac{c_{A^-}\, c_{H_3O^+}}{c_{AH}\, c_{H_2O}} \ . \tag{2.99}$$

Similarly, the thermodynamic dissociation constant

$$K_D = \frac{a_{A^-}\, a_{H_3O^+}}{a_{AH}\, a_{H_2O}} = K_D' \frac{\gamma_{A^-}\, \gamma_{H_3O^+}}{\gamma_{AH}\, \gamma_{H_2O}} \tag{2.100}$$

could have been obtained by a rigorous derivation, using activities of the equilibrium participants and activity-related reaction rate constants in (2.94) and (2.95). This, however, is of no concern here. Rather, it was to be shown that the derivation leads to the important conclusion that the dissociation constant is given by the standard free energies of activation of dissociation and association,

$$K_D' = e^{(\Delta G_a^{0\neq} - \Delta G_d^{0\neq})/RT} \ , \tag{2.101}$$

whose relative magnitude thus determines whether the compound concerned is a strong or a weak acid. The qualitative plot in Fig. 2.13, for instance, characterizes an acid with a dissociation constant smaller than unity. It is also emphasized that the homogeneous reaction rates, (2.94) and (2.95), can neither be controlled nor influenced from the outside, except by changes in pressure and temperature and by changing the solvent, which, however, means treating a completely different system.

Heterogeneous Equilibria

The kinetics of equilibria at glass–solution interfaces can be treated in much the same way but must observe some special features of the two-phase system. The interfacial dissociation of silanol groups at a glass surface

$$\equiv \text{SiOH(s)} + \text{H}_2\text{O(sol'n)} \rightleftharpoons \equiv \text{SiO}^-(\text{s}) + \text{H}_3\text{O}^+(\text{sol'n}) \qquad (2.102)$$

is again taken as an example. The conditions are explained in Fig. 2.14, where the energy of the proton is plotted as a function of the vertical distance from the glass surface. Curve 1 characterizes the purely chemical energy barrier and is, for example, valid at the very first contact between glass surface and solution when no charges have yet developed at the interface. In correspondence to Fig. 2.13, curve 1 is determined by the standard *chemical* free energies of activation in the dissociation and association direction of the surface groups. The dissociation of the silanol group, however, generates opposite charges at the interface and thus an electric field that is directed into the glass, hinders further dissociation, and, in turn, supports the (reverse) association. The corresponding *electrical* energy, which is plotted as curve 2 as a function of distance for the equilibrium case, must thus be added to the chemical energy (curve 1) in order to obtain the *electrochemical* energy curve 3. As seen in Fig. 2.14, this causes the proton energy at the summit of curve 1 to change by a fraction of the total electrical energy $F\varepsilon_e$, which is determined by the so-called *symmetry factor* α. This important quantity is, for instance, obtained geometrically,

$$\alpha = \frac{\varphi_s - \varphi_{\text{OHP}}}{\varphi_{\text{gl}} - \varphi_{\text{OHP}}} , \qquad (2.103)$$

where φ is the potential, indices gl and s stand for glass and summit, respectively, and the index OHP means *outer Helmholtz plane* and denotes the distance between the plane where the surplus of charge-carrying ions of the solution is located and the glass surface. It is also called the plane of closest approach of the ions because the space between the OHP and the glass surface is occupied by adsorbed water molecules and is not accessible to the ions at equilibrium. The centre plane of the adsorbed water molecules, incidentally, is called the *inner Helmholtz plane*. As shown by the plots of Fig. 2.14, the difference

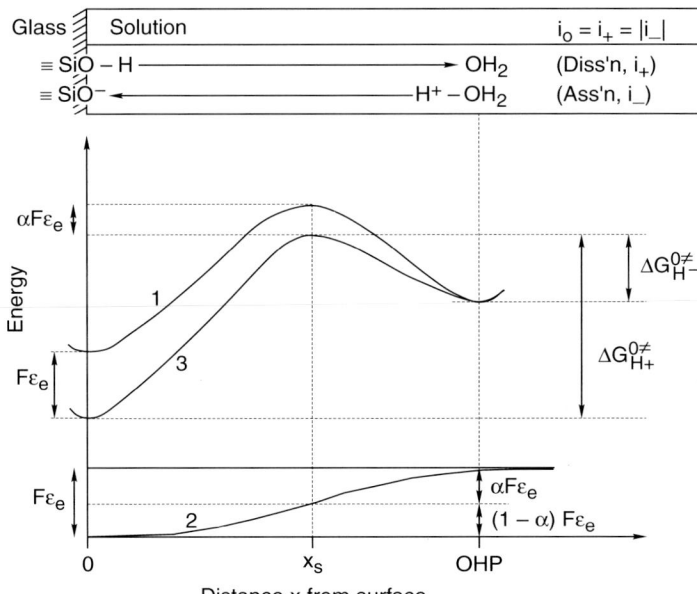

Fig. 2.14. Schematic energy–distance plots of the proton for the heterogeneous dissociation of a weakly acidic (e.g. silanol) group at a glass surface in aqueous solution and the (reverse) association reaction. Curve 1: chemical energy corresponding to Fig. 2.13 and valid at the very first glass–solution contact before the generation of charges. Curve 2: electrical energy due to the charges at the interface caused by the dissociation at equilibrium. Curve 3: electrochemical energy consisting of chemical (curve 1) and electrical energy (curve 2) at dissociation equilibrium. ε_e is the equilibrium potential of the glass membrane, $\Delta G_{H+}^{0\neq}$ and $\Delta G_{H-}^{0\neq}$ are standard free energies of activation of (anodic) dissociation and (cathodic) association, respectively, OHP is the outer Helmholtz plane, and x_s denotes the distance of the energy summit from the glass surface

$$(1-\alpha) = \frac{\varphi_{gl} - \varphi_s}{\varphi_{gl} - \varphi_{OHP}} \tag{2.104}$$

is of equal importance for the derivation as α. Thus, at equilibrium, where the potential difference between glass and solution is ε_e, the standard free energy of activation of silanol dissociation is given by

$$\Delta G_{H+}^{0\neq} = \Delta G_d^{0\neq} - (1-\alpha)F\varepsilon_e \tag{2.105}$$

and that of the (reverse) association by

$$\Delta G_{H-}^{0\neq} = \Delta G_a^{0\neq} + \alpha F\varepsilon_e \, , \tag{2.106}$$

where the negative sign of the potential ε_e of the glass relative to that of the solution

$$\varepsilon_e = (\varphi_{gl} - \varphi_{sol'n}) < 0 \, , \tag{2.107}$$

which reverses the sign of the electrical terms of (2.105) and (2.106), is taken into account. The electrochemical rate constant of the interfacial silanol dissociation, consequently, is given by

$$k'_{H+} = \frac{kT}{h} e^{-(\Delta G^{0\neq}_{H+}/RT)} = k_{H+}\, e^{[(1-\alpha)F\,\varepsilon_e)]/RT} \;, \tag{2.108}$$

where k_{H+} is the chemical rate constant corresponding to k_d in (2.94), so that the reaction rate of silanol dissociation, if expressed as (anodic) current density, is represented by

$$i_{H+} = Fv_{H+} = Fk_{H+}\, c'_{SiOH}\, c_{H_2O}\, e^{[(1-\alpha)F\,\varepsilon_e)]/RT} \;. \tag{2.109}$$

In the same way, the rate constant of the (reverse) interfacial siloxy association with protons is obtained,

$$k'_{H-} = \frac{kT}{h} e^{-(\Delta G^{0\neq}_{H-}/RT)} = k_{H-}\, e^{-(\alpha F\,\varepsilon_e/RT)} \;, \tag{2.110}$$

and yields the silanol dissociation rate

$$i_{H-} = -Fv_{H-} = -Fk_{H-}\, c'_{SiO-}\, c_{H_3O+}\, e^{-(\alpha F\,\varepsilon_e/RT)} \tag{2.111}$$

if it is expressed as (cathodic) current density.

The phase boundary equilibrium is characterized by equal cathodic and anodic current densities, which are called the (equilibrium) exchange current density,

$$i_{0,H} = i_{H+} = Fk_{H+}\, c'_{SiOH}\, c_{H_2O}\, e^{[(1-\alpha)F\,\varepsilon_e)]/RT} = |i_{H-}|$$
$$= Fk_{H-}\, c'_{SiO-}\, c_{H_3O+}\, e^{-(\alpha F\,\varepsilon_e/RT)} \;, \tag{2.112}$$

whose rearrangement, finally, yields the equilibrium potential difference between the OHP and the glass surface as a function of the hydronium concentration of the solution,

$$\varepsilon_e = -\frac{RT}{F} \ln \frac{k_{H+}}{k_{H-}} + \frac{RT}{F} \ln \frac{c'_{SiO-}\, c_{H_3O+}}{c'_{SiOH}\, c_{H_2O}} \;. \tag{2.113}$$

Equation (2.113) is equal to (2.22) of Sect. 2.5.2, which represents the Galvani voltage of the glass membrane as a function of pH, because

- the ratio of the rate constants of a reaction is equal to its equilibrium constant, (2.99),
- activity coefficients are contained in the rate constants, (2.99) and (2.100), and
- Galvani voltage and standard Galvani voltage are independent of the so-called zeta potential, the potential difference between the OHP and the bulk of the solution, which merely adds a constant term to the equilibrium potential according to (2.113), whose absolute value is immeasurable anyway.

The Galvani voltage of the glass membrane as a function of pH at 100% pH selectivity has thus been derived on a purely kinetic basis. The pM function of glass electrodes and the electrode response in the transition pH range can be determined by an analogous procedure. The derivations are not given here but can easily be carried out by the interested reader.

Heterogeneous Equilibria in Electric Fields

When the phase boundary equilibrium is subjected to a polarizing or overvoltage, η, the anodic and cathodic current densities are no longer equal but differ by an amount that is easily derived from the foregoing equations. This is demonstrated by Fig. 2.15 for a negative polarization of a glass membrane, which thus has the same sign as the equilibrium potential. The polarizing voltage is simply added to the equilibrium potential in (2.109) and (2.111), which results in the anodic (dissociation) current density

$$i_{H+} = F k_{H+} \, c'_{SiOH} \, c_{H_2O} \, e^{[(1-\alpha)F(\varepsilon_e + \eta)]/RT} \tag{2.114}$$

and the cathodic (association) current density

$$i_{H-} = -F k_{H-} \, c'_{SiO^-} \, c_{H_3O^+} \, e^{-[\alpha F(\varepsilon_e + \eta)]/RT} \,, \tag{2.115}$$

whose sum is the net current density through the interfacial equilibrium,

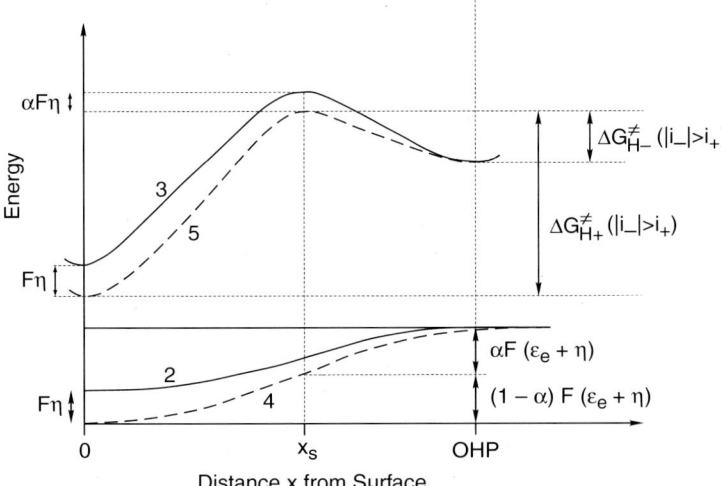

Fig. 2.15. Energy–distance plots of a proton for heterogeneous dissociation and association of a glass surface group in aqueous solution. Curves 2 and 3: at equilibrium, as curves 2 and 3 in Fig. 2.14. Curves 4 and 5: electrical and electrochemical energy, respectively, when the interfacial equilibrium is subject to a negative overvoltage (polarization)

$$i_H = i_{H+} + i_{H-} = i_{0,H}\left(e^{[(1-\alpha)F\eta]/RT} - e^{-(\alpha F\eta)/RT}\right), \qquad (2.116)$$

which is caused by the polarizing voltage. At cathodic polarization as assumed in Fig. 2.15, the net current density transfers protons from the interfacial equilibrium into the glass. Polarization due to ionic diffusion from the solution to the OHP, so-called diffusion overvoltages [2.40], can be neglected with good approximation because of the small net current densities caused at glass membranes even if large voltages are applied.

By extending the exponential expressions of (2.116) and retaining the first two terms for small polarizing voltages, a simple linear relationship between current density and polarization is obtained,

$$i_H = \frac{i_{0,H}F\eta}{RT}, \qquad (2.117)$$

whose rearrangement and derivation yields the "polarizability" or "reaction resistance"

$$\left(\frac{\delta\eta}{\delta i}\right)_{c,T} = \frac{RT}{Fi_{0,H}} = \rho_{gl/sol'n} \qquad (2.118)$$

of the interfacial equilibrium. Equation (2.118) shows that the polarizability of the phase boundary equilibrium is the smaller, the larger the exchange current density. The exchange current density thus is a significant quantity in that it characterizes the interfacial equilibrium with respect to its polarizability. Because estimates yielded exchange current densities of the order of $mA\,cm^{-2}$, the polarizability of interfacial equilibria between electrode glasses and solutions is of the order of $25\,mV\,cm^2\,mA^{-1}$. If, for instance, a $1\,cm^2$ large glass membrane with a resistance of $500\,M\Omega$ is subjected to a voltage of $3500\,V$, which causes a current density of $7\,\mu A\,cm^{-2}$, the interfacial equilibrium is polarized by as little as $0.18\,mV$. These small values justify the quantitative treatment of ionic transfer experiments and electrolyses described in Sect. 2.6, which were always conducted under even less demanding conditions and were thus not impaired by polarization.

Surprisingly, the large exchange current density of the interfacial equilibrium is opposed by an extremely small concentration of negative (e.g. siloxy) surface groups. An estimate on the basis of electrostatics yielded a siloxy concentration of the order of $10^{-17}\,mol\,cm^{-2}$ for a potential difference $0.1\,V$, compared to an estimated $10^{-9}\,mol\,cm^{-2}$ silanol concentration. The resulting lifetime of a siloxy group is thus of the order of $10^{-9}\,s$, and the exchange frequency amounts to $10^9\,s^{-1}$. These extreme magnitudes can only be understood by the structure of the double layer at the glass surface. Obviously, silanol groups (and siloxy groups where present) at the glass surface adsorbed water molecules in the IHP, and water molecules (and hydronium ions where present) in the OHP are ideally lined up due to hydrogen bridge formation, thus favouring a chain mechanism of proton exchange without the necessity of major reorientation of the particles, as indicated by the reaction bars above the plots in Fig. 2.14. Besides, this orientation of the double layer favours tunnelling of the protons, which is basically not excluded because of the DeBroglie wavelength of approximately 0.4 nm at an assumed velocity of 10^5 cm s^{-1} [2.113]. If tunnelling determines the proton exchange rate to a major extent, however, the exchange current density of the interfacial equilibrium involving cations other than protons, which generally do not tunnel, must be expected to be smaller than that involving protons. In order to gain more detailed kinetic information about the glass–solution interface, the measurement of standard exchange current densities of glass electrode membranes is needed and has thus been proposed [2.71, 114].

2.6 Field-Driven Transport of Ions Into and Within Glasses

2.6.1 Electric Contacts

This section is concerned with the field-driven transport of ions into and within glasses, particularly under direct current (d.c.) conditions. Ions, for instance protons, are introduced into lithium silicate and aluminosilicate membrane glasses, replacing from zero to 100% of the lithium ions. The concentration-dependent conductances of these "protonated" glasses, together with concentration profiles measured by means of IBSCA, yield the concentration-dependent mobilities of the "guest protons" and the lithium ions of the glass. Moreover, the electric field is reversed, protonated glasses are "lithiated", and attempts have been made to transfer ions other than protons and lithium ions into glasses.

All of these experiments depend on a basic, but in no way trivial problem. The glasses investigated must somehow be coupled to the measuring circuit via phases acting as ion sources and sinks [2.115], which supply or take up the appropriate ion species. This is because the measurement of d.c. conduction, in a sense, extends anodically and cathodically across the surfaces of the investigated solid phase, or, more correctly, d.c. conduction involves ions that do not originally belong to the glass phase.

Consequently, by far not all ion-supplying and ion-accepting electrically coupling phases are electrodes (if defined as locations where the conduction changes from electronic to ionic and vice versa), but in the majority of cases represent so-called electric or ohmic contacts which supply the ions to or accept them from the glass. On the other hand, electrodes are nevertheless needed for contacting the ion-supplying phases. This is immediately obvious when, instead of the cell part of immediate interest,

$$(+)\text{cation source}|\text{glass}(-)\ ,\qquad\qquad(2.\text{XII})$$

the complete cell used for d.c. measurements is considered, for instance

$$(+)\text{positive electrode}|\text{cation source}|\text{glass}|\text{cation sink (or anion source)}$$
$$|\text{negative electrode}(-) , \qquad (2.\text{XIII})$$

where the metal terminals are electrodes as defined above.

In cell (2.XIII), the electrodes are denoted positive and negative because electrodes must not simply be called anodes or cathodes because of their relative sign. The terms anodic and cathodic are reserved for the definition of the current flowing through the electrode/electrolyte interface, an anodic current carrying negative charge into, or positive charge out of, an electrode and a cathodic current transporting positive charge into, or negative charge out of, the electrode [2.116]. If electrodes are nevertheless frequently termed anodes or cathodes, this must strictly be done on the basis of the current flowing through their surface. The immediate function of the cell employing the electrodes must thus be clear because whether, for instance, the positive electrode represents an anode or a cathode depends on whether the electrochemical reaction is externally driven or is a spontaneous process. For example, the anode during charging of a battery (when negative charge flows from the electrolyte into the electrode) becomes the cathode during the subsequent discharge of the battery (when the electrode accepts positive charge from, or supplies negative charge to, the electrolyte), the electrode being the positive electrode during both processes.

In Sect. 2.6, mainly electrolysis cells, as represented by cell (2.XIII), are described so that the (left) positive electrode supplying positive charge to the cation source of the glass may be called the anode. Because the glass is always treated as what it really is, a (solid) electrolyte between two electric contacts, its surface facing the anode is termed the anodic glass surface despite the transport of positive charge (cations) from the electric contact into the glass. A corresponding terminology applies to the cathodic glass surface.

The ions supplied by the anode to the electric contact may differ from those introduced from the electric contact into the glass. If this is likely to disturb the transfer process, the (left) anodic electric contact may consist of two solutions employing an intermediate diaphragm. Also, the positive electrode and the adjacent electric contact of cell (2.XIII) may be replaced by an oxidizable metal, which is then called a non-blocking electrode. The cations of the metal generated by the external electric field are offered to the anodic glass surface, where they are or are not transferred into the glass, depending on their size relative to the available ion sites in the glass. For example, protons generated by a positive Pt,H_2 electrode (often called a protode) are transferred into any glass (see Sect. 2.6.2); the transfer of silver ions generated at a silver anode depends on the glass composition [2.117, 118]; and lead ions generated at a positive lead electrode are practically not transferred into any glass but form lead oxide with oxide ions, for instance, from borate glass networks at the glass surface [2.119] because of their ionic radius and, possibly, their two-fold positive charge.

Salt melts have frequently been used as electric (ohmic) contacts in the past, see for instance [2.120]. However, they require rather high temperatures of the electrolyses. This was different during the work described in this section. Due to the high depth resolution of IBSCA (see Sect. 2.4) applied for analysing the electrolysed glass, it was possible to conduct the electrolyses at temperatures below 100 °C, which means that aqueous (and non-aqueous) solutions could serve as electric contacts [2.54, 56, 121] and the interpretation of the results did not require transformation to application temperatures of glass electrodes. It was thus also found during the experiments that the interfacial equilibrium established between the solutions and the glass controlled the ratio of the different ions simultaneously transferred into the glasses [2.54, 120]. Because this heterogeneous equilibrium depends on the solution composition, the ion transfer could be predetermined and controlled very accurately by the choice of the solutions applied as cation sources for the glass specimens, so that the generation of homogeneously modified glass layers and the study of their transport properties were possible [2.54, 120].

The combination of electrolysis and depth-resolving subsurface analysis by IBSCA represents a modification of the well-known moving boundary method for determining ionic transport numbers in electrolyte solutions [2.93]. We have thus called the technique the modified moving boundary method (m.m.b.m.). It not only yields the concentration profile of the moving boundary at the end of an electrolysis but also informs about its development by repeating the electrolysis under identical conditions for various periods of time, see Sects. 2.4 and 2.6.5–7. The m.m.b.m. thus gives more and basically different information compared to the original method. A puzzling complication was at first presented by the sterical hindrance of the ion transfer caused by the relative radii of cations and anionic sites. This phenomenon, however, was later studied in some detail and yielded information about the glass structure [2.55]. The first (high-temperature) application of the moving boundary method to solids was mentioned in 1922 by *Kraus* and *Darby* [2.122], who followed the shift of the boundary between sodium ions and transferred silver ions by observing the spread of the translucency of the silver-containing ion-exchanged glass microscopically.

The cathodic glass surface is of minor significance in these experiments as it releases an amount of cations equivalent to that taken up by the anodic surface. A significant exception was the electrolyses conducted to prove the anodic uptake and migration of protons and thus to exclude the transfer of hydronium ions into lithium silicate glasses [2.54, 55], see also Sect. 2.6.2.

Protons as well as ions of other elements have been transferred anodically also from a glow discharge into glasses [2.123, 124]. Temperatures were near 300 °C, and the d.c. voltage was increased during each experiment from a few volts up to 75 V to keep the d.c. current constant, the increase rate depending on discharge atmosphere, pressure, and alternating current (a.c.) voltage. However, this procedure represents an ion injection rather than a glass electrolysis, and, correspondingly, the aim of the experiments was not to create glass layers with certain electrochemical properties, for instance ion mobilities, but to modify the glass properties in general, for example to achieve an improved adhesion of metal layers to the glass surfaces [2.124].

In contrast to electric (or ohmic) contacts and non-blocking electrodes, so-called blocking electrodes neither accept ions from, nor supply ions to, neighbouring electrolytes [2.115]. Because, however, the d.c. measurement and blocking electrode exclude each other by definition, blocking electrodes are mainly used for measurements under a.c. conditions. An exception is the application of blocking electrodes to the so-called plateau method for measuring the d.c. conductance of glasses [2.125–127]. It is based on the idea that the time between the onset of the electric field and the onset of electrode polarization by the discharge of ions represents a (very short) period characterized only by the d.c. conductance of the glass. Although the results of such measurements "are not unreasonable", *Ernsberger*, on the grounds of a literature search and personal experience, concludes "that the 'plateau' is a variable and somewhat subjective effect" and doubts "that a plateau necessarily exists" [2.128].

At any rate, really blocking electrodes are difficult to achieve experimentally. Indeed, the only realizable blocking electrode seems to be a positive non-porous, inert electrode, for instance platinum, evaporated onto a fresh, unchanged glass plate or membrane, in an inert atmosphere, for instance argon. If, in addition, the applied voltage is sufficiently small to ensure that the oxides (or other negative ions) are not detached from the network and do not migrate to the anode, the electrode can be assumed to "block" any electron exchange with constituents of the glass.

As also discussed by *Ernsberger* [2.128], these conditions are, if at all, extremely difficult to realize because the glass surface is nearly always modified ("leached" in some way) by ion exchange and hydration, see Sect. 2.7. This leads either to a transport of protons or hydronium ions, depending on the glass composition, into the glass or the generation of oxygen, if the electrode is positive, or to an extraction of alkali ions, hydrogen generation, and alkali hydroxide formation, if it is negative. It is thus also in doubt whether the polarization of blocking electrodes as observed by the measurement of potential profiles in glass plates [2.126, 127] is caused by the creation of space charge, which only rarely has a chance to build up [2.128], or whether its real cause is the generation and accumulation of polarizing material at the metal electrode [2.128–132].

2.6.2 Electromigration of "Guest Protons" and "Guest Deuterons"

The simplest case of ion transfer and subsequent migration is given by 100% protonation [2.54, 121, 133]. It is secured by anodic dilute acid solutions. The interfacial equilibrium of the hydrogen ions of the solution with the surface groups R^- of the glass (R = SiO, [AlOSi], or [BOSi]) converts the groups into their acidic form RH, except for the negligible part of negatively charged entities involved in the equilibrium (see Sect. 2.5.2) so that only protons are supplied to the underlying glass when an electric field directed into the glass is applied to the membrane. The ionic transfer is thus purely protonic and free of alkali ions.

Transport Properties of Lithium Ions

Figure 2.16 gives two examples that are typical of all oxidic lithium silicate, aluminosilicate, borosilicate, and boroaluminosilicate membrane glasses investigated. The concentration profiles measured after various migration periods show that the lithium ions are progressively displaced from the anodic towards the cathodic surface of the membrane [2.54]. Their rear boundaries are distinct but develop a slightly differently decreasing sharpness during the ionic migration [2.121]. The amount of lithium drifted away from the anodic membrane surface, which is calculated from their profile and the lithium content of the glass, is equal to the lithium amount given off by the membrane to the cathodic solution and analyzed by spectroscopy, as well as to the

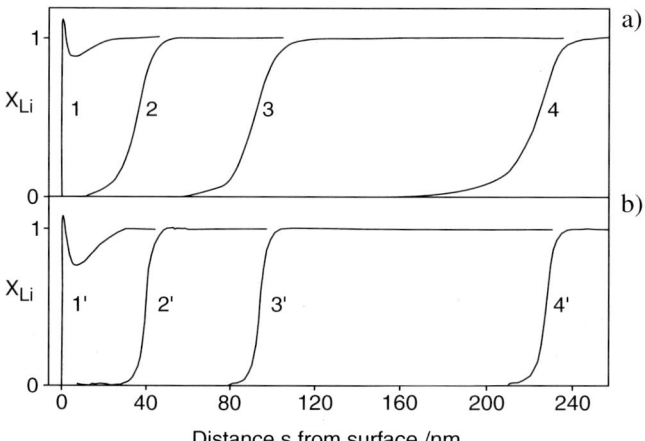

Fig. 2.16. Lithium concentration profiles below the anodic surface of lithium-based oxidic glass membranes after various periods of electrolysis with anodic dilute aqueous sulphuric acid. (**a**) Lithium silicate glass (1) before and after (2) 30 min, (3) 80 min, and (4) 190 min of electrolysis. (**b**) Lithium aluminosilicate glass (1′) before and after (2′) 26 min, (3′) 64 min, and (4′) 152 min of electrolysis

equivalent charge q transported through the membrane during the electroysis [2.54],

$$(-)n_{\text{Li,anode}} = -n_{\text{Li,cathode}} = \frac{q}{F} = \frac{1}{F} \int\limits_0^t i(t)\mathrm{d}t \ , \tag{2.119}$$

where F is the Faraday constant, $i(t)$ is a time-dependent current, t is the electrolysis time, the minus sign indicates a loss by the membrane, and the bracketed minus sign denotes a shift away from the surface indicated. Equation (2.119) is valid for ions with unit charge. The average uncertainty of these measurements is below $\pm 0.02\%$ if leached layers (see Sect. 2.7.1) are taken into account. The validity of (2.119) confirms that the average mobility of the lithium ions can be obtained from the resistivity ρ and the lithium concentration c_{Li}^0 of the glass,

$$\bar{u}_{\text{Li}} = (\rho \, c_{\text{Li}}^0 \, F)^{-1} \ , \tag{2.120}$$

as well as from the migration rate of the lithium ions at the beginning of the electrolysis, $v_{\text{Li}}(t \to 0) = \mathrm{d}s/\mathrm{d}t(t \to 0)$, and the overall electric field strength applied, $\boldsymbol{E}_{\text{total}} = U^0/d$, where d is the membrane thickness and U^0 is the overall voltage,

$$\bar{u}_{\text{Li}} = \frac{\boldsymbol{E}_{\text{total}}}{v_{\text{Li}}(t \to 0)} \ . \tag{2.121}$$

Because the ionic mobility in glasses is subject to a certain distribution, $u_{\text{Li}} = f(x_{\text{Li}})$, the mobilities obtained by (2.120) and (2.121) are average mobilities of the lithium ions [2.55], as represented by

$$\bar{u}_{\text{Li}} = \frac{1}{x_{\text{Li}}} \int\limits_0^{x_{\text{Li}}=1} u_{\text{Li}}(x_{\text{Li}})\mathrm{d}x_{\text{Li}} \ , \tag{2.122}$$

where x_{Li} is the mole fraction and u_{Li} the individual mobility of the lithium ions. The sharpness of the migrating boundary in Fig. 2.16 is mainly determined by the mobility distributions of the lithium and the following ions [2.54], see also below. Note that the lithium concentration profiles below the cathodic membrane surface do not change during the electrolyses.

Average mobilities of lithium ions determined according to (2.120) and (2.121) depend on the thermal history of the glasses [2.121], and it must be assumed (but has not yet been verified) that also mobility distributions are functions of the thermal history of the glass. Thus, average mobilities have been found to differ by a factor of up to 6 for quenched glasses, which have larger mobilities than rate-cooled glass membranes [2.121]. In order to obtain maximum reproducible transport data, glass membranes must thus be subjected to a certain cooling rate from above a certain minimum temperature.

A cooling rate of $5\,K\,min^{-1}$ from $5\,K$ above T_g resulted in reproducibilities within $\pm 0.5\%$ of the mobilities, independent of type and composition of the membrane glass. All glass specimens reported here were subjected to this cooling procedure, which, in addition, was carried out in a dry argon atmosphere in order to avoid interaction of the glasses with oxygen and water vapour at high temperatures and to exclude irreversible changes in the surface regions of the membranes, see also Sect. 2.7.2.

Proof of Proton and Deuteron Transfer

The lithium ions drifting towards the cathodic membrane surface are replaced by hydrogen ions from the anodic solution. Their concentration profile cannot be measured by IBSCA but must be obtained by NRA [2.54, 72, 87]. Besides, hydrogen ions form their own indicators within the glass by combining with non-bridging oxides, yielding OH groups that exhibit a strong tailing infrared absorption band with a maximum at $3390\,cm^{-1}$ $(2.95\,\mu m)$ [2.54, 55, 121]. The increasing intensity of this band during an electrolysis of a lithium silicate glass with anodic dilute sulphuric acid is presented in Fig. 2.17a. Figure 2.17b shows the increasing band of OD groups with a maximum at $2470\,cm^{-1}$ $(4.05\,\mu m)$ during the electrolysis of the same glass using an anodic deutero-sulphuric acid [2.55, 135]. The spectra demonstrate that deuterium ions enter the glass membrane anodically, as do hydrogen ions. Anodic lithium concentration profiles of deuterated glass membranes are also practically identical to those of protonated glasses, Fig. 2.16.

The following three experiments proved that the hydrogen species entering lithium-based glasses in electric fields is protons and not hydronium ions nor a mixture of protons and hydronium ions [2.54].

(1) Protonated surface layers generated by electrolyses with anodic non-aqueous protonic solvents, which were extremely dry and thus able to supply only protons to the glass, exhibited infrared absorption bands and absorptivities that were identical with those generated by electrolyses with anodic aqueous solutions.

(2) Electrolyses of glass membranes covered with removable vapour-deposited, sputtered-on or painted-on platinum electrodes in extremely dry hydrogen atmosphere, which can only inject protons, also yielded infrared absorption bands identical to those observed after electrolyses with aqueous acid solutions.

(3) Application of NRA, by which, in contrast to IBSCA, proton concentration profiles can be measured (see Sect. 2.4), showed a one-for-one replacement of lithium ions by protons in membranes electrolysed with aqueous (and non-aqueous) solutions. Leached layers formed during the electrolyses were discerned from the purely protonated parts of the glass (see Fig. 2.58) and were also taken into account with the measurements described under (1) and (2).

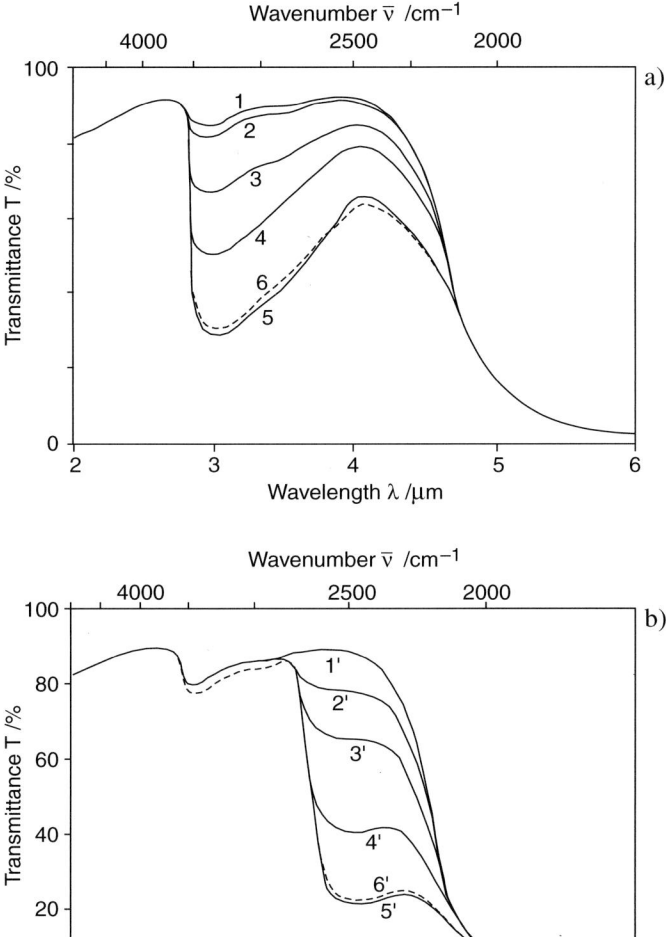

Fig. 2.17. Change of infrared absorption of lithium silicate glass membranes during electrolyses with anodic acid solution in water (**a**) and heavy water (**b**). Charge densities transferred into (a): (1) untreated, (2) leached, (3) 2.0×10^{-6}, (4) 5.9×10^{-6}, (5) 15.2×10^{-6} equ. cm^{-2}; (6) after (5) and 10 days in dilute D$_2$SO$_4$ at $75°$. Change $(5)\rightleftarrows(6)$ (releaching) is reversible. Charge transferred into (b): (1′) untreated, (2′) 2.0×10^{-6}, (3′) 5.9×10^{-6}, (4′) 15.2×10^{-6}, (5′) 28.4×10^{-6} equ. cm^{-2}; (6′) after (5′) and 10 days in dilute H$_2$SO$_4$ at $75°$C. Change $(5')\rightleftarrows(6')$ (releaching) is reversible

These results are in agreement with the argument that it is highly improbable for hydronium ions (H_3O^+) to enter lithium-based oxidic glasses for spatial reasons. Their ionic radius ($r_{hydr} = 140\,pm$, estimate [2.136]) is comparable to that of potassium ions ($r_K = 138\,pm$, [2.137]), which were found to be entirely excluded sterically from penetrating lithium-based glasses [2.97] ($r_{Li} = 76\,pm$ [2.137]), see also Sect. 2.6.4. In addition, it is unlikely for energetic reasons that the entity H_3O^+ enters any glass network in an undissociated form.

On the other hand, the replacement of lithium ions by protons generates void space in the glass network so that protonated and partly protonated layers are under some "internal" as well as under an "external" tensile stress, which is exerted by the underlying membrane. Both effects influence the ionic mobilities in the layers but must be distinguished from each other. Internal stress is an intrinsic property of protonated glasses which cannot be released by cooling because high temperatures cause condensation of the OH groups of protonated glass layers and subsequent reactions (see Sect. 2.6.7) [2.55]. Ionic mobilities in protonated glass layers are thus properties of glasses which are necessarily under an internal tensile stress even if the layers were separated from the underlying membranes. External tensile stress, in contrast, could be released by separating the protonated layers from the membranes if this is feasible. Its effect on the ionic mobilities, however, seems insignificant because concentration-dependent mobilities of lithium ions and protons in partly protonated glasses (see Sect. 2.6.5), which were determined on layers with rather different thicknesses, agreed within the limits of experimental error. Vice versa, the compressive stress exerted by the thin protonated layer on the underlying much thicker membrane certainly decays with depth within a short distance and is not expected to affect, on the average, the mobility of the lithium ions of the unchanged glass.

According to the experiments mentioned above, the field-driven amount of protons $n_{H,anode}$ entering the glass membrane anodically is equivalent to the amount of lithium ions leaving the anodic for the cathodic surface, to that of the lithium ions leaving the membrane cathodically, and to the electric charge transported through the membrane during the electrolysis [2.54],

$$+n_{H,anode} = (-)n_{Li,anode} = -n_{Li,cathode} = \frac{q}{F} = \frac{1}{F}\int_0^t i(t)dt \ . \qquad (2.123)$$

The same result was obtained for deuterons. It is thus justified to call the described electrolyses "protonation" and "deuteration" of the glass and the transferred ions "guest protons" and "guest deuterons", respectively. Protons and deuterons present a rare case of ionic migration in solids [2.138, 139].

Transport Properties of Guest Protons and Deuterons –
Qualitative Description

During field-driven protonation, the resistance of the membrane increases continuously, as already reported by *Haugaard* [2.99, 101], *Sendt* [2.140], and others. Deuteration also increases the membrane resistance. The average mobility of the transferred protons and deuterons, consequently, is smaller than that of the lithium ions of the glass,

$$(\overline{u}_H \text{ and } \overline{u}_D) < \overline{u}_{Li} , \tag{2.124}$$

which is contrary to what should be expected from the relative size of the ions [2.141] and must thus be caused by the strong polarizing effect of the "small" protons and deuterons on the oxygens. Because, in addition, the migrating boundary between the lithium ions and the following protons (and deuterons) remains distinct during long periods of electrolysis and thus for long distances travelled by the boundary, the individual mobilities of the different ions do not overlap either. The maximum individual mobility of guest protons and guest deuterons, consequently, is smaller than the minimum individual mobility of the lithium ions,

$$(u_{H,max} \text{ and } u_{D,max}) < u_{Li,min} . \tag{2.125}$$

Contrary to a statement in the literature [2.142], the above-mentioned slight decrease of sharpness of the migrating boundary with migration time is only partly caused by interdiffusion. This was shown by introducing long intermittent periods (up to 24 h, 50 °C) between migration and the analysis by IBSCA, which did not cause considerable changes of the lithium concentration profiles. The slight loss in distinctness with migration distance is probably caused by the mobility distributions of the different ionic species, which are both present within the range of the migrating boundary, and by the strong depth dependence of the electric field that is caused by the different mobilities of protons and lithium ions in this membrane region [2.54], see also Sect. 2.6.6. Because of the continuous change of composition and properties within the migrating boundary, even if it occurs within a small distance (see Fig. 2.16), and because of the thickness change of the boundary region during electrolysis, protonated layers do not form phases in the thermodynamic sense.

Lithium ions can be completely removed by 100% protonation, independent of composition and thermal history of the glass, at least after long migration periods. This sheds some doubt on the existence of "impasses" of the network for ionic migration, which are thought to retain cations because their structure is impermeable in certain directions [2.143, 144], at least in lithium-based oxidic glasses.

Optical Properties of Protonated and Deuterated Glasses

Figure 2.17 shows that protons and deuterons replacing lithium ions in an oxidic glass form OH and OD groups, respectively, with non-bridging oxides. Protons and deuterons thus generate their own indicators because the OH and OD groups exhibit strong infrared absorption with characteristic absorption maxima [2.145, 146]. Figure 2.18 shows maximum absorbances as functions of charge transported through the membrane for protonation and deuteration. A lithium silicate glass containing 25 mol% Li_2O was, respectively, 100% protonated and 100% deuterated with defined charges under manifold conditions. Charges were between 0.1 C and 2.4 C at total overall electric fields between 3.5 kV cm^{-1} and 70 kV cm^{-1}. One to four absorbances per membrane and totals of 24 (protons) and 14 (deuterons) absorbances were measured. Several of the membranes were quenched, the others were rate-cooled. However, independent of these rather different conditions, the absorbances measured are linear functions of the charge transported during the electrolyses.

Table 2.5 presents numerical results. The "coulometric absorptivity" defined by

$$\varepsilon_q = \frac{\mathbf{A}A}{q} = \frac{\mathbf{A}A}{nF} , \qquad (2.126)$$

and the molar absorptivity [2.147] according to

$$\varepsilon_n = \frac{\mathbf{A}A}{n} \qquad (2.127)$$

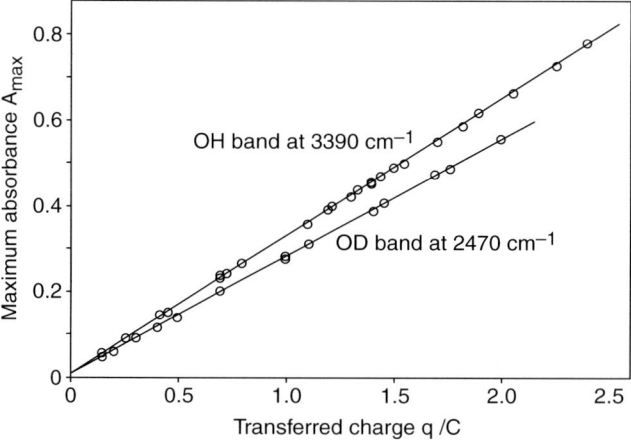

Fig. 2.18. Maximum absorbance of infrared spectra of a lithium silicate glass after electrolyses with anodic H_2SO_4 (OH) and anodic D_2SO_4 (OD) as functions of transferred charge q

Table 2.5. Maximum "coulometric" ε_q, molar ε_n, and integral "coulometric" $\varepsilon_{I,q}$ and molar $\varepsilon_{I,n}$ absorptivities of 100% protonated and 100% deuterated lithium (25 mol%) silicate pH glass membranes

Absorption band:	OH (maximum)	OD (maximum)
Wavenumber, $\bar{v}/\mathrm{cm}^{-1}$	3 390	2 470
Wavelength, $\lambda/\mu\mathrm{m}$	2.95	4.05
$\varepsilon_q/\mathrm{cm}^2\ \mathrm{C}^{-1}$	0.323 ± 0.001	0.276 ± 0.0015
Number of measurements	24	14
Correlation factor	0.998	0.995
$\varepsilon_n/\mathrm{cm}^2$ equiv.$^{-1}$	$31\,176 \pm 96\ (\sim 32\,500)$ [a]	$26\,640 \pm 145$
$\varepsilon_{I,q}/\mathrm{cm}\ \mathrm{C}^{-1}$	267	
$\varepsilon_{I,n}/\mathrm{cm}$ equiv.$^{-1}$	25.8×10^6	

[a] Maximum molar absorptivity of water-containing lithium (20 mol%) silicate glass as measured by *Scholze* [2.145] but referred to OH groups instead of H_2O content (average of two reported values)

are presented for the protonated and the deuterated glass. **A** is the absorbance, A the membrane cross section, q the charge, n the respective number of moles of OH and OD groups, and F is the Faraday constant. The molar absorptivity was calculated from the coulometric absorptivity by

$$\varepsilon_n = F\,\varepsilon_q \ , \tag{2.128}$$

which is valid if the charge of entering and replaced ions is unity, as is the case with protons, deuterons, and lithium ions.

As expected and indicated by the linear functions $\mathbf{A}_{\mathrm{max}} = f(q)$ in Fig. 2.18, the coulometric absorptivity ε_q is constant and has a correlation factor of 0.998 for OH and 0.995 for OD, Table 2.5. The small experimental error of these data, which is a consequence of the high accuracy of coulometric measurements and can certainly still be reduced, suggests using "coulometric absorptivities" as standard absorptivities for the exact determination of "water" (as OH groups) in lithium-based oxidic glasses. This procedure would eliminate the necessity and the uncertainties introduced by glasses that contain defined amounts of water through introduction into the melt. It is also supported by the independence of the coulometric absorptivity of the conditions under which the electrolyses are carried out. However, it rests on the assumption that the intensity maximum of the infrared spectrum is independent of the position and surrounding of the OH and OD groups in the glass structure, that is, it makes no difference whether the proton or deuteron replaces a lithium ion and is bound to a non-bridging oxygen or whether the OH and OD groups have been introduced into the network by the melt. Thus, the verification of the proposal needs more basic and experimental investigations. But a comparison of the molar absorptivity of the 25 mol% lithium silicate glass obtained from the coulometric absorptivity ($31\,176 \pm 96\,\mathrm{cm}^2$ equiv^{-1}) with the molar absorptivity of 20 mol% lithium silicate glass as reported by *Scholze* [2.145] ($\sim 32\,500\,\mathrm{cm}^2$ equiv^{-1}) looks promising. This coulometric standardization on the basis of aqueous anodic solutions would be restricted to lithium-based glasses because sodium-based glasses, for instance, have been reported to admit water together with protons, probably as hydronium ions, during electrolyses with anodic aqueous solutions [2.140, 148]. However, the use of aqueous solutions can be circumvented by applying removable, vapour-deposited anodic Pt,H$_2$ electrodes in very dry hydrogen atmosphere or by using extremely dry protonic solvents as anodic contacts for the electrolyses. At any rate, this procedure, if feasible, would yield considerably more exact standard absorptivities than the use of glasses with defined water contents introduced via the melt.

Membrane Glass Polarization

After removal of the electrolysing voltage, glass membranes exhibit a polarization that is characterized by the following properties.

(1) The direction of the polarization is equal to the direction of the electrolysing field.
(2) The polarization decays with time.
(3) The decay rate is independent of whether or not the cell containing the membrane is short-circuited, which means that the decay is not connected with a current through the membrane.
(4) In agreement with (3), the decay rate is independent of the electrical contact of the membrane used for the measurements, for example aqueous or non-aqueous electrolyte solutions or metals, such as mercury.
(5) The decay rate depends strongly on temperature. The polarization can be frozen-in for months at sufficiently low temperatures, for instance in liquid nitrogen.
(6) During the decay, the anodic and the cathodic surfaces of the glass membrane exhibit a correct glass electrode response. The electrode slope measured before the electrolysis is maintained, and the standard emf of the cell containing the membrane shifts towards the value measured before the electrolysis.
(7) The magnitude of the polarization depends on the mole fraction of protons by which the membrane glass is protonated (see Sect. 2.6.3). Protonation with $x_H = 1.0$ yields the largest polarization and thus the longest time of decay, for instance 12 h between 1.8 V and 0.03 V at 50 °C, whereas electrolysis with $x_H = 0$ (and thus $x_{Li} = 1.0$) causes close to zero polarization, which is below 0.1 V and decays within 10 min at 50 °C.
(8) Changes in the concentration profiles of anodic and cathodic subsurface ranges and of the migrating boundary region by the polarization are not observed.

Tomozawa has reported a polarization effect [2.149], which he calls the charging current during polarization and discharging current during decay because both increase and decrease of the polarization were connected with the flow of (opposite) electric currents. He moreover distinguishes between an anomalous charging current, which is observed with nearly every experiment, and an electrode polarization, which is contributed to the electric contacts used; these contacts consisted of sputtered-on and painted-on gold layers, were not defined thermodynamically, and gave rather different results. However, being connected with currents, these effects cannot explain the polarization described above. On the other hand, neither such charging nor discharging currents have been observed during our migration experiments.

The polarization described here is not a "battery effect" because no electrolysed material, apart from anodically introduced protons and cathodically dismissed lithium ions, is detected at or below the glass surface, see (8) above. The alignment of dipoles or oppositely charged ions is also excluded as the cause of the polarization because it would result in a direction of polarization opposite to the direction of the electrolysing voltage. Besides, the effect is not caused by polarization of one or both of the interfacial equilibria because they are found to exhibit an intact electrode functioning during the decay, see point (6) above.

One explanation, however, is in agreement with the characterizations of the polarization as given above. The locus of the polarization seems to be the range of the migrating boundary, where protons and lithium ions with rather different mobilities overlap and cause a drastic change of the field strength with depth. Thus, the diffusion voltage in this region must be subject to depth-dependent polarization, which decays after the electrolysing field has been turned off. In addition, the depth-dependent electric field must be assumed to influence the depth distribution of the ions having different mobilities, and their rearrangement after removal of the electric field may add to the decay of the polarization. Ionic shifts connected with these effects are minute (probably in the 0.1 nm range [2.142]) and are not detectable by IBSCA. Independent of the origin of the described polarization, however, glass membranes were kept until the polarization had decayed before further analysis by IBSCA if the aim of the investigation allowed doing so.

2.6.3 Interfacial Equilibria in Electric Fields – Transfer of Different Cations Across Glass–Solution Interfaces

The Experimental Phenomenon

After studying 100% protonated glasses we were interested in investigating also the optical and transport properties of partly protonated glass layers and therefore tried to introduce protons and lithium ions simultaneously by means of electrolyses with anodic solutions containing both ion species. However, the solutions initially applied and arbitrarily containing an activity ratio of lithium to hydronium ions of about 10, corresponding to (pH − pLi) = 1, caused a lithium mole fraction of only $x_{Li} = 0.08$ in a lithium aluminosilicate pNa glass and of $x_{Li} = 0$ in a lithium silicate pH glass. As expected, the lithium mole fraction which was transferred into the pNa glass increased when the relative activity of the lithium ions in the anodic solution was increased and became 100% at $(pH - pLi) = 4$. Surprisingly, however, the pH glass was still 100% protonated after application of this solution. On increasing the relative lithium activity further, this glass started to take up small mole fractions of lithium ions when the difference $(pH - pLi)$ was as large as 9, and was finally 100% "lithiated" at an activity ratio of lithium to hydronium ions of $10^{12.5}$ and above. In order to generate a protonated layer with equal mole fractions of lithium ions and protons, $x_H = x_{Li} = 0.5$, in the pH glass, the lithium activity of the anodic solution had to be $10^{10.5}$ times that of the hydronium ions!

This tremendously large activity ratio and the large difference of the activity ratios necessary for introducing lithium ions into the pNa and pH glasses suggested strongly that, for the simultaneous transfer of alkali ions and protons, the glass surface does not act as an inert plane that is crossed by the ions simply according to their competing activities in the solution, but that the transfer must be controlled by some mechanism involving certain glass properties. In addition, the numbers measured hinted at the possibility that the ionic transfer might be controlled by the interfacial equilibrium between glass surface groups and ions in the solution. Indeed, this idea could be verified and resulted in a transfer mechanism which was based on phase boundary equilibria and explains not only the observed transfer phenomena quantitatively [2.54, 55, 97] but also supports the dissociation mechanism of the glass electrode response, which rests on the same equilibria [2.72, 87]. This transfer mechanism will be explained in detail in the following.

Control of Ionic Transfer by Interfacial Equilibria

A quantitative equation for the coverage of the glass surface with different ion species is obtained by expressing the interfacial equilibrium in terms of mole fractions. According to Sect. 2.5.4, the combined dissociation and association equilibrium is given by [2.54, 55, 72, 87]

$$RH(s) + Li^+(sol'n) + H_2O(sol'n) \rightleftarrows RLi(s) + H_3O^+(sol'n) , \qquad (2.129)$$

where R stands for any surface group, for instance R = SiO, [AlOSi], (s) and (sol'n) denote glass surface and solution, respectively, and the minute surface concentration of negatively charged surface groups, which cancels in (2.129) but is still involved and determines the potential of the glass, is not taken into consideration. The equilibrium constant of (2.129) is the product of the dissociation constant, $K''_{D,H}$, and the association constant, $K'_{A,Li}$, and is also termed the selectivity product [2.72, 87].

$$K''_{D,H} K'_{A,Li} = \frac{c'_{RLi}\, a_{H_3O^+}}{c'_{RH}\, a_{Li^+}} , \qquad (2.130)$$

where $K''_{D,H}$ contains the activity of water and both constants involve appropriate activity constants of the activities of the surface groups (see Sect. 2.5.4) and c' and a are surface concentration and activity in solution, respectively, of the species indicated. Expressing surface concentrations by mole fractions x' as given by

$$x'_{RH} = 1 - x'_{RLi} = \frac{c'_{RH}}{c'_{RLi} + c'_{RH}} = \frac{1}{\frac{c'_{RLi}}{c'_{RH}} + 1} , \qquad (2.131)$$

and combining (2.130) and (2.131) results in an expression for the mole fraction of protons bound to surface groups, RH,

$$x'_{RH} = 1 - x'_{RLi} = \frac{1}{K''_{D,H} K'_{A,Li} \frac{a_{Li^+}}{a_{H_3O^+}} + 1} , \qquad (2.132)$$

and of lithium ions bound to the functional groups, RLi,

$$x'_{RLi} = 1 - x'_{RH} = \frac{1}{\frac{1}{K''_{D,H} K'_{A,Li}} \frac{a_{H_3O^+}}{a_{Li^+}} + 1} , \qquad (2.133)$$

as functions of the ratio of the ionic activities in the solution [2.54, 97]. Equation (2.132) for a lithium silicate pH glass, were R = SiO, at three temperatures is presented as solid lines in Fig. 2.19 [2.72, 87, 97]. The position of these curves was fixed by equating the abscissae of their inflection points to the selectivity products, which had been measured potentiometrically by means of appropriate glass electrode cells. The justification of this step is given by the expression

$$K''_{D,H} K'_{A,Li} = \left(\frac{a_{H_3O^+}}{a_{Li^+}} \right)_{x'_{RLi=0.5}} , \qquad (2.134)$$

which results from introducing the ordinate of the inflection point, $x'_{RLi} = 0.5$, into (2.132) or (2.133).

The curves in Fig. 2.19 separate the areas of glass surface coverage with lithium ions and protons. They can also be looked at as titration curves of the weak acid RH at the glass surface with the strong base LiOH under formation of the salt RLi [2.87], which, in contrast to salts in homogeneous solutions, dissociates only to a minute extent because of the strong counter-effect of the potential of the glass generated by the negative surface charge of the anionic groups R$^-$ present [2.72].

Besides the solid lines representing the glass surface coverage, Fig. 2.19 contains experimental results of electrolyses of the same glass. These data points (open and closed circles, triangles) denote mole fractions of lithium, $x_{Li} = 1 - x_H$, which were measured below the glass surface by IBSCA after electrolyses with anodic solutions containing various relative lithium and

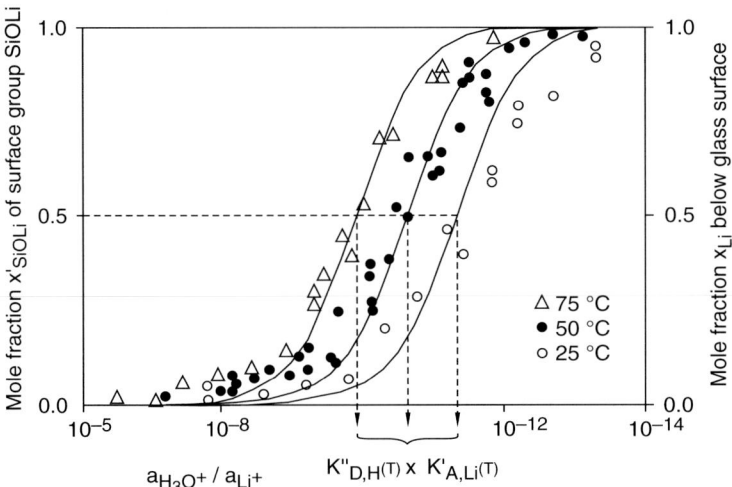

Fig. 2.19. Dependence of surface state and of electrolytic transfer of a rate-cooled lithium silicate pH glass membrane on the solution composition at various temperatures. Solid lines and left ordinate: calculated equilibrium mole fractions of SiOLi surface groups whose horizontal positions were adjusted by potentiometrically measured selectivity products. Dots and right ordinate: lithium mole fractions below anodic glass surface as measured by IBSCA after electrolyses with appropriate anodic solutions

hydronium activities. Figure 2.20 shows that these concentration profiles consist of concentration plateaus which are constant within $x_{Li} = 0.005$ under constant conditions, independent of electrolysis time, and that the partly protonated glass layers have uniform composition [2.97]. At each temperature, the experimental data points and their position with respect to the abscissa agree fairly well with the calculated curves in Fig. 2.19, which represent the equilibrium coverage of the glass surface [2.72, 87]. This agreement indicates that the ion transfer is controlled by the surface coverage, in particular because the measurements yielding the identical position of the different types of curves are completely independent of each other and of a different nature. The left ordinate in Fig. 2.19, which denotes mole fractions of surface groups, $x'_{SiOLi} = 1 - x'_{SiOH}$, is thus valid for both sets of data, which, consequently, are also characterized by the same selectivity constant due to the identical inflection point,

$$\left(\log \frac{a_{H_3O^+}}{a_{Li^+}} \right)_{x_{Li}=0.5} = \left(\log \frac{a_{H_3O^+}}{a_{Li^+}} \right)_{x'_{RLi}=0.5} = -p(K''_{D,H}K'_{A,Li}) . \quad (2.135)$$

This conclusion is supported by the extension of the concentration plateaus in Fig. 2.20 up to the glass surface [2.54]. (The small deviations within 1–2 nm below the surface were traced back to insufficient conditions during the experimentally difficult transfer of the specimens between electrolyses and

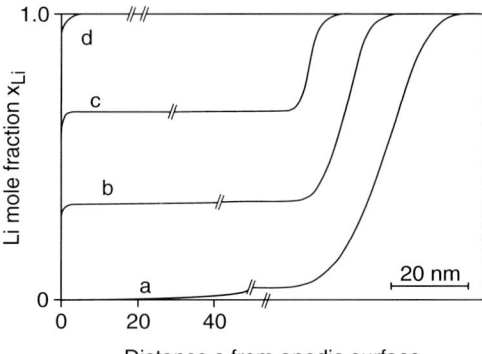

Fig. 2.20. Lithium concentration profiles measured by IBSCA below the anodic surface of a rate-cooled lithium silicate glass membrane after electrolyses with anodic solutions containing (a) $(a_{Li^+}/a_{H_3O^+}) < 10^8$, (b) $(a_{Li^+}/a_{H_3O^+}) = 10^{10.3}$, (c) $(a_{Li^+}/a_{H_3O^+}) = 10^{10.9}$, and (dS) $(a_{Li^+}/a_{H_3O^+}) > 10^{12.5}$; 5–70 kV cm^{-1}, 50 °C

analysis by IBSCA.) These profiles can thus be looked at as an extension of the equilibrium concentrations at the glass surface into the third dimension, the depth of the glass membrane, which is brought about by the electrolyses. They also show that the relative average mobilities of the different ions are practically independent of depth and are not significantly changed by a slight uptake of water by the glass during the electrolyses either (see Fig. 2.58 of Sect. 2.7).

The Mechanism of Ion Transfer

The experimental observations can be understood by the following transfer mechanism. Cations, for instance lithium ions and protons, which are attached to functional groups at the glass surface by the dynamic interfacial equilibrium, represent the outermost particles of the glass. As such, they "spend" a large part of their charge by their bond to the surface groups and thus belong to the glass phase as, for instance, any other particle positioned at the immediate glass surface. They are also far less hydrated (in general: solvated) than they would be in solution.

Thus, when a glass membrane between two solutions is subjected to an overall voltage U^0, the electric field strength generated across the membrane will extend up to the surface groups. The cations attached to the groups will follow the driving force of the electric field (if it is directed into the glass), and their first successful jump from the surface position into the next available site in the glass will be mainly a jump within the glass phase, as is that of any alkali ion within the bulk glass. The relative jump frequency of two different ions leaving the surface groups, and thus their relative mobility through the glass surface, their "transfer mobility", will be nearly equal to the relative mobility they would have in the bulk glass. This means that their transport number through the interface or their "transfer number",

$$n'_H = 1 - n'_{Li} = \frac{x'_{RH} \, \overline{u}'_{RH} \, (x'_{RH})}{x'_{RH} \, \overline{u}'_{RH} \, (x'_{RH}) + x'_{RLi} \, \overline{u}'_{RLi} \, (x'_{RLi})} \ , \tag{2.136}$$

will be nearly equal to their transport number in the glass,

$$n_H = 1 - n_{Li} = \frac{x_H \, \overline{u}_H \, (x_H)}{x_H \, \overline{u}_H \, (x_H) + x_{Li} \, \overline{u}_{Li} \, (x_{Li})} \tag{2.137}$$

where x' and x are mole fractions at the surface and in the glass, $\overline{u}(x')$ is the mole fraction-dependent average mobility through the glass surface, also called the "transfer mobility", and $\overline{u}(x)$ is the mobility in the glass of the ions indicated.

Thus, during the electrolysis of a glass membrane, a partly protonated glass layer is generated whose increasing thickness is determined by the slow migration of protons. Simultaneously, lithium ions are transferred and migrate through the layer according to their larger mobilities. They slow down when they leave the layer, thus satisfying the electrochemical continuity condition, which requires transport of equal charges through any cross section of the glass (see also Sect. 2.6.5).

After leaving the surface groups, the ions are replaced by appropriate ions from the solution. However, the interfacial equilibrium is maintained only if the current density i_{el} imposed by the electrolysis is sufficiently smaller than the exchange current density, $i_{0,H}$ and $i_{0,Li}$, of either ion taking part in the equilibrium (see Sect. 2.5.5). Only under this condition, the polarization $\eta = |\varepsilon - \varepsilon_0|$ of the equilibrium, that is, the shift of the potential relative to the equilibrium potential ε_0 caused by the current density i_{el} is negligible, as indicated by the general expression (2.138) for small polarization [2.150], see Sect. 2.5.5,

$$\eta = \frac{R T \, i_{el}}{F \, i_0} \ . \tag{2.138}$$

A sufficiently small polarization can always be attained by choosing an appropriate current density of the electrolysis. For example, an electrolysis with a total voltage of 3500 V of a 0.05 cm thick membrane (field strength: 70 kV cm^{-1}) with a cross section of 1 cm^2 of a glass having a resistance of $10^{10} \, \Omega$ cm at 25° causes a polarization of $\eta = 1.8$ mV if an exchange current density of 10^{-4} A cm^{-2} [2.71] is assumed. The polarization is probably even smaller because the exchange current density of glasses, which has not been exactly determined yet because of principle difficulties [2.71, 114. 151], is probably larger than assumed here. Under the given conditions, the ratio of the electrolysis to the exchange current densities is 14.3. It can be shown that both this current density ratio i_{el}/i_0 and the polarization η are nearly independent of temperature. The example demonstrates also that the objection that the application of "high" electric field strengths to glass membranes changes significantly the "conditions at and below their surfaces" is not justified. The critical quantity of glass electrolyses to be kept small is the polarization of the interfacial equilibrium and not the electric field strength applied across the glass membrane.

Figure 2.19 indicates that transfer experiments yield relatively exact temperature-dependent selectivity products and thus thermodynamic data of interfacial equilibria if the transfer of the alkali ion of interest is not hindered sterically (see Sect. 2.6.4). An interesting example is the measurement of such data of a lithium silicate pH glass in a rate-cooled (4.5 K min^{-1} from 5 K above T_g) and quenched state (> 1000 K min^{-1} from the melt) [2.152]. Table 2.6 shows the results. Independent of temperature, the selectivity products of the rate-cooled glass are about six times as large as those of the quenched glass resulting in standard free energies ΔG^0 of the equilibrium which are larger by 7–8% for the quenched glass. The most significant difference, however, is shown by the (negative) standard free entropy ΔS^0 of the equilibria, which, for the quenched state, is nearly twice that of the rate-cooled state. This remarkable difference is obviously connected with the "order" of the environment of the surface groups (see also Sect. 2.6.4). The difference is also shown in Fig. 2.21, which compares the transfer properties of a lithium silicate pH and a lithium aluminosilicate pNa glass [2.55]. The shape of the curves that separate lithium and proton transfer is identical for the different glasses and can be described by expression (2.133), whereas their position on the abscissa differs by eight orders of magnitude for the rate-cooled state of the different glasses and by nine orders of magnitude for the rate-cooled pNa and the quenched pH glass, as expected from the large differences of the selectivity products of silicate pH and aluminosilicate pNa glasses [2.72, 87].

Table 2.6. Thermodynamic data of the interfacial equilibria between a rate-cooled and quenched lithium silicate pH glass and solutions with various pH and pLi (ΔG^0, ΔH^0, and $\Delta H^0_{average}$ given in kJ mol^{-1}; ΔS^0 and $\Delta S^0_{average}$ in J K^{-1} mol^{-1})

Thermal history	Temp. °C	$p(K''_{D,H}K'_{A,Li})$	ΔG^0	ΔH^0	$\Delta H^0_{average}$	ΔS^0	$\Delta S^0_{average}$
Rate-cooled	25	+11.3	+64.5	+59.0		−18.5	
	50	+10.5	+65.0	+60.3	+59.6	−14.5	−16.5
	75	+ 9.8	+65.3	+59.6		−16.4	
Quenched	25	+12.1	+69.1	+59.3		−32.9	
	50	+11.3	+69.9	+60.3	+59.7	−29.7	−31.5
	75	+10.6	+70.7	+59.6		−31.9	

Significance, Application, Remarks

The concentration profiles in Fig. 2.20 demonstrate that homogeneously protonated glass layers with any proton mole fraction can be prepared by electrolysis, and Figs. 2.19 and 2.21 show the composition of the anodic solutions
that must be chosen if glass layers with particular proton contents are to be
obtained. This means that oxidic glasses are not only alkali ion conductors,
as which they are generally known, but can also be made mixed-alkali ion-
proton and pure proton conductors (see also Figs. 2.16 and 2.17). These findings modify *Schwabe's* and *Dahms'* [2.47] and *Hammond's* [2.48] conclusion
from their extended electrolyses with tritium-labelled anodic solutions that
"oxidic glasses are not permeable to hydrogen ions". However, they would
have never observed proton migration through glass membranes under their
experimental conditions (electrolysis time, membrane thickness, voltage applied, temperature) anyway [2.49, 50] because of the small proton mobility,
and fast protons, whose existence in any oxidic glass has been claimed [2.49],
do not exist in silicate glasses, as was shown experimentally (see Sect. 2.6.7).

Two applications of protonated glasses are to be mentioned at this point.

(1) Homogeneous, protonated glass layers with various proton contents
were prepared in order to investigate the transport properties of such modified
glasses. Thus, concentration-dependent mobilities of protons and lithium ions
were measured and will be reported in Sect. 2.6.5, and the conditions for the
stability of migrating boundaries in glasses were studied and are discussed in
Sect. 2.6.6. This work resulted also in the detection of proton traps in glasses
with low proton contents, which will be described in Sect. 2.6.7.

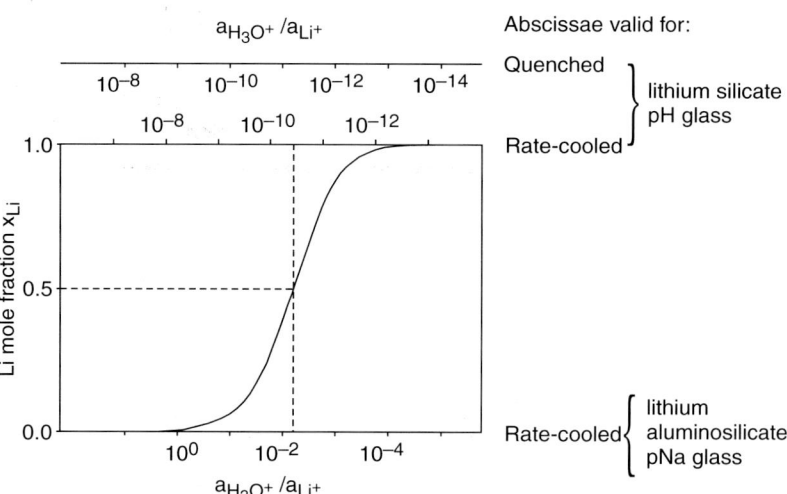

Fig. 2.21. Comparison of transfer properties of a lithium aluminosilicate pNa glass
and a lithium silicate pH glass with various thermal histories at $50\,^\circ$C

(2) The experimental verification of the equilibrium coverage of glass surfaces by electrolyses according to Figs. 2.19 and 2.20 and (2.132) and (2.133) and the detection that glass electrodes with protonated membranes show an unchanged glass electrode slope in alkali-free buffers prove the dissociation mechanism of glass electrodes (see Sect. 2.11).

Finally we would like to quote a historical remark which demonstrates that experimental observations are of little value if they cannot be evaluated and interpreted for lack of adequately sophisticated methods. Already in 1928, *Quittner* concluded from his glass electrolyses with anodic solutions containing various ionic species that "the number of ions present (in the anodic solution) can thus not be the only cause responsible for how many of these ion species migrate into the glass" (translation from German) [2.153]. Quittner intuitively grasped that there was a certain mechanism at work which controlled the field-driven transfer of ions into glasses, without being able to explain it by his results or explore it by means of the experimental tools available to him. His observations had to wait for an understanding until, about half a century later, sufficiently sensitive methods had been developed.

2.6.4 Sterical Effects

Direct current (d.c.) glass electrolysis with electrolyte solutions as non-blocking anodic contacts presents the possibility for offering cations with various properties (e.g., ionic radius, polarizing power, and polyvalency) to the glass surface; their transfer into the glass as well as their subsequent migration, if they are transferable, and thus the effect of the glass structure upon these processes can be studied. The only precondition is the knowledge of the thermodynamic data of the corresponding surface equilibria. If unknown, these data must be obtained from, or determined by, separate experiments because it must be ensured for the transfer that the glass surface is completely covered with the ions of interest. Experiments of this kind with several different aspects have been performed [2.54, 56, 133, 135] and have been summarized in a brief overview [2.55]. The work concerned with ion radius-connected transfer and migration is reported here. Other glass structure-related electrolyses will be found in Sects. 2.6.5 and 2.6.7.

Whether or not a monovalent cation is transferred across the interface between anodic solution and glass under the action of an electric field depends mainly on its radius relative to the radius distribution of the vacancies of the glass network possibly available for its accommodation. The simplest case is the electrolysis of sodium ions (ionic radius (six-coordinated): 0.102 nm [2.137]) into a lithium (ionic radius (six-coordinated): 0.076 nm [2.137]) silicate glass, for which the selectivity product and thus the pH_{tr} value is known from potentiometric measurements.

Figure 2.22 presents subsurface lithium and sodium concentration profiles measured by means of IBSCA after electrolyses of a lithium silicate pH glass

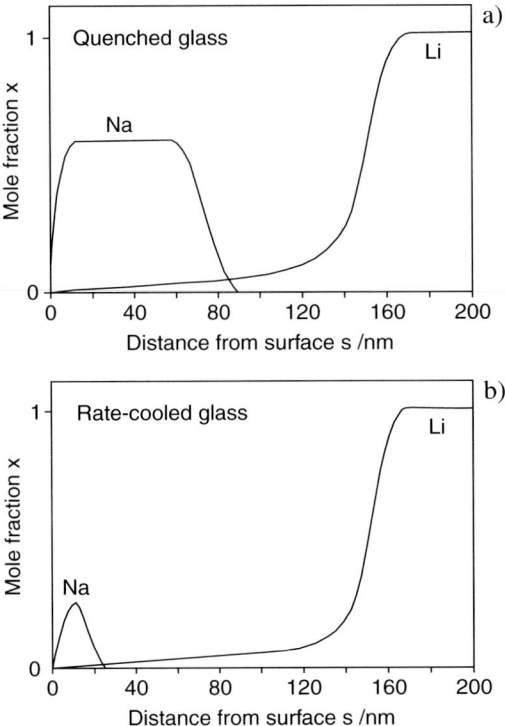

Fig. 2.22. Different structures of a lithium silicate pH glass due to different thermal histories as observed by electrolysis and subsequent concentration profiling with IB-SCA (steric effect). (**a**) Sodium ions are admitted to the quenched glass, whereas (**b**) no sodium is found in the rate-cooled glass after the electrolysis although complete coverage of the anodic surface with SiONa was secured by potentiometry

with anodic solutions containing sodium ions at pH > (pH$_{tr}$ + 2), that is, at complete sodium coverage of the glass surface (50 °C). Figure 2.22b shows that no sodium is admitted into the rate-cooled glass except for a time-independent small peak below the surface, which is probably caused by glass hydrolysis. In contrast, sodium is transferred into the quenched glass, but only up to an upper mole fraction of 0.65, Fig. 2.22a, which could neither be exceeded by increasing the activity ratio of sodium to hydronium ions in the anodic solution nor by changing other parameters, for instance the temperature. In both cases, it was verified that protons are transferred into the glass instead of sodium ions, although proton transfer is obvious from the movement of the migrating boundary, which was similar to the moving boundaries observed during 100% protonation, Fig. 2.16. Figure 2.23 gives an overview of the sodium transfer as functions of the activity ratio of hydronium to sodium ions and demonstrates again the different transfer properties of the glass in its different thermal states.

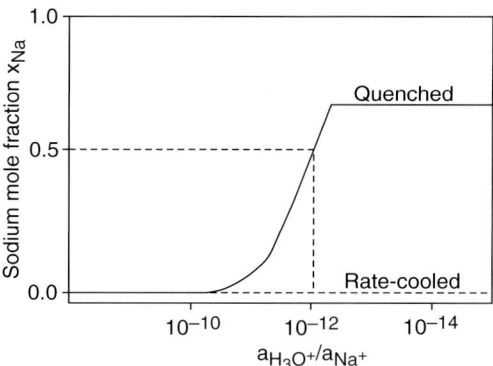

Fig. 2.23. Overview of the transfer properties to sodium ions of the lithium silicate pH glass shown in Fig. 2.22 in the quenched and rate-cooled state

The transfer of protons is surprising at first sight because proton admission should be excluded at 100% sodium coverage of the glass surface. This effect, however, is easily explained. The sodium ions, bound to the siloxy surface groups but unable to follow the electric field and to enter the glass, block the glass surface to the transfer of any ions. As a consequence, the electric field across the surface groups increases so that the surface equilibrium is polarized more strongly than in the case of, for instance, proton transfer. The polarization increase, which is connected with a shift of the equilibrium, continues until some protons appear at the surface groups, which, not being hindered sterically, are transferred easily into the glass at the relatively high polarization of the SiOH groups. They enter the glass either as the only ions from the solution, Fig. 2.22b, or together with sodium ions, for instance at their limiting mole fraction $x_{Na} = 0.65$, Fig. 2.22a.

The average size of the vacancies determining the transfer of the ions across the interface is obviously also characteristic of their subsequent migration in the glass. This is demonstrated by the quenched glass, Fig. 2.24, which demonstrates the increasing thickness of the "sodiated" subsurface glass layers. Figure 2.24a shows the growth of the layer thickness while sodium ions are constantly supplied by the same, unchanged anodic solution. A partly "sodiated" protonated layer is generated directly below the glass surface; its thickness increase is determined by the average mobility of the sodium ions. In addition, a completely protonated layer is formed between the partly sodiated layer and the bulk glass, whose boundary is not contained in Fig. 2.24a because of figure dimensions but is shown in Fig. 2.22. In contrast, Fig. 2.24b demonstrates the dilution, (2) and (3), of primarily introduced sodium ions (1), while only protons are supplied to the glass by a sodium-free, acidic solution. The migration rate of the sodium ions and thus the increase of the layer thickness are determined by the concentration dependence of the sodium ion

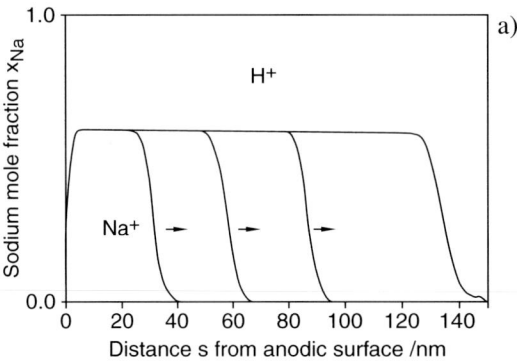

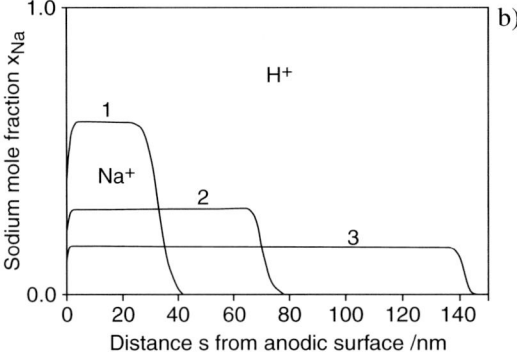

Fig. 2.24. Sodium concentration profiles measured by IBSCA demonstrating the change of the glass layer with maximum sodium content (as in Fig. 2.22) during electrolyses of a quenched lithium silicate pH glass. (**a**) Constant sodium content of the anodic solution leads to a uniform layer with increasing thickness. (**b**) Replacement of the primary sodium-containing solution by dilute acid causes dilution of the sodium ions by protons

mobility. The total number of sodium ions in the glass is constant during the dilution.

Figures 2.22 and 2.24a inform also about the relative magnitude of the average mobilities of the ion species, $\overline{u}_{\mathrm{Li}}(x_{\mathrm{Li}} = 1) > \overline{u}_{\mathrm{H}}(x_{\mathrm{H}} = 1) > \overline{u}_{\mathrm{Na}}(x_{\mathrm{Na}} = 0.65)$ and of the extreme individual mobilities of the migrating ions, $u_{\mathrm{Li,min}} > u_{\mathrm{H,max}}$ and $u_{\mathrm{H,min}} > u_{\mathrm{Na,max}}$, which, however, are caused by different effects. The lithium ion mobility is determined by the size of the vacancies of the original glass network and partly by their polarization of the oxygens, and the mobility of the guest protons is mainly given by their polarizing action on the oxygens, whereas the mobility of the sodium ions is mainly due to their large radius relative to the vacancies in the network. In the rate-cooled glass, in contrast, the relatively smaller radius of the vacancies suppresses transfer and migration of sodium ions altogether. The different thermal states of the lithium silicate glass are thus not only characterized by the thermodynamic data of the interfacial equilibria, in particular the standard entropy [2.152], as reported in Sect. 2.6.3, see Table 2.6, but also by the size of the vacancies in the network. Table 2.7 compares thermodynamic and spatial data.

Different from lithium silicate pH glasses, lithium aluminosilicate pNa glasses allow the complete replacement of their lithium by sodium ions in both the quenched and the rate-cooled state. This is shown by the lithium and sodium concentration profiles in Fig. 2.39 of Sect. 2.6, which were measured after the electrolyses of a rate-cooled lithium aluminosilicate glass at 100% sodium transfer. A stable migrating boundary is not observed because the mobilities of the ions obviously overlap. The structural conditions of aluminosilicate glasses are of great interest for application as pNa electrode glasses and will be treated in Sect. 2.7.3, where we show that the glass structure determines the conditions under which alkali-selective electrodes must be applied, pretreated, and stored if reproducible pNa and pK measurements are to be conducted. It will also be shown in Sect. 2.7.3 that the structure of lithium silicate glasses, which are ideal pH glasses, is so seriously changed by small additions of certain compounds that they are practically useless as electrode glasses.

2.6.5 Simultaneous Electromigration of Ions and "Guest Ions"

After its development, the modified moving boundary method involving the subsurface-analytical technique IBSCA as the indicating means presented a new way of investigating transport processes in glasses [2.54–56, 88, 121, 127, 133, 135]. It is of continuing interest because it shows directly the time-dependence of the depth-dependent concentrations of the migrating ions, whereas all other methods, whether based on direct or alternating current, depend on the conclusion of the ion movement from overall, for instance electrical or optical, effects [2.99, 101, 140, 154]. Its application is thus of particular interest for the investigation of electrode membrane glasses [2.72, 87]. Simultaneous migration of lithium ions of the glass and protons which replace lithium ions, so-called "guest protons", was the first example studied [2.135]. In addition to its basic interest for electrode glasses, the lithium ion-"guest

Table 2.7. Comparison of standard entropies of interfacial equilibrium (see Table 2.6) and of spatial conditions of a lithium silicate pH glass with different thermal histories

	Rate-cooled	Quenched	Literature
$\Delta S^0/\mathrm{J\,K^{-1}\,mol^{-1}}$	-16.5	-31.5	[2.152]
Vacancies with radius r, in %			
$0.133\,\mathrm{nm} > r > 0.095\,\mathrm{nm}$	0	65	[2.55]
$r < 0.095$	100	35	

proton" system promised some insight into the mixed-alkali effect because, in contrast to mixed-alkali glasses usually investigated, the various concentration ratios of lithium ions and protons are contained in the same glass network, and neither ion is sterically hindered. In addition, this ion couple forms distinct moving boundaries, see Fig. 2.20, due to the favourable relative mobilities of the ions, see for instance (2.125), and is thus ideally suited for an investigation by the moving boundary method.

Subsequent to the reported work, the principle of the modified moving boundary method with solutions as anodic contacts was applied by *Doremus* and coworkers to the investigation of ionic migration processes in a soda-lime silicate glass [2.148]. Indicator methods were NRA for hydrogen and Rutherford backscattering for other elements. Regrettably, none of our published general results were observed by these authors so that several basic experiments were repeated. Another paper [2.155] reporting on glass electrolyses using several (vacuum-deposited) metals as anodes and NRA as indicator means for hydrogen, Rutherford backscattering for other elements, and resonant backscattering for oxygen profiles represents an example of the application of (at least partly) blocking electrodes (150 °C), in contrast to non-blocking solutions. The current is carried by positive ions (Na^+, H^+, Ca^{2+}) and, in extremely high fields, by electrons, but not by oxide ions. The different character of blocking anodes is made particularly obvious by the appearance of current fluctuations (at constant voltage) at certain times whereas a smooth decrease of the current is observed at other times. This effect has never been observed with non-blocking solutions as contacts.

The Electrochemical Continuity Condition

The relative migration rates of ions in glass are regulated by the electrochemical continuity condition, whose basis is the condition of electroneutrality. It states that, at any given moment, equal charges per unit time are transported through any cross section of the electrolyte perpendicular to the direction of the electric field, as expressed by

$$i = \frac{dq}{dt} = \sum_j \frac{dq_j}{dt} = \sum_j i_j = \text{const.} , \tag{2.139}$$

where, respectively, i and q are total current and charge, i_j and q_j are partial current and charge carried by ion species j, and "const." refers to the position in the glass and not necessarily to time. For an overview, Fig. 2.25 presents a schematic with the denotation of the quantities for 100% and partial protonation. Equation (2.139) may be expressed by the more detailed relationship

$$\sum_j x_j \bar{u}_j(x_j) c^0 z_j F A \boldsymbol{E} = \sum_j i_j = i , \tag{2.140}$$

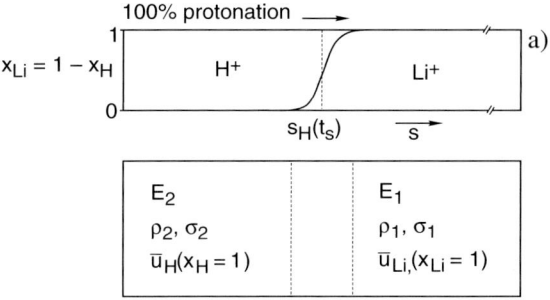

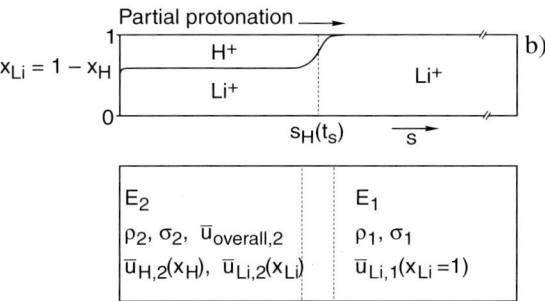

Fig. 2.25. Systematic overview of denotations used for the description of the electric quantities of the glass (1) and the protonated layers (2) during and after 100% (**a**) and partial protonation (**b**). Mole fraction x, electric field strength E, resistivity ρ, conductivity σ, average ionic mobility $\bar{u}$, overall mobility $\bar{u}_{\text{overall}}$, and distance from anodic surface s

where x, $\bar{u}(x)$, and z are mole fraction, mole fraction-dependent average mobility, and ionic charge of the ion indicated, c^0 is the total concentration of sites in the glass, A is the cross section of the glass membrane, and E is the electric field strength. For equally charged ions and constant cross section, (2.140) changes to

$$\sum_{j} x_j \bar{u}_j(x_j) E = \frac{i}{c^0 z_j F A} , \qquad (2.141)$$

which, for 100% protonation of a glass with mobile lithium ions, (2.141) reduces to

$$\bar{u}_{\text{Li},1}(x_{\text{Li},1} = 1) E_1 = \bar{u}_{\text{H},2}(x_{\text{H},2} = 1) E_2 = \frac{i}{c^0 F A} , \qquad (2.142)$$

where indices 1 and 2 indicate the glass and the protonated layer, respectively, Fig. 2.25. If the glass is only partly protonated so that the layer contains lithium ions in addition to protons, (2.142) changes to

$$[x_{H,2}\overline{u}_{H,2}(x_{H,2}) + x_{Li,2}\overline{u}_{Li,2}(x_{Li,2})]\, \boldsymbol{E}_2 = \overline{u}_{Li,1}(x_{Li,1} = 1)\boldsymbol{E}_1 = \frac{i}{c^0 F A} \tag{2.143}$$

or to

$$\overline{u}_{\text{overall}}\boldsymbol{E}_2 = \overline{u}_{Li,1}(x_{Li,1} = 1)\boldsymbol{E}_1 = \frac{i}{c^0 F A} \tag{2.144}$$

if the bracket on the left side of (2.143) is defined as the overall mobility [2.55],

$$\overline{u}_{\text{overall}} = x_{H,2}\overline{u}_{H,2}(x_{H,2}) + x_{Li,2}\overline{u}_{Li,2}(x_{Li,2}) \ . \tag{2.145}$$

According to (2.142) and (2.144), the ratio of the electric fields in glass and layer is inversely proportional to the corresponding mobilities. Thus, for 100% protonation

$$\frac{\boldsymbol{E}_2}{\boldsymbol{E}_1} = \frac{\overline{u}_{Li,1}(x_{Li,1} = 1)}{\overline{u}_{H,2}(x_{H,2} = 1)} \tag{2.146}$$

and for partial protonation

$$\frac{\boldsymbol{E}_2}{\boldsymbol{E}_1} = \frac{\overline{u}_{Li,1}(x_{Li,1} = 1)}{\overline{u}_{\text{overall},2}} \tag{2.147}$$

holds.

Expression (2.143) allows the ionic migration in the glass to be visualized. After transfer, the lithium ions and protons migrate through the partially protonated range with rates corresponding to their concentration-dependent mobilities. The front of the protons forms the measurable migrating boundary between the glass and the slowly extending layer, whereas the faster lithium ions, after leaving the layer, slow down and continue their migration with a rate that is given by their mobility in the unchanged glass. Determining concentration-dependent ionic mobilities, that is, the proportionality constant between ionic rate and electric field strength, thus means measuring the rate of both ions in, and the field strengths across, the partly protonated layer at various ion concentrations. We will now show how this can be accomplished and, based on the data obtained, propose a migration mechanism for lithium ions and guest protons [2.55]. For simplicity, the following text frequently uses the term proton instead of guest proton unless the latter is to be especially emphasized.

Electric Fields in Protonated Layer and Bulk Glass During Protonation

Figure 2.26 shows schematically that, during 100% or partial protonation of a glass membrane with thickness d, the total voltage U_0 applied is divided

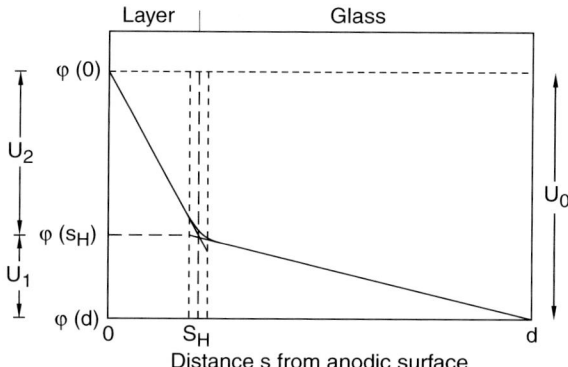

Fig. 2.26. Schematic potential distribution $\varphi(s)$ within a glass membrane with thickness d during protonation with total voltage U_0, after the protonated layer has reached the thickness s_H. $\varphi(0)$, $\varphi(d)$, and $\varphi(s_H)$ are the potentials of the anodic and cathodic surface and of the migrating boundary, respectively

into two parts, one, U_1, across the yet unchanged glass membrane and the other, U_2, across the protonated layer,

$$U_0 = U_1 + U_2 \ . \tag{2.148}$$

If the protonation has proceeded to a thickness s_H of the layer so that $(d-s_H)$ is the thickness of the unprotonated glass, the electric field strength in the layer is given by

$$\boldsymbol{E}_2 = \frac{U_2}{s_H} \tag{2.149}$$

and that across the rest of the membrane by

$$\boldsymbol{E}_1 = \frac{U_1}{(d - s_H)} \ . \tag{2.150}$$

The total voltage can thus also be expressed by the electric field strengths in the two glass regions,

$$U_0 = \boldsymbol{E}_1(d - s_H) + \boldsymbol{E}_2 s_H \ . \tag{2.151}$$

Further insight into the condition of the electric fields within the membrane is gained by considering the resistivities ρ (or conductivities σ) within the unprotonated glass,

$$\rho_1 = \frac{1}{\sigma_1} = \frac{1}{Fc^0 \overline{u}_{Li,1}(x_{Li,1} = 1)} \ , \tag{2.152}$$

and within the layer during 100% protonation,

$$\rho_2 = \frac{1}{\sigma_2} = \frac{1}{Fc^0 \overline{u}_{H,2}(x_{H,2} = 1)} , \tag{2.153}$$

and during partial protonation,

$$\rho_2 = \frac{1}{\sigma_2} = \frac{1}{Fc^0 \overline{u}_{\text{overall},2}} . \tag{2.154}$$

Combination of (2.146), (2.152), and (2.153) as well as that of the expressions (2.147), (2.152), and (2.154) show that the electric field strengths in the membrane during protonation are proportional to the resistivities,

$$\frac{E_2}{E_1} = \frac{\rho_2}{\rho_1} , \tag{2.155}$$

and combining (2.155) with (2.151) finally yields the equations for the normalized electric fields in the protonated layer,

$$\frac{E_2}{\frac{U_0}{d}} = \frac{1}{\frac{\rho_1}{\rho_2}\left(1 - \frac{s_H}{d}\right) + \frac{s_H}{d}} , \tag{2.156}$$

and the unchanged glass,

$$\frac{E_1}{\frac{U_0}{d}} = \frac{1}{\frac{s_H}{d}\left(\frac{\rho_2}{\rho_1} - 1\right) + 1} , \tag{2.157}$$

as functions of the normalized thickness of the protonated layer, s_H/d [2.56]. (The expression "overall electric field strength" for U_0/d is misleading and should strictly be avoided.) Knowledge of the expressions (2.156) and (2.157) yields also the dependence of the electric field strengths on time because the time dependence of the layer thickness is given by

$$s_H(t) = \left[\frac{2U_0}{c^0 F(\rho_2 - \rho_1)}t + \left(\frac{\rho_1 d}{\rho_2 - \rho_1}\right)^2\right]^{1/2} - \frac{\rho_1 d}{(\rho_2 - \rho_1)} . \tag{2.158}$$

The electric field strengths according to (2.156) and (2.157) are plotted as functions of $\log(s_H/d)$ for several resistivity ratios ρ_2/ρ_1 of layer and glass in Figs. 2.27 and 2.28, respectively. The basically similar plots differ at large values of $\log(s_H/d)$, where the normalized field $E_2/(U_0/d)$ across the protonated layer approaches unity, whereas that across the unchanged glass, $E_1/(U_0/d)$, approaches the ratio of the resistivities of glass and protonated layer, ρ_1/ρ_2. This is in accordance with the ratio of the electric field strengths, E_2/E_1, which is independent of the layer thickness, as also shown in Fig. 2.27. Because, however, the measurement of concentration-dependent mobilities is conducted on relatively thin anodic subsurface layers of membranes, such thick protonated layers are not of practical interest here.

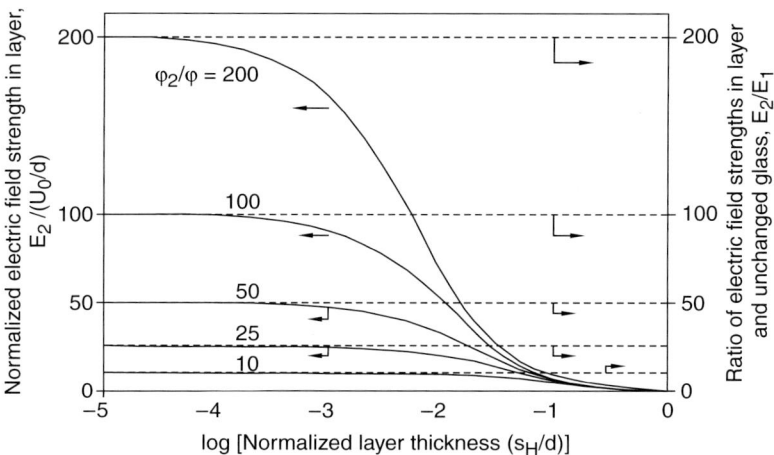

Fig. 2.27. Normalized electric field strength $E_2/(U_0/d)$ within a protonated layer during protonation of a glass membrane with thickness d as functions of the logarithm of the normalized distance s_H/d of the migrating boundary from the anodic glass surface. ρ_2/ρ_1 is the resistivity ratio of layer and glass. Also given are the (constant) ratios E_2/E_1 of the electric field strengths within layer and bulk glass

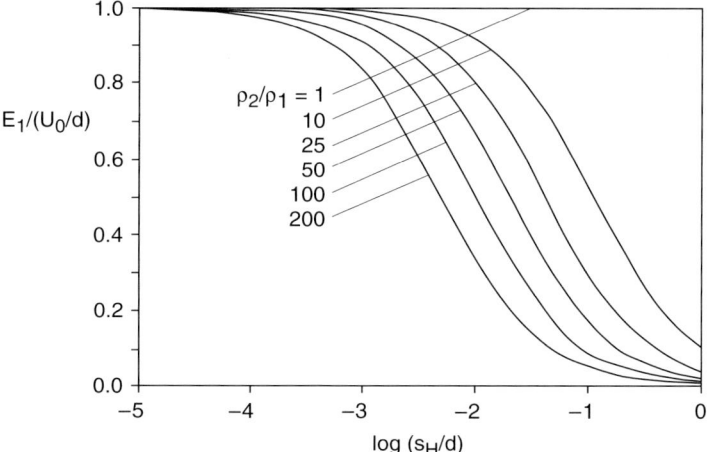

Fig. 2.28. Normalized electrical field strength $E_1/(U_0/d)$ within the bulk glass during protonation of a membrane with thickness d as functions of the normalized distance s_H/d of the migrating boundary from the anodic glass surface. ρ_2/ρ_1 is the ratio of layer and glass resistivities, respectively

In addition, in Fig. 2.29 the electric potential of the migrating boundary relative to the potential of the anodic glass surface $\varphi(s_H)/\varphi(s_H = 0)$ is plotted as a function of the logarithm of the normalized layer thickness. The curves are for several ratios ρ_2/ρ_1 and were obtained by means of

$$\frac{\varphi(s_H)}{\varphi(s_H = 0)} = \frac{U_1}{U_0} = \frac{(d - s_H)}{s_H \left(\frac{\rho_2}{\rho_1} - 1\right) + d} , \qquad (2.159)$$

which is derived similarly to the expressions for the electric fields. Different from the electric field strengths according to (2.156) and (2.157), the potential approaches zero as the migrating boundary approaches the cathodic membrane surface. However, in thin layers, whose thicknesses are of interest for the following, the electric field strengths and potentials have nearly constant values, from which they deviate at the larger thicknesses the smaller the resistivity ratio ρ_2/ρ_1 of layer and unchanged glass is, Figs. 2.27–2.29.

The potential distribution (2.159), Fig. 2.29, was derived as the basis for the determination of ion mobilities in protonated glasses with various proton contents, which was enabled by the anodically and cathodically non-blocking contacts represented by the electrolyte solutions used. As also mentioned in Sect. 2.6.1, several papers, particularly the detailed work by *Proctor* and *Sutton* [2.127], were carried out with the completely different aim to study the behaviour of glasses under the influence of blocking electrodes, for instance aluminium. Temperatures were at and above the annealing point of the glass (383 °C), and narrow aluminium probes evenly distributed in the glass plates yielded the potential distribution in the glass. Accordingly, completely different results were obtained. Largely due to the development of a space charge at the positive electrode, a steady-state or, more exactly, a quasi-static state of the potential distribution developed. It was distinguished by a large (space charge-caused) potential drop at the anode, a much smaller potential drop

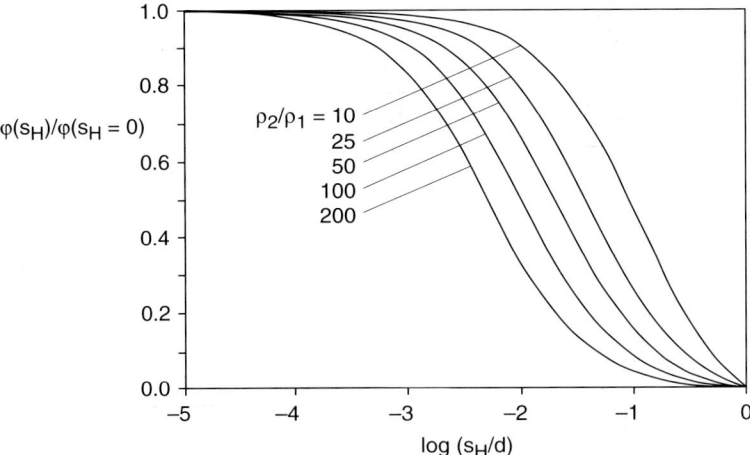

Fig. 2.29. Potential $\varphi(s_H)$ of the migrating boundary referred to the potential of the anodic surface $\varphi(s_H = 0)$ during protonation of a glass membrane with thickness d as a function of the normalized distance s_H/d of the migrating boundary from the anodic glass surface. ρ_2/ρ_1 is the ratio of layer and glass resistivities, respectively

at the cathode, which was probably caused by a pile-up of cations, and, dependent on the extension of the specimen relative to the voltage applied, a nearly linear potential distribution with a small potential gradient in the central part of the glass. Despite, or because of, the clear results of the experiments, the paper tempts one to doubt the existence of really blocking electrodes for glasses and to discuss their usefulness as electrodes for electrochemical measurements.

Determination of Concentration-Dependent Mobilities

Concentration-dependent mobilities of lithium ions and guest protons were determined in a lithium silicate pH glass with a total lithium content of nearly $25 \, mol \, dm^{-3}$ by the arrangement sketched in Fig. 2.10 of Sect. 2.4. A flat, about 0.5 mm thick membrane of the rate-cooled glass melted to a shaft glass tube with high resistivity was contacted by appropriate solutions and subjected to a total voltage U_0 ranging from 350 V to 3500 V. The mole fractions of protons to be achieved in the protonated glass were predetermined by means of the relative pH and pLi values of the external, anodic solutions. During the electrolyses, the current through the membrane was monitored as a function of time, thus yielding the initial current $i(t = 0)$, the current at the end of the protonation $i(t_s)$, and the total charge $q = \int_0^{t_s} i \, dt$ transported through the membrane. After electrolysis and decay of some polarization, see Sect. 2.6.2, the membrane was subjected to infrared spectroscopy (an IR spectrum of the blank was always measured before the treatment) and subsequently to concentration profiling by IBSCA, yielding the exact mole fractions of protons and lithium ions in the protonated layer and the distance s_H travelled by the migrating boundary.

Mobility of Lithium Ions in Unchanged Glass

The initial current $i(t = 0)$ of the electrolysis yielded the resistivity of the unchanged glass according to

$$\rho_1 = \frac{U_0 \, A}{d \, i(t = 0)} \, , \tag{2.160}$$

which, in turn, gave the mobility of the lithium ions in the yet unprotonated glass by the expression

$$\overline{u}_{Li,1}(x_{Li,1} = 1) = \frac{1}{\rho_1 \, c^0 \, F} \, . \tag{2.161}$$

Proton Mobility in 100% and Overall Mobility in Partly Protonated Glass

The resistivity of 100% and of partly protonated glass layers was obtained from the total resistance $R(t_s)$ of the membrane measured at the end of the electrolysis and from the thickness s_H of the protonated layer as obtained by IBSCA,

$$\rho_2 = \frac{R(t_s)A}{s_H} - \frac{\rho_1(d - s_H)}{s_H} . \tag{2.162}$$

The proton mobility at 100% protonation as well as the overall mobility at partial protonation was thus determined by

$$\overline{u}_{H,2}(x_{H,2} = 1) \qquad \text{or} \qquad \overline{u}_{\text{overall},2} = \frac{\rho_1 \overline{u}_{Li,1}(x_{Li,1} = 1)}{\rho_2} , \tag{2.163}$$

where ρ_2 denotes the respective resistance of 100% or partly protonated layers.

Proton Mobility in Partly Protonated Glass

The mobility of protons in partly protonated glass is obtained from

$$\overline{u}_{H,2}(x_{H,2}) = \frac{s_H}{t_s \overline{\boldsymbol{E}}_2(s_H)} , \tag{2.164}$$

where $\overline{\boldsymbol{E}}_2(s_H) \cong \overline{\boldsymbol{E}}_2(t_s)$ is the approximate time average of the electric field strength that acted on the protonated layer during the electrolysis time. Its exact value can be obtained by means of (2.156) and (2.158) above. However, as shown by a plot of $\boldsymbol{E}_2/(U_0/d)$ versus (s_H/d) (not reproduced), the arithmetic mean

$$\overline{\boldsymbol{E}}_2(s_H) = 0.5 \left[\boldsymbol{E}_2(s_H = 0) + \boldsymbol{E}_2(s_H) \right] \tag{2.165}$$

agrees with the exact value within a few percent, even at large resistivity ratios, for instance $\rho_2/\rho_1 = 200$, as long as (s_H/d) is not larger than approximately 10^{-3}. This condition corresponds to a layer thickness of 500 nm in an 0.5 mm thick membrane and was always given during the experiments.

Lithium Ion Mobility in Partly Protonated Glass

The lithium ion mobility in partly protonated glasses is obtained from the migration rate of the lithium ions under the influence of the time average of the electric field strength across the layer. Figure 2.30 sketches the conditions. During electrolysis time t_s, that is, while the protonated layer with thickness s_H is generated, lithium ions migrate through the protonated layer and form the lithium-for-lithium exchanged layer between s_H and s_{Li}. The boundary between lithium of the glass and penetrated lithium, which is not measurable by IBSCA, is assumed distinct, and its distance s_{Li} from the anodic surface is obtained from the total charge transported:

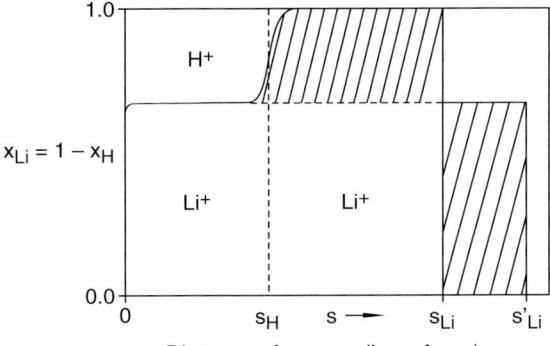

Fig. 2.30. Lithium concentration profile of a lithium silicate glass membrane partly protonated with mole fraction $x_H = 1 - x_{Li}$ up to the migrating boundary s_H. Lithium ions from the solution have replaced lithium ions of the glass up to the calculated, distinct boundary s_{Li}, which allows the calculation also of boundary s'_{Li} for determining the lithium ion mobility at mole fraction x_{Li}

$$s_{Li} = \frac{1}{c^0 F A} \int_0^{t_s} i \, dt \ . \tag{2.166}$$

Consequently, the distance s'_{Li} that would have been migrated by the lithium ions, if they had continued their migration at the mole fraction x_{Li} in the protonated layer, is determined by means of

$$s'_{Li} = s_{Li} + \frac{x_{H,2}}{x_{Li,2}} (s_{Li} - s_H) \ , \tag{2.167}$$

so that the lithium ion mobility is either given by

$$\overline{u}_{Li,2}(x_{Li,2}) = \frac{s'_{Li}}{t_s \overline{E}_2(t_s)} \ , \tag{2.168}$$

or by

$$\overline{u}_{Li,2}(x_{Li,2}) = \frac{s'_{Li}}{s_H} \overline{u}_{H,2}(x_{H,2}) \ . \tag{2.169}$$

Mole fraction-dependent mobilities of both lithium ions and protons can thus be determined in the entire concentration range $0 \le x_H \le 1$.

Concentration-Dependent Mobilities

Figure 2.31 presents average mobilities of guest protons and lithium ions and the conductivity of a lithium silicate pH glass as functions of the proton mole fraction, $x_H = (1 - x_{Li})$ [2.54]. The different data are referred to each other by

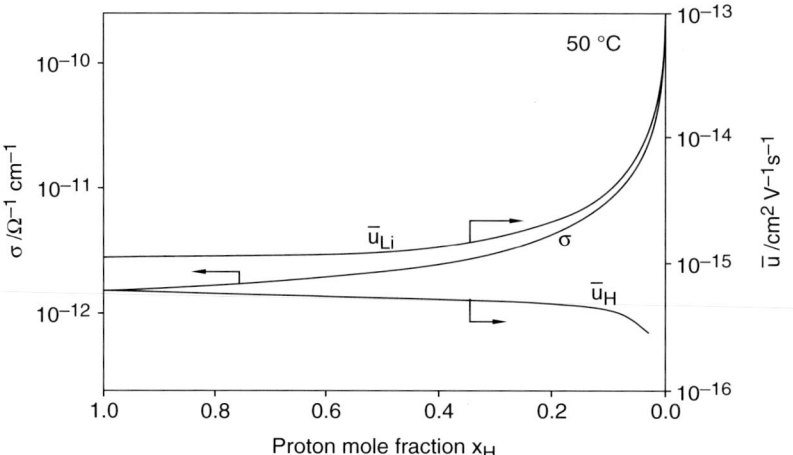

Fig. 2.31. Experimental average mobilities of lithium ions and protons and conductance of the protonated glass as functions of proton mole fraction $x_H = 1 - x_{Li}$, 50 °C. The different ordinates are referred to each other by $\sigma(x_{Li} = 1)/\overline{u}_{Li}(x_{Li} = 1) = F c^0 = 2172 \, \text{A s cm}^{-3}$

$$\frac{\sigma(x_{Li} = 1)}{\overline{u}_{Li}(x_{Li} = 1)} = F c^0 = 2172 \, \text{A s cm}^{-3} , \qquad (2.170)$$

where $c^0 = 22.5 \, \text{mol dm}^{-3}$ is the site concentration in the glass. Similar to the lithium ion mobility, the conductance decreases with increasing proton mole fraction, the change being greatest at low proton contents. The "equitransport glass", which is the glass with an equal transport number of lithium ions and protons, $x_{Li}\overline{u}_{Li}(x_{Li}) = x_H\overline{u}_H(x_H)$, contains a proton mole fraction $x_H = 0.68$,

$$\frac{\overline{u}_{Li}(x_{Li} = 0.32)}{\overline{u}_H(x_H = 0.68)} = 2.13 . \qquad (2.171)$$

Although strongly deviating from the "ideal" linear connection of the conductivities of the original unprotonated and the fully protonated glass, the conductivity does not exhibit the typical feature of the mixed-alkali effect, a conductance minimum, which is generally found with mixed-alkali glass series containing two alkali ion species with constant total but changing relative alkali mole fractions, see for instance [2.156]. This result agrees with observations on sodium silicate glasses containing ion-exchanged potassium [2.157, 158] and silver [2.157]. Also, another typical feature of the mixed-alkali effect, a cross-over of mobility-mole fraction and diffusivity-mole fraction curves of the ions [2.159–161] is not observed with protonated glasses. If, despite its controversial discussion [2.162], the conclusion is allowed that ion-exchanged glasses do not generally exhibit a mixed-alkali effect, this would mean that the mixed-alkali effect is at least strongly connected with the different sizes of the vacancies generated by the different alkali ions during cooling of the melt. Obviously, not all vacancies are available to both alkali ion species in mixed-alkali glasses, whereas ion-exchanged glasses, including protonated glasses, are distinguished by vacancies which are sterically accessible to both ions, for instance lithium ions and ion-exchanged protons in the lithium silicate glasses investigated. The results thus emphasize that mixed alkali glasses are not just glasses with different concentration ratios of alkali ions but are also distinguished by the variation of their glass networks whose structure is determined by the steric conditions of their alkali ions. It seems that this effect has not always been adequately taken into account when the mixed-alkali effect was discussed in the past.

Proposal of a Migration Mechanism

The results shown in Fig. 2.31 can be understood by the following simple mechanism [2.135]. The only assumptions are (a) that the negative sites are uniformly distributed in the network and formation of clusters is neglected [2.163] and (b) that the negative sites are equally available sterically to both lithium ions and protons. The field-driven migration of an ion, that is, the probability of an ion to leave a certain site for a neighbouring site, will then depend on (1) the strength of its own bond to the site it occupies, (2) the site concentration, that is, the average distance between the sites, and (3) the bond strength of the surrounding ions to the sites they occupy. For a given site concentration, consequently, the following two extreme situations may be visualized.

(a) A cation strongly bound to the site it occupies and surrounded mainly by the other ion species whose bond strength to their sites is much weaker, will preferably leave its position according to its own bond to the site. This situation is obviously given by a guest proton surrounded mainly by lithium ions or, in terms of protonation, by protons at low proton contents. The strong bond of the proton to the SiO^- group, compared to that of lithium ions, is shown by the low mobility ratio of protons to lithium ions at low proton contents, $\overline{u}_H(x_H \to 0/\overline{u}_{Li}(x_{Li} \to 1)$, which is smaller than 6×10^{-3}, Fig. 2.31. (The exact proton mobility at low proton mole fractions, $x_H \to 0$, is unknown because of the formation of proton traps [2.54, 55], see Sect. 2.6.7.)

(b) The jump probability of a cation that is only weakly bound to its site but surrounded mainly by ions with a much stronger bond to their sites, in contrast, will be determined by the bond of the surrounding ions. This situation is approached by a lithium ion surrounded mostly by protons, that is, by lithium ions in a highly protonated glass. The immobilization of the lithium ions is obvious from the mobility ratio of lithium ions to protons in a highly protonated glass, $\overline{u}_{Li}(x_{Li} \to 0)/\overline{u}_H(x_H \to 1) = 1.9$, as compared to the mobility ratio of lithium ions in unprotonated, original glass to protons in highly protonated glass, $\overline{u}_{Li}(x_{Li} = 1)/\overline{u}_H(x_H = 1) = 67$, Fig. 2.31, which demonstrates the weak bond of unhindered lithium ions, compared to the strong bond of protons, to SiO$^-$ sites.

Also, the concentration dependence of the transport data is understood by this mechanism. An exchange of protons for lithium ions at high proton mole fractions is expected to have relatively little effect on the mobilities of both lithium ions and protons as observed, Fig. 2.31. The mobility of lithium ions will not change significantly as long as they are surrounded mainly by immobilizing protons, and the proton mobility remains basically unchanged because of the strong bond of the proton to the sites. On the other hand, replacement of lithium ions by protons at high lithium mole fractions means reducing the concentration of the unhindered, more mobile lithium ions of the glass, which is generally accompanied by a relatively large conductivity decrease [2.164], whereas this change will not much influence the mobility of the protons because of their strong bond to the sites. This mechanism is in agreement with the experimental results, Fig. 2.31, the only exception being the decreasing proton mobility at proton mole fractions below ~ 0.15, where the structural effect of proton traps becomes detectable and is the more obvious the smaller the proton mole fraction [2.54, 55], see Sect. 2.6.7.

Displacement of Electric Field Profiles During Protonation

Combination of the concentration profile $x(s_H)$ measured after protonation of a glass membrane, see for instance Figs. 2.16 and 2.20, and the concentration-dependent resistivity $\rho(x)$ of the glass as shown in Fig. 2.31 yields the depth-dependent resistivity $\rho(s_H)$ and, with (2.155), the depth-dependent electric field strengths $\boldsymbol{E}(s_H)$ [2.54, 56], called the profile of the electric field or, briefly, the field profile. An example is given in Sect. 2.6.6. Several field profiles measured after different migration times and thus different distances s_H of the migrating boundary from the anodic glass surface in the same glass exhibit systematic changes [2.55], which reflect the constant ratio of the electric fields, $(\boldsymbol{E}_2/\boldsymbol{E}_1) = \text{const.}$, in layer and glass during protonation, as shown Fig. 2.27. They are discussed in the following because they simplify the determination and discussion of field profiles during protonations.

As an example, Fig. 2.32 sketches two field profiles obtained after migration of the boundary by $s_{H,1}$ and $s_{H,2}$ from the anodic surface. $\boldsymbol{E}_1$ and $\boldsymbol{E}_2$ are the electric field strengths in the unchanged glass and the protonated layer, respectively. Ohm's law,

$$\log \boldsymbol{E} = \log \rho + \log \frac{i}{A} \,, \tag{2.172}$$

whose validity is a precondition of the following derivation, can thus be applied to the protonated layer and to the bulk glass in both cases. Doing this yields the result that the difference of the logarithms of the electric fields $\boldsymbol{E}_2$ and $\boldsymbol{E}_1$ is independent of the distance of the migrating boundary from the anodic glass surface and of the (time-dependent) current,

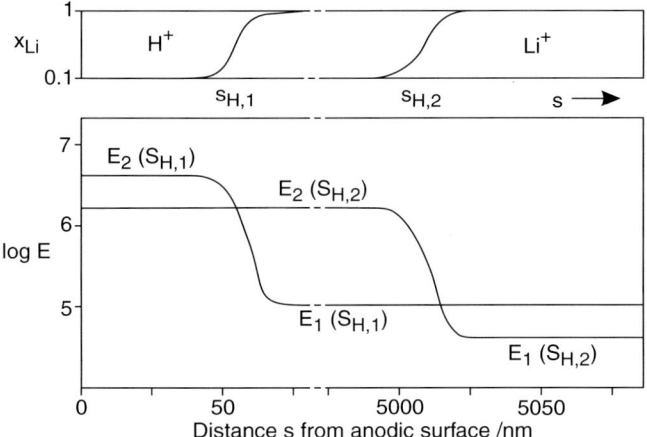

Fig. 2.32. Schematic field profiles in logarithmic presentation, $\log \boldsymbol{E} = f(s)$, within two membranes of a glass during protonation up to the distances $s_{H,1}$ and $s_{H,2}$. $\boldsymbol{E}_2$ and $\boldsymbol{E}_1$ are electric field strengths in layer and glass, respectively

$$\log \boldsymbol{E}_2(s_{H,1}) - \log \boldsymbol{E}_1(s_{H,1}) = \log \boldsymbol{E}_2(s_{H,2}) - \log \boldsymbol{E}_1(s_{H,2})$$
$$= \log \frac{\rho_2}{\rho_1} \neq f(i) \ . \tag{2.173}$$

Expressed in another way, the difference of the logarithms of the electric fields $\boldsymbol{E}_2$ in protonated layers with, for example, thicknesses $s_{H,1}$ and $s_{H,2}$, is equal to the difference of the logarithms of the electric fields $\boldsymbol{E}_1$ in the glass under these conditions,

$$\log \boldsymbol{E}_1(s_{H,1}) - \log \boldsymbol{E}_1(s_{H,2}) = \log \boldsymbol{E}_2(s_{H,1}) - \log \boldsymbol{E}_2(s_{H,2})$$
$$= \log \frac{i_1}{i_2} \neq f(\rho) \ . \tag{2.174}$$

It is obvious that (2.174) expresses the same fact as (2.173) in a different way. Both equations can be helpful when the development of an electric field profile during protonation of a glass membrane is discussed. It is emphasized, however, that the field profile *within* the migrating boundary depends on individual conditions, for instance distance s_H and glass composition, see Fig. 2.16. Because the extension of the migrating boundary is only a small part of its distance from the anodic glass surface, it is generally not of interest. If it is, its field profile must be determined individually in each case, as indicated above and described in Sect. 2.6.6.

2.6.6 The Moving Boundary Between Different Electromigrating Ions

The modified moving boundary method (m.m.b.m.) can be applied for quantitative measurements of transport data only if the distances travelled by the ions can be clearly identified. This is possible if the boundary between the different ions is stable, where a stable migrating boundary shall be understood to separate two glass regions, for instance the original, unchanged glass and an ion-exchanged layer, with depth- and time-independent concentrations and transport numbers [2.54, 56]. This definition applies to potentiostatic and amperostatic experiments. Despite the importance of boundary stability for field-driven replacement as well as diffusional moving boundaries, for which theoretical analyses are available [2.165, 166], experimental reports are scarce [2.167, 168]. The most recent publication treats the morphological instability of the field-driven interface between monocrystalline KCl and AgCl [2.169] and thus demonstrates the wide range of application of the subject. The application of boundary stability to electrolysed glasses is best discussed on the basis of the original moving boundary method (m.b.m.) as applied for measuring transport numbers in solutions. The comparison with the application to glasses will then show common points and differences of the two techniques.

The Original Moving Boundary Method in Solutions

The principle of the original moving boundary method is sketched in Fig. 2.33, see for instance [2.93]. In a vertical tube, the solutions of two uni-uni-valent salts with concentrations c_1 and c_2 having one ion (e.g., anion A^-) in common, are placed one above the other and subjected to a total voltage U_0, which causes anions and cations to migrate in opposite directions. The leading cation is the ion under investigation, the other is the so-called indicator ion, which is detectable in some way, for example by light absorption. The position of the boundary between the cations can thus be followed during its movement along the tube towards the cathode. After the boundary has been shifted by a certain distance s, for instance from cross section a–b to cross

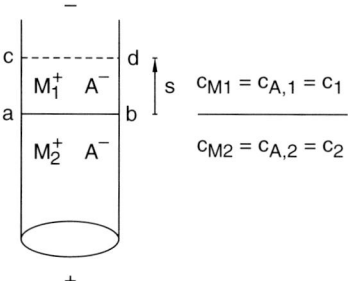

Fig. 2.33. Principle of the original moving boundary method in solution. Solutions of two uni-uni-valent salts with a common anion forming a common boundary are subjected to an electric voltage, which moves cations and anions in opposite directions. The distance s travelled by the boundary relative to the total charge transported yields the transport number of the leading cation M_1^+

section c–d, all of the leading cations M_1^+ within the volume $A\,s$ have crossed cross section c–d so that their transport number, that is the fraction of the total charge q carried by the particular ion during this period, is given by

$$t_1 = \frac{c_1 s\, AF}{q} \, .\tag{2.175}$$

The measurement of transport numbers is the main object of the original moving boundary technique.

However, certain experimental conditions must be met for a successful application of the method, and these are mainly concerned with the stability of the boundary between the migrating cations. Thus, for mechanical stability, the density of the lower solution must be larger than that of the upper one, independent of which solution contains the leading ion. For electrical stability, the mobility u_1 of the leading ion, for example of M_1^+, must be larger than the mobility u_2 of the following ion M_2^+:

$$\frac{u_1}{u_2} > 1 \, .\tag{2.176}$$

This rather simple condition has several consequences for the properties of the solutions applied, which should be known to realize the complexity and the complicated character of the seemingly simple method.

The transport numbers t_1 and t_2 can be expressed by the respective mobilities u_1 and u_2 of the leading and the indicating cation, respectively,

$$t_1 = \frac{c_1 u_1}{c_1 u_1 + c_1 u_-} = \frac{u_1}{u_1 + u_-}\tag{2.177}$$

and

$$t_2 = \frac{c_2 u_2}{c_2 u_2 + c_2 u_-} = \frac{u_2}{u_2 + u_-} \, ,\tag{2.178}$$

where u_- is the mobility of the common anion. Equations (2.177) and (2.178) yield the ratio of the transport numbers

$$\frac{t_1}{t_2} = \frac{u_1(u_2 + u_-)}{u_2(u_1 + u_-)} > 1 \, ,\tag{2.179}$$

which is larger than unity because of (2.176). The transport number of the leading cation must thus be larger than that of the following cation if the moving boundary is to be stable. Further, a stable migrating boundary means an equal migration velocity of the leading and the indicator ion and thus the transport number t_2 of the indicator ion M_2^+ must be given by an expression

$$t_2 = \frac{c_2\, s\, A\, F}{q}\tag{2.180}$$

that corresponds to (2.175). Consequently, the concentration ratio of the salts must also be larger than unity,

$$\frac{t_1}{t_2} = \frac{c_1}{c_2} > 1 ,\tag{2.181}$$

because of (2.179). The relationship, (2.181), is called Kohlrausch's regulating function. Finally, the resistances ρ_1 and ρ_2 of the solutions containing the leading and the indicator ion are given by

$$\rho_1 = \frac{1}{c_1 F (u_1 + u_-)}\tag{2.182}$$

and

$$\rho_2 = \frac{1}{c_2 F (u_2 + u_-)} ,\tag{2.183}$$

respectively, yielding the ratio of the resistances

$$\frac{\rho_1}{\rho_2} = \frac{c_2 (u_2 + u_-)}{c_1 (u_1 + u_-)} < 1 ,\tag{2.184}$$

which is smaller than unity because of (2.176) and (2.181). Because, in addition, the electric field $\boldsymbol{E}$ in the solutions is proportional to the resistance at an identical current density i, also

$$\frac{\boldsymbol{E}_1}{\boldsymbol{E}_2} = \frac{\rho_1}{\rho_2} < 1\tag{2.185}$$

must be valid, which means that a stable moving boundary is characterized by a smaller field strength in front of, and a larger electric field strength behind, the moving boundary.

The method can be modified for directly measuring the mobility u_1 of the leading ion. The information needed is the electric field strength $\boldsymbol{E}_1$ within the solution of this ion, which can be obtained by inserting two electrodes reversible to one of the ions present (or two Pt electrodes if large total voltages are applied) through the wall of the tube at a known distance and by measuring the voltage between them during the transport experiment. The mobility of the leading ion is then given by

$$u_1 = \frac{s_1}{t_1 \boldsymbol{E}_1} ,\tag{2.186}$$

where s_1 is the distance travelled by the boundary in time t_1.

If the conditions derived above are met, the moving boundary between the different ions is stable but, nevertheless, tends to lose some of its sharpness with increasing migration distance because of diffusion. Fortunately, there is a mechanism at work that counteracts the diffusional spreading of the boundary. It is caused by the large gradient of the electric field at the immediate boundary between the ions, Fig. 2.34a. If, by any mischance, one of the leading ions M_1^+ at the boundary falls back into the region of higher field strength E_2, its comparatively high mobility moves it forward again along the relatively large potential gradient. Conversely, a slower ion M_2^+ that finds itself in the region of lower field strength E_2 ahead of the boundary is slowed down due to its relatively low mobility and is thus also returned to the boundary. This so-called "self-regulating effect" keeps the boundary relatively sharp. Its effectiveness, however, depends on the gradient of the electric field at the boundary and rarely eliminates the effect of diffusion completely.

The number of conditions to be met for a successful application of the moving boundary method sometimes make it rather laborious to find appropriate salts with common anions before the actual measurements. While, however, there is almost always a combination that can favourably be applied, the modified moving boundary method as applied to glasses is restricted to few exchanged cations if quantitative measurements are to be carried out, as will be seen in the following discussion.

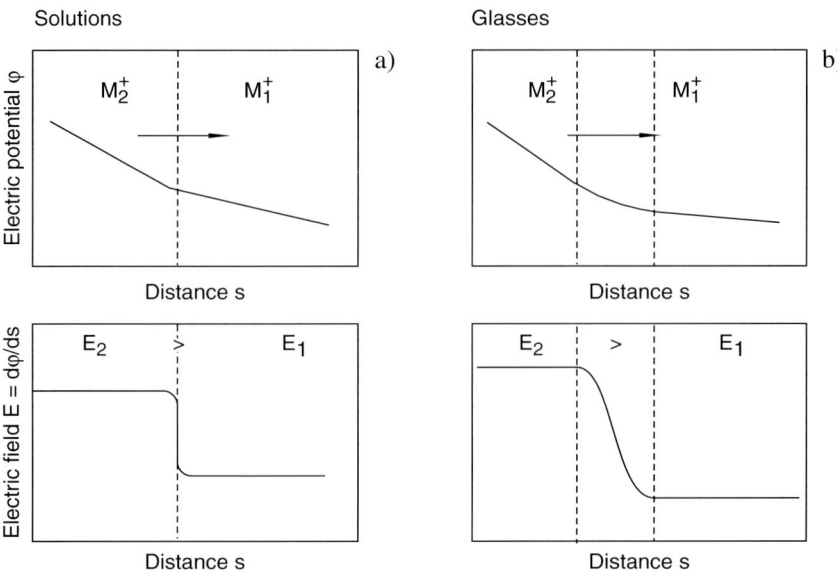

Fig. 2.34. Comparison of the original moving boundary method for solutions (**a**) and its modification as applied to glass membranes (**b**). The boundary in solutions is kept distinct by the large electric field gradient ("self-regulating effect"), whereas the boundary in glasses tends to broaden with time because of the concentration-dependent mobilities of the ions. In both media the boundary is stable if the resistivity (and electric field) gradient in the direction of migration is negative

The Modified Moving Boundary Method for Glasses

The modified moving boundary method (m.m.b.m.) for glasses is based on the same principles as the original method. However, two basic differences are (1) that only cations migrate within the glass network (at least in all cases of a quantitative application of the method) and (2) that the total concentration c^0 of the mobile ions is constant in these cases. Consequently, three different types of migrating boundary must be distinguished, Fig. 2.35.

(1) 100% of the leading cations M^+ originally present in the glass are replaced by less mobile cations N^+, which are, for instance, introduced from a solution, Fig. 2.35a, so that

$$c_{N,2} = c_{M,1} = c^0 , \tag{2.187}$$

where indices 1 and 2 denote the original glass and ion-exchanged layer, respectively. Stability of this kind of boundary is characterized by

$$\frac{\overline{u}_{N,2}}{\overline{u}_{M,1}} < 1 \tag{2.188}$$

and thus also by the condition (2.185) as in the case of the original method.

(2) Fewer than 100% of the ions M^+ in the glass are replaced by less mobile ions N^+,

$$c^0(x_{N,2} + x_{M,2}) = c_{M,1} = c^0 , \tag{2.189}$$

where x is the mole fraction of the ions indicated, see Fig. 2.35b. A stable boundary is formed by the front of the less mobile ions penetrating the glass if the condition

$$\frac{u_{\text{overall},2}}{\overline{u}_{M,1}} < 1 \tag{2.190}$$

and thus also (2.185) are met. u_{overall} is the overall mobility of the ions in layer 2 as defined by (2.145). During the electrolysis, the more mobile ions, which are also supplied by the anodic solution, migrate through the ion-exchanged layer according to their larger mobility.

(3) As sketched in Fig. 2.35c, the glass or glass layer to be subjected to electromigration is already partly ion-exchanged, for instance by electrolysis, and contains the mole fraction $x_{N,1}$ of the less mobile ion, which is transferred at a larger mole fraction during the second electrolysis. These concentration conditions are expressed by

$$(x_{N,2} + x_{M,2}) = (x_{N,1} + x_{M,1}) = 1 . \tag{2.191}$$

The stable moving boundary between the regions containing the ions at different concentrations is characterized by the condition

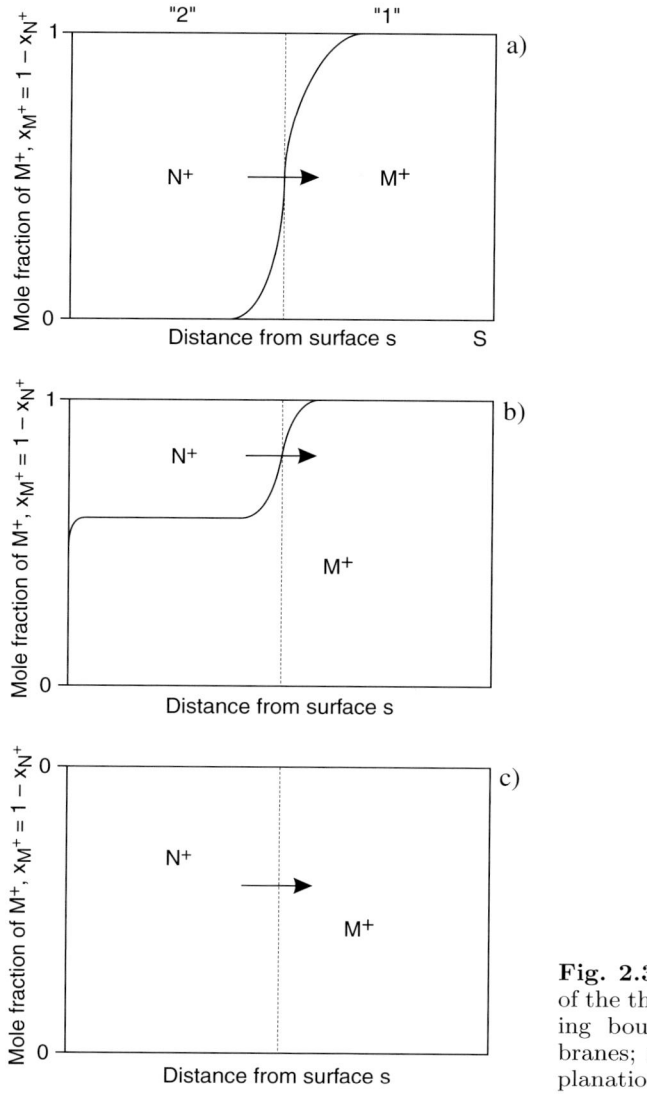

Fig. 2.35. Schematic diagrams of the three types of stable moving boundaries in glass membranes; see text for explicit explanation

$$\frac{u_{\text{overall},2}}{u_{\text{overall},1}} < 1 \, , \tag{2.192}$$

which also yields the stability condition (2.185) for the ratios of the resistivities and the electric fields in the corresponding glass ranges. The overall mobilities in (2.190) and (2.192) are functions of the mole fractions of the different ions according to Fig. 2.32, so that these equations are actually more complicated than the corresponding condition, (2.176), for stable moving boundaries in solutions.

Equations (2.188), (2.190), and (2.192) apply to the homogeneous regions of the respective glass and ion-exchanged layer. However, the concentration dependence of the ionic mobilities causes a certain distribution of the migrating ions in the boundary range and, consequently, a finite extension of the moving boundary, which is not caused by diffusion, Fig. 2.34b. Concentrations, mobilities and overall mobilities, and, consequently, resistivities and electric fields within the range of the moving boundary are thus complicated functions of distance in the direction of migration. Quite generally, however, a stable boundary region is characterized by the condition that the overall mobility of the ions must be an increasing function of distance, so that the stability condition for the entire membrane is given by

$$\frac{1}{c^0 F}\left[\frac{d\left(\frac{1}{u_{\text{overall}}}\right)}{ds}\right] = \frac{d\rho}{ds} = \frac{1}{i(s)}\left(\frac{dE}{ds}\right) < 0 , \qquad (2.193)$$

where $i(s)$ is the electrolysis time-dependent and thus migration distance-dependent current density. The validity of (2.193) was shown experimentally by means of lithium silicate and aluminosilicate glasses, as described in the following.

Figure 2.36b presents resistivity profiles $\rho(s)$ in a lithium silicate pH glass after various degrees of potentiostatic protonation. They were obtained by combining the concentration profiles $c(s)$ given in Fig. 2.36a with concentration-dependent resistivities $\rho(c)$ according to Fig. 2.31. The larger resistivities of the protonated layers relative to the resistivity of the glass

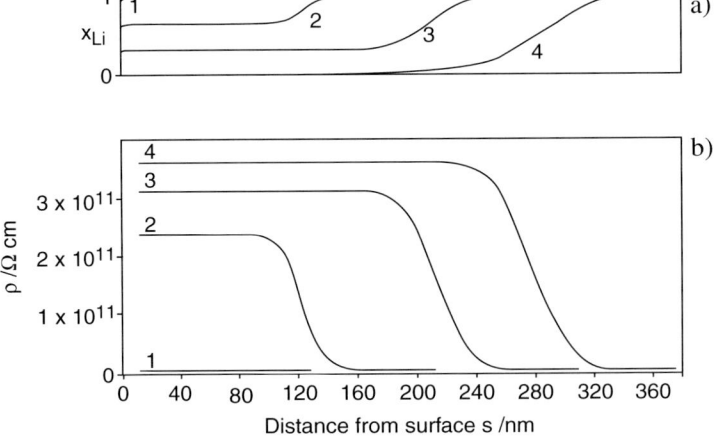

Fig. 2.36. Results obtained by applying the m.m.b.m. to a lithium silicate pH glass membrane for various degrees of protonation: 1 = 0%, 4 = 100%. (**a**) Anodic lithium concentration profiles, (**b**) resistivity profiles calculated from the concentration profiles and concentration-dependent resistances, 50 °C, see Fig. 2.31. Both plots indicate stability of the moving boundary

and the negative resistivity gradients throughout the moving boundary region clearly indicate that the boundaries are stable at any possible proton concentration.

Figure 2.37b shows profiles of the electric field at three different depths within a membrane of the same glass during 100% protonation at constant total voltage. They were obtained from the concentration profiles $x_H(s)$ given in Fig. 2.37a, which were measured in membranes of the same glass after different periods of protonation, current densities $i(s_H)$ at the end of the respective protonation periods, and concentration-dependent resistivities $\rho(x)$,

$$E(s) = i(s_H)\rho(x_H(s)) \ . \tag{2.194}$$

Although this moving boundary is also stable as expected, it becomes less sharp with increasing distance from the surface. At first thought, this broadening could be attributed to diffusion during migration. This would correspond to experimental results by *Chemla* [2.170], who showed by means of radioactive tracers for monocrystalline salts that electromigration and diffusion are independent processes. *Cooper* [2.142], who analysed the amperostatic mode of field-induced ionic migration in glasses, also concluded from these results that the spreading of the moving boundary under potentiostatic conditions, as shown in Fig. 2.37a, was caused by diffusion. However, diffusion is ruled out by the simple fact that the shape of the concentration profiles is

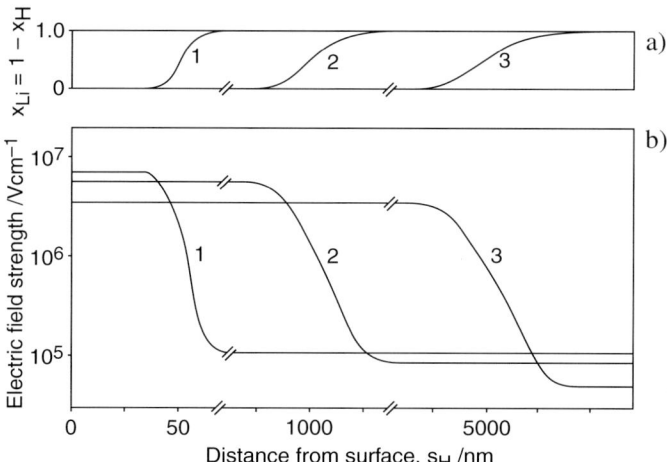

Fig. 2.37. Results obtained by applying the m.m.b.m. to a lithium silicate pH glass for the generation of 100% protonated layers with 50 nm, 1000 nm, and 5000 nm thickness. (**a**) Anodic lithium concentration profiles, (**b**) electric field profiles calculated from the concentration profiles, concentration-dependent resistances, and current densities at the end of the electrolyses. The plots indicate stability of the moving boundaries. Total voltage applied: 3500 V, membrane thickness: 0.5 mm, 50 °C

independent of the time between electrolysis and analysis by IBSCA, which was extended up to several days.

It is thus concluded that the spreading of the moving boundary under potentiostatic conditions is unavoidable since it is caused primarily by the concentration dependence of the ionic mobilities. The relative concentrations within the boundary range are coupled to each other by the locally changing electric field, and any "self-regulation" by the field gradient initiates concentration changes along the entire line of the migrating ions in the direction of migration and thus changes also the electric field profile. Obviously, the self-regulating effect cannot cope with this situation because of the small field gradients compared to those in solutions, see Fig. 2.34.

These results show also that the independent migration and diffusion observed by Chemla could be obtained only because of two favourable experimental conditions. (1) The migrating boundary indicated by a thin layer of radioactive tracers was formed of identical ions, which guaranteed zero gradient of the electric field and absence of a self-regulating effect. (2) The ionic mobilities in the monocrystalline salts obviously showed no detectable distribution and caused no inherent spread of the boundary during migration. It must thus be assumed that different conditions would have resulted in a coupling of diffusion and electromigration as in the case of glass protonation.

Figure 2.38 presents lithium concentration and resistivity profiles obtained by transferring various amounts of lithium ions into the 100% protonated layer of a lithium silicate pH glass. It corresponds to a reversal of the field after protonation but has the advantage of a discontinuous boundary between the ions at the 100% $\equiv$SiOLi-covered glass surface. Besides, protons of

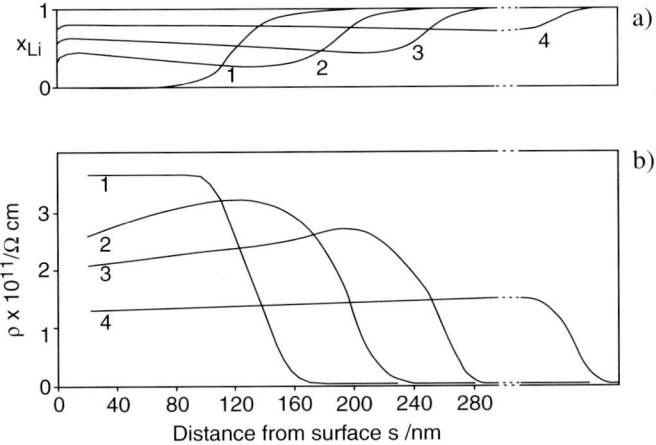

Fig. 2.38. Time-dependent unstable lithium concentration (**a**) and resistivity profiles (**b**) during field-driven anodic transfer (2, 3, 4) of more mobile lithium ions into a 100% protonated layer (1) of a lithium silicate glass. The number of protons is constant during their dilution

the layer are not lost to the solution during the migration experiments. The moving boundary is unstable as indicated by the positive resistivity gradients in the protonated range. While lithium ions penetrate the protonated layer and dilute the protons, thus decreasing resistivity and electric field in this region, the replaced protons migrate into the yet unprotonated glass and increase layer thickness and glass resistivity. As shown in Table 2.8, the overall resistance R_L of the membrane increases during this electrolysis because the resistivity decrease caused by the decreasing proton concentration is overcompensated by the thickness increase of the layer. Unstable moving boundaries have been observed after field reversal also at high temperatures [2.171, 172].

Figure 2.39 shows a different example of boundary instability. 100% sodium ions are transferred into a lithium aluminosilicate glass by an electric field. The sodium ions, which are not hindered sterically, penetrate the glass and replace and overtake the lithium ions progressively. Because the overall resistance of the membrane decreases slightly during this process, it must be concluded that the mixed-alkali glass formed exhibits no typical mixed-alkali properties. Obviously, ion-exchanged glasses generated by electrolysis are not subject to the mixed-alkali effect if the sites accommodate both ion species. This agrees with observations on protonated glasses and also with literature reports on diffusionally ion-exchanged glasses [2.157, 158, 173]. Figure 2.40, finally, shows a more complicated situation, where sodium ions penetrate a protonated layer of the lithium aluminosilicate glass mentioned, which leads to the destabilization also of the stable boundary between lithium ions and protons.

2.6.7 Proton Traps and Heat Treatment-Induced Network Changes

Protons introduced electrolytically into oxidic glasses are bound to oxides, thus forming hydroxyl groups which exhibit absorption in the infrared of the

Table 2.8. Calculated resistance R_L of a 100% protonated layer on a lithium silicate pH glass during dilution of the protons by electrolytically introduced lithium ions

q_2/q_1 [a]	$\overline{x}_H$ [b]	ρ [c] $10^{11}\,\Omega\,cm$	s [d] nm	R_L $10^6\,\Omega$
0.0	1.0	6.7	100	6.7
0.7	0.5	4.6	200	9.1
2.9	0.25	3.0	400	12.1
18.0	0.13	1.7	800	13.8

[a] Ratio of charges transferred during proton dilution, q_2, and during primary protonation, q_1

[b] Average proton mole fraction of the layer

[c] Specific resistance of the protonated layer according to Fig. 2.32

[d] Thickness of the protonated layer as obtained by IBSCA

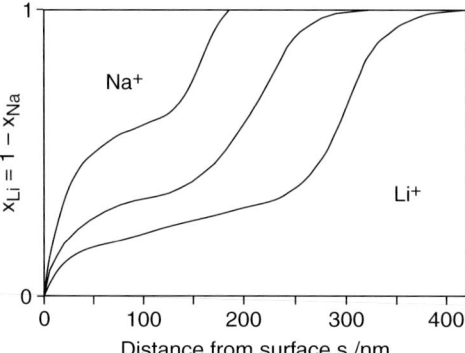

Fig. 2.39. Unstable lithium–sodium distribution during field-driven introduction of sodium ions into a lithium aluminosilicate glass membrane, 50 °C

spectrum. Because the absorption bands are relatively sensitive to the kind of bonding and to the environment of the OH groups, protons can be used to gain information about the glass structure and about structure changes, which are, for instance, caused by heat treatment. Two particularly striking examples, the formation of proton traps [2.54–56] and the rearrangement of the glass network of protonated glasses by heat treatment [2.55], are presented in this section.

Proton Traps

An obvious feature of Fig. 2.31, which cannot be explained by the proposed or any other known migration mechanism, is the strong mobility decrease of electrolytically introduced protons with decreasing proton concentration at proton mole fractions below approximately 0.15. Such layers can be pre-

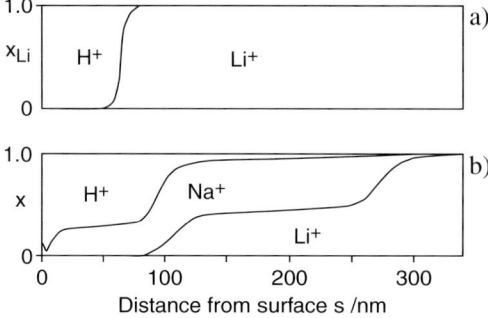

Fig. 2.40. Stable proton–lithium distribution during 100% field-driven protonation of a lithium aluminosilicate glass (**a**) and proton–sodium–lithium distribution during the introduction of sodium ions into the protonated glass causing two unstable boundaries (**b**)

pared either by direct field-driven protonation of glass membranes, see, for example, Fig. 2.36, or by diluting high proton concentrations by electrolytically introducing 100% lithium ions, see Fig. 2.38, which has the advantage of a constant total proton content of the glass membranes during protonation and dilution and a simple performance of quantitative measurements. Due to the low proton mobility at small concentrations, however, their concentration profile practically does not change after dilution to approximately 10%.

This effect is partially caused by the electric field strengths in glass and layer, which attain increasingly similar magnitudes with decreasing proton concentration, see Fig. 2.38. The main reason, however, is given by the glass structure as is obvious from infrared spectra. Figure 2.41a shows the IR spectra of a 100% protonated lithium silicate pH glass before (1) and after dilution of the protons to nearly 10% (3). Obviously, Beer's law is not valid. Compared to the spectrum of the 100% protonated membrane, that of the glass after dilution exhibits smaller absorbances between 2.8 μm and 3.5 μm

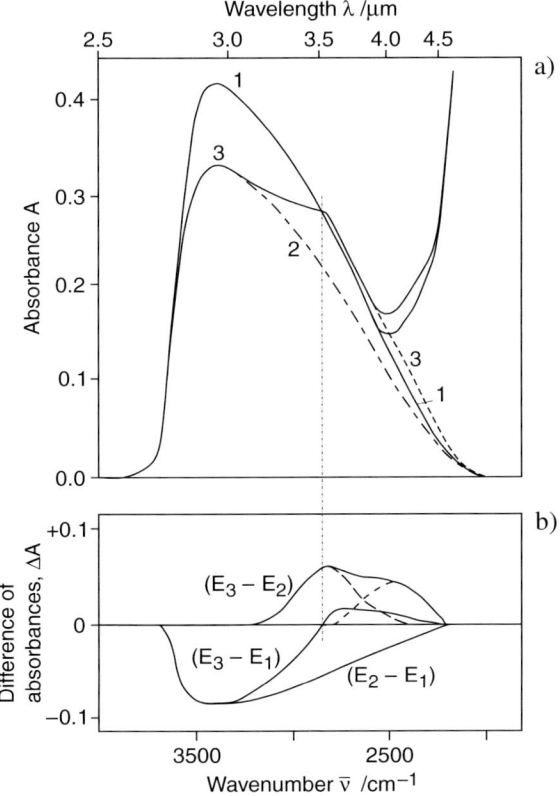

Fig. 2.41. Infrared spectra of a protonated lithium silicate glass showing the effect of proton dilution by electrolytically introduced lithium ions, see text

and larger absorbances between 3.5 μm and 4.5 μm, which is particularly obvious by the difference $(E_3 - E_1)$ of the spectra in Fig. 2.41b. The smaller absorbances cover the range of absorption bands which are attributed to single and twin silanol groups with maxima between 2.75 μm and 2.95 μm and at 2.9 μm, respectively [2.145, 146, 174]. The larger extinctions are positioned in the range of the infrared bands characteristic of hydrogen bridges, whose maxima are located between 3.35 μm and 3.85 μm and at 4.25 μm if the hydrogen bridge involves an SiO_4 tetrahedron, respectively [2.145, 146]. Low proton concentrations characterized by spectrum (3) thus indicate high relative concentrations of hydrogen bridges also called "mixed twin siloxy groups" and symbolized by the formula $[\equiv SiOH \cdots OSi \equiv]^- Li^+$. This is particularly obvious from the difference $(E_3 - E_2)$ between spectrum E_3 after proton dilution and spectrum E_2 of the 100% protonated glass with correspondingly lower total proton content, whose absorbance at 2.95 μm is equal to that of spectrum E_3 at the same wavelength.

In terms of ion migration, these results are explained by the hold up probability of the proton at the different sites. The scheme in Fig. 2.42 shows arbitrary relative bond energies of lithium ions and protons to single and twin siloxy groups together with the respective ionic exchange processes and wavelengths of the maxima of the IR absorption bands. Protonation of a single siloxy group (lithium–proton exchange step 1,1′) and first and second protonation of a twin siloxy group (exchange steps 3,3′ and 4,4′) take place under

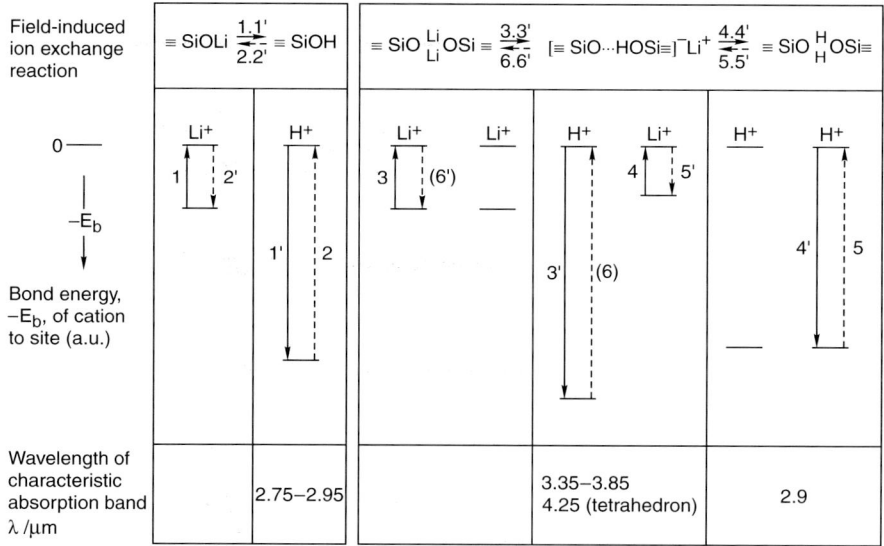

Fig. 2.42. Scheme showing arbitrary relative bond energies $-E_b$ of lithium ions and protons to single and twin siloxy groups, corresponding exchange reactions, and wavelengths of the characteristic absorption bands

comparable energy loss and must be expected with nearly equal probability. These exchanges characterize the protonation of lithium silicate glasses under formation of stable migrating boundaries as discussed in Sect. 2.6.6. Equally, re-exchanging the proton of a single silanol group (exchange step 2,2′) and one of the protons of a twin silanol group for a lithium ion (exchange step 5,5′), for which the energy is supplied by the electric field, is not expected to be hindered in any way and should also be equally probable although this cannot be observed during ion migration because of the instability of the moving boundary due to the violation of the stability condition. Deprotonation of a mixed twin silanol group (exchange step 6,6′), however, although comparably possible energetically, will be relatively improbable because the lithium ion present in or at this group is much less strongly bound to this site than the proton. The lithium ion of the mixed twin silanol group is thus much more likely to exchange for a different migrating lithium ion than is the more strongly bound proton.

Consequently, protons of mixed twin silanol groups must be expected to have larger hold up probabilities and smaller individual mobilities than protons of single and twin silanol groups,

$$u_H(\text{mixed}) < u_H(\text{single}), \, u_H(\text{twin}) \, . \tag{2.195}$$

Mixed twin silanol groups thus act as proton traps. Their relative concentration $x_H(\text{mixed})/x_H(\text{total})$ increases with decreasing total proton concentration

$$\frac{d\left(\frac{x_H(\text{mixed})}{x_H(\text{total})}\right)}{dx_H(\text{total})} < 0 \, , \tag{2.196}$$

and this results in a decreasing average proton mobility with decreasing total proton mole fraction at small proton concentrations,

$$\left(\frac{d\bar{u}_H}{dx_H}\right)_{x_H \to 0} > 0 \, , \tag{2.197}$$

as observed. For this reason, the plot $\bar{u}_H(x_H)$ in Fig. 2.31 is not extrapolated at proton concentrations below approximately $x_H = 0.05$. The average proton mobility, which approaches the individual mobility when the concentration approaches zero, may even assume extremely small values in this concentration range.

The effect of mixed twin silanol groups was observed with all lithium silicate glasses investigated that contained more than about $20 \, \text{mol}\%$ Li_2O. It was not found, however, with lithium aluminosilicate glasses containing a concentration ratio of lithium oxide to aluminum oxide of approximately unity, probably because of the large concentration of $[\equiv SiOAl \equiv]^-$ sites, which are present in such glasses [2.175] and which exclude the generation of observable concentrations of twin silanol groups able to form mixed twin silanol groups or proton traps during deprotonation.

Heat Treatment-Induced Network Changes

Drastic changes of the infrared spectrum are observed when protonated glasses are heat-treated. As an example, Fig. 2.43 presents the IR spectra of a lithium silicate pH glass membrane before (1) and after 100% protonation (5). When this membrane is heated to 300 °C, the spectrum changes to spectrum (4) within a relatively short time, which depends on the thickness of the protonated layer. Thereafter spectrum (4) is constant, even during prolonged treatment at 300 °C. However, if the protonated membrane before or after heating to 300 °C is subjected to the higher temperature 490 °C, the IR spectrum changes to spectrum (3), which is nearly identical to that observed with the original glass containing dissolved water that was introduced into the melt, spectrum (2). Further protonation of the heat-treated glass results in spectra similar to spectrum (5) found after primary protonation.

The change from spectrum (5) to spectrum (4) at 300 °C indicates the loss of most or all of the twin silanol groups. Their equilibrium with molecular water,

$$\equiv SiOH \cdots HOSi \equiv \ \rightleftarrows \ \equiv SiOSi \equiv + H_2O , \tag{2.198}$$

is obviously shifted to the right (water) side with increasing temperature, so that water molecules will leave the glass by diffusion. Surprisingly, formation and diffusion of water seem to be possible without major changes in the glass

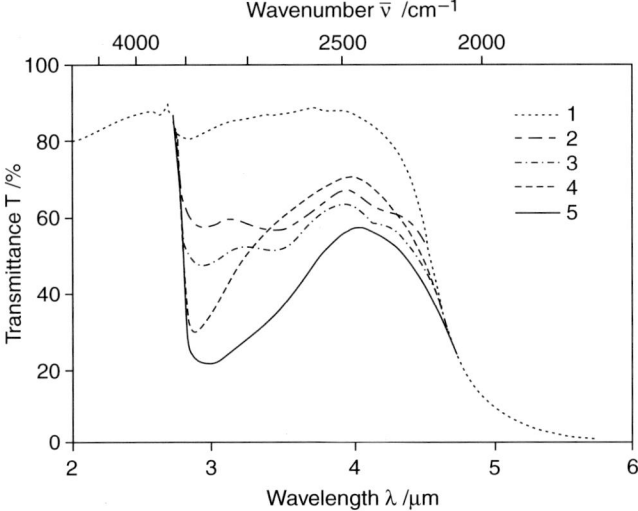

Fig. 2.43. Changes of infrared spectrum of 100% protonated lithium silicate glass caused by heat treatment (1) before, (5) after protonation, (4) after heating protonated glass to 300 °C for various periods, (3) after heating protonated and protonated and 300 °C-treated protonated glass to 490 °C; (2) spectrum of the original glass containing dissolved water

network. The single silanol groups are obviously unaffected at $300\,°C$ as is concluded from the OH band with maximum at $2.95\,\mu m$, which shows strong tailing, spectrum (c). The absorptivity of this band remains constant after attaining its final value. However, temperatures near the transformation temperature of the glass, which is $508\,°C$, obviously rearrange the glass network and cause an equilibrium distribution of the remaining silanol groups that is equal or at least similar to that of water-containing glass. Interdiffusion of lithium ions and protons between the protonated layers and the bulk glass is obviously negligible during these experiments because the character of the spectra practically does not depend on the thickness of the layers. According to IBSCA measurements, the protonation of heat-treated glass membranes results in the formation of normally and 100% protonated glass *below* the modified protonated layers and thus in infrared spectra similar to spectrum (b) if the time of the protonation is sufficiently extended.

2.6.8 A Different Situation: Electrolysis of Borate Glasses

The electrolyses treated so far are characterized by the replacement of the original cations of the glass by different (or the same) cations, which are supplied by anodic non-blocking contacts, that is, by electrolyte solutions or by anodic metals or hydrogen, which are oxidized and release their ions into the glass network. Strictly speaking, however, these experiments are not electrolyses of the glasses in the electrochemical sense, as the glass network, which remains unchanged during the process, merely serves as the matrix for the transfer and migration of the travelling cations. They are therefore more appropriately called transfer or migration experiments.

However, there are also observed actual electrolyses in the strict sense of the word, where the cations travelling towards the cathode are not replaced by other cations but where the charge balance is met by a change, for instance a decomposition, in the glass network and a corresponding discharge of appropriate anions at the anode. Examples are alkali borate glasses, which, under certain conditions, show a cathodic discharge of alkali ions as with silicate glasses and an equivalent anodic discharge of oxide ions. The conditions of this "real" electrolysis are: (1) There must be a mechanism available by which the anions are liberated at the rear boundary of the cations leaving for the cathodic surface and by which they are transported through the anodic, electrolysed glass region, and (2) the electric field within this anodic region must be sufficiently large to facilitate the transport. In other words, competing processes that need smaller driving forces than the anion transport, for instance ionic migration, must be excluded. Such "real" electrolyses of alkali borate glasses were investigated by means of IBSCA and discussed on the basis of optical basicity as obtained from UV spectra of the glass specimens before, during and after the electrolyses [2.119, 176, 177]. The $0.5\,mm$ thick, plane-parallel glass samples were coated with the electrode metals (Pt, Ag, Pb) by vapour deposition or were merely contacted by the metals if this was appropriate (Hg, Tl(Hg)). The experiments were carried out in extremely dry, inert atmospheres because of the hygroscopicity of the borate glasses.

Figure 2.44a shows the sodium concentration profile below the surface of a blank specimen of a sodium borate glass with 35 mol% Na_2O as measured by IBSCA before (dashed line) and after mechanical removal of a thin layer of sodium carbonate (solid line) that had formed during storage, handling, and polishing. Figure 2.44b presents sodium and lead concentration profiles at and below the anodic glass surface after the electrolysis of a specimen in the cell

$$(+)Pb \mid glass \mid Hg(-) , \qquad\qquad (2.XIV)$$

during which the charge of 0.049 C, corresponding to a charge density of $0.0245\,C\,cm^{-2}$, was transported through the sample, and after subsequent removal of the remaining lead by dissolution in mercury [2.119, 177]. As in

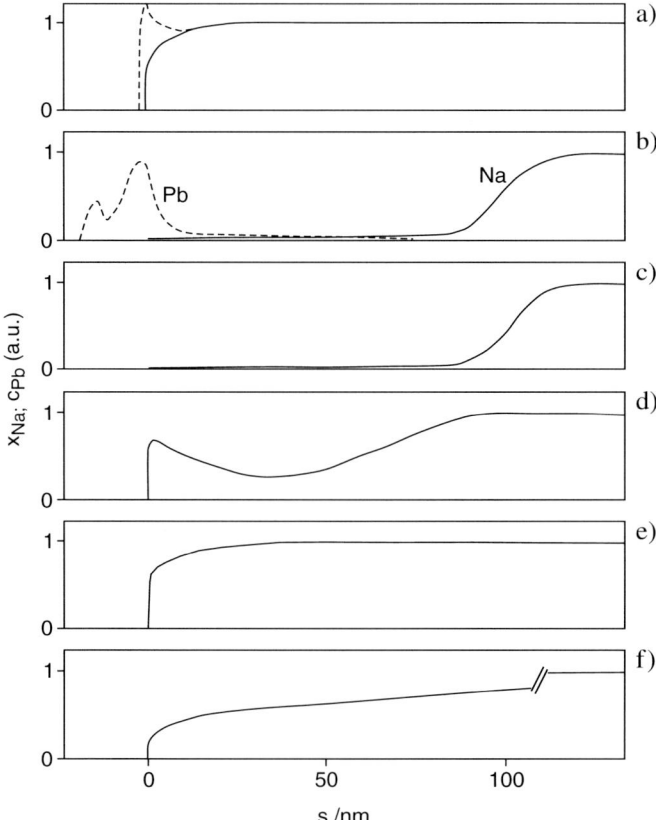

Fig. 2.44. Sodium and lead concentration profiles measured by IBSCA below the surface of a sodium borate glass after electrolyses under various different conditions, see text

the case of 100% protonation of lithium silicate glasses (see Fig. 2.16), the sodium ions of the borate glass drifted towards the cathodic surface under formation of a sharp, stable rear boundary. However, they were neither replaced by protons, as shown by infrared spectroscopy, nor by an equivalent amount of lead ions, as demonstrated by the concentration profile of lead in Fig. 2.44b and by UV spectrum (1) in Fig. 2.45. This spectrum rather indicates anodic formation of lead oxide, as is demonstrated by comparing with the spectrum of a 40 nm thick PbO layer deposited on a borate glass sample, spectrum (2) of Fig. 2.45. Spectrum (3) of Fig. 2.45, which was measured after the removal also of the PbO layer by ion ablation, is typical of lead ions in borate glasses and indicates that, indeed, very small amounts of Pb^{2+} ions have migrated into the glass. The sodium concentration profile below the cathodic glass surface equals that of the blank specimen (solid line in Fig. 2.44a), demonstrating a simple discharge of sodium ions and dissolution of the sodium atoms formed in the cathodic mercury. Apart from the anodic replacement of the small amount of the sodium by lead ions, the electrolytic cell reaction thus consists mainly of the removal of nearly the total Na_2O content of the glass from the region below the anodic surface, the formation of anodic lead oxide and a cathodic discharge of sodium ions

$$Pb + Na_2O(B_2O_3)_n + (Hg) \rightarrow PbO + nB_2O_3 + 2Na(Hg) \ . \qquad (2.199)$$

The electrolysis involves a decomposition of the glass network and is thus basically different from the field-driven anodic cation replacement as observed with silicate glasses, during which the glass network merely serves as the matrix and remains unchanged, even when blocking electrodes are applied (although the possibility of an electrolytic oxygen displacement cannot basically be excluded under certain conditions [2.178], particularly at elevated temperatures).

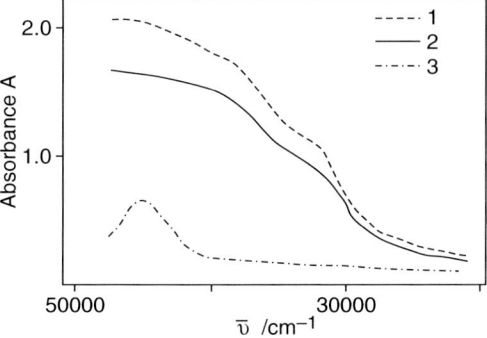

Fig. 2.45. UV spectra of sodium borate glass. (1) After electrolysis with lead anode, see Fig. 2.44b; (2) sodium borate glass with 40 nm thick vapour-deposited PbO layer; (3) after electrolysis and removal of PbO layer by ion sputtering

A similar sodium concentration profile is observed when the sodium borate glass is electrolysed by means of a mercury cathode and an inert metal anode, for instance platinum, in inert atmosphere,

$$(+)\text{Pt} \mid \text{glass} \mid \text{Hg} \ (-) \ , \tag{2.XV}$$

see Fig. 2.44c [2.177]. Again, the sodium ions are neither replaced by protons, as shown by infrared spectra, nor by other cations because of the absence of corresponding oxidizable elements. Thus, the deficit in sodium below the anode must have been balanced by the removal of an equivalent amount of oxide, and the electrolytic cell reaction must be formulated by

$$\text{Na}_2\text{O}(\text{B}_2\text{O}_3)_n + (\text{Hg}) \rightarrow 1/2\text{O}_2 + n\text{B}_2\text{O}_3 + 2\text{Na}(\text{Hg}) \ , \tag{2.200}$$

similar to the reaction involving a lead anode, (2.199). The modified glass layer below the platinum anode has thus lost essentially the entire sodium content and approximately $17.7\,\text{mol}\%$ of its oxide.

The remaining B_2O_3 layer shows neither macroscopic nor microscopic cracks. Its structure must be closely related to, or must even be the same as, that of the original glass because a prolonged reverse electrolysis, during which the platinum electrode is used as an oxygen-flooded cathode, restores the original sodium concentration profile, Fig. 2.44d and 2.44e. (The surplus of Na_2O formed at the platinum cathode due to the large charge transported during the prolonged electrolysis had been removed, together with the platinum electrode before the analysis by IBSCA). The structure of the anodic Na_2O-free glass region has not been determined but would be of high interest because of a possible application of such, perhaps modified, layers as reversible oxide- and alkali ion-storing materials in active layer systems, for example, in electrochromic devices [2.179].

The sodium ions travelling towards the cathodic surface during the electrolysis of the sodium borate glass leave behind oxide whose charge is primarily unbalanced, and the question arises how these oxides are transported through the nearly Na_2O-depleted glass layer to the anode where they are discharged. Individual oxide ions, O^{2-}, do certainly neither exist in the glass nor are they transported as such through the layer. Therefore, a mechanism has been proposed, which is based on the ability of the glass former boron to exist in two different coordination states, that is, in threefold and fourfold coordination. This process, which is called "structure switching (oxide) anion transport mechanism" [2.119, 177], can be described in the following way. After withdrawal of two sodium ions, the remaining oxide is primarily bound to two four-coordinated boron atoms according to the formal stoichiometric equation

$$[\text{O}(\text{BO}_2)\text{O}(\text{BO}_2)\text{O}]\,\text{Na}_2 \rightarrow [\text{O}(\text{BO}_2)\text{O}(\text{BO}_2)\text{O}]^{2-} + 2\text{Na}^+ \ , \tag{2.201}$$

which, however, does not present any structural details as, for instance, BO_4 units are rarely, if at all, linked to each other. One of the four-coordinated borons may then change its coordination number to three by "handing over", or shifting, an oxide to a nearby three-coordinated boron under the action of the electric field, which, by this change, becomes four-coordinated and can itself hand one of its oxides over to the next three-coordinated boron, and so forth,

$$BO_3 + [O(BO_2)O(BO_2)O]^{2-} \rightarrow [O(BO_2)O(BO_2)O]^{2-} + BO_3 \ . \quad (2.202)$$

By extending this basic process, it is seen how the oxides are shifted from the rear side of the leaving sodium ions, where they are liberated, through the B_2O_3 glass layer to the anodic glass surface, where they are discharged. All that is required is the continuous interchange between threefold and fourfold coordination of the boron atoms of the layer with an appropriate rotation or other slight oscillation of the borate units.

A necessary condition for the appearance of the structure-switching oxide transport is the absence of competing processes that need smaller driving forces than the shift of the oxides through the electrolysed layer. For instance, lead acts as a blocking anode because the twofold positively charged lead ions, except for small concentrations, are obviously hindered sterically and by their twofold positive charge to penetrate the glass and thus enable the structure-switching oxide transport. Inert metal electrodes such as platinum in dry inert atmosphere do not offer any competing ionic processes either, and even a proton-injecting anode, Pt,H_2, does not exclude the shift of oxides, obviously, because protons are little mobile also in borate glasses. In this case, however, the oxides, on their arrival at the anode, react with the protons to form water, which combines with the generated B_2O_3 to form boric acid,

$$6H^+ + 3O^{2-} + B_2O_3 \rightarrow 2H_3BO_3 \ , \quad\quad\quad\quad (2.203)$$

which is a bad glass former that crystallizes. This explains also why, unlike silicate glasses, the investigated borate glass cannot be protonated.

Electrodes supplying ions with relatively large mobilities, in contrast, prevent structure-switching oxide transport. Sputtered-on silver anodes, for example, form silver ions, whose ionic radius is smaller (67 pm) than that of the replaced sodium ions (99 pm) and which replace up to 100% of the sodium ions of the glass under formation of an unstable moving boundary that extends progressively into the glass during the electrolysis [2.117]. Thallium amalgam anodes supply thallous ions [2.176], which also form an unstable migrating boundary with the sodium ions of the glass, Fig. 2.44f, and obviously have mobilities comparable to those of sodium ions despite their larger ionic radius (151 pm). The unit transport number of sodium ions in unchanged borate glass, which follows from the concentration profile in Fig. 2.44a (solid line) under an oxygen-free mercury *cathode*, can also be viewed under this aspect. Obviously, the mobility of the sodium ions is sufficiently large to exclude structural changes so that the borate glass network is maintained and serves as the matrix for the migrating sodium ions.

The UV spectra of the p-block metal ions Pb^{2+} and Tl^+ contained in the electrolysed glass allow one to calculate the optical basicity, that is, the electron donor power of the oxides of the sites occupied by these ions [2.180–182] (for details, see Sect. 3.3). Electrolysis experiments as carried out here are of particular interest for these investigations because the electrolytically introduced thallous ions and traces of lead ions are in sites of sodium ions which have drifted away towards the cathode, while they probably occupy other sites if they are introduced by the glass melting process. Surprisingly, both thallous and lead ions monitor larger optical basicities than would be expected from the average composition of the surrounding anodic glass matrix. Thus, thallous ions indicate an optical basicity of 0.66–0.67, corresponding to an Na_2O content of the surrounding of 60 mol% [2.176, 177], although the electrolysed glass contains less than 35 mol% Na_2O, which would correspond to an optical basicity of 0.53. Similarly, the lead ions in the electrolysed glass layer with less than 3 mol% Na_2O, corresponding to an average optical basicity of 0.44, monitor an actual optical basicity of 0.49 and thus an Na_2O content of the layer of 18 mol% [2.119, 177]. These results can be understood on the basis of a recent paper, which reports that the sodium ions of sodium borate glasses reside in two locally different sites, one with a larger and the other with a smaller value than the average optical basicity [2.183]. Both lead and thallous ions thus obviously replace sodium ions in sites with the larger optical basicity, that is, with 0.49 and 0.65, respectively, which eliminates the necessity to assume modifications of the sites by the probe ions [2.177].

2.7 Interfacial Equilibria Under the Influence of Subsurface Concentration Gradients

The phase boundary equilibrium at the glass/solution interface as described in Sect. 2.5 involves ions of the solution, functional groups at the glass surface to which ions from the solution are attached, and a minute concentration of dissociated, negatively charged surface groups, which give the glass a negative potential and are the basis of the glass electrode response. The glass surface thus has what may be called an electrochemical structure [2.184] that depends on the solution composition and differs considerably with respect to structure and composition from the bulk glass. In this way, the glass surface represents a two-dimensional phase between glass and solution, whose atomic thickness, in addition to the large exchange current density of the interfacial equilibrium, ensures short response times and a high stability of the potential of the glass [2.72, 87]. Indeed, response times of glass electrodes of the order of tens of milliseconds have been reported [2.66, 76, 77, 185]. With respect to the glass/solution interface, a glass membrane is an ideal sensing element of glass electrodes.

However, the glass surface is nearly never in equilibrium with the underlying glass because either the ionic species attached to the surface groups or their concentration differ from that of the alkali ions of the bulk glass [2.72, 87]. The resulting concentration gradients cause an interdiffusion of the ions in subsurface regions of the glass, which, being a solid-state diffusion, is a rather slow process. The rate of the ions leaving the interfacial equilibrium for the glass interior, if expressed as a current density, is much smaller than the equilibrium exchange current density of the interfacial equilibrium, so that the phase boundary equilibrium is practically undisturbed and supplies a constant concentration ratio of ions to the subsurface glass region as long as the contacting solution is not changed [2.71, 72, 87]. The glass surface groups act as terminals for the penetrating ions in the same way as they do during a field-driven transfer of ions (see Sect. 2.6). The rate of the interdiffusion is practically undisturbed by the charge present at the glass surface, as was estimated and can also be concluded from a theoretical study of double layers [2.186].

The driving force of the interdiffusion is the time-dependent concentration gradient of the ions below the glass surface and not simply the free energy difference of the ions in the solution and the glass, as was assumed when the glass surface was still believed to be an inert interface between the different phases. For example, for the irreversible interdiffusion of protons from silanol surface groups versus lithium ions of a lithium silicate glass according to

$$\equiv SiOH(s) + \ \equiv SiOLi(gl) \rightarrow \equiv SiO^-(s) + \ \equiv SiOH(gl) + Li^+(sol'n) \ , \tag{2.204}$$

whose electrochemical free energy is negative, $\Delta \overline{G}_{diff} < 0$, the protons are supplied by the $\equiv$SiOH groups at the surface. The generated lithium siloxys dissociate immediately because the siloxy group takes part in the interfacial dissociation equilibrium of silanol, which is thus practically not disturbed,

$$\equiv SiOH(s) + H_2O \rightleftarrows \equiv SiO^-(s) + H_3O^+ \ , \tag{2.205}$$

and whose electrochemical free energy remains zero during the interdiffusion, $\Delta \overline{G}_{eq'm} = 0$. The change of the electrochemical free energy $\Delta \overline{G}_{total}$ of the overall reaction

$$\equiv SiOLi(gl) + H_3O^+(sol'n) \rightarrow \equiv SiOH(gl) + H_2O(sol'n) + Li^+(sol'n) \ , \tag{2.206}$$

in which surface groups do not appear, is thus equal to the electrochemical free energy of the interdiffusion,

$$\Delta \overline{G}_{total} = (\Delta \overline{G}_{diff} - \Delta \overline{G}_{eq'm}) = \Delta \overline{G}_{diff} < 0 \ . \tag{2.207}$$

Consequently, neither the overall reaction (2.206) nor its energy change indicate that the exchange of lithium ions of the glass versus protons from the solution occurs by two distinct reactions, the attachment of the proton to the surface siloxy group according to equilibrium (2.205) and its subsequent exchange for subsurface lithium ions, which are immediately released into the solution, (2.204). Statements such as: "The surface SiOH groups are not involved in this equilibrium exchange between the bulk of the glass and the solution" and: "The surface groups are at equilibrium with both solution and bulk glass" [2.148], can only result from not knowing that the mechanism described here has been shown experimentally by field-driven anodic transfer of alkali ions and protons within the transition range of the interfacial equilibrium, see Sect. 2.6.3. The treatment of the ion exchange omitting the interfacial dissociation equilibrium according to [2.148], however, cannot explain that reaction (2.206) takes place also if the solution contains large concentrations of lithium ions [2.152] because they should shift (2.206) to the left side or even reverse it. Contrary to expectation, reaction (2.206) proceeds to the right side even at activity ratios as high as $(a_{Li^+}/a_{H_3O^+}) = 10^{10}$, for instance, when the glass is a pH silicate glass [2.97, 152]. A corresponding observation was reported by *Scholze* for a sodium calcium silicate glass and neutral solutions with high concentrations of potassium ions [2.187].

It is surprising how close researchers came to a correct understanding of the surface processes. The most obvious example is a paper by *Schwabe* and *Dahms*, who already in 1960 concluded from their radioactive tracer experiments that "we must assume that the Na:H ratio, which corresponds to the exchange equilibrium, forms immediately only at the interface swollen surface layer/solution after the glass has been immersed in the solution" [2.9]. They thus "felt" that there was an equilibrium at the glass/solution interface which was distinguished from the interdiffusion below the glass surface by a different rate. However, misled by the then generally accepted, highly influential ion exchange concept into assuming ion exchange to be the cause of the electrode potential, they failed to draw the correct, rather obvious conclusion.

The coverage of the glass surface, which is given by the solution composition, determines the ion species which enter the glass and interdiffuse with its alkali ions. Three cases must be distinguished [2.87, 188].

- At 100% pH selectivity of the glass, where the glass surface is covered with the acidic form of the surface groups, for instance with silanol (SiOH), protons interdiffuse versus the alkalis of the glass.
- At pure alkali (pM) selectivity, where the glass surface bears only the salt form of the surface groups, for example [AlOSi]M, alkali ions from these groups interdiffuse against the alkali ions of the glass.
- In the transition range with mixed pH and pM sensitivity, where the glass surface contains both the acidic and the salt form of the surface groups, both protons and alkali ions interdiffuse versus the alkalis of the glass.

The penetrating and interdiffusing ion species are not determined by the surface groups *per se*, but by the equilibrium form of the surface groups present and thus primarily the solution composition. For instance, the entity [AlOSi], which is typical of alkali-selective glasses, offers alkali ions to the glass only if its salt form [AlOSi]M covers the glass surface, that is, at pH values pH $\geq$ (pH$_{tr}$ + 2). At pH $\leq$ (pH$_{tr}$ − 2), in contrast, the acidic form, [AlOSi]H, is present at the glass surface and supplies protons to the glass interior, and in the transition range at approximately (pH$_{tr}$ − 2) $\leq$ pH $\leq$ (pH$_{tr}$ + 2), both protons and alkali ions interdiffuse with the alkali ions of the glass. Because the different penetrating ions trigger different subsequent reactions, for instance network hydrolysis, the glass surface coverage also determines the kind of corrosion to which glasses are subjected. Examples are given below.

A further consequence of subsurface ionic interdiffusion is the formation of diffusion potentials [2.189, 190]. They are significant because they are included in the emf of glass electrode cells although their major part is generally cancelled by the diffusion potential below the opposite glass membrane surface, which usually has a similar magnitude and an opposite sign. Fortunately, subsurface diffusion potentials are often constant in the pH or pM range of practical interest. In certain cases, however, special treatment and storing of glass electrodes are required to secure constant diffusion potentials and thus meaningful results. Application of a special technique allowed the measurement of diffusion potentials of pH glass electrodes within the entire pH range, including the sodium error range [2.190]. Also, some potential drifts of pH-sensitive and alkali-sensitive glass electrodes were traced back to diffusion processes [2.87, 96]. These findings will be described in this section within the appropriate context.

As with field-driven ion transfer, the relative sterical conditions of ions and glasses are significant [2.55, 121, 191] in that they can drastically change the "normal" transfer of ions into, as well as their interdiffusion within, glasses. Indeed, sterical hindrance has often obscured these processes and thus hindered the understanding of glass electrode processes in the past. Examples will also be reported in this section.

2.7.1 Proton/Alkali Ion Interdiffusion: Glass Leaching

We now treat the interdiffusion of protons from the acidic form of glass surface groups versus alkali ions of the glass, which is confined to 100% proton coverage of the glass and thus to solutions with pH $\leq$ (pH$_{tr}$ − 2) (−2 corresponding to 99% proton coverage is arbitrarily chosen as the lower limit of the transition range but can be easily changed to different numbers and coverages) [2.86, 87, 192]. Because of the release of alkali ions of the glass and their replacement by penetrating protons, which trigger an additional uptake of more or less water, this process is called leaching and the resulting layers leached layers [2.193]. These expressions are to be preferred to other frequently used terms, for instance swollen surface layer or gel layer [2.97], because the properties, particularly the mechanical strength, of leached layers are glass-like [2.86] and thus unlike those of gels. Words such as swollen and gel evoke completely wrong connotations, and as it is both a surprising and a regrettable fact that, in glass electrode research, words with unclear, faulty or double meanings have considerably delayed the understanding of the physical phenomena at the glass membrane, such expressions should be strictly avoided in the future. However, experience shows that attempts at replacing old terminology by a more correct one take a long time to succeed.

Glass leaching has been investigated almost exclusively in the past (see for instance [2.194]). This is not surprising because most glasses of practical interest are silicate glasses with large transition (pH_{tr}) values, which cause surface coverage with silanol groups even at rather large pH and alkali ion activities and thus in the majority of practically relevant solutions. Indeed, because of the rare occurrence of glass/solution interaction at $pH \geq (pH_{tr}+2)$ resulting in alkali/alkali ion interdiffusion, glass leaching is generally believed to be the only kind of glass corrosion, and *Eisenman's* [2.10] and others' extensive measurements of alkali/alkali exchange between electrode glasses and solutions, for instance by means of radioactive tracers [2.195–200], have gone unnoticed by corrosion researchers or have been ignored as being a particular reaction in the "special field" of electrode membrane glasses.

In the following, we demonstrate the formation and structure of leached surface layers by means of a practical example, and then show how the conditions of the solution determine the structure of leached layers.

Formation and "Structure" of Leached Layers

Leached Layer Formation. The formation of a leached layer below the surface of a new multi-component lithium silicate pH glass membrane is shown in Fig. 2.46 [2.86]. The leaching solution, 0.1 molal sulphuric acid, excluded the presence of any ions bound to the glass surface groups except protons. Figure 2.46a presents time-dependent lithium concentration profiles, and in Fig. 2.46b the lithium loss per unit surface area as a function of the square root of leaching time is plotted. The primarily large change rate of the concentration profile at first decreases at a rather large rate, and the linear dependence of the lithium loss on the square root of time during this period indicates diffusion control. The loss rate, however, soon becomes smaller and finally approaches zero: the final concentration profile (g) developed in approximately one day no longer changes with time. Infrared spectra showed that the lithium ions lost to the solution are replaced by hydrogen ions, and additional NRA measurements, resulting in the subsurface hydrogen concentration profile shown in Fig. 2.47, yield a one-for-one replacement of lithium ions by protons in the internal part of the leached layer (at approximately $x_{Li} \geq 0.3$) and an additional hydrogen content in the outer part. Despite diffusion control, the lithium concentration profiles, including profile (g), exhibit

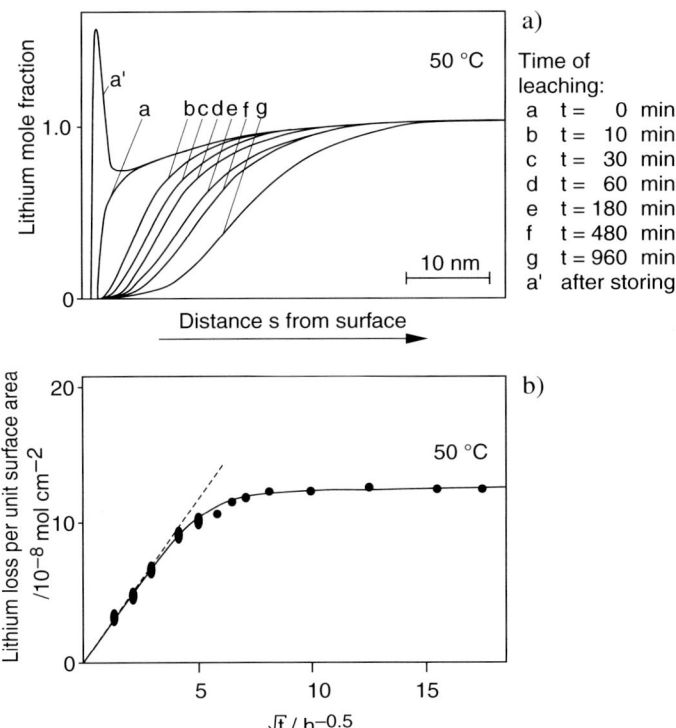

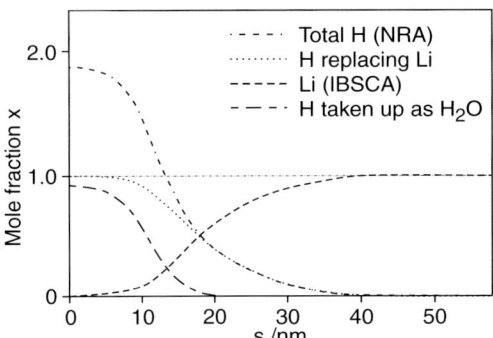

Fig. 2.46. Development of a leached layer below the surface of a multi-component lithium silicate pH glass in 0.05 molar sulphuric acid. (**a**) Time-dependent lithium concentration profiles measured by IBSCA. (**b**) Time dependence of lithium deficit of the layer

Fig. 2.47. Concentration profiles in steady-state leached layer (g) of Fig. 2.46a. Lithium as measured by IBSCA, total hydrogen taken up as measured by NRA, hydrogen replacing lithium 1:1, and hydrogen taken up as water obtained as the difference of the former profiles

a sigmoid shape, which indicates a type of interdiffusion that is different from that typically observed in a uniform material.

These experimental results explain the formation of leached layers in some detail. The primary process is an interdiffusion of protons from the acidic form of the surface groups versus lithium ions of the glass. The invading protons obviously trigger or catalyse an uptake of small amounts of water, which is restricted to the outer layer part with approximately $x_H \leq 0.3$. However, it is improbable that the water enters the glass as hydronium ions, that is, bonded to the entering protons, (a) because of steric conditions (the ionic radii of the hydronium ion compares to that of the potassium ion, which is hindered sterically to enter lithium silicate glasses, for instance in electric fields [2.54, 152, 191], see Sect. 2.4.6), (b) because the existence of silanol (SiOH) groups with oxygen-bonded protons at silicate glass surfaces is well known [2.201], and (c) because protons and not hydronium ions enter lithium silicate glasses also during field-driven migration (see Sect. 2.6.2). The uptake of water, however, seems to be connected in some way to the penetration of protons into silicate glasses because it is not observed when alkali ions penetrate aluminosilicate glasses during alkali/alkali ion interdiffusion [2.188], see Sect. 2.7.3.

The penetrating water hydrolyses the glass network and "prepares" its dissolution. At the beginning of the leaching process, network dissolution takes place with an increasing rate simultaneously with lithium/proton interdiffusion with a decreasing rate, which is indicated by the decreasing gradient of the lithium concentration profile. After some time, which is about one day in the example of Fig. 2.46, the dissolution rate v_d and the interdiffusion or layer formation rate v_D become equal, and the leached layer assumes a steady state, profile (g) of Fig. 2.46a. This type of glass corrosion, which is characterized by a release of glass components from different depths, is called incongruent dissolution in contrast to congruent dissolution, where the unchanged glass composition is dissolved layer by layer. A steady state of the subsurface glass range was first suggested by *Schwabe* and *Dahms* [2.9] and concluded from chemically measured sodium concentration profiles of a leached sodium calcium silicate glass by *Boksay* [2.74], who also treated his experimental results mathematically [2.75, 202]. The large initial exchange rate of lithium ions and protons during the generation of leached layers explains the so-called formation period of pH glass electrodes, during which the sensors exhibit considerable potential drifts [2.87]. During this period, the ionic current densities through the glass surface are comparable to the exchange current densities of the interfacial equilibrium, which thus causes the polarization of the potential [2.196].

Leached Layer "Structure". The term "structure" as used in the following does not mean that the depth-dependent atomic structure of the leached layer including that of the glass network has been evaluated. "Structure" is restricted here to qualitative changes in the subsurface range, for instance by hydrolysis and with respect to ion mobilities, as far as they can be concluded from concentration profiles and electrochemical measurements, for example from ionic transfer experiments and glass electrolyses as reported in Sect. 2.6.3. "Structure" thus just means "construction" in the sense of the German word "Aufbau". Nevertheless, even this restricted knowledge is quite valuable because it explains the mode of formation and the properties of leached layers and allows significant conclusions, for instance with respect to reactions of membrane glasses and, in some cases, also of other glasses.

Due to the depth-dependent hydrolysis of the glass network, the interdiffusion coefficient of lithium ions and protons is not only given by its concentration dependence, see Fig. 2.31, but depends also on the depth-dependent state of the glass network. This explains the sigmoid shape of the lithium concentration profile as well as the resistivity profile perpendicular to the membrane surface, which was evaluated by combining profile and concentration-dependent resistivities [2.86]. Surprisingly at first sight, the local resistivity within the leached layer range $x_{Li} \cong 0.5$ to $x_{Li} \cong 0.8$ is about 300 times the resistivity of the bulk glass, and *Wikby*, by applying a special pulse technique [2.78], reported even a factor of thousand for a lithium silicate glass with a similar composition [2.203]. The cause is the overlapping of two effects. On the one hand, the proton mobility in the unchanged glass network is smaller than the lithium ion mobility and is thus rate-determining in the entire concentration range [2.54]. On the other hand, the mobility of both ions increases with increasing degree of network hydrolysis. Thus, starting out in the bulk glass, the local resistivity increases in the inner range of the leached layer, which has a nearly unchanged glass structure, and then decreases in the outer leached layer region with increasing degree of hydration, see Fig. 2.47. The resistivity maximum does thus not indicate a mixed-alkali effect between lithium ions and lithium-replacing protons, as could be concluded, but is caused by the depth dependence of both proton mobility and degree of network hydration. This agrees with mobility measurements as described in Sect. 2.6.5 and also with results reported for alkali/silver [2.157, 173] and alkali/alkali ion-exchanged glasses [2.158]. The high resistivity of leached layers is a significant property because it protects lithium silicate pH glasses from forming much thicker layers and from more serious corrosion.

The glass shown in Figs. 2.46 and 2.47 can be viewed as a model glass for demonstrating the different processes of leached layer formation. This is particularly obvious from the sodium calcium silicate glass used by *Boksay* when he, for the first time, applied chemical concentration profiling by means of fractional glass dissolution in hydrofluoric acid and subsequent analysis of the fractions [2.74, 75, 202]. Figure 2.48a shows that this glass exhibits much less balanced processes in that initial interdiffusion is considerably faster than dissolution so that much thicker leached layers are formed. Even after 145 h, the sodium concentration step moves further into the bulk glass with a nearly unchanged shape, and it must be expected that the thickness of the steady-state leached layer will be in the $100\,\mu\mathrm{m}$ range or even larger if the layer does not separate and peel off before it reaches the steady state (see also Sect. 2.7.5). Similar results were reported by *Scholze*, who confirmed the basic findings reported here also for sodium-containing silicate glasses and particularly stated that the leaching rate is controlled by proton diffusion and not by sodium or hydronium ion diffusion [2.97]. In contrast, potassium-containing glasses investigated by *Boksay* [2.74] and by *Scholze* [2.97] exhibit dissolution rates in the applied solutions that are larger than the initial inter-diffusion rate and do not attain a leached layer, see Fig. 2.48b. These glasses thus dissolve nearly congruently under the prevailing conditions.

Leached Layers Under Various Conditions

The "thickness" of a steady-state leached layer depends on the prevailing leaching conditions. *Boksay* has given an equation for the time-dependence of the steady state [2.75, 202], which was further developed by introducing concentration-dependent diffusion coefficients by *Doremus* [2.136]. For the present treatment, however, a simple semiquantitative expression will suffice to explain the features of such layers and their changes due to changing conditions. Figure 2.49 sketches a steady-state leached layer after two different leaching times. The expression is based on the condition of the steady state, which is equal dissolution and formation rate of the layer. The formation rate is expressed by the lithium loss rate

$$-\left(\frac{dn_{\mathrm{Li}}}{dt}\right)_D = A\overline{D}_{\mathrm{eff}}(T)\left(\frac{dc_{\mathrm{Li}}}{ds}\right)_{D,\mathrm{eff}} \qquad (2.208)$$

of the glass, where A is the glass surface area, $\overline{D}_{\mathrm{eff}}(T)$ is the temperature-dependent effective interdiffusion coefficient, and $(dc_{\mathrm{Li}}/ds)_{D,\mathrm{eff}}$ is the effective or maximum concentration gradient, and the dissolution rate is given by the loss rate or counter-diffusion rate of protons

$$-\left(\frac{dn_{\mathrm{H}}}{dt}\right)_d = A\,c^0\left(\frac{ds}{dt}\right)_d. \qquad (2.209)$$

Equality of these rates then yields

$$\left(\frac{dx_{\mathrm{Li}}}{ds}\right)_{D,\mathrm{eff}} = -\frac{1}{\overline{D}_{\mathrm{eff}}(T)}\left(\frac{ds}{dt}\right)_d, \qquad (2.210)$$

which presents the relative conditions of the steady state of a leached layer if the lithium concentration is expressed by its mole fraction, $x_{\mathrm{Li}} = c_{\mathrm{Li}}/c^0$, where c^0 is the total concentration of cationic sites in the glass.

It is essential for the discussion of (2.210) that the formation rate of the leached layer, which is mainly given by the temperature-dependent in-terdiffusion coefficient of lithium ions and protons, can be controlled only

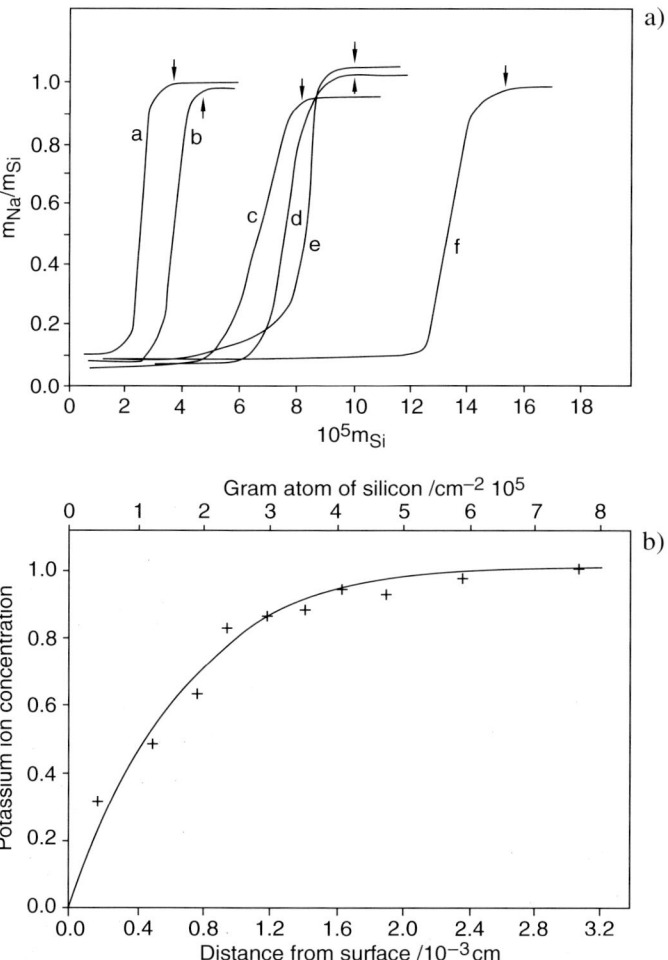

Fig. 2.48. Alkali concentration profiles obtained by leaching alkali strontium silicate glasses in water according to *Boksay* [2.74]. (**a**) Alkali = sodium, leaching times: (a) 12, (b) 20, (c) 51, (d) and (e) 72, (f) 145 hours. (**b**) Alkali = potassium (time unknown), 40 °C. Arrows mark the depth of most probable layer separation

by the temperature, whereas the dissolution rate depends also on solution parameters, for instance pH, glass network-dissolving compounds such as hydrofluoric acid, and hydrodynamic conditions [2.192]. An exception to this rule is leaching in isotopic solutions, for instance in heavy water (D_2O), where both the dissolution rate and the layer formation rate are determined by the isotopic solvents and ions, respectively. Because both effects are expected to have the same sign (slower dissolution of the layer in D_2O than in H_2O and a smaller interdiffusion coefficient of D^+/Li^+ than of H^+/Li^+), the isotopic effect of leaching is not expected to be detectable, as reported [2.204]. Nev-

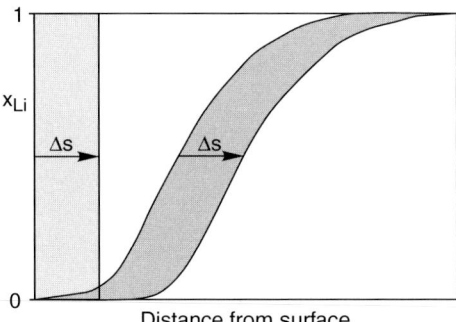

Fig. 2.49. Schematic of the time-dependent shift Δs of the glass surface and the steady-state leached layer shown in Figs. 2.46 and 2.47

ertheless, careful experiments yielded slightly steeper concentration profiles in the transition range of the steady-state leached layer in deuterium oxide than in ordinary water solutions so that a slightly different magnitude of the isotopic effect of the two processes can be concluded [2.205]. A further condition for the development of a steady state is that the dissolution rate is smaller than the initial layer formation rate.

These rules are demonstrated by comparing Figs. 2.50 and 2.51. They present steady-state lithium concentration profiles in the lithium silicate pH glass already shown in Fig. 2.46, which were formed in light water solutions at constant temperature (50 °C). Figure 2.50 demonstrates that steady-state leached layers of this glass are generated within the entire pH range if the solution does not contain additional glass-corroding compounds. (The layer formed at the surface in NaOH was a porous deposit which could be removed mechanically or by dissolution in dilute acids). The profiles in this figure show also that the dissolution rate is larger at larger pH, where the concen-

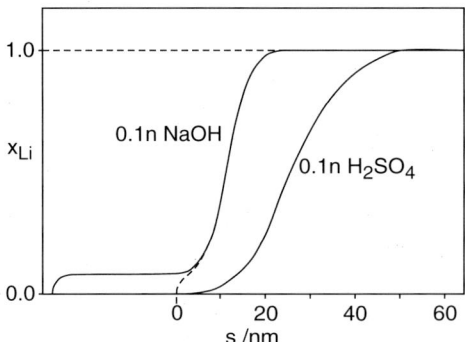

Fig. 2.50. Lithium concentration profiles in steady-state leached layers of a multicomponent lithium silicate pH glass generated in 0.1 molar NaOH and in 0.05 molar H_2SO_4 at 50 °C

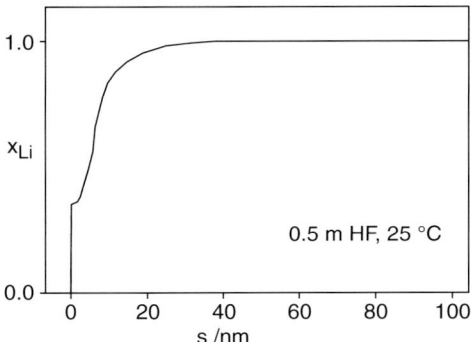

Fig. 2.51. Steady-state lithium concentration profile in the subsurface layer of a lithium silicate glass formed in 0.5 molar HF solution, 25 °C

tration profile in the layer formation region must be steeper in order to cause a formation rate that meets the dissolution rate. As a result, the steady-state leached layer at the higher pH extends less into the glass (is "thinner") than that at the low pH. In Fig. 2.51 this trend is intensified by increasing the dissolution rate nearly beyond the initial formation rate by adding hydrofluoric acid to the corroding solution. Merely a small remainder of a layer is observed while most of the glass is dissolved congruently.

Figure 2.52 shows lithium concentration profiles of two leached layers which were formed in a quiet and in a stirred, unbuffered, carbon dioxide-free, neutral solution. Contrary to expectation, the quiet solution causes a slightly thinner leached layer than the stirred solution. The reason is that the glass dissolution causes a small increase of pH at the immediate glass surfaces which is larger in the stagnant, unstirred solution than in the thin adhering layer of the stirred solution. A corresponding pH difference of $\Delta pH = 2–3$ in front of the electrodes was measured after several days.

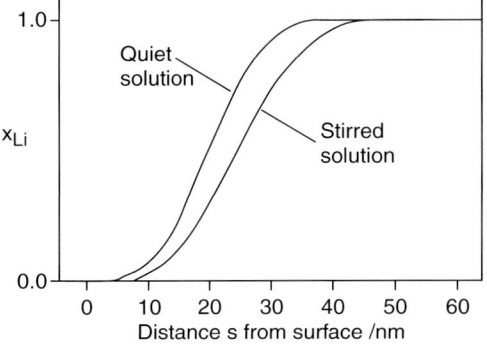

Fig. 2.52. Lithium concentration profiles of a lithium silicate pH glass generated in unbuffered neutral solution at different hydrodynamic conditions, 50 °C

Steady-state leached layers formed in dilute sulphuric acid (and thus at nearly equal pH) at four different temperatures are shown in Fig. 2.53. The layer thickness increases with increasing temperature, which indicates a stronger temperature dependence of the formation rate (and interdiffusion coefficient) than of the dissolution rate. Temperature changes, however, yielded a surprising result. As expected, the new steady-state concentration profiles formed within periods comparable to the primary formation period when the temperature had been increased, whereas nearly no change in the profiles could be detected after temperature decreases, even after periods as long as one month. The simple explanation of this unexpected effect is that the thick layers formed at the high temperature must be reduced in thickness by dissolution at the small dissolution rate given by the lower temperature, which takes a considerably longer time than layer formation at any temperature. In addition, the formation of the layer continues while the surplus layer thickness is being removed at the low temperature, which additionally increases the dissolution period. The "quasi-irreversible" glass state at the lower temperature can last for years after an appropriate temperature decrease [2.192].

Because all glasses must be subject to this effect, this observation is of general significance. Thus, glasses after contact with solutions at high temperature exhibit corrosion at low temperatures which differs from that of the original, untreated glasses. Sensitive glass applications, consequently, make the knowledge of the individual glass history mandatory in order to guarantee reproducible product quality. In glass coating, for instance, only a constant surface state (and history) of the substrate glass can secure reproducible layer systems [2.206]. Standardized glass testing procedures may give information about the chemical durability, which is different for samples that had been in contact with solutions at high temperature. Time-saving high-temperature tests developed for economic reasons can thus never completely replace real-time experiments and are often meaningless.

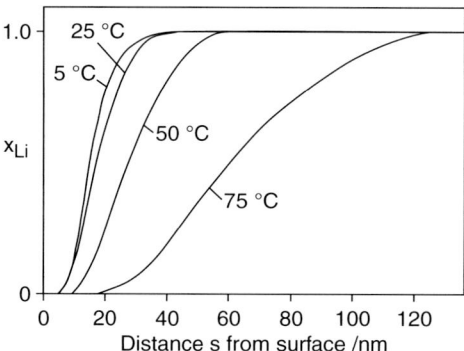

Fig. 2.53. Subsurface steady-state lithium concentration profiles of a lithium silicate pH glass formed in 0.05 molar H_2SO_4 at various temperatures

2.7.2 Interaction of Glasses with Humid Atmospheres

It is not always realized in practice that glasses react also with water vapour, for instance in humid atmospheres. However, the mechanism of the glass/water vapour reaction differs basically from that of the glass/liquid water interaction.

(a) There is no interfacial equilibrium to control the surface coverage and the ion species transferred into the glass.

(b) The glass is not dissolved, and protons from the surface interdiffuse versus the alkali ions of the glass in an unhydrated, mostly unchanged, uniform glass network, thus yielding "normal" concentration profiles.

(c) The alkali ions of the glass arriving at the surface are not removed by dissolution but react with gaseous or adsorbed compounds and form reaction products, which deposit either as bulky or crystalline materials or porous layers or as compact and uniform, non-porous layers.

(d) In the case of compact layers, the reaction rate is determined either by the ionic interdiffusion rate in the glass or by the diffusion rate of the ions or the reacting compounds (or both) in the layer.

(e) In the latter case, concentration profiles in the glass are expected to have a time-independent, rather discontinuous shape similar to those of the original glass and to differ from those as expected under (b). The most interesting case of this rather large number of possibilities was observed with a lithium silicate pH glass. The main results are briefly presented in the following paragraph. A thorough description of the investigation is given in [2.207].

The lithium concentration profiles in Fig. 2.54 demonstrate the growth of a lithium carbonate layer during contact of the lithium silicate glass with a

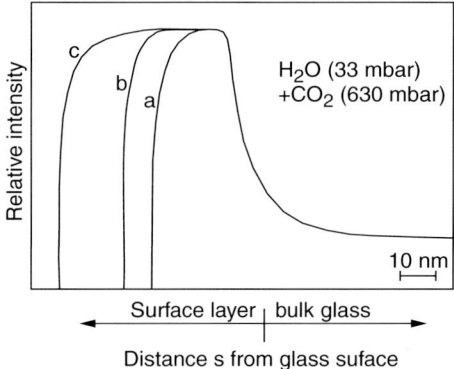

Fig. 2.54. Formation of lithium carbonate layers at the surface of a lithium silicate glass in contact with humid carbon dioxide at 120 °C. Contact periods: (a) 48 h, (b) 66 h, (c) 96 h

mixture of water vapour and carbon dioxide at $120\,°C$ (where lithium bicarbonate is not a stable compound). The approximate position of the glass/layer interface and the rather similar depth scales of both phases are also indicated. Figure 2.55 presents lithium concentration profiles which were measured in the glass after removing the lithium carbonate layers that had formed at $120\,°C$ and $35\,°C$. It is most interesting that the different mechanisms mentioned above under (b) and (e) are indeed observed with this one glass at both temperatures. At $35\,°C$, time-dependent "normal" concentration profiles are found (profiles b, c, d) and indicate rate control by interdiffusion in the glass, whereas the time-independent concentration profile observed at $120\,°C$ (profile a) demonstrates control of the reaction rate by diffusion of either lithium ions or carbon dioxide in the surface layer. In addition, the profiles demonstrate compactness of the layers at both temperatures, which was additionally verified by electron microscopy [2.207].

The example demonstrates impressively that slightly varying conditions can have considerable consequences for the reaction of a glass and can even change the mechanism and the resulting glass product. Newly developed surface processes must therefore be investigated extremely thoroughly before their technical application if undesired results are to be excluded. As shown by the example, temperature differences, which are not extreme in glass production and are thus probably applied without further inquiries by production staff to accelerate an industrial process for economic reasons, can result in a glass with completely undesired properties. In addition, the failure may not easily be detected in an early production state and can go unnoticed, at worst, until the product is applied by the customer. The example has also some practical significance because it demonstrates that sulphur dioxide, which is often used together with water vapour and oxygen to dealkalize

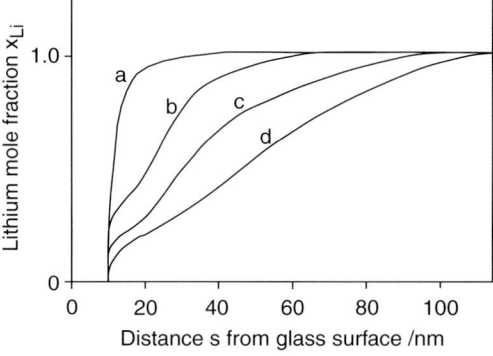

Fig. 2.55. Subsurface lithium concentration profiles generated during reaction of a lithium silicate glass with humid carbon dioxide and measured after removal of the lithium carbonate layers formed. Reaction conditions: (a) $120\,°C$ after $120\,h$, and $35\,°C$ after (b) $120\,h$, (c) $144\,h$, (d) $216\,h$

glass surfaces for better chemical durability [2.207, 208], can be replaced, in principle, by a non-poisonous mixture of water and carbon dioxide.

2.7.3 Alkali/Alkali Ion Interdiffusion

If the interfacial equilibrium causes a coverage of the glass surface with the salt form of the glass surface groups, which is the case at pH $\geq$ (pH$_{tr}$ + 2), and if the alkali ions at the surface are not identical with those of the glass, the concentration gradients below the interface result in an interdiffusion of the different alkalis. A further condition is that the ions are not hindered sterically, i.e., by their ionic radius, to be accommodated by the glass network. With silicate glasses, alkali/alkali ion exchange is observed only at extreme solution compositions because of the large selectivity products of these glasses, see Table 2.2 in Sect. 2.5.4. For this reason, silicate glasses are not applied as alkali-sensitive membrane glasses. Aluminosilicate glasses exhibit alkali/alkali ion exchange if (pH − pM) $\geq$ 2, i.e., in much more "normal" solutions, because their selectivity product is much larger. These conditions were indirectly noticed by *Quittner* during his glass electrolyses as early as in 1928 [2.153], and they also controlled *Eisenman's* ion exchange experiments using radioactive tracers [2.10], as a re-evaluation of his data on the basis of the dissociation mechanism has shown [2.188].

Figure 2.56 presents lithium and sodium concentration profiles, as obtained by IBSCA, below the surface of a lithium aluminosilicate glass membrane after three different periods of lithium/sodium ion exchange using a sodium-containing solution with pH $\geq$ (pH$_{tr}$ + 2) [2.188]. The exchange is purely diffusion-controlled for at least 360 h, as demonstrated by the linear dependence of both lithium loss and sodium uptake on the square root of time in Fig. 2.57. Protons do not take part in the interdiffusion, neither is water taken up by the glass because (a) the respective amounts of exchanged sodium and lithium ions are equal during the entire exchange period, Fig. 2.57, and (b) infrared spectra and NRA confirmed the absence of hydrogen in the glass after long periods of diffusional lithium/sodium ion exchange (35 days, 50 °C) and field-driven lithium/sodium replacement (64 h, 50 kV cm^{-1}, 50 °C). Different from a proton/alkali ion exchange of silicate glasses, the dissolution of the glass is not "prepared" by an uptake of water and a connected hydration, but the glass is dissolved congruently. The dissolution rate is 0.04 nm h^{-1} (25 °C), as determined by the shift of the glass surface, Fig. 2.56, which was obtained from the change of the sum (Li + Na) of the lithium (Li) and the sodium (Na) concentration profiles. The rate of interdiffusion defined as [2.10]

$$v_1 = \frac{n_{\text{Li}}^2}{A^2\, t} = \frac{n_{\text{Na}}^2}{A^2\, t} \tag{2.211}$$

(where n is the number of moles of the ions indicated, A is the surface area, and t is time) is of the order of 10^{-20} mol^2 cm^{-4} s^{-1} (25 °C), is not too differ-

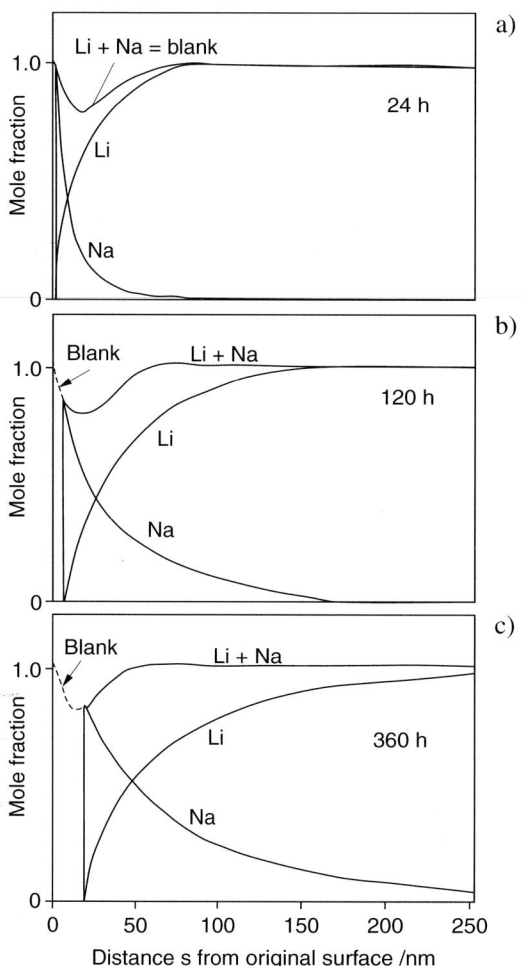

Fig. 2.56. Time dependence of 1:1 sodium/lithium ion exchange of a lithium aluminosilicate pNa glass in sodium-containing solution with pH $\geq (pH_{tr} + 2)$, 25 °C

ent from that resulting from Eisenman's experiments $(10^{-17}\,\mathrm{mol^2\,cm^{-4}\,s^{-1}})$ [2.188], particularly if the different glass compositions are taken into account.

The absence of hydrogen in the glass after long contacts with solutions causing 100% sodium coverage of the glass surface and after long-time electrolyses with such solutions as anodic contacts allows an interesting conclusion. The glass contained an atomic ratio of Al:Si:Li = 3.0:7.0:2.7. This gross composition, however, yields no information about the structural entities formed by silicon and aluminium and thus about the relative amounts of SiO and [AlOSi] groups at the glass surface. In addition, it is primarily unknown whether or not bridging oxygens, SiOSi, which are expected to be

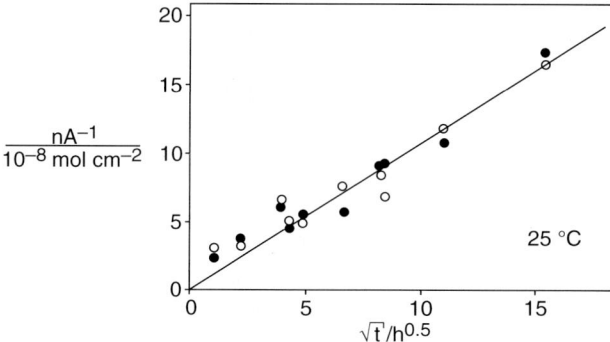

Fig. 2.57. Lithium loss (open circles) and sodium uptake (closed circles) of a lithium aluminosilicate pNa glass in sodium-containing solution with pH $\geq$ (pH$_{tr}$+ 2) as a function of the square root of time, 25 °C

present, undergo hydrolysis and thus form surface silanol groups, SiOH, upon contact with solutions. However, the confirmed absence of hydrogen in the glass after the experiments means that SiOH had neither been present originally at the glass surface nor had they formed from surface SiOSi groups by hydrolysis because the sodium-containing solutions applied were characterized by pH $\leq$ (pH$_{tr}$ (SiO)-2), which would have left silanol groups, if present, unchanged. (pH$_{tr}$(SiO) is the transition pH of SiO groups or silicate glasses.) This, in turn, would have led to the detection of protons by surface analyses after the transport processes, which was not the case. Because, in addition, it can be assumed that the concentrations of SiO and [AlOSi] entities are the same in the glass and at the glass surface, it is further concluded that the bulk glass practically contains [AlOSi] units and no appreciable amounts of non-bridging oxygens. The corresponding equilibria, for instance

$$\equiv SiO^- + \equiv Al \rightleftarrows [\equiv SiOAl \equiv]^- \ , \tag{2.212}$$

in the glass melt, from which the glass was formed, must thus be far on the right, that is, on the $[\equiv SiOAl \equiv]^-$ side. This fact seems to explain why only sodium ions and no protons enter the glass by interdiffusion and migration and also verifies that they do so exclusively via [AlOSi]Na surface groups in equilibrium with sodium ions in the contacting solution.

2.7.4 Diffusion Potentials in Subsurface Glass Layers

Interdiffusion of different ions in any medium, and thus also in the subsurface range of glasses as discussed above, causes an interdiffusion voltage or briefly (and less accurately) a diffusion potential owing to the different diffusion coefficients (or mobilities) of the ions. Because in an electrochemical cell involving a glass electrode the subsurface diffusion potentials are arranged in series with the other galvanic voltages (or potentials), they are part of the emf measured. Fortunately, there is a diffusion potential below either of the opposite surfaces of the electrode membrane, which cancel each other to a large extent because of their opposite sign. However, the diffusion potential at the reference side of the membrane is constant, whereas the one below the measuring glass surface, in principle, can be subject to changes in the interfacial equilibrium and thus in the measuring solutions. The investigation of subsurface diffusion potentials in membrane glasses and their dependence on the interfacial equilibria is thus of high theoretical as well as practical interest.

For this reason, a special method was developed that allows the measurement of diffusion potentials of pH glass membranes at pure pH selectivity and in the sodium error range [2.87, 96]. In the following, we first report on this method and then discuss diffusion potentials in alkali-selective glasses. With specific examples we demonstrate how pM glass electrodes must be treated in order to secure reproducible measurements and also show what can happen if fundamental knowledge is ignored (or not obtained) during the development of pH electrode glasses.

Determination of Diffusion Potentials in Leached Layers

Because of the large difference of the mobilities of lithium ions and lithium ion-replacing protons ("guest protons") in lithium silicate glasses, see Fig. 2.31 of Sect. 2.6.5, rather large diffusion potentials must be expected in leached layers of such pH glass membranes. Figure 2.58a shows the lithium and the hydrogen concentration profiles in the steady-state leached layer of a lithium silicate pH glass as also presented in Fig. 2.47. For the determination of the diffusion potential, the leached layer can roughly be divided into two parts. The inner part (transition range) consists mainly of the original glass network which is unchanged by hydration. It extends from approximately $x_{Li} \cong 0.3$ to $x_{Li} = 1$ and shows a large change in the lithium (and replacing hydrogen) concentration. The diffusion potential in this range is denoted by $\varepsilon_{j,tr}$. In contrast, the network of the outer part of the leached layer is changed by hydration (or hydrolysis), extends from the glass surface to $x_{Li} \cong 0.3$, and exhibits much smaller lithium concentration gradients. The diffusion potential in this layer region is termed $\varepsilon_{j,L}$. The total diffusion potential of the leached layer is thus equal to the sum, $\varepsilon_j = \varepsilon_{j,tr} + \varepsilon_{j,L}$.

Outer Part of the Leached Layer. Inner and outer leached layer ranges can be experimentally separated, as a first approximation, by field-driven protonation, which displaces the unchanged transition range and the corresponding diffusion potential $\varepsilon_{j,tr}$ into a deeper glass region, whose distance from the glass surface can be chosen by the electrolysis conditions (electric field, electrolysis time, and temperature). This is shown by the concentration profiles measured after the protonation of a pH glass membrane in Fig. 2.58b. The outer part of the leached layer, however, is changed by the protonation: the

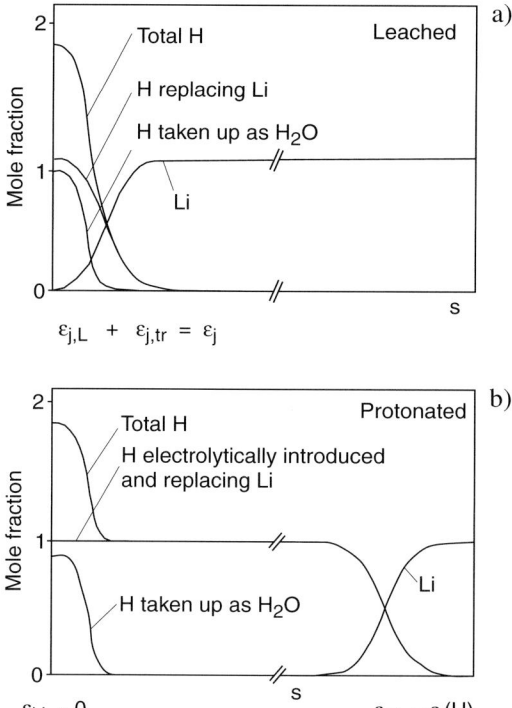

Fig. 2.58. Determination of the diffusion potential $\varepsilon_{j,L}$ in the outer, hydrated part of a leached layer of a lithium silicate pH glass by field-driven protonation. The diffusion potential $\varepsilon_{j,tr}$ of the transition layer is maintained, whereas $\varepsilon_{j,L}$ is zero in the protonated glass (lower plot)

interdiffusion of lithium ions and protons is eliminated so that the diffusion potential of the outer layer range of the protonated membrane is zero, $\varepsilon_{j,L} = 0$. The difference of the emfs of a cell involving a glass electrode in the original state, E, and in the (subsequently) protonated state, E_H, consequently, yields the diffusion potential in the outer leached layer range,

$$E - E_H = \varepsilon_{j,L} , \tag{2.213}$$

and the pH functions of the original and the protonated glass electrode result in the pH dependence of the outer diffusion potential,

$$E(\text{pH}) - E_H(\text{pH}) = \varepsilon_{j,L}(\text{pH}) . \tag{2.214}$$

The hydrogen concentration profiles in Fig. 2.58 show that the "hydration state" of the membrane is practically not changed during the protonation. Incidentally, it was shown by calculations and experiments that the protonation neither changed the pH of the internal buffer solution of the glass electrodes used nor shifted the potential of the internal reference electrode. Figure 2.59 shows the experimental results.

(a) At constant surface coverage with silanol groups and thus at 100% pH selectivity of the membrane, the unprotonated and the protonated electrodes exhibit an identical response,

$$\frac{dE}{d\,pH} = \frac{dE_H}{d\,pH} = -k' , \qquad (2.215)$$

where k' is the practical slope factor. The diffusion potential is constant under these conditions, $\varepsilon_{j,L} = $ const., which agrees with *Eisenman's* prediction that the diffusion potential in an ion exchange membrane is constant as soon as the boundary conditions are fixed [2.10]. The negative value $\varepsilon_{j,L} = (-26 \pm 2)\,mV$ for the glass investigated means that also in the hydrated network

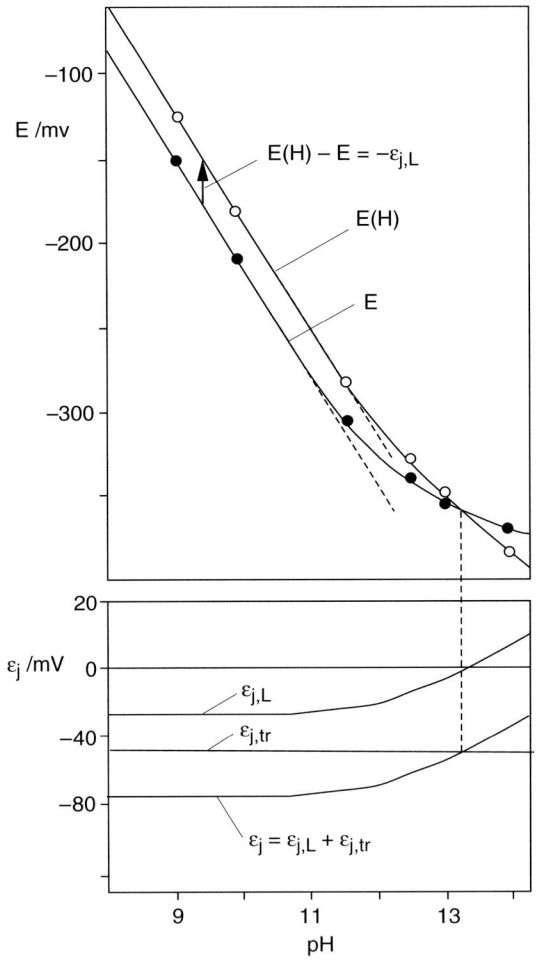

Fig. 2.59. emfs of a cell with a pH glass electrode with leached (E) and subsequently protonated membrane $(E(H))$, subsurface total ε_j, and partial diffusion potentials $\varepsilon_{j,L}$ and $\varepsilon_{j,tr}$ of the membrane as functions of pH

of the outer part of the leached layer the mobility of lithium ions is larger than that of protons. The glass is negative with respect to the solution. The finding expressed by (2.215), which is valid also when alkali-free solutions are used, is highly significant as it excludes the necessity of two different ions to participate in the potential formation and, consequently, eliminates the ion exchange theory of glass electrodes experimentally.

(b) In the sodium error range of the glass electrode, where both sodium ions and protons are present at the glass surface, the outer diffusion potential depends on pH and pNa and changes increasingly with increasing surface concentration of sodium ions. This is demonstrated, for instance, by $(d\varepsilon_{j,L}/dpH)_{pNa = const} > 0$ in Fig. 2.59, where it even changes its sign at high pH. This is surprising because the rate-cooled membrane glass excludes the penetration of alkali ions other than lithium because of steric conditions (see Fig. 2.22 in Sect. 2.6.4). The effect seems to be caused by the sodium ions at the glass surface, which block part of the sites to protons and influence the subsurface interdiffusion between protons and lithium ions without taking part in the interdiffusion. It is, consequently, called the *indirect sodium error*. The effect is significant as it causes the deviation of the experimental pH response of pH glass electrodes in the sodium error range from the theoretical response predicted by equations for the phase boundary potential. *Eisenman* took care of this deviation by the potential ion exchange constant, $K_{ij}^{pot} = (u_j/u_i)^n K_{ij}$ [2.209], which consists of the reversible ion exchange constant K_{ij}, the irreversible mobilities u_j and u_i, and a variable n, and can thus fit the theory to any experimental values [2.63].

Inner or Transition Part of the Leached Layer. Unlike the diffusion potential in the outer leached layer region, the diffusion potential $\varepsilon_{j,tr}$ in the transition range between approximately $x_{Li} = 0.3$ and $x_{Li} = 1$ remains constant during the protonation. This was verified by the independence of both the lithium (and hydrogen) concentration profile and the emf of the cell containing the protonated glass electrode, of the displacement of the transition layer. $\varepsilon_{j,tr}$ was determined on the basis of an equation given by *Conti* and *Eisenman* [2.10, 209, 210] for calculating the diffusion potential V_D across a cation exchanger membrane,

$$V_D = -\frac{n\,RT}{F}\ln\frac{c_i(d) + \frac{u_j}{u_i}c_j(d)}{c_i(0) + \frac{u_j}{u_i}c_j(0)}\;, \tag{2.216}$$

where c is the concentration and u the ion mobility of the ions indicated, d is the thickness of the membrane, and $n = (d \ln a/d \ln c)$ is a parameter describing the non-ideality of the system. Preconditions for applying (2.216) are (a) uniform chemical composition of the ion exchanger, (b) constant mobilities of the interdiffusing ions, and (c) fixed concentrations at the membrane boundaries.

The determination of the diffusion potential $\varepsilon_{j,tr}$ was carried out by subdividing the concentration profile of the displaced transition layer into several subconcentration profiles, Fig. 2.60, and by applying the differential form

$$\Delta\varepsilon_{j,tr} = -\frac{RT}{F} \ln \frac{x_{Li}(s_n + \Delta s_n) + \frac{u_H(s_n + 0.5\Delta s_n)}{u_{Li}(s_n + 0.5\Delta s_n)} x_H(s_n + \Delta s_n)}{x_{Li}(s_n) + \frac{u_H(s_n + 0.5\Delta s_n)}{u_{Li}(s_n + 0.5\Delta s_n)} x_H(s_n)} \qquad (2.217)$$

of (2.216) to each of the sublayers thus generated. $x(s_n)$ and $x(s_n + \Delta s_n)$ are mole fractions of the ions indicated at the sublayer boundaries s_n and $s_{n+1} = (s_n + \Delta s_n)$, respectively, and $u(s_n + 0.5\Delta s_n)$ is the average mobility of the ions indicated within the sublayer extending from s_n to s_{n+1}. The numerical values of the average mobilities of lithium ions and protons were taken from Fig. 2.31 of Sect. 2.6.5. Besides, ideality of each sublayer was assumed, which means that n in (2.216) is equated to unity. The entire diffusion potential in the transition range of the leached layer was then obtained by summation of the differential diffusion potentials according to

$$\varepsilon_{j,tr} = \sum_{x_{Li}=0.3}^{x_{Li}=1} \Delta\varepsilon_{j,tr} . \qquad (2.218)$$

The application of the differential form (2.217) to the sublayers of the lithium/proton boundary within the protonated glass obviously meets the preconditions for (2.216). Besides from theoretical considerations, this was concluded from results obtained for different numbers (seven, ten, twelve) of sublayers with different extensions Δs_n, resulting in total diffusion potentials $\varepsilon_{j,tr}$ which agreed to within 3%. The value determined for the glass described here is $\varepsilon_{j,tr} = (-48 \pm 2)$ mV. Consequently, the total diffusion potential in the steady-state leached layer amounts to $\varepsilon_j = \varepsilon_{j,L} + \varepsilon_{j,tr} = (-26 \pm 2 - 48 \pm 2)$ mV $= (-74 \pm 4)$ mV at 100% pH selectivity of the glass and changes as $\varepsilon_{j,L}$ changes in the sodium error range.

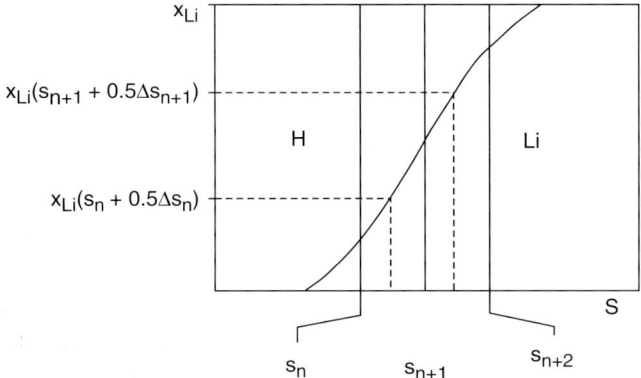

Fig. 2.60. Section of a subsurface lithium concentration profile of a protonated layer demonstrating the division into sublayers for determining the diffusion potential $\varepsilon_{j,tr}$

The total diffusion potential ε_j, which is larger than the Nernst slope $(RT \ln 10)/F$ at all practical temperatures of glass electrode application, is caused, in part, by the large difference of the mobilities of lithium ions and (guest) protons in the unchanged network of the glass [2.54] in the transition range, as expected. Surprisingly, however, the mobilities of the ions in the hydrated layer also seem to be rather different and thus contribute a similarly large value to the total diffusion potential. Total and partial diffusion potentials are plotted as functions of pH in the entire pH range in the lower part of Fig. 2.59, which shows that, unlike $\varepsilon_{j,L}$, the total diffusion potential ε_j is negative also in the higher sodium error pH range.

Diffusion Potentials Caused by Alkali Ion Interdiffusion

Diffusion potentials caused by an interdiffusion of the alkali ions of glasses versus different alkali ions from solutions are mainly of interest for alkali-selective electrode membrane glasses because such aluminosilicate glasses show relatively little sterical hindrance, whereas hydrogen-selective silicate glasses, which usually contain lithium, are generally not "open" even to sodium ions in the sodium error pH range (see Sect. 2.6.4). An exception will be treated at the end of this subsection.

Except for very few cases [2.10], mobilities of alkali ions, in particular of replacing ("guest") alkali ions in "host" glass networks, have not been reported, as the mobility ratios given at temperatures between 350 °C and 740 °C accumulated by *Doremus* [2.211] are not relevant to glass electrodes. Thus, in the following, two effects of alkali/alkali ion interdiffusion will be reported which are relevant for the special treatment of pM glass electrodes before, during, and between pM measurements and for general aspects concerning the development of pH electrode glasses.

Alkali-Selective Glass Electrodes. The first case is concerned with pNa-sensitive glass electrodes, and a lithium aluminosilicate glass is chosen as example. Figure 2.61 shows that this glass, if leached in acid solution, exhibits 100% pH response (plot c). The pH function extends up to high pH values (not shown in the figure), the limit being given by pH $\cong (\mathrm{pH_{tr}} - 2)$ and the transition pH being determined by the impurity alkali contents of the solutions. Subsequently brought into a solution causing 100% glass surface coverage with [AlOSi]Na groups (point a), the electrode exhibited a potential drift, which shifted the entire pH and pNa characteristics to more positive values by about $+25\,\mathrm{mV}$ within 24 h (potential change a→b, see also inset). Subsequent storage of the sensor in acidic solution reversed the drift, but only partly.

This observation indicates the following processes at and below the membrane surface. The first contact of the new glass with acid solution caused the formation of a leached layer, as was confirmed by IBSCA. The connected diffusion potential (interdiffusion of lithium ions of the glass and protons from

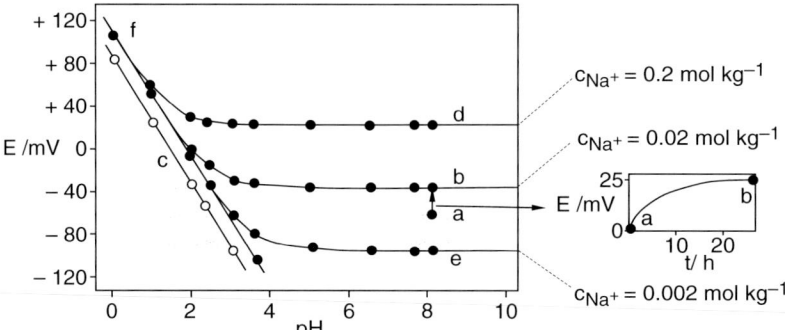

Fig. 2.61. pH function (c) of a new pNa glass electrode and subsequent potential shift (a→b) of the entire electrode function (b–f, d–f, e–f) by a diffusion voltage caused by sodium uptake

the surface) was independent of pH as long as the glass surface contained only [AlOSi]H groups and was thus undetectable when the pH function was measured. The subsequent coverage of the glass surface with sodium in solution (a) caused a complicated three-particle interdiffusion (protons, sodium, and lithium ions) resulting in the positive potential drift from a to b, which attained a steady state after about a day. This steady state was maintained, as verified by IBSCA, as long as the glass surface contained only [AlOSi]Na groups. The subsequent return to a surface coverage with [AlOSi]H caused different interdiffusion conditions which resulted in a negative potential drift.

The example demonstrates that, independent of the alkali ion of the membrane glass, alkali-selective glass electrodes ideally should be conditioned and stored before and between uses in solutions containing the alkali ion to be measured and having a pH value that guarantees a continuous coverage of the membrane glass surface with the measuring ion, pH $\geq$ (pH$_{tr}$ + 2). This is necessary because aluminosilicate glasses, which are usually applied as alkali-selective membrane glasses, exhibit neither sterical hindrance to sodium nor to potassium ions, even if the glass structure is determined by small ions such as lithium in the reported experiments. This treatment excludes potential drifts by varying diffusion potentials during measurements because it allows for such drifts before the application if an electrode has been accidentally immersed in a solution not meeting the strict conditions. The generally recommended application of 0.1 or 1 molar sodium chloride solution [2.212–214] safely meets these conditions. As demonstrated, however, the reason for this recommendation is not the development and preservation of the alkali response as suggested in the literature but the protection of the glass from penetrating inappropriate ions.

pH Glass Electrodes. Different from alkali-selective aluminosilicate membrane glasses, silicate pH glasses are generally protected from penetrating alkali ions. (a) Within the range of pH selectivity, their functioning necessitates 100% surface coverage with silanol groups, which causes the formation of leached layers and, at the same time, excludes alkali ions from being bonded to the surface and from entering the glass by interdiffusion. (b) In the sodium error range, the surface is partly covered with alkali siloxy groups so that the glass would be prone to entering alkali ions. The steric conditions of silicate glasses, however, block their penetration, especially because they usually contain lithium ions, which generate small cationic sites. Thus, in general, pH glasses are by far better protected from potential drifts and are more practicable than pM glasses.

Figure 2.62, however, demonstrates what happens when the protecting sterical hindrance in the sodium error pH range of a pH glass is not given. The glass shown is one of a group of glasses that had been patented "because of small sodium errors" [2.215] and which, for this reason, contained various amounts of tantalum oxide. Obviously, the large tantalum atoms impaired the steric conditions and "opened" the surface of the leached, but otherwise new glass electrode for sodium ions in the sodium error range. The result is a large, time-dependent, positive diffusion potential reflected by a corresponding change of the cell emf when the electrode is exposed to solutions with pH $\geq (\mathrm{pH_{tr}} - 2)$. The total drift was $+150\,\mathrm{mV}$ within two days at $50\,^\circ\mathrm{C}$. Figure 2.62 shows also that the glass electrode exhibited a

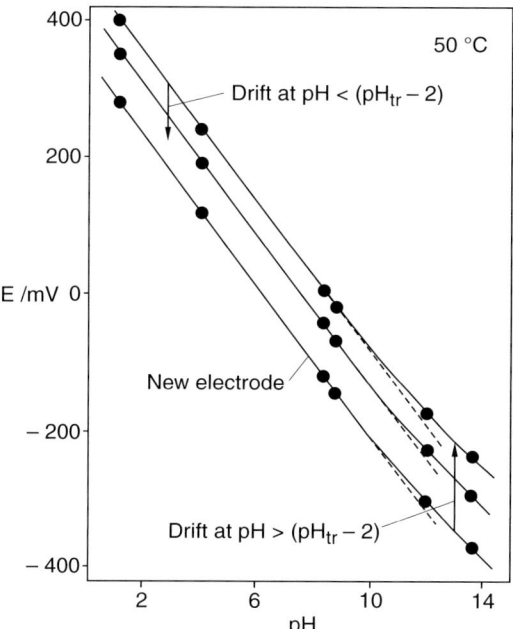

Fig. 2.62. Unusual shift of pH function of a pH glass electrode caused by interdiffusion in the membrane glass, which is not protected sterically from sodium ion penetration

normal response within the entire pH range during the slow drift and thus that the diffusion potential is indeed the cause of the potential drift. At pH values outside the sodium error pH range, that is at pH $\leq$ (pH$_{tr}$ − 2), the effect is reversed, although not completely because the conditions below the glass surface have changed from a two-particle (protons, lithium ions) to a three-particle interdiffusion (protons, lithium, sodium ions), which is a rather complicated situation. IBSCA revealed that the sodium ions, in addition to their diffusion back towards the surface and into the acid solution, continued to diffuse further into the glass, as is to be expected from their concentration gradients. The complete liberation of the glass from the introduced sodium may thus take months, if not years. Potassium ions had a similar effect as sodium, drift rate and maximum drift being only slightly smaller. The effect is called the *time-dependent sodium error* of pH glass electrodes [2.96]. A similar observation has recently been reported on silver ions in a sodium aluminosilicate glass [2.216]. Sensors subject to the described deficiency must never be exposed to solutions in the sodium error range, and even contact with potassium-containing solutions at high pH must be excluded. These extreme restrictions make practical application impossible.

2.7.5 Additional Surface Reactions

There are several reactions at glass surfaces which impair the correct response of glass electrodes or prevent their functioning altogether. Two of these deleterious reactions are briefly reported here because they are caused by inappropriate membrane glass compositions, in one case even by a minor glass component. This is another example demonstrating that good knowledge about all properties of a newly developed glass is an absolute necessity because even small changes in the glass composition, made, for instance, to improve a certain glass property, can result in an impairment of the main functioning of the glass which may even be difficult to detect.

Separation of Leached Surface Layers

The first example is concerned with glasses, in particular sodium calcium silicate pH glasses, that form leached layers of the type mentioned in Sect. 2.7.1. Such glasses, especially with large sodium contents and during application at high temperatures, form rather thick leached layers, which, due to an appreciable water uptake and a high degree of hydrolysis, are under mechanical stress and tend to crack and peel off the bulk glass. For instance, the arrows in Fig. 2.48a denote the depth where the surface layers of the investigated glasses tended to separate from the underlying glass [2.74]. As long as even thick leached layers are connected with the glass through the transition region with continuously varying composition (and properties), they do not present separate phases in the physico-chemical sense and thus do not impair the correct functioning of the glass membrane. However, on their separation, which is often not easily detected, they become new phases that disturb the correct electrode response. For example, cavities formed between the cracked and partly separated but still mechanically adhering layers are filled with stagnant solution, whose pH increases because of the continuing ion exchange with the fresh glass surface. The electrode, consequently, attains a mixed potential of numerous local potentials and the electrode potential of interest (if the measuring solution still contacts the membrane). Unfortunately, the connected slow potential drifts are easily overlooked during routine applications of glass electrodes, at least at the beginning of the impairment.

Phosphate and Fluoride Error of pH Glass Electrodes

Another case of layer formation is the precipitation of compounds at the glass surface, which result from a reaction of glass components released during the slow glass dissolution with compounds in the solution. Two examples recently detected and elucidated are the formation of lanthanum orthophosphate and lanthanum fluoride from small lanthanum contents of the membrane glasses, as recommended by *Perley* [2.217, 218], with phosphate and fluoride ions, respectively, of the solution [2.219]. These reactions are particularly interesting because the formation of various stages of the precipitates can be followed by characteristic changes of the electrode potential although only minute amounts of the salt are generated. To give an impression, the solubility product of lanthanum orthophosphate is $K_{s0} = 10^{-22.4} \, mol^2 \, dm^{-2}$ at $25 \,°C$ [2.220], which means that a saturated $LaPO_4$ solution contains $10^{-11.2} \, mol \, dm^{-3}$ of the salt if the solution is otherwise free of phosphate and as little as $10^{-21.4} \, mol \, dm^{-3}$ of lanthanum ions if, for example, the total phosphate concentration is $0.1 \, mol \, dm^{-3}$. The condition for these numbers is pH ≥ 13, which indicates another interesting aspect. The solubility of lanthanum phosphate depends strongly on the solution pH because the orthophosphate ion concentration determining the concentration of lanthanum ions is itself a strong function of pH [2.221]. For example, at a given total phosphate concentration, the concentration of lanthanum ions increases by nearly six orders of magnitude when the pH is lowered from pH $= 7$ to pH $= 4$. The physico-chemical basis of these phenomena is thoroughly discussed in [2.219].

The observation is surprising indeed. As demonstrated by Fig. 2.63, a glass electrode with a glass membrane containing small amounts of lanthanum attains the correct potential, for instance, in a phosphate buffer solution (pH $\cong 7$), which is constant for a rather long time. However, after a certain period, which is called the induction time t_i and which lasts approximately 55 min at $50 \,°C$ with the example given, the potential starts to drift towards more negative values with a constant drift rate and stabilizes after a period, which is termed the drift time t_d and is approximately 2.6 times the induction period. Potential drifts observed with various other types of glass electrodes are between $-4 \, mV$ and $-18 \, mV$ and can be as large as $-35 \, mV$ if the electrode

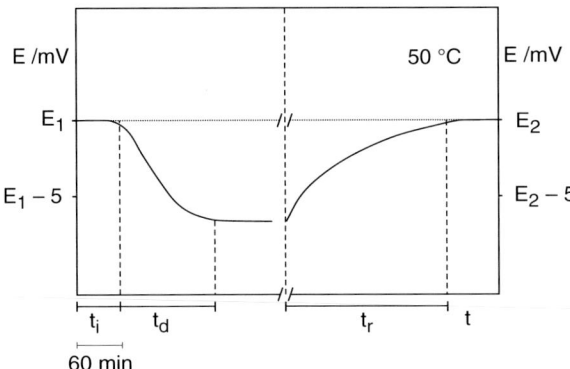

Fig. 2.63. Phosphate error of a pH glass electrode with lanthanum-containing membrane. t_i = induction time, t_d = drift time, t_r = recovery or return time

has been heat-treated in (phosphate-free) solutions, for instance for sterilization, before the measurements. When the glass electrode is subsequently transferred into phosphate-free solutions with smaller pH, for example, into a phthalate buffer with pH = 4, the potential returns asymptotically to its (pH-dependent) correct value during the so-called restoring time t_r. Potential change and induction, drift, and restoring time depend on glass composition, temperature, and pH. An example of the temperature dependence is shown in Fig. 2.64, which also demonstrates by parallel measurements on a glass electrode with a lanthanum-free membrane in the same solution that the "phosphate error" described is indeed caused by the lanthanum content of the membrane glass.

The phosphate error is taken into account by introducing a time-dependent term $\Delta\varepsilon(t)$ into the glass electrode characteristic,

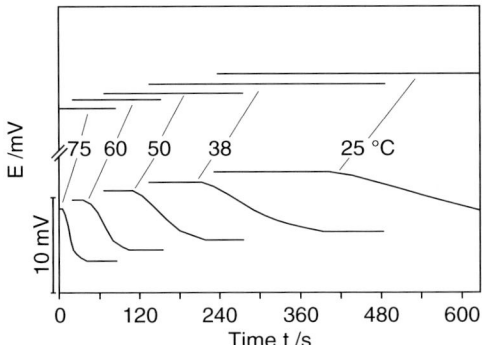

Fig. 2.64. Temperature dependence of phosphate error of a pH glass electrode with lanthanum-containing membrane as compared to the potential formation of a glass electrode with lanthanum-free membrane (upper constant potentials)

$$E(t) = E^0 - \Delta\varepsilon(t) - k'\mathrm{pH} \; , \tag{2.219}$$

which makes the emf a function of time. $\Delta\varepsilon(t)$ and its derivative $(\mathrm{d}\Delta\varepsilon(t)/\mathrm{d}t)$ can be negative, positive, or zero, depending on individual conditions of the measurement, such as sequence and composition of, and time of contact with, each of the solutions. The phosphate error can thus result in incorrect slopes, standard potentials, and zero-potential pH values. However, owing to the small change rate of the potential, in particular at low temperatures (see Fig. 2.64), the deficiency is often not realized.

Figure 2.65 shows the phosphate error as caused by an internal phosphate reference buffer of a glass electrode. It shifts the electrode potential to more positive values and can be compensated by the external phosphate error. This, however, does not yield the correct glass electrode potential as indicated by the potential drift during the subsequent restoring time. Because the construction of glass electrodes does not allow the internal phosphate error to be eliminated, the electrode will never attain its correct potential E_1 but will exhibit an asymmetry potential $(E_2 - E_1)$. Probably, this phenomenon appeared frequently during the development of new membrane glass compositions in the past and either went unnoticed or led to the rejection of the new glass composition, or, if appearing at the internal membrane surface, was interpreted as an asymmetry potential. In any of these cases, the phosphate error may have caused serious set-backs in electrode glass research, whose extent will remain unknown for ever. A mechanism for the development of the phosphate error has been proposed [2.219] and is briefly sketched in the following paragraph.

Assume a lanthanum-containing glass membrane in contact with a phosphate-containing solution. The lanthanum, being a network former [2.10], is released to the solution at the glass dissolution rate, which is constant if the leached layer of the silicate glass is in a steady state. It reacts with phosphate ions immediately after the release and forms lanthanum phosphate, which, in

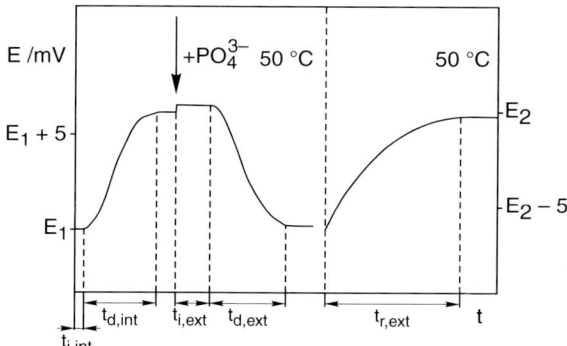

Fig. 2.65. Phosphate error of internal membrane surface of pH glass electrode $(t_{i,\mathrm{int}}, t_{d,\mathrm{int}})$ causing an irreversible asymmetry potential of the electrode

the beginning, precipitates as small, unconnected, probably non-crystalline groups or patches. As long as the precipitate does not represent a phase able to generate an independent potential, the membrane potential is solely given by the pH response of the glass and is constant, which is the case during the induction time t_i. However, when the amount of the salt present at the glass surface is large enough to form crystals at the end of t_i, the larger, by now crystalline patches act as a phase from now on and develop potentials with the glass and the solution that are thermodynamically fixed by the lanthanum content of the glass and the small lanthanum ion concentration of the LaPO$_4$-saturated solution, respectively, and are thus constant. The glass membrane attains a mixed potential which changes due to the following increase of the glass surface coverage with crystalline lanthanum phosphate and causes a potential drift of the glass during the drift time t_d. Finally, the potential attains a constant value when the glass surface is completely covered with a monolayer of the crystalline salt, which probably consists of crystalline unit cells, and this is the case at the end of the drift time t_d. Because hexagonal LaPO$_4$ has a rather open structure and is transparent to water molecules [2.222, 223] and thus to hydronium ions, the salt layer does not isolate the glass surface from the solution but allows the continued existence of the potential of the glass in addition to its own potential so that the membrane exhibits a constant mixed potential, which is the "faulty" potential observed after the drift time t_d. Further formation of LaPO$_4$ can only increase the thickness of the salt layer but does not influence the membrane potential. Unlike the formation, the dissolution of the LaPO$_4$ layer at low pH (and higher solubility of the salt) is not subject to certain characteristic steps and thus proceeds continuously during restoring time t_d as observed, Figs. 2.63 and 2.64. This mechanism is quantitatively supported by electrochemical and thermodynamic measurements, crystallographic data, and by analyses using SEM, EDX, and IBSCA. Details are reported in [2.219], where also a theoretical equation corresponding to the empirical expression (2.219) is presented.

2.7.6 Significance of Interfacial Equilibria for Glass Corrosion

As has been shown, the significance of electrode membrane glasses is not restricted to their functioning as electrode glasses; they also represent excellent model glasses for studying the initiating steps of glass corrosion. This role is particularly obvious from the fundamental, general basis, the phase boundary equilibrium, which unifies all types of glass corrosion and whose significance has been discovered during these studies.

In the past, glass corrosion was generally equated with glass leaching. The exchange of alkali ions of the glass for hydrogen ions of contacting solutions was assumed to be driven simply by the free energy difference of the ions between the phases, and the phase boundary was believed to be an inert interface. However, the glass surface is not an inert plane but is electrochemically structured in that it is the locus of an interfacial equilibrium involving functional surface groups which gives rise to a potential difference between glass and solution and to the presence of alkali ions and/or protons at the glass surface. The ions bound to the surface groups enter the glass because of the concentration gradients in the solid and exchange for the alkalis of the glass. The exchange of ions between solution and glass thus proceeds via surface groups and not across inert network parts of the glass surface. The interfacial equilibrium is maintained due to its large exchange current densities despite the diffusional disturbance (except for the very beginning of the interaction). The interface can be looked at as and reacts like a two-dimensional "inter-phase" intermediate between the solution, with which it is in equilibrium, and the glass, with which it is not. The composition of the interphase depends on the solution composition relative to the thermodynamic properties of the glass surface groups. These conditions initiate two basically different types of glass corrosion.

(a) One is the uptake of protons initiated by the acidic forms of the surface groups, which is generally connected with an uptake of water. It is termed leaching and represents what was believed to be the only kind of glass corrosion. Of the numerous publications on this topic we only cite a few characteristic ones. *Hench*, for example, has divided glasses into five groups according to their characteristic surfaces formed by leaching [2.194], which, in part, had been shown by the described measurements on electrode glasses. *Boksay* has proposed [2.75, 202], and *Doremus* has further developed [2.136], an equation describing the steady state of leached surface layers. *Scholze* [2.97] and later *Ernsberger* [2.224] have shown the existence of molecular water in leached layers of sodium calcium silicate glasses. *Scholze* has further applied IR spectroscopy [2.97] and *Schwabe* radiochemical tracers [2.9, 66] to study glass leaching, and *Shul'ts* and coworkers have extensively investigated leaching of electrode glasses [2.223, 225]. In addition, a new interpretation of reflectance spectra of surface layers, for instance of leached layers, has recently been given [2.227]. These and all other studies clearly reflect the type of glass corrosion that takes place under the basic experimental condition $pH \leq (pH_{tr} - 2)$.

(b) The other kind of glass corrosion is an exchange of alkali ions from the glass surface versus alkalis of the glass, in which protons are not involved. It has been reported by various researchers, see for instance [2.9, 199], and was extensively investigated by means of radioactive tracers by *Eisenman* [2.10]. Their findings, however, did not fit the general experience with glass corrosion at that time and were thus more or less ignored by corrosion researchers or believed to be a special property of electrode membrane glasses. A re-evaluation of Eisenman's and other scientists' data and our own measurements [2.188] yielded the condition $pH \geq (pH_{tr} + 2)$ for this type of glass–solution interaction. If it is not met, the glass is subject to leaching with all of its typical phenomena (see above) or to the complicated mixed leaching/alkali exchange interaction in the transition pH range, even if it is an alkali-selective glass [2.188, 226].

Contrary to this treatment, a recent review comes to the conclusion that, independent of the kind of glass (alkali silicate, alkali borosilicate, and aluminosilicate glasses), in most cases the rate at which water enters the glass structure controls the kinetics of the other glass–solution reactions (ion exchange, dissolution, etc.) [2.228]. According to the examples given here, however, this is certainly too general a statement. Rate control by water penetration may be significant in appliance glasses, for instance window and container glasses, whose alkali content is much smaller than that of the glasses treated here. As a result of the low alkali content, the concentration of surface groups is also much smaller so that water may enter the glasses through inert surface regions and may thus determine the rate of leaching.

A complication which has certainly caused some confusion in the investigation of corrosion initiated by leaching as well as alkali/alkali exchange is sterical hindrance [2.55, 152]. For glass electrodes, however, it is also a practically important phenomenon, which resulted in the development of lithium-containing membrane glasses for pH electrodes which can be used in the sodium error range without interference by interdiffusion. A remaining effect of the sterically blocked alkali ions at the pH glass surface is an unavoidable deviation of experimental from theoretical potentials called the *reversible indirect sodium error*, which has been taken into account in the past by a mixture of reversible and irreversible processes [2.10, 63, 209]. Lack of sterical hindrance, on the other hand, necessitates certain treatments of, and storing solutions for, alkali-selective glass electrodes.

Interdiffusion of different ions necessarily causes a diffusion potential. It is of no concern to glass leaching but significant for glass electrodes. Although, and despite its large magnitude of the order of the Nernstian slope, a large part of the diffusion potential is generally compensated by the diffusion potential below the opposite surface of the membrane, it impairs emf measurements if it depends on pH or pM as, for instance, in the sodium error range of pH glass electrodes. It is again emphasized that interdiffusion and diffusion potentials are a necessary consequence of the interfacial equilibria and are thus inseparably connected with the dissociation mechanism, a fact which obviously has not always been understood [2.229].

Glass corrosion can give rise to extremely thin layers formed from components of glasses during their dissolution. The generation of monolayers of disturbing $LaPO_4$ and LaF_3 is a striking example [2.219] that fits into the three-layer glass group as described by *Hench* [2.194]. Moreover, it should be realized that glass corrosion is generally concerned with extremely small amounts, often traces, of material. For example, $1\,m^2$ of a steady-state leached layer such as layer (g) shown in Fig. 2.46a contains as little as approximately $1\,mg$ of protons [2.206], and even a fraction of this amount can easily impair the products of a glass coating line, as glass coatings in particular are most sensitive to minute changes of the glass surface and thus also to the first steps of glass corrosion.

Further progress of glass corrosion depends on individual conditions during the exposure of the glass to the corroding medium. Only one of a large number of examples may be indicated. Because leaching never causes a decrease, as erroneously stated in [2.230], but always an increase of the pH of the leaching solution [2.231], the dissolution rate increases during the glass/solution contact, particularly if the solution, as generally is the case, is unbuffered. A critical number is the ratio glass surface area to solution volume [2.194]. It is extremely large, for instance, when stacks of glass plates are stored and no measures are taken to keep condensing water off the glass. Condensed water may lead to extremely non-uniform corrosion and inhomogeneous glass surfaces and even to a generation of pits within short periods of time because the pH increase of the water is an auto-catalysed process: the more alkaline the solution, the larger is the dissolution rate and the faster, in turn, is the pH increase of the solution. This is one of the few examples which justifies the term glass corrosion, whereas quite generally the term glass durability should be preferred because of the high chemical resistivity of glasses. For example, dynamic interaction of glass surfaces with water, for instance by weathering or car washing, causes no remarkable corrosion of the glasses [2.232].

Hwang and *Han* [2.233] recently investigated "hydration" (leaching) of lithium silicate pH glass electrode membranes by means of complex impedance measurements, infrared spectroscopy, and scanning electron microscopy ($80\,^{\circ}$C). Their results agree completely with what has been reported above. (a) The bulk resistance of the membrane was nearly unchanged at pH = 7 but decreased with time at pH = 10 because of the larger rate of network dissolution at high pH. (b) The resistance of the leached layer increased markedly at pH = 7 with leaching time because of the formation of a thin region of protonated glass between the leached layer and the bulk glass. This effect was much smaller at pH = 10, where the leached layer and the protonated range are considerably thinner than at the lower pH. The authors confirmed these results by infrared spectra. (c) The electron micrograph of the cross section of a membrane after leaching at pH = 7 shows a rather bulky or crystalline surface layer, whereas that of a membrane leached at pH = 10 exhibits an uncovered smooth glass surface. If the "commercial buffer solution" with pH = 7 used by the authors was a phosphate buffer and the buffer solution with pH = 10 did not contain phosphate, as can be assumed with high probability, the layer found at pH = 7 consisted probably of lanthanum phosphate because the investigated membrane glass contained, *inter alia*, a small amount of lanthanum, which in phosphate solutions gives rise to the formation of $LaPO_4$ deposits and the phosphate error [2.219]. This is also supported by the larger scatter of the time-dependent layer resistance at pH

= 7. Unfortunately, however, the authors did not measure emfs during the "hydration" processes and thus could not detect this effect. Their publication (as well as the work reported above) would have benefited from a comparison between the findings and published results, in particular because they are in excellent agreement [2.86, 192, 193, 206].

2.8 Hydrogen-Sensitive Platinum-Covered Glass Electrode Membranes

This brief section reports measurements on glass electrodes whose membranes with 100% protonated external surface were covered with platinum layers and contacted with hydrogen-containing atmospheres. Experiments with silica glass disks carrying platinum layers in hydrogen atmospheres have been reported before. *Jorgensen* and *Norton* [2.234] electrolysed Corning 7940 silica glass at 1000 °C, and *Bazán* [2.235] measured emfs of platinum-covered Suprasil® between 700 °C and 900 °C. The aim of both studies was to obtain information about the transport number of protons in vitreous silica. Different from their work, the measurements reported here were carried out at application temperatures of glass electrodes, i.e., from ambient temperature to 95 °C, and the purpose was to get better knowledge about the potential-forming electrode reaction at the interface solid electrolyte glass/deposited Pt,H_2 electrodes. 100% externally protonated glass membranes were chosen because the potential-forming equilibrium was expected to involve protons and to be possibly disturbed by a second (e.g. alkali) ion. These expectations were confirmed by preliminary measurements on unprotonated membranes, which indeed did not yield sufficiently constant potentials.

A rather large number of platinum layers with different structures and arrangements were tested, the best-suited being a 50 nm thick smooth vapour-deposited layer covered by electrolysed micro-crystalline, porous, black platinum (1.2×10^{-4} mol cm^{-2}) with large internal surface area, which is known to be a good catalyst for hydrogen dissociation [2.236]. Contact to the layers was made by stripes of platinum paint along the electrode shafts. The internal reference electrode of the glass electrode was a Thalamid® electrode (see Sect. 2.12) in 3.5 molar KCl solution with pH = 7. Potential measurements were carried out in a thermostated jacketed glass vessel under streaming hydrogen-containing gases. The exact composition of the hydrogen–argon mixtures were determined by gas chromatography.

The electrochemical cell formed by the platinum-covered glass electrodes is represented by the cell scheme

$$\text{int. ref. electrode} | \text{int. ref. buffer} | \text{glass, protonated glass} | Pt, H_2(p_{H_2}) ,$$
(2.XVI)

and the emf of this cell is given by

$$E_C = \sum \varepsilon_i + \varepsilon_{gl/Pt} , \qquad (2.220)$$

where $\sum \varepsilon_i$ is the sum of all internal potentials of the glass electrode, and $\varepsilon_{gl/Pt}$ is the potential at the protonated glass/platinum interface. The difference of the emfs measured by the same cell at two hydrogen partial pressures is thus solely given by the potentials of the protonated glass/Pt,H_2 electrodes,

$$\Delta E = E_2 - E_1 = \varepsilon_{gl/Pt,2} - \varepsilon_{gl/Pt,1} , \qquad (2.221)$$

and is independent of all other potentials of the cell, including the diffusion potentials within the glass membrane.

As an example, Fig. 2.66 presents temperature-dependent emfs of cell (2.XVI) at four hydrogen partial pressures between 10^{-4} and 1 bar, which exhibit linear dependence of the emf on temperature. Table 2.9 compares the hydrogen functions of the protonated membrane glass/Pt,H_2 electrode measured at three temperatures with the corresponding theoretical values, which agree within $\pm 0.5\%$, and Table 2.10 presents the temperature dependence of the electrode slopes, which show agreement with theory within $\pm 2\%$. The Pt,H_2 electrode in contact with the solid electrolyte-protonated glass thus shows the same hydrogen response as the platinized platinum/hydrogen electrode in aqueous solutions, where it is one of the most reliable and exact electrodes [2.236, 237]. The results show in particular that, during and in between the isothermal measurements, the protonated glass did not change its properties as an electrolyte despite the rather wide range of hydrogen partial pressures applied (1–10^{-4} bar) and that, during the non-isothermal measurements, the electrolytic properties of the glass changed reversibly within the entire temperature range from $20\,°C$ to $95\,°C$. This behaviour also corresponds exactly to that of the Pt,H_2 electrode in aqueous solutions.

The following electrode reaction is proposed for the potential formation of the Pt,H_2 electrode in contact with protonated glass. Owing to its catalytic activity, the platinum catalyses the dissociation of adsorbed hydrogen molecules into atoms and that of hydrogen atoms into electrons and protons. These react with siloxy groups at the glass surface, which are good proton acceptors, and form silanol groups. In this way, an interfacial equilibrium between the different forms of glass surface groups, protons, and negatively charged platinum is generated, which is represented by

$$\frac{1}{2}H_2(Pt) + SiO^-(s) \rightleftarrows SiOH(s) + e^-(Pt) , \qquad (2.222)$$

where the bracketed (s) indicates the glass surface. Equation (2.222) meets the well-known condition for the formation of an interfacial potential, which is the separation of charges between the contacting phases due to a combining electrochemical reaction. This process was also proposed for the potential formation at the silica glass/Pt,H_2 interface at high temperatures by *Bazán* [2.235], who pointed out its resemblance to our dissociation mechanism of the glass electrode [2.192], see also [2.72] and Sect. 2.5.2.

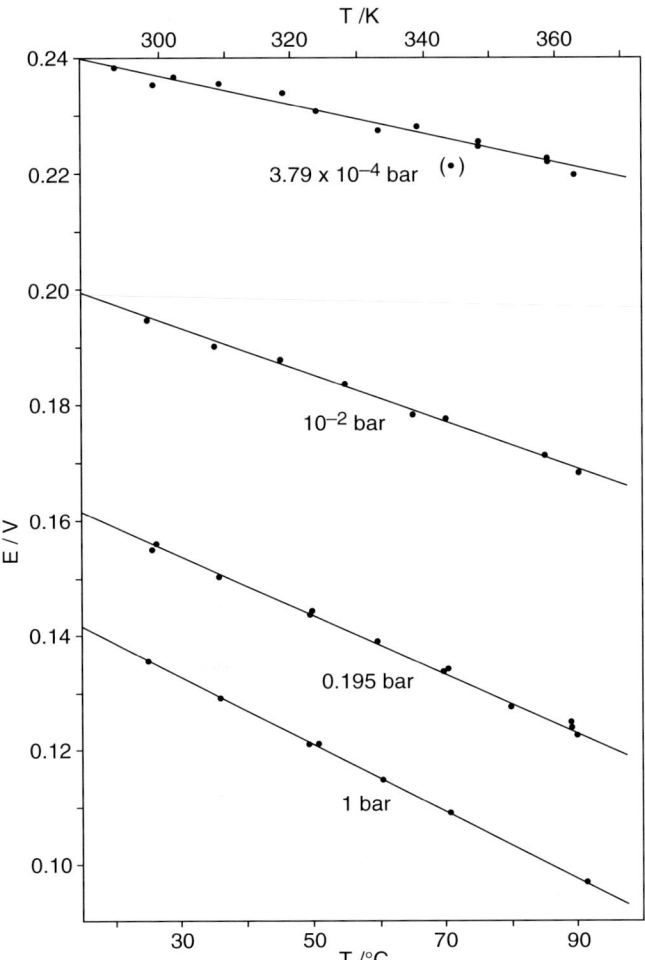

Fig. 2.66. Emf of glass electrode cell (2.XVI) with protonated and platinum-covered membrane surface in hydrogen-containing atmospheres at various hydrogen partial pressures as functions of temperature

Reaction (2.222) is very probable as the potential-forming equilibrium, in particular because a different reaction can hardly be imagined and three principally possible processes are essentially excluded.

(a) Alkali ions had been removed from the glass surface by electrolytic protonation, see Sect. 2.6.2. The absence even of traces of alkali below the membrane surface up to a depth of at least 200 nm was shown by IBSCA measurements, see Sect. 2.4. The participation of ions other than protons in the equilibrium is thus excluded.

(b) An interfacial reaction involving water, for instance according to

Table 2.9. Comparison of measured hydrogen functions ("hydrogen slopes") of the electrode-protonated membrane glass/Pt,H_2 at various temperatures with the theoretical value $-2.303RT/2F$; hydrogen partial pressure in bar

T K	$\frac{\Delta \varepsilon_{gl/Pt}}{\Delta \log p_{H_2}}$ mV	$-\frac{2.303RT}{2F}$ mV
300	−29.62	−29.76
330	−32.80	−32.74
360	−35.75	−35.71

Table 2.10. Temperature dependence of the "hydrogen slope" of the electrode protonated membrane glass/Pt,H_2 in various temperature ranges and the theoretical value $-2.303R/2F$; hydrogen partial pressure in bar

T_1 K	T_2 K	$\frac{\Delta\left(\Delta\varepsilon_{gl/Pt}/\Delta \log p_{H_2}\right)}{\Delta T}$ mV K^{-1}	$-\frac{2.303R}{2F}$ mV K^{-1}
300	330	−0.106	
300	360	−0.102	
330	360	−0.098	
Average		−0.102	−0.0992

$$\frac{1}{2}H_2(Pt) + H_2O(s) \rightleftarrows H_3O^+(s) + e^-(Pt) \tag{2.223}$$

is essentially excluded because siloxy groups are known to be much stronger proton acceptors than water molecules [2.72]. In addition, only traces of water molecules, if any at all, will be present at the glass surface so that the equilibrium

$$SiO^-(s) + H_3O^+(s) \rightleftarrows SiOH(s) + H_2O(s) \tag{2.224}$$

would be positioned on the extreme right side, and protons would be nearly completely converted to silanol groups. This conclusion is supported by the observation that measurements employing extremely dry hydrogen-containing gases yielded the most constant potentials.

(c) Potential formation purely on the basis of proton adsorption to the glass surface is extremely improbable as long as siloxy groups are present, because the chemical bond of protons to the acceptor groups SiO^- is much stronger than the adsorption bond of protons to an undefined locus of a glass surface.

The thermodynamic treatment of the electrode reaction (2.222) is of interest as it reveals some relationship of platinum-covered with solution-contacted glass electrode membranes. The electrochemical free energy change of the equilibrium (2.222) is given by

$$\Delta \overline{G}_{\text{gl/Pt}} = \Delta G^0_{\text{gl/Pt}} + RT \ln \frac{a'_{\text{SiOH}}}{a'_{\text{SiO}^-} \sqrt{p_{\text{H}_2}}} - F\varepsilon_{\text{gl/Pt}} = 0 \; , \tag{2.225}$$

where $\Delta G^0_{\text{gl/Pt}}$ is the standard chemical free energy change, a' is the surface activity of the groups indicated, $\varepsilon_{\text{gl/Pt}}$ is the potential of the glass|Pt,H$_2$ electrode, and the number of electrons exchanged is $z = 1$. (In order to meet *IUPAC Recommendations* [2.238], the potential $\varepsilon_{\text{gl/Pt}} = \varphi_{\text{Pt}} - \varphi_{\text{gl}}$, where φ is the internal electric potential of the phases indicated, is taken negative in (2.225) because the direction of reaction (2.222) is the reverse of that of cell (2.XVI).) Rearrangement of (2.225) yields

$$\varepsilon_{\text{gl/Pt}} = -\varepsilon^0_{\text{gl/Pt}} + \frac{RT}{F} \ln \frac{a'_{\text{SiOH}}}{a'_{\text{SiO}^-}} - \frac{RT}{2F} \ln p_{\text{H}_2} \; , \tag{2.226}$$

and differentiation of (2.226) with respect to $\ln p_{\text{H}_2}$ gives

$$\frac{\mathrm{d}\varepsilon_{\text{gl/Pt}}}{\mathrm{d}\ln p_{\text{H}_2}} = \frac{RT}{F} \frac{\mathrm{d}}{\mathrm{d}\ln p_{\text{H}_2}} \left(\ln \frac{a'_{\text{SiOH}}}{a'_{\text{SiO}^-}} \right) - \frac{RT}{2F} \tag{2.227}$$

or, in decadic logarithms and with the abbreviation $2.303RT/F = k$,

$$\frac{\mathrm{d}\varepsilon_{\text{gl/Pt}}}{\mathrm{d}\log p_{\text{H}_2}} = -\frac{k}{2} \left[2 \frac{\mathrm{d}}{\mathrm{d}\log p_{\text{H}_2}} \left(\log \frac{a'_{\text{SiO}^-}}{a'_{\text{SiOH}}} \right) + 1 \right] \; . \tag{2.228}$$

Because, in addition, the surface activity of the siloxy groups is certainly much smaller than that of the silanol surface groups, $a'_{\text{SiO}^-} \ll a'_{\text{SiOH}}$ [2.72, 95], see Sect. 2.5.3, and thus the expression

$$\left| \frac{\mathrm{d}\log a'_{\text{SiO}^-}}{\mathrm{d}\log p_{\text{H}_2}} \right| \gg \left| \frac{\mathrm{d}\log a'_{\text{SiOH}}}{\mathrm{d}\log p_{\text{H}_2}} \right| \tag{2.229}$$

holds, (2.228) may finally be approximated by

$$\frac{\mathrm{d}\varepsilon_{\text{gl/Pt}}}{\mathrm{d}\log p_{\text{H}_2}} = -\frac{k}{2} \left[2 \frac{\mathrm{d}\log a'_{\text{SiO}^-}}{\mathrm{d}\log p_{\text{H}_2}} + 1 \right] \; . \tag{2.230}$$

Equation (2.230) shows that, strictly speaking, the thermodynamically correct p_{H_2} dependence, the "hydrogen gas function", of the interfacial potential $\varepsilon_{\text{gl/Pt}}$ is smaller than the ideal slope $k/2$ because an increase of the hydrogen partial pressure causes a decrease of the siloxy activity at the glass surface so that

$$\frac{\mathrm{d}\log a_{\text{SiO}^-}}{\mathrm{d}\log p_{\text{H}_2}} < 0 \; , \tag{2.231}$$

and the sum of the terms in the square brackets of (2.230) is below unity. Thus, in principle, also the potential at the protonated glass/Pt,H$_2$ electrode exhibits a thermodynamic subideal response, as does the potential of the glass electrode membrane in contact with solutions [2.95], see Sect. 2.5.3. However, the effect seems to be too small to be detected by our measurements with an uncertainty of $\pm 0.5\%$, see Table 2.9, whereas pH glass electrodes exhibit the thermodynamic subideal response by the sub-Nernstian response, which is well detectable experimentally [2.95, 237]. It is thus justified to use the approximate slope

$$\frac{\mathrm{d}\varepsilon_{\mathrm{gl/Pt}}}{\mathrm{d}\log p_{\mathrm{H}_2}} = -\frac{k}{2} = -\frac{2.303RT}{2F} \tag{2.232}$$

of the hydrogen function of glass/Pt,H_2 electrodes. The temperature dependence of the slope is thus given by

$$\frac{\mathrm{d}}{\mathrm{d}T}\left(\frac{\mathrm{d}\varepsilon_{\mathrm{gl/Pt}}}{\mathrm{d}\log p_{\mathrm{H}_2}}\right) = -\frac{k}{2T} = -\frac{2.303R}{2F} = -9.92 \times 10^{-2}\,\mathrm{mV\,K^{-1}} \tag{2.233}$$

and is expected to be independent of temperature. Both expressions are in good agreement with measured data, see Tables 2.9 and 2.10.

It was found during the search for optimal platinum layers that, in contrast to glass electrodes, the most frequent problem with Pt,H_2 electrodes on glasses is an insufficient stability of potential. This difference is probably explained in the following way. Glass electrodes are characterized by large exchange current densities [2.71, 87] and intimate contact of their membrane surface with the contacting solution, which result in large exchange currents and insensitivity of their potential to mechanical and electric disturbances. In glass|Pt,H_2 electrodes, in contrast, the contact of the glass with platinum is rather patchy because of the crystallinity of the metal, which allows only small parts of the surface area to generate the interfacial equilibrium (2.222) and thus causes a sensitive potential between glass and platinum. Heat treatment for improving the contact was not applied because, owing to the low T_g of the glass, it would have easily impaired the spherical cap-like shape of the only 300 μm thick membranes. It would probably have also resulted in the growth of the platinum crystals, fewer grain boundaries, and a reduction of proton transfer rates through the platinum layer. The optimum layers were finally obtained by vapour deposition at high deposition rates. They consisted of extremely fine crystals of the metal, which obviously generated a sufficiently large contact area with the glass surface and also enough grain boundaries for a sufficient proton diffusion rate. The additional electrolytically deposited, catalytically active platinum on the platinum layer met the condition of supplying enough protons for the interfacial reaction.

The main result of these experiments is that the membrane glass obviously participates in the formation of phase boundary potentials with its functional surface groups, independent of whether the contacting phase is a liquid electrolyte or a solid metal. This is best seen by comparing the equations representing the interfacial equilibria for the solution [2.72],

$$\mathrm{SiOH(s)} + \mathrm{H_2O(sol'n)} \rightleftarrows \mathrm{SiO^-(s)} + \mathrm{H_3O^+(sol'n)}\,, \tag{2.234}$$

and for the Pt,H_2 electrode,

$$\mathrm{SiOH(s)} + \mathrm{e^-(Pt)} \rightleftarrows \mathrm{SiO^-(s)} + \frac{1}{2}\mathrm{H_2(Pt)} \ . \tag{2.235}$$

In both cases, the proton acceptor at the glass surface is the siloxy group. The proton donor of the contacting phase is the hydronium ion of the solution in the case of glass electrodes, and catalytically dissociating hydrogen atoms with the Pt,H_2 electrode. Both reactions result in a charge separation and thus a galvanic voltage between the phases – independent of the kind of material that contacts the glass – and generate a thermodynamically correct response of the electrode potentials.

2.9 The Dissociation Mechanism of Glass Electrodes – Discussion

The preceding sections presented a detailed treatment of various electrochemical phenomena at and below glass surfaces in contact with solutions and gases, and within glasses. Many of these properties are observed with ion-sensitive membranes of glass electrodes and form what is called the *dissociation mechanism* of these sensors. Thus, phase boundary (or interfacial) equilibria leading to the activity-dependent potential of glass membranes have been treated in Sect. 2.5, subsurface ionic interdiffusion between glass surface and glass interior and the connected formation of subsurface glass layers and diffusion voltages were presented in Sect. 2.7, and Sect. 2.6 treated field-driven ionic migration in glasses, which, in several cases, is applied to verify the dissociation mechanism experimentally. These processes and equilibria will be discussed in some detail in the following. In some instances we will refer to equations derived earlier in the text without repeating them; in others, where their immediate presence seems necessary, they will be repeated or briefly derived.

2.9.1 Interfacial Equilibrium

It has been a surprising discovery that the interfacial equilibrium between functional groups at glass surfaces and cations in contacting solutions is the cause of the electric response of glass membranes and thus of the functioning of glass electrodes. In addition, it was interesting to detect that the interfacial equilibrium is also quantitatively responsible for practically all secondary reactions, for instance subsurface interdiffusion, the formation of diffusion potentials, and the type of glass corrosion, and it even determines the ionic species which are transferred anodically into glasses when electric fields are applied.

Origin of the Glass Membrane Potential

Figure 2.67 demonstrates the interfacial equilibrium by a general equation of the (zero) electrochemical free energy change [2.72, 108], in which the activities a_i of the ions in the solution and the activities a_j of the various forms of the glass surface groups are given in two separate terms. Because the standard free energy change ΔG^0 of the heterogeneous reaction is fixed by the glass composition and the kind of solution, the ionic activities, for instance $a_{H_3O^+}$ and/or a_{M^+}, which, besides temperature and pressure, are the only free variables of the glass surface/solution system, determine the activities of the various kinds of surface groups and the potential of the glass in the following way. Besides the acidic form RH and the salt form RM, the interfacial equilibrium involves the anionic form R^- of the surface groups, which represents a negative charge at the glass surface and, consequently, the potential ε_m of the glass membrane relative to that of the solution. This demonstrates the electrochemical character of the interfacial equilibrium, in which a chemical and an electric driving force balance each other [2.72, 87, 108]. For example, an increase of one (or both) of the cation activities in the solution increases the activity of the appropriate form of the surface groups, RH or RM, and necessarily decreases the activity of the anionic form R^-, but only to an extent which is given by the chemical driving force resulting from the increased ionic activity and the counteracting electric driving force of the increased charge of the created negative groups. The increase of the potential ε_m of the membrane, which reflects the balanced decrease of the charge at the glass surface, represents the well-known response of glass electrodes to changes of

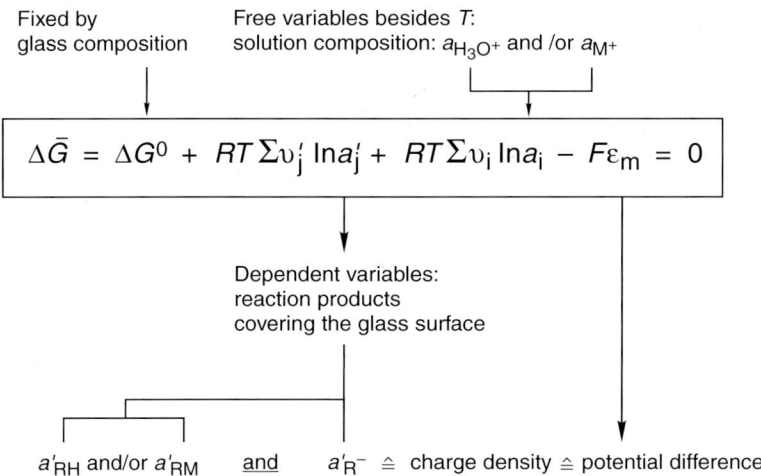

Fig. 2.67. General equation of electrochemical free energy change of interfacial equilibrium between glass and solution, demonstrating the interconnection of solution composition, glass surface coverage, and electric potential of the glass

pH and/or pM. In short, the necessary participation of (negatively charged) anionic surface groups in the heterogeneous electrochemical equilibrium between glass and solution is the basis of the glass electrode functioning.

It can be imagined that, at extremely low pH values, the acidic form of the surface groups may take up an additional proton according to

$$RH(s) + H_3O^+(\text{sol'n}) \rightleftarrows RH_2^+(s) + H_2O(\text{sol'n}) \tag{2.236}$$

although experimental evidence of this equilibrium has not been presented. We have not expressly tried to detect this form of surface equilibrium, in particular because the positively charged "doubly protonated surface group" is undetectable by glass electrolyses because the necessarily very small pH of the anolyte will always result in 100% protonation [2.97, 239], independent of whether the surface contains RH or RH_2^+ groups. Besides, the modified form (2.236) of the equilibrium, in principle, will not change the basic mechanism of potential formation, except that the glass membrane will have a positive potential relative to that of the solution in contrast to the negative potential it has in the case of negative surface groups.

The "Thickness" of the Interface

We have concluded that the surface groups that take part in the heterogeneous equilibrium are positioned at the direct surface of the membrane glass so that the "phase" between glass and solution, which is responsible for the potential formation, is a two-dimensional plane with monoatomic thickness, the "interface" [2.72, 87]. The main reason for this conclusion is the kinetic properties of the interfacial equilibrium, which, in addition, correspond well to the kinetic derivation of the equation for the membrane potential [2.71]. Estimated exchange current densities have an order of magnitude that is comparable with those of many metal electrodes under similar conditions [2.71]. They enable response times of glass membranes in the range of tens of milliseconds [2.66, 76, 77], which were not feasible if solid-state diffusion, even within distances of only a few atomic dimensions, would be involved. Besides, an interfacial equilibrium that yields well-defined activity-dependent potentials is physically conceivable only if the involved functional groups of the solid are in direct contact with the solution containing the dissolved ions. It cannot be imagined how glass groups positioned within the static environment of subsurface glass regions might take part in the interfacial equilibrium.

It is mentioned at this point that, also for kinetic reasons, *Pungor* in several recent papers [2.185, 240] rejects ion exchange between solutions and ion-selective electrode (ISE) membranes as the process responsible for the potential formation of ISEs. He also emphasizes that a membrane potential can only be created by a charge separation between the different phases. He proposes chemisorption of ions at the membrane surface as the underlying reaction, whereas we have shown the specific dissociation and association processes of functional surface groups to be the reactions responsible for bringing about the required charge separation at glass membrane surfaces, see equilibria (2.17) and (2.29) for light water solutions, (2.326) and (2.327) for deuterium oxide solutions, and (2.222) for the charge separation between membrane and Pt,H$_2$ electrode.

Subsurface interdiffusion of ions leaving the surface groups versus ions of the glass interior is a completely different matter, see also below. It proceeds independently and disturbs the interfacial equilibrium only if the interdiffusion rate is comparable with the formation rate of the equilibrium. Expressed in quantitative terms, the subsurface interdiffusion polarizes the equilibrium only if the ionic current density representing the rate of the ions interdiffusing into and out of the glass is of the same order of magnitude as the exchange current density of the equilibrium, which indicates the rate with which the ions attached to the surface groups are interchanged for ions of the same kind from the solution and thus characterizes the dynamic stability of the equilibrium [2.71]. Most glasses meet this condition, particularly if their subsurface layer is in or near a steady state so that the ionic interdiffusion rate is small. Exceptions are "soft" glasses with extremely high dissolution and interdiffusion rates and most glasses during the initial period of subsurface layer formation when the rate of the interdiffusing ions is still high because of initially large concentration gradients. The period of first contact of glass electrodes with solutions is called the "formation period" [2.86, 241], during which the electrodes are useless because the time-dependent disturbance of the interfacial potential, probably in addition to the time-dependent diffusion potential, is reflected by drifts of the electrode potential.

A Well-Known Principle of Potential Formation

Dissociation of interfacial groups of glasses represents a separation of charges between the contacting phases, the glass being negatively and the solution positively charged. Charge separation is the precondition of potential formation, also for metal electrodes [2.71]. This is obvious when, for instance, the dissociation equilibrium of acidic glass surface groups

$$RH(s) + H_2O(sol'n) \rightleftarrows R^-(s) + H_3O^+(sol'n) \tag{2.237}$$

and the corresponding potential of the glass

$$\varepsilon_m = \varepsilon_m^0 + \frac{RT}{F} \ln \frac{a'_{R^-} \, a_{H_3O^+}}{a'_{RH} \, a_{H_2O}} \tag{2.238}$$

are, respectively, compared with the equilibrium between a Pt,H$_2$ electrode and hydronium ions in a solution,

$$\frac{1}{2}H_2(Pt) + H_2O(sol'n) \rightleftarrows e^-(Pt) + H_3O^+(sol'n) \,, \tag{2.239}$$

and the potential of the Pt,H$_2$ electrode,

$$\varepsilon_{\text{Pt,H}_2} = \frac{RT}{F} \ln \frac{a'_{\text{e}^-\,(\text{Pt})}\, a_{\text{H}_3\text{O}^+}}{\sqrt{p_{\text{H}_2(\text{Pt})}}\, a_{\text{H}_2\text{O}}} \ . \tag{2.240}$$

(The standard potential of the Pt,H$_2$ electrode is defined to be zero at all temperatures.) The characteristic process at both electrodes is a dissociation under formation of protons, which leave the electrode for the solution and hydrate, and electrons, which remain at the electrode. The difference is that, in the case of glass membranes, the electrons are localized in the valence band of surface groups, mostly of oxygen, for instance SiO$^-$, whereas in the case of metals, here platinum, they are delocalized in the conduction band of the metal [2.71]. The comparison of the given equations nevertheless demonstrates that the potential formations at the glass and the metal surface are comparable processes and that, consequently, the principle of the glass electrode response has been well known for a long time. A difference of minor significance is that an effect such as the formation of "jellium", a finite distribution of electrons across the metal/solution interface [2.242], obviously cannot be expected at the glass/solution interface because the electrons are localized in the valence band of the surface groups.

The Interface Between Protonated Glass and a Pt,H$_2$ Electrode

It was interesting with respect to the dissociation mechanism to find that the potential formation at the interface between an electrolytically protonated glass and a platinum layer in hydrogen-containing atmospheres is also based on functional surface groups of the glass, see Sect. 2.10. The electrode glass(H$^+$)|Pt,H$_2$ responds to hydrogen as does the Pt,H$_2$ electrode in aqueous solutions with constant pH, showing thermodynamically correct hydrogen slopes at hydrogen partial pressures between at least 1 and 10^{-4} bar and their correct temperature dependence from ambient to 95 °C. In addition, the interfacial equilibrium

$$\frac{1}{2}\text{H}_2(\text{Pt}) + \text{SiO}^-(\text{s}) \rightleftarrows \text{SiOH}(\text{s}) + \text{e}^-(\text{Pt}) \tag{2.241}$$

of the glass(H$^+$)|Pt,H$_2$ electrode, which was proposed as the potential-determining process, yields a subideal response as does the interfacial equilibrium between glass membranes and solutions, although the effect seems to be too weak to be detected experimentally. The functioning of the glass(H$^+$)|Pt,H$_2$ electrode in hydrogen, consequently, shows that siloxy (and probably also (SiOAl)$^-$ and related groups) is a strongly proton-accepting glass surface group not only in aqueous solution but acts as such also in contact with solid platinum in hydrogen atmospheres. This is a significant generalization of the basis of the dissociation mechanism.

Earlier Attempts to Explain Glass Electrode Functioning by Interfacial Equilibria

Electrically charged entities attached to the surface of membrane glasses have rarely been taken into account as the cause of the glass electrode potential. Two authors, however, must be mentioned at this point. Their assumptions are surprisingly close to the processes which were actually found to take place at the glass surface. However, both authors failed to describe correctly the electrode potential because of fundamental errors in their treatments.

(a) Already in 1943, *Tendeloo* put forward the idea that the equilibrium responsible for glass electrode functioning is the dissociation of weak acids (e.g., silicic acid) and salts attached to the glass surface [2.243]. He thus suggested an explanation of potential formation which we could later verify experimentally as the dissociation mechanism. However, his thermodynamic treatment of the assumed dissociation equilibrium suffered from two serious errors. (1) Tendeloo used the chemical instead of the electrochemical equilibrium to derive an equation for the pH-dependent glass electrode potential, an error, incidentally, frequently reported [2.244]. (2) Although he at first quite correctly introduced the anions at the surface into the equations, he eliminated them at a later stage of the derivation, not realizing the significance of their negative charge for the membrane potential and thus obviously not knowing how to treat the equations containing these "disturbing" particles. Tendeloo's treatment thus resulted in equations which, like those of all other past theories, describe the pH and pNa functions of membrane glasses purely thermodynamically and do not explain them mechanistically. His proposal has thus been forgotten.

(b) In 1962, *Eisenman* developed a theory which predicts the selectivity order of silicate and aluminosilicate glasses for protons and alkali ions [2.62, 63]. He based this work on the change of the standard free energy of the ion exchange between solution and glass surface and, for this reason, calculated bond energies of protons and alkali ions to the different negative sites $(SiO^-, [AlOSi]^-, SiOSiO^-)$ of the glasses. Regrettably, however, he neglected two fundamental preconditions for his model. (1) He calculated the electrostatic energies and omitted the chemical parts although the electrochemical treatment would have demanded the sum of both. (2) He treated the sites at the glass surface as if they were separate singly charged anions in homogeneous solution, and not as sites positioned at a negatively charged glass surface although it is known today that the potential of the glass changes the bond strength of each site considerably [2.72] (see Sect. 2.5.4 and explanations below). As a result, Eisenman obtained extremely negative and positive, respectively, standard free energies of the ion exchange for silicate and aluminosilicate glasses, which, on conversion, yield selectivity constants with unrealistic orders of magnitude. Examples for the silicate groups are: $\log K_{H,Li} = -98.6$ and $\log K_{Na,K} = -8.6$ and for the aluminosilicate group: $\log K_{H,Li} = +45.1$, and $\log K_{Na,K} = +11.8$. That, nevertheless, the selectivity orders obtained by Eisenman reflect the experimentally determined selectivity orders seems to be caused by the nature of the model he used. Because it is based on the relative radii assumed for the anionic sites and on the radii of the cations, the selectivity orders are obviously *a priori* built into the model and do not depend on the absolute magnitude of the standard free energy differences calculated.

Experimental Verification of the Interfacial Equilibrium

The dependence of the membrane potential on the hydronium and alkali ion activities in the transition region is described by

$$\varepsilon_m = \varepsilon_H^0 + k' \log(a_{H_3O^+} + K_{D,H}'' K_{A,M}' a_{M^+}) , \qquad (2.242)$$

where $k' = \alpha k$ is the subideal or practical slope of the membrane glass. (This equation has been derived in Sect. 2.5.4 (2.82).) The product of dissociation and association constant in the logarithm is called the selectivity product [2.87] and characterizes the selectivity of glasses, in the example of (2.242) to hydronium and alkali ions M^+. Small selectivity products mean a relatively strong influence of hydrogen ions upon the potential and are thus characteristic of pH glasses, whereas large selectivity products indicate a strong effect of the particular alkali ion on the potential and characterize pM glasses.

Thus, for a specific membrane glass in solutions with a specific alkali activity, the relative magnitudes of the two terms in the logarithm indicate which of the ionic species preferably determines the potential. Therefore, the logarithm of the second term was defined the transition pH, see Sect. 2.5.4,

$$\text{pH}_{tr} = - \log(K_{D,H}'' K_{A,M}' a_{M^+}) , \qquad (2.243)$$

so that at $\text{pH} < \text{pH}_{tr}$ the influence of the hydronium ions on the membrane potential is stronger than that of the alkali ions; at $\text{pH} > \text{pH}_{tr}$ the reverse is the case; and at $\text{pH} = \text{pH}_{tr}$, i.e., at the transition pH of the solution, both ions determine the membrane potential equally [2.72, 87, 108].

The same terms, $a_{H_3O^+}$ and $(K_{D,H}'' K_{A,M}' a_{M^+})$, determine also the relative mole fractions of the acidic and the salt forms of the glass surface groups which are present at the membrane surface [2.97, 239], see (2.132) and (2.133) in Sect. 2.6.3. For example, the mole fraction of the salt form RM of the surface group R is given by

$$x_{RM}' = 1 - x_{RH}' = \left(\frac{a_{H_3O^+}}{K_{D,H}'' K_{A,M}' a_{M^+}} + 1 \right)^{-1} . \qquad (2.244)$$

Thus, at $\text{pH} < \text{pH}_{tr}$, the mole fraction of the acidic is larger than that of the salt form of the surface groups; at $\text{pH} > \text{pH}_{tr}$ the reverse is true; and at $\text{pH} = \text{pH}_{tr}$ the surface contains equal mole fractions of the surface group modifications. Consequently, potential formation and surface coverage of electrode glass membranes always appear in parallel, not only in the extreme cases of pure pH and pM selectivity but also in the transition range, where both forms of surface groups are present in analytically detectable amounts. The experimental verification of this thermodynamic conclusion was thus of high significance for proving the general validity of the dissociation mechanism. For this reason, the following experiments were conducted [2.72, 87, 108]. As described in Sect. 2.6.3, the relative mole fractions of the acidic and salt forms of the surface groups at the membrane surface were measured by electrolyses employing anolytes with various relative lithium and hydronium ion activities and by subsequently analysing the anodic subsurface ranges of the membranes for their relative concentrations of protons and lithium ions by IBSCA. During the electrolyses, the different ions migrated into the glass and generated uniform protonated glass layers whose concentrations were depth-independent within $x = 0.005$, see Fig. 2.20. These subsurface ranges represent "extensions" of the equilibrium surface mole fractions into the glass and thus yield the mole fractions of the different forms of the surface groups. (For an explanation of the transfer and subsequent migration and for additional justification of this conclusion, see Sect. 2.6.3). Plotting the mole fractions thus obtained as functions of the difference (pLi – pH) of the anolyte applied during the electrolyses resulted in a sigmoid curve (see Fig. 2.19), whose shape was identical to that of a theoretical curve representing mole fractions of the surface groups which had been calculated by (2.244) as functions of (pLi – pH). After adjusting the horizontal position of the calculated to that of the measured curve, the "best" point of inflection of the plots (at $x_H = x_{RH} = x_{Li} = x_{RLi} = 0.5$) yielded the selectivity product $(K''_{D,H} K'_{A,M})_{migr}$, where the index migr indicates that the origin of the data is migration experiments. It agreed well with a selectivity product $(K''_{D,H} K'_{A,M})_{pot}$ that had been obtained by potentiometric measurements on cells containing glass electrodes with the same membrane glass. This result was found with various lithium silicate pH and lithium aluminosilicate pNa glasses having vastly different selectivity products, see Table 2.2 of Sect. 2.5.4, and at various temperatures, Fig. 2.19. Because in all of these experiments the different selectivity products are the result of two basically different kinds of experiment, their agreement means that surface coverage and potential of the membrane glass are caused by the same interfacial equilibrium, which is the central equilibrium of the dissociation mechanism.

Estimate of Heterogeneous Dissociation and Association Constants

The magnitude of the selectivity product is easily accessible, whereas a method for the determination of the single heterogeneous dissociation and association constants of the interfacial equilibria, of which it is composed, has still to be developed. As described in Sect. 2.5.4, however, possible combinations of the single constants have been obtained by basic considerations about the difference between heterogeneous and homogeneous dissociation and association constants of functional surface groups and related dissolved acids and salts, respectively [2.72]. For example, it was estimated that the maximum heterogeneous dissociation constant of silanol groups at the surface of a lithium silicate pH glass is $K''_{D,H} = 10^{-16}\,\mathrm{mol\,kg^{-1}}$, whereas the minimum heterogeneous association constant of SiOLi groups of the same glass is $K'_{A,Li}=10^4\,\mathrm{kg\,mol^{-1}}$, see Sect. 2.5.4, Table 2.3. However, the dissociation constant is probably even smaller and the association constant larger than these values.

Even if these estimated data differ considerably from the correct magnitudes, it is obvious that the heterogeneous dissociation constant of the silanol group, $\equiv$SiOH, is smaller than the first dissociation constant of silicic acid, $(HO)_3SiOH$, in homogeneous solution by *at least* six orders of magnitude. The estimate thus confirms the strong effect of the negative potential of the glass membrane on the dissociation of acidic surface groups.

The corresponding effect of the negative potential of the glass on the salt form of the surface groups is even more striking. According to the estimate, siloxy lithium and in particular aluminosiloxy lithium groups (Sect. 2.5.4, Table 2.4) have rather large association constants and are present at the glass surface at considerable mole fractions under favourable conditions in the solution, whereas dissolved alkali silicates, like most salts in homogeneous solution, are practically completely dissociated under any condition. Thus, the response of membrane glasses to alkali ions is enabled by the negative potential of the glass membrane: it causes primarily the formation of surface salt groups, for instance SiOM and [AlOSi]M, the equilibrium of whose incomplete association is then the basis of the alkali error of pH membrane glasses and the alkali selectivity of pM membrane glasses. The above estimate also confirms a fact often asserted in the preceding discussions, namely that the concentration of dissociated surface groups is minute (see, e.g., Sect. 2.5.3, in particular Table 2.1).

Rigorous Treatment of the Interfacial Equilibrium

The derivation of the glass membrane potential as a function of hydronium ions yielded (2.39) in Sect. 2.5.3, which can be expressed in a general form by

$$\varepsilon_m = -k \log K_{D,H} + k \log \frac{a'_{R^-}\, a_{H_3O^+}}{a'_{RH}\, a_{H_2O}} \tag{2.245}$$

or, after rearrangement, by

$$\varepsilon_m = -k \log K_{D,H} + k \log \frac{a'_{R^-}}{a'_{RH}\, a_{H_2O}} - k\,\mathrm{pH} \ . \tag{2.246}$$

Equation (2.246) contains the unknown activities of surface groups, whose further treatment presented some problems. The usual way is to get rid of quantities that are only approximately constant by combining the corresponding term with a constant term and to define their sum as a constant quantity, which, in the example at hand, would be the standard potential of the interfacial equilibrium [2.87]

$$\varepsilon_m^{0'} = -k \log K_{D,H} + k \log \frac{a'_{R^-}}{a'_{RH} \, a_{H_2O}} \, . \tag{2.247}$$

However, the standard potential defined in this way is not strictly constant because a change in the pH in (2.246) and thus a change in the hydronium activity in (2.245) must necessarily change also the other quantities taking part in the equilibrium so that the problem only seems to be solved.

A Note on Presentations in Textbooks

This treatment, incidentally, is also often used in textbooks when electrode potentials are derived by means of the Nernst equation, see for instance [2.245–247]. For example, the rigorous equation for the potential of a metal electrode (e.g. silver) electrode with the electrode reaction

$$Ag(m) \rightleftarrows Ag^+(sol'n) + e^-(m) \tag{2.248}$$

is first derived by means of electrochemical free energies

$$\varepsilon_{Ag/Ag^+} = \varepsilon_{Ag/Ag^+}^0 + k \log \frac{a_{Ag^+(sol'n)} \, a_{e^-(m)}}{a_{Ag(m)}} \tag{2.249}$$

and subsequently treated by equating the activities of pure silver and electrons in the metal to unity and by including them into the standard potential

$$\varepsilon_{Ag/Ag^+}^{0'} = \varepsilon_{Ag/Ag^+}^0 + k \log \frac{a_{e^-(m)}}{a_{Ag(m)}} \, , \tag{2.250}$$

which reduces (2.249) to the simple Nernst equation

$$\varepsilon_{Ag/Ag^+} = \varepsilon_{Ag/Ag^+}^{0'} + k \log a_{Ag^+(sol'n)} \, . \tag{2.251}$$

Textbooks often neither mention that (2.249) expresses an equilibrium between ions in the solution and electrons in the metal, nor do they present (2.250). To avoid any misunderstanding: this derivation is not faulty, and there is probably no other way of handling activities of pure metals and electrons in metals. Changes in the electron activity as a function of changes in the silver ion activity are immeasurable (except by changes in the potential). What is highly unsatisfactory, however, is that (2.251), in spite of being correct, does not explain to a student why a change in the activity of the silver ions in the solution changes the potential of the silver metal. This information is lost by hiding the electron activity in the standard potential. Certainly, thermodynamics can basically not inform about the mechanism of (electrode) reactions. However, if such information is known, it should not be hidden as in (2.251) without explanation, at least not in textbooks.

The Subideal Response

We have initially chosen a corresponding treatment of the potential of glass membranes by defining the "modified standard potential" of the interfacial equilibrium by (2.247) despite serious misgivings: "The term $\left(c'_{\mathrm{SiO^-}}/c'_{\mathrm{SiOH}}\right)$ is *practically* constant and is, consequently, introduced into the modified standard potential. Rigorously, however, the surface concentration of the siloxy groups actually determining the potential of the glass membrane changes as a function of the hydrogen ion activity" [2.87]. We subsequently verified this statement by finding that the pH-dependent activity of the negatively charged anionic surface groups reduces the ideal to the subideal response of electrode glass membranes [2.72, 95].

The fact that a pH change is necessarily coupled to changes of the other participants in the equilibrium in (2.246) dictates that the activity term on the right side of (2.246) must be included in the slope and not in the standard potential [2.95]. Thus, the rigorous, thermodynamically correct slope is given by

$$
\frac{\mathrm{d}\varepsilon_{\mathrm{m}}}{\mathrm{pH}} = \frac{\mathrm{d}}{\mathrm{dpH}}\left(\log \frac{a'_{\mathrm{R^-}}}{a'_{\mathrm{RH}}\, a_{\mathrm{H_2O}}}\right)k - k \tag{2.252}
$$

and, after rearrangement, by

$$
\frac{\mathrm{d}\varepsilon_{\mathrm{m}}}{\mathrm{pH}} = -\left[1 - \frac{\mathrm{d}}{\mathrm{dpH}}\left(\log \frac{a'_{\mathrm{R^-}}}{a'_{\mathrm{RH}}\, a_{\mathrm{H_2O}}}\right)\right]k = -\alpha k = -(1-n)k \;. \tag{2.253}
$$

In this way, the standard potential remains constant,

$$
\varepsilon_{\mathrm{H}}^0 = -k \log K_{\mathrm{D,H}} \;, \tag{2.254}
$$

and the rigorous equation for the potential is given by

$$
\varepsilon_{\mathrm{m}} = \varepsilon_{\mathrm{H}}^0 - \left[1 - \frac{\mathrm{d}}{\mathrm{d\,pH}}\left(\log \frac{a'_{\mathrm{R^-}}}{a'_{\mathrm{RH}}\, a_{\mathrm{H_2O}}}\right)\right]k\,\mathrm{pH} \;. \tag{2.255}
$$

α is the electromotive efficiency [2.248], and n is the electromotive loss factor [2.95]. The second term (the derivative) in the square brackets of (2.253) and (2.255) is a positive quantity, so that the slope is always smaller than unity. Because

$$
\left|\frac{\mathrm{d}}{\mathrm{dpH}}\left(\log a'_{\mathrm{R^-}}\right)\right| \gg \left|\frac{\mathrm{d}}{\mathrm{dpH}}\left(\log a'_{\mathrm{RH}}\, a_{\mathrm{H_2O}}\right)\right| \;, \tag{2.256}
$$

the fractions in the logarithms of (2.253) and (2.255) can be equated to the nominator $a'_{\mathrm{R^-}}$, so that

$$
\frac{\mathrm{d}}{\mathrm{dpH}}\left(\log \frac{a'_{\mathrm{R^-}}}{a'_{\mathrm{RH}}\, a_{\mathrm{H_2O}}}\right) = \frac{\mathrm{d}\log a_{\mathrm{R^-}}}{\mathrm{dpH}} = -\frac{\mathrm{dpR}}{\mathrm{dpH}} \;, \tag{2.257}
$$

which demonstrates more clearly the determining role of the anionic groups in (2.253) and (2.255) and also makes these equations less clumsy. For further simplification, the definition $pR \equiv -\log a_{R^-}$ corresponding to that of pH was introduced into (2.257) [2.95].

Thermodynamics yields also that, apart from possibly small influences of activity coefficients, the electromotive efficiency $\alpha = (1 - n)$ is the same for all surface groups R (and thus for membrane glasses), and for the hydronium, deutonium, and alkali responses of glass membranes [2.95, 108]. For the pD and pM functions, however, this thermodynamic conclusion remains to be verified experimentally.

Subideal and Sub-Nernstian Response

The rigorous thermodynamic treatment of the interfacial equilibrium yields the unexpected result that the slope of glass electrode membranes cannot be "ideal" ($k = (RT \ln 10)/F$) but must necessarily be subideal ($k' < k$) although it does not *a priori* give the magnitude of k'. It was thus of high interest that glass electrodes had repeatedly been reported to exhibit electromotive efficiencies in the range $\alpha = 0.995$–0.998 (see [2.95]), a deficiency which is called the sub-Nernstian response. We confirmed these reports by careful measurements on several pH glass electrodes, which yielded $\alpha = 0.997$–0.998 at $25\,^{\circ}\mathrm{C}$ and $50\,^{\circ}\mathrm{C}$ [2.95]. Obviously, the magnitude of the subideal response is indeed given by the reported electromotive efficiency $\alpha = (1 - n)$, which, in turn, has thus found an explanation by the thermodynamics of the interfacial equilibrium.

Mechanistic Details of the Interfacial Equilibrium

The information obtained by the reported rigorous treatment yields remarkable details of the interfacial equilibrium. What is generally known is the overall practical slope

$$\frac{d\varepsilon_m}{dpH} = -k' = -\alpha\, k = -(1 - n)k \ , \tag{2.258}$$

which is in the range 58.5–59.0 mV. The electromotive loss factor subdivides the overall response into (a) the pH dependence of the activity of the negative surface groups,

$$\frac{d \log a'_{R^-}}{dpH} = -\frac{dpR}{dpH} = n = 1 - \alpha \ , \tag{2.259}$$

and (b) the dependence of the membrane potential on the activity of the negative surface groups,

$$\frac{d\varepsilon_m}{d \log a'_{R^-}} = -\frac{d\varepsilon_m}{dpR} = -\frac{1 - n}{n}k = \frac{\alpha}{1 - \alpha}k \ . \tag{2.260}$$

As a result of the magnitude of α (and n), dpR/dpH is of the order of several thousands. This confirms our repeated statement that the change in the activity of negative surface groups caused by pH changes is minute, see Table 2.1 in Sect. 2.5.3. The qualitative statement was at first derived from the strong influence of the potential of the glass upon the dissociation of the acidic groups [2.87], whereas (2.259) now presents a quantitative measure. To give an impressive example: a pH change by -14, corresponding to a change in the hydronium activity by a factor of as large as 10^{14}, causes a change of the activity of the negative surface groups by only -8% if the average electromotive efficiency of 0.9975 is assumed as the basis.

In contrast, $d\varepsilon_m/dpR$ is of the order of several tens of volts per unit pR, see Table 2.1 in Sect. 2.5.3. It reflects the strong relationship between the potential of the glass and the charge density at its surface, see Fig. 2.67. Because of this most direct response of the potential to an activity, it is called the "internal slope" of the electrode response. An example may also demonstrate the magnitude of this term for $\alpha = 0.9975$: the membrane potential changes by as much as $\pm 23.6\,\mathrm{V}$ when the activity of the negative surface groups is changed by only $\pm 10\%$.

A Different Presentation of Interfacial Equilibria

Integration of (2.257) and observing (2.259) yield the version

$$\frac{a'_{R^-}\, a^n_{H_3O^+}}{a'_{RH}\, a_{H_2O}} = \text{const.} \tag{2.261}$$

of the heterogeneous interfacial equilibrium. It represents an interesting modification of the equation as obtained from (2.246),

$$\frac{a'_{R^-}\, a_{H_3O^+}}{a'_{RH}\, a_{H_2O}}\, 10^{-(\varepsilon_m/k)} = K_{D,H} \,, \tag{2.262}$$

as it demonstrates the effect of the electromotive loss factor n upon the dissociation of acidic surface groups. It is also interesting to compare (2.261) with the corresponding homogeneous dissociation equilibrium

$$\frac{a_{R'^-}\, a_{H_3O^+}}{a_{R'H}\, a_{H_2O}} = K_{D,\text{sol'n}} \tag{2.263}$$

of a corresponding dissolved acid R'H, which reveals that the characteristic difference between heterogeneous and homogeneous dissociation is the power n of the hydronium activity for the heterogeneous dissociation. In addition, (2.261) again demonstrates that changes of the activity of the negative surface groups caused by pH changes are minute because the electromotive loss factor n is of the order of a thousandth.

2.9.2 Subsurface Interdiffusion

Origin

The acid and salt forms of the glass surface groups, whose dissociation and association equilibria, respectively, cause the response of the glass membrane, are also the source of a serious complication for the application of glasses as electrode membranes. Because the ions attached to the surface groups differ nearly always in their kind and/or their concentration from the alkali ions of the glass interior, concentration gradients are set up and lead to subsurface interdiffusion. Although ionic interdiffusion between glasses and solutions had been known for a long time, it was unknown before the detection of the dissociation mechanism that it is the glass surface in equilibrium with the solution which supplies the ions that diffuse into the glass versus the ions of the glass interior. Until this discovery, the glass surface had been tacitly assumed to be an inert plane through which cations diffused under the driving force of the difference of their free energies in the two phases.

Systematic Occurrence

The knowledge that the interfacial equilibrium determines the coverage of the glass surface and thus the ion species available for an interdiffusion changed the situation [2.72, 87, 108]. The above statement about the basically parallel occurrence of selectivity and surface coverage must be extended to subsurface interdiffusion: pH selectivity is connected with the presence of the acidic form of the surface groups and, in addition, with an interdiffusion of the protons of the surface groups versus the cations of the glass, whereas pM selectivity always appears together with a coverage of the surface with the salt form of the surface groups and also with an interdiffusion involving their alkali ions. In the transition range, both states overlap to an extent which is given by the relative presence of the acidic and salt forms at the surface. Besides, this change in the subsurface glass composition is the first step of glass corrosion, whose nature is thus also determined by the selectivities of membrane glasses.

Characteristic Composition Changes of Subsurface Glass Ranges

The transport properties of the ions in the glass and the possible participation of water in the interdiffusion processes determine whether glasses dissolve congruently or incongruently, and by doing so also determine the nature of the subsurface ranges generated. In lithium silicate pH glasses [2.71, 188], for instance, the small diffusion coefficients of protons and the participation of only small amounts of water lead to incongruent dissolution and to the formation of so-called leached layers, whose formation rate is determined by the proton mobility, which is smaller than that of the lithium ions of the glass by two [2.86, 192] to three [2.79] orders of magnitude. After a certain time, these layers attain a steady state and thus a constant thickness, which is determined by equal rates of layer formation and dissolution [2.86, 192, 241]. The independence of the concentration profiles of leached layers from mechanical friction confirms that they consist mainly of ion-exchanged glass [2.86] and do not represent "swollen surface layers" as usually assumed. The alkali/alkali interdiffusion in lithium aluminosilicate pM glasses, on the other hand, is connected with congruent dissolution of glass layers whose alkali ions have been exchanged for alkali ions from surface groups and whose thickness increases for long periods. Protons and water molecules are excluded from this exchange [2.87, 188]. Alkali/alkali-exchanged layers will also certainly attain steady states after extremely long exchange times. The development of constant layer thicknesses, however, has not yet been observed under these conditions, even after extended reaction periods. Alkali/alkali-exchanged glass layers were also found to be mechanically stable and thus to consist of an unchanged glass network, which contains the exchanged alkali ions, and not of a gelled glass.

Consequences

The various interdiffusion processes are important for the application of glass electrodes and for the development of new electrode membrane glasses. Mainly two consequences are to be mentioned.

(a) *Disturbance of Interfacial Equilibria.* In principle, cations that leave surface groups for the glass disturb the prevailing interfacial equilibrium and thus polarize the phase boundary potential. This effect is negligible if the current density, representing the rate by which the ions leave the surface group, is much smaller than the exchange current density of the equilibrium, which is a kinetic measure of the stability of the potential [2.71], see Sect. 2.5.5. "Good", modified lithium silicate pH electrode glasses, which have small dissolution rates and take up only small amounts of water, always meet this condition after they have been in contact with solutions for a certain time. However, the condition is not met during the initial time after their first contact with solutions because, owing to the large concentration gradients during this period, the ionic interdiffusion rates are still large. So-called soft glasses, for instance potassium silicate glasses, which are distinguished by large dissolution rates and by an uptake of large amounts of water, do not meet the condition either and are thus not suited as electrode membrane glasses.

(b) *Subsurface Diffusion Voltages.* Ionic interdiffusion always causes diffusion voltages, which add to the total potential of glass membranes and, consequently, to the emf of the glass electrode-containing cells. Thus diffusion voltages in glass membranes are unavoidable and glass electrodes are therefore ready for measurements only as long as the diffusion potentials in their membranes are constant. This condition is automatically met by pH glass membranes after their leached layers have nearly formed, whereas it is not met during the first stages of the leaching process. Changing diffusion voltages and large interdiffusion current densities (see explanations under (a)) both cause potential drifts during the initial leaching process, a phenomenon empirically acknowledged for a long time and called the "formation period" [2.86, 241]. In contrast, diffusion voltages in pM-selective membranes must be kept constant by preconditioning and storing the electrodes in solutions that guarantee a permanent coverage of the glass surface with the alkali ion to be measured [2.213], that is, their pH must be considerably larger than the transition pH, $pH > (pH_{tr} + 2)$ [2.96], see Sect. 2.5.4. The purpose of this treatment is thus not to generate and maintain certain selectivities of the membrane glass, as suggested in the literature [2.212–214], but to exclude potential drifts during measurements [2.96].

Sterical Effects: the Reason Why Modern pH Glasses are Lithium Silicate Glasses

A precondition for the participation of ions in subsurface interdiffusion is their transfer from the surface group to a cationic site in the glass. This seems to be trivial. However, cations can penetrate glasses only if they "fit" the glass network, or, in other words, if their radius is not appreciably larger than the radius of the pathways and sites in the glass. We were the first to prove experimentally that "sterical hindrance" is indeed a frequent phenomenon [2.55, 97, 239], see Sect. 2.6.4; empirically it had been the guiding idea in the development of lithium silicate-based electrode membrane glasses [2.218]. Lithium silicate glasses are generally "closed" to cations larger than lithium and are thus not subject to potential drifts caused by ionic interdiffusion. For this reason, practically all pH electrodes today contain membranes made of more or less modified lithium silicate glasses.

However, the mechanism of potential formation has meanwhile revealed that the sterical hindrance of these glasses operates only in the pH range of the sodium error, where alkali ions are attached to surface groups and thus have a chance to penetrate the glass. At pure pH selectivity, in contrast, the surface is covered with proton-bearing silanol groups, SiOH, which "block" the glass surface, independent of the alkali ion of the glass [2.87, 96].

A striking example for the operation of sterical hindrance is lithium silicate glasses that contain tantalum oxide [2.215], see Sect. 2.7. The integration of tantalum atoms in the glass network obviously widens the glass structure and creates larger vacancies than are present in the pure silicate glass. Sodium and even potassium ions penetrate these glasses in the pH range of the sodium error, generating large potential drifts [2.87, 96], see Fig. 2.62. At pure pH selectivity, however, these glasses also show the protecting effect of the proton-bearing silanol groups. Glass electrodes containing such membranes are thus applicable only at 100% pH selectivity, for which, however, there is not much use.

Determination of Subsurface Diffusion Potentials

The range of subsurface interdiffusion in pH membrane glasses can be displaced to greater depths by means of electrolytic protonation [2.239], as described in detail in Sect. 2.7.4. This possibility yielded the following significant results.

(a) The diffusion voltage within the outer part of the leached layer, which has a partly hydrated structure, was determined experimentally to be typically $\varepsilon_{j,L} = (-26 \pm 2)$ mV (glass negative) [2.72, 241].

(b) Calculations as described in detail in Sect. 2.7.4 yielded also diffusion voltages in the transition parts of the leached layer, which has a nearly unchanged glass structure. The value obtained is typically $\varepsilon_{j,tr} = (-48 \pm 2)$ mV.

(c) The sum of the values determined under (a) and (b) yields the total diffusion voltage of the leached layer, which amounts to $\varepsilon_j = (-74 \pm 4)$ mV. The magnitude of these values [2.72] is not generally expected.

(d) The diffusion potential $\varepsilon_{j,L}$ of the outer part of the leached layer is independent of pH at pure pH selectivity of the membrane glass but changes in the sodium error pH range to less negative values and even changes its sign at high pH values [2.96], see Fig. 2.59. Because alkali ions do not participate in the subsurface interdiffusion owing to their sterical hindrance, this change in the interdiffusion is obviously caused by their presence at the glass surface, where they replace a certain amount of protons and block their participation in the interdiffusion with lithium ions. The effect has been measured; its magnitude, however, can hardly be obtained theoretically because of the complicated conditions of the interdiffusion.

The Three Sodium Errors of Glass Electrodes

It is common knowledge that the potential of glass electrodes shows positive deviations at large pH and large alkali concentrations. They are termed the sodium (or alkali) error and are explained by the onset of an alkali sensitivity of the membrane glass. Much less known is the fact that this potential deviation is actually the sum of two deviations, both of which result from the presence of alkali ions at the glass surface.

- The alkali ions attached to some surface groups contribute directly to the interfacial potential because the pH is in the onset of the transition range of the membrane glass. Accordingly, we have termed this part of the sodium error the *direct reversible sodium error*.

- The alkali ions at the surface hinder some of the protons from penetrating the network and thus influence the lithium ion–proton interdiffusion in the leached layer, although they do not participate themselves in the interdiffusion because of their sterical hindrance. This leads to an indirect pH-dependent change of the subsurface diffusion voltage, which also vanishes when the glass comes into contact with solutions outside the sodium error range. We have thus called it the *indirect reversible sodium error*. The effects were separated by potential measurements on protonated and unprotonated glass electrodes, see Sect. 2.6.2.

- A third kind of sodium error has already been described above. It is caused by the diffusion of sodium and even potassium ions into tantalum (and other) oxide-containing lithium silicate glass membranes and is characterized by large potential drifts to more positive values. This "error" is slowly, and only partly, reversible in solutions outside the sodium error pH range. We have thus termed it the *time-dependent, irreversible sodium error*.

Experimental Disproof of the Ion Exchange Theory

The potonation experiments discussed yielded an additional result, which is most significant for the mechanism of the glass electrode response. The potential slope of protonated (alkali-free) pH glass membranes, $(\mathrm{d}\varepsilon_\mathrm{m}(\mathrm{H})/\mathrm{dpH})$, in alkali-free buffer solutions is equal to the slope $(\mathrm{d}\varepsilon_\mathrm{m}/\mathrm{dpH})$ of the membranes before their protonation:

$$\frac{\mathrm{d}\varepsilon_\mathrm{m}(\mathrm{H})}{\mathrm{dpH}} = \frac{\mathrm{d}\varepsilon_\mathrm{m}}{\mathrm{dpH}} = -k' \ , \tag{2.264}$$

see Fig. 2.59. This result dismisses an exchange of different ions as the origin of the pH response of electrode glass membranes and disproves experimentally all glass electrode theories based on an ion exchange [2.86, 192].

No Combined Equation for Interfacial and Diffusion Potentials

The foregoing discussion has shown that diffusion potentials in glass membranes are caused by the glass surface coverage and, primarily, by the interfacial equilibrium. Although they usually represent complications, they can often qualitatively explain deviations of experimental from expected results. However, it is not helpful and often not even possible to combine the equation of the interfacial with an equation of the diffusion potential, as exemplified by *Eisenman* for the ion exchange theory [2.62, 63], because the potential-determining ions often differ from the interdiffusing ions. The most general example is the response of a lithium silicate pH glass in a sodium-containing solution in the sodium error pH range, where sodium and hydronium ions generate the interfacial potential, whereas interdiffusing lithium ions and protons are responsible for the diffusion potential. Besides, the dependence of the individual ionic diffusion coefficients on concentration and glass structure, which would be necessary for such calculations, is unknown, except for the treatment reported in Sect. 2.7.4.

An example for such a combination is Eisenman's equation [2.62]

$$E = E^0 + n\frac{RT}{F}\ln\left[a_i^{1/n} + \left(K_{ij}^{pot}a_j\right)^{1/n}\right] , \tag{2.265}$$

which combines Nicolsky's equation [2.57], (2.6) of Sect. 2.3, and an expression for the diffusion potential [2.209, 249], (2.216). The constant in the brackets on the right side of (2.265) is given by

$$K_{ij}^{pot} = \left(\frac{u_j}{u_i}\right)^n K_{ij} \tag{2.266}$$

and consists of

- Nicolsky's ion exchange constant K_{ij}, which is a reversible quantity;
- the ratio of the mobilities u of the ions i and j, being irreversible quantities, whose structure-dependent and concentration-dependent magnitudes are generally unknown; and
- the adjustable parameter n originating from the assumption that glass is "a perfect cation exchanger obeying n-type non-ideal behaviour (i.e., $d\ln a/d\ln c = n$)" [2.10].

Equation (2.266) allows (2.265) to be fitted to any experimental results [2.63], independent of the physical meaning of the parameters. This suffices as long as only agreement between the describing equation (2.265) and experiments is required, and it represents an excellent basis for practical applications. No mechanistic information, however, can be expected from Eisenman's equation because the basis is the purely thermodynamic ion exchange theory, see Sect. 2.3. The dissociation mechanism, on the other hand, explains the glass electrode response by experimentally verified physico-chemical reactions. It thus represents a physical understanding of their functioning, and consequently involves no equations that must be fitted to the experiments, such as (2.265) and (2.266).

Is the Glass Electrode Potential a Reversible Quantity?

This question may be unexpected, but it is not trivial. The total potential $\varepsilon_{gl.el.}$ of a glass electrode is the sum of five potentials (see Fig. 2.68a)

$$\varepsilon_{gl.el.} = \varepsilon_{ref} - \varepsilon_{m,i} - \varepsilon_{j,m,i} + \varepsilon_{j,m,o} + \varepsilon_{m,o} , \tag{2.267}$$

the reversible potential ε_{ref} of the internal reference electrode, two reversible interfacial potentials, $\varepsilon_{m,i}$ at the internal and $\varepsilon_{m,o}$ at the outer membrane surface, and two irreversible subsurface diffusion potentials, $\varepsilon_{j,m,i}$ below the internal and $\varepsilon_{j,m,o}$ below the outer membrane surface. The diffusion potentials below the opposite surfaces are generally almost equal, for instance when both surfaces have developed a leached layer at 100% pH selectivity, and nearly cancel due to their opposite sign. Although, consequently, the resulting total potential $\varepsilon_{\mathrm{gl.el.}}$ of glass electrodes is close to the sum of its reversible parts, it is not an entirely reversible quantity because it contains also small irreversible parts. However, these extend only over thin subsurface regions, have opposite signs, and thus add little to the absolute value of the total potential. The irreversibility does not affect the pH and pM response of glass electrodes because the electrode functioning is entirely determined by the reversible interfacial potential at the outer membrane surface, except for disturbances by the diffusion potential within the transition range.

The conditions in ion exchanger membranes are different. For comparison, Fig. 2.68b shows an arrangement for a cation exchanger membrane and a bi-ionic system after *Mackay* and *Meares* [2.250], which corresponds to the glass electrode arrangement in Fig. 2.68a. The total potential $\varepsilon_{\mathrm{M.el.}}$ of this "membrane electrode" is made up of six potentials,

$$\varepsilon_{\mathrm{M.el.}} = \varepsilon_{\mathrm{ref}} - \varepsilon_{\mathrm{Do,i}} + \varepsilon_{\mathrm{j,M}} + \varepsilon_{\mathrm{Do,o}} - \varepsilon_{\mathrm{i,sol'n,i}} + \varepsilon_{\mathrm{j,sol'n,o}} . \qquad (2.268)$$

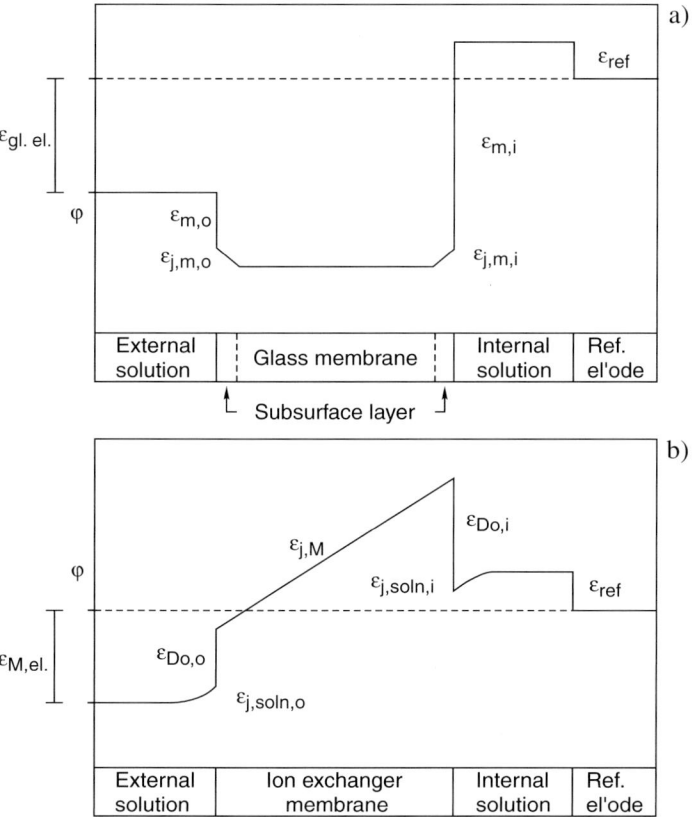

Fig. 2.68. Systematic potential distribution within (**a**) a glass electrode and (**b**) an ion exchanger electrode

As with the glass electrode, the potential ε_{ref} of the reference electrode is reversible. The so-called Donnan potentials $\varepsilon_{\text{Do,i}}$ and $\varepsilon_{\text{Do,o}}$ at the internal and external membrane surfaces, respectively, also are reversible potentials. They result from the fact that cations, but no anions, can penetrate the membrane/solution interface. Due to the transparency of the membrane to cations and solvent molecules, the counterdiffusion of cations with different mobilities across the membrane causes the diffusion voltage $\varepsilon_{\text{j,M}}$, which is irreversible. The two diffusion voltages $\varepsilon_{\text{j,sol'n,i}}$ and $\varepsilon_{\text{j,sol'n,o}}$ within the solutions, which are caused by the cation transport from the solutions to the membrane surfaces and vice versa, must also be added to the total potential. It thus results that, in contrast to glass electrodes, the potential of an ion exchanger membrane electrode contains a considerable irreversible part, which extends through the entire membrane thickness [2.250] and does not cancel. Also different from glass electrodes, its response is partly irreversible because one of its origins is the pH-dependent and pM-dependent irreversible diffusion potential $\varepsilon_{\text{j,M}}$, whose magnitude is determined by the pH- and/or pM-dependent Donnan potential at the external membrane surface also outside the transition range (the internal Donnan potential is fixed by the internal reference buffer solution).

Glass Membranes are No Ion Exchanger Membranes

In conclusion, the treatment of glass electrodes as perfect ion exchangers [2.62, 63] is mechanistically faulty because neither cations (or anions) nor solvent molecules diffuse through their membranes. The irreversible parts of their potentials are small, extend only over a few tens of nanometres below the surfaces, and often nearly cancel because of their equal magnitudes and opposite signs. In contrast, the diffusion potential of ion exchanger membranes is generally considerable and extends through the entire membrane because cations diffuse through the ion exchanger material. Strictly speaking, (2.265), which combines the equations for the phase boundary and the diffusion potential, is thus restricted to real (organic and inorganic) ion exchangers and does not correspond to the physical chemistry of glass electrodes, although it is certainly well suited to describe measurements by fitting the involved parameters. Interestingly, Eisenman has experimentally demonstrated the condition for transforming a glass membrane into an ion exchanger membrane by showing that a $1\,\mu\text{m}$ thick membrane of NAS 27-4 glass, which had been hydrated (leached) through its entire thickness, behaved exactly like an ion exchanger membrane [2.251]. With a membrane having a thickness of tens of millimetres, this observation would not have been made because its complete hydration is impossible.

Unjustified Criticism

The above paragraphs on ionic interdiffusion in glass membranes demonstrate that our studies included the potential-determining interfacial equilibrium as well as subsurface interdiffusion, and even yielded the magnitude of diffusion potentials. We found that the diffusion potential, in spite of being part of the total potential, does not contribute to the activity-indicating potential but frequently acts as a disturbing quantity and must therefore be carefully considered. In the light of these (thoroughly published) results, Doremus' statement that "in this theory (i.e., the dissociation mechanism) the equilibrium of ions at the glass surface (Eq. 12) is included, but not the diffusion potential" [2.252], seems quite inexplicable. A key to a possible explanation of his remark is perhaps given by the inappropriate mention of Nicolsky's equation (!) (Eq. 12) in this connection and by his additional statement that "the internal diffusion potential must be included to obtain agreement between theory and experiment". It is obviously difficult to free oneself from thinking in terms of an accustomed, but merely thermodynamically founded, theory and start to study, and perhaps finally accept, a new mechanism that has been experimentally verified.

2.9.3 pD Response of Glass Membranes

Glass electrodes have long been known to respond to deutonium ions and alkali ions in heavy water solutions in a similar way as they respond to hydronium and alkali ions in ordinary water [2.253, 254]. In Sect. 2.11 we show that the functioning in both solvents reflects the same mechanism [2.108], and we also show that this was confirmed by experiments which corresponded exactly to those that had proved the dissociation mechanism in light water.

Two cases of isotopic difference must be distinguished. (a) The heterogeneous dissociation of deuterated surface groups, which is the basis of the pD response, involves the isotopic ion, i.e., the deuteron, instead of the proton,

$$RD(D_2O) + D_2O \rightleftarrows R^-(D_2O) + D_3O^+ , \tag{2.269}$$

and is thus termed a direct isotope effect [2.108]. (b) The association of surface groups with alkali ions, which is the origin of the pM(D_2O) response, differs only by the solvation of the alkali ions (and surface groups),

$$R^-(D_2O) + M^+(D_2O)_n \rightleftarrows RM(D_2O) + nD_2O , \tag{2.270}$$

and therefore is called an indirect isotope effect. The thermodynamic treatment of these interfacial equilibria yields expressions for the pure pD and pNa(D_2O) dependencies of the interfacial potential and their combination an equation for the potential in the transition range. The relationships correspond to those in light water and are also characterized by subideal slopes, which, however, remain to be verified experimentally.

The knowledge that glass electrodes function by the same type of mechanism in the isotopic solvents yields a quantitative explanation for the empirical determination of pD and pNa(D_2O) values, a method that has been applied for a long time but was never explained, see for instance [2.253–255]. It comprises the calibration of a pH glass electrode cell in pH standard buffer solutions in light water, its application to the unknown solution in heavy water, and the addition of a correction term δ_{glass} to the apparent (operational) pH^D thus obtained:

$$pD = pH^D + \delta_{glass} .\tag{2.271}$$

The same procedure, carried out with a pNa glass electrode cell and pNa standard solutions in light water, yields the pNa(D_2O) of an unknown solution in heavy water, where the correction $\delta_{glass,Na}$ is added according to

$$pNa(D_2O) = pNa(H_2O)^{D_2O} + \delta_{glass,Na} .\tag{2.272}$$

Details will be presented in Sect. 2.11. For this discussion it is of interest that both correction terms reflect properties of the membrane glass. Thus, the difference between pD and pH^D involves the heterogeneous dissociation constants $K_{D,H}$ and $K_{D,D}$ of the glass surface groups and (to a lesser degree) the subsurface diffusion voltages $\varepsilon_{j,m,H}$ and $\varepsilon_{f,m,D}$ of the correspondingly leached electrode membrane:

$$\delta_{glass} = \log \frac{K_{D,H}}{K_{D,D}} + \frac{\varepsilon_{j,m,D} - \varepsilon_{j,m,H}}{k} .\tag{2.273}$$

This explains why correction terms reported for various glass electrodes do not represent one common value but are scattered over a small but significant range [2.255], see Sect. 2.11, Fig. 2.74. The correction term δ_{glass} is no universally constant number but a specific quantity of membrane glasses. This is also confirmed by rather different correction terms δ_{glass} obtained for pH silicate and pNa aluminosilicate glasses and their temperature dependencies.

A corresponding expression results for the correction term for empirical pD(Na) measurements,

$$\delta_{glass,Na} = \log \frac{K_{A,Na(D_2O)}}{K_{A,Na(H_2O)}} ,\tag{2.274}$$

which, however, does not involve diffusion voltages because alkali/alkali interdiffusion in subsurface glass layers is independent of the solvent as long as the membrane surface is completely covered with the salt form of the surface groups.

Perhaps the most surprising result is that the sodium error of glass electrodes is slightly, but detectably, smaller in heavy than in light water solutions. Expressed in different terms, the sodium-selective range of glass electrodes is slightly wider in light than in heavy water. This is a consequence of the rather different magnitudes of the pD and pNa(D_2O) correction terms ($\delta_{glass}/\delta_{glass,Na} \approx 4.6$), which result in transition (pH,pD) values in the isotopic solvents that differ by ($pD_{tr} - pH_{tr}) \approx 0.3$.

2.9.4 Potential-Impairing Reactions

There are several reactions at electrode glass membranes that interfere with the interfacial equilibrium and disturb or obscure the correct functioning of glass electrodes. Examples have been given in Sect. 2.7.5. Now we will briefly discuss the formation of lanthanum phosphate and fluoride layers at membrane glasses [2.219]. It is a remarkable reaction and an extremely insidious one because of its delayed appearance and imperceptibly small amounts of material involved. Therefore, this reaction probably often occurred during empirical membrane glass development and is still occurring with glass electrodes rather frequently without being noticed. Although most glasses – but not all – react with orthophosphate ions and, in parallel fashion, with fluoride ions, the following discussion focuses on the effect of phosphate, which was studied in more detail.

Since 1948/49, when *Perley* first recommended small contents of lanthanum oxide in electrode glasses because of several favourable effects [2.217, 218], researchers involved in glass development have used this ingredient. However, it was not until 45 years later that the reaction of the lanthanum of membrane glasses with orthophosphate ions, for example in phosphate buffer solutions, was detected [2.219]. The solubility of $LaPO_4$ in aqueous solution is so small that the lanthanum ions, on reaching the glass surface by the slow dissolution of the glass, react instantaneously with orthophosphate ions. The resulting lanthanum orthophosphate molecules remain at the surface, accumulate, and form patches of a monomolecular layer which slowly grow, eventually crystallize, and extend over the entire membrane surface. Only after complete coverage of the glass surface does the layer grow in the direction perpendicular to the glass surface.

This sequence of layer formation in a phosphate-containing solution is reflected by the development of the membrane potential (see Fig. 2.63), which during the primary formation of $LaPO_4$ molecules is constant. Depending on temperature and glass composition, this "induction period" can last several hours (see Fig. 2.64), a long span which is usually assumed to be indicative of "good" behaviour of glass electrodes. After this period, the potential suddenly starts to drift towards more negative values, and the "drift period" lasts 2.5 times as long as the induction period. The drift extends over a range of –4 mV to –18 mV, with heat-treated membranes even up to –32 mV. The change is caused by the formation of more extended patches of freshly crystallized $LaPO_4$ that act as thermodynamic phases and develop their own potentials, which add to the potential of the glass membrane in a proportion given by the surface areas of the two materials exposed to the solution. After complete coverage of the membrane surface with $LaPO_4$, the electrode potential becomes finally constant, but is faulty. This is demonstrated by a continuous positive potential drift towards the original value after the electrode is brought into acidic solutions, which dissolve the deposit according to the pH-dependent solubility of $LaPO_4$.

This mechanism has been confirmed by electrochemical measurements at various temperatures and pH values and by considering the strong pH dependence of the solubility of orthophosphate [2.219]. The deposits in their various stages of development were also analysed by IBSCA, SEM, and EDX. An equation was derived that describes the time-dependent pH response of glass electrode potentials during the development of the phosphate error.

It is almost certain that, due to the extremely insidious character of the reaction, many membrane glasses were developed which had excellent properties but were subject to the phosphate error because of a lanthanum oxide content. In this case, their faulty potentials were either detected in routine tests during glass development or during subsequent electrode applications in phosphate-containing solutions, for instance phosphate buffers. In both instances, the "new" electrodes were probably discarded and the reason for their faulty behaviour remained unexplained. In a worse case, however, the potential drifts were falsely interpreted as solution changes, with serious consequences. An example is the pH of phosphate-containing solutions with bacteria for biochemical reactions. A sudden pH change of such solutions often indicates a disturbance of the process or the end of the active bacteria lifetime. Because biochemical processes are usually very costly, the phosphate effect may thus have seriously reduced their economic efficiency in both cases.

2.10 Electrometric pH Measurements

Glass electrodes are applied for measuring activities or concentrations of alkali ions in aqueous and sometimes non-aqueous solutions. By far the most frequent application, however, is to pH measurements. Indeed, pH measurement by means of glass electrodes is the most often conducted analytical procedure in nearly all fields of chemistry, that is, in research and development and for process control, in medicine, pharmacology, environmental analysis, food production, and so forth. However, the subject of pH is actually rather complicated and has been controversially discussed worldwide for the last two decades. The complication arises from the fact that the concept of pH is unique amongst chemical, physical, and physico-chemical quantities. (a) Electrochemical cells applied to its measurement unavoidably involve a liquid junction between the solution of interest and the reference electrode solution which introduces a principally unknown liquid junction potential into the measured emf. (b) The proton – or, more exactly, its hydrated form, the hydrogen or hydronium ion, H_3O^+ or $H_3O^+(H_2O)_n$ – is a single ion, and single ion activities, in principle, are immeasurable. Nevertheless, the significant role of the proton necessitates the definition of an appropriate measure of its activity, the pH, and a measuring procedure that guarantees traceability of the measured pH values to this definition. Because, judging from experience, this field is generally not well understood, it seems appropriate to outline the fundamentals of pH measurement and thus to provide the reader with the knowledge needed for carrying out meaningful pH measurements and for avoiding serious errors during the experimental work and in the interpretation of the obtained pH values.

2.10.1 Definition of pH

Different from other cations, which, perhaps in hydrated or otherwise complexed form, are present in aqueous (and non-aqueous) solutions at fixed concentrations, hydrogen or hydronium ions are always in equilibrium with hydroxyl ions. Stated in a different way, water is dissociated to a small degree into (solvated) protons and hydroxyl ions,

$$H_2O \rightleftarrows H^+ + OH^- \ . \tag{2.275}$$

Equilibrium (2.275) is deliberately formulated with a proton and not with a hydronium ion, as in several textbooks and monographs, because the corresponding equilibrium

$$H_2O + H_2O \rightleftarrows H_3O^+ + OH^- \tag{2.276}$$

tends to suggest that each proton is attached to only one water molecule, which is not the case in reality. The process is called autodissociation and is controlled by the (thermodynamic) autodissociation constant

$$K_D = \frac{a_{H_3O^+}\, a_{OH^-}}{a_{H_2O}} = \frac{m_{H_3O^+}\, m_{OH^-}}{m_{H_2O}} \frac{\gamma_{H_3O^+}\gamma_{OH^-}}{\gamma_{H_2O}} \ , \tag{2.277}$$

where a is the activity, m the molality, and γ the activity coefficient of the ionic species indicated. In not too highly concentrated solutions, the molality of water is practically constant and close to that of pure water, $m_{H_2O} = 55.51\,\mathrm{mol\,kg^{-1}}$, and can be included in the autodissociation constant,

$$K_D' = K_D m_{H_2O} = \frac{a_{H_3O^+}\, a_{OH^-}}{\gamma_{H_2O}} = m_{H_2O^+}\, m_{OH^-} \frac{\gamma_{H_3O^+}\gamma_{OH^-}}{\gamma_{H_2O}} \ . \tag{2.278}$$

Besides, the activity coefficient of water is approximately unity under these conditions so that for dilute solutions

$$K_W = K_D\, a_{H_2O} = a_{H_3O^+}\, a_{OH^-} \tag{2.279}$$

and for infinitely diluted (or ideal) solutions by

$$K_W \cong m_{H_3O^+}\, m_{OH^-} \ . \tag{2.280}$$

K_W is called the ion product of water. Incidentally, autodissociation or auto-protolysis is a characteristic property of all protonic solvents. For methanol, for instance, the autodissociation equilibrium is given by

$$2CH_3OH \rightleftharpoons CH_3OH_2^+ + CH_3O^- \ . \tag{2.281}$$

The above equations demonstrate two characteristic properties of hydrogen ions in solution. (a) Their concentration (and activity) is not fixed *per se* but depends on the composition of the solution, particularly on the concentration of acidic and basic components. It must, consequently, be measured *in situ*, that is, by a method which does not cause any concentration or activity changes. This requirement is met by electrochemical means, for example by a potentiometric glass electrode cell dipping into the solution and yielding a potential difference, referred to that of a reference electrode, which characterizes the hydrogen activity. (b) The possible hydrogen concentration (and activity) covers a wide range extending from approximately $10 \, \mathrm{mol \, kg^{-1}}$ up to $10^{-15} \, \mathrm{mol \, kg^{-1}}$ at room temperature. This range of nearly 16 decades cannot be expressed in a direct numerical manner if a tremendous number of decimal places is to be avoided. Sørensen first realized this fact and consequently introduced the symbol pH (originally p_H) by defining it as the negative logarithm of the hydrogen ion *concentration*, $pH = - \log c_{H+}$ [2.25]. Fifteen years later, when further development of the theory of electrolyte solutions had led to the definition of the ionic activity, he changed the definition to the negative logarithm of the hydrogen ion *activity*, $pH = - \log a_{H+}$ [2.256]. Because only dimensionless numbers can be expressed by logarithms, it is now customary to express the definition by

$$pH = - \log \frac{a_H}{m_0} = - \log \frac{m_H \, \gamma_H}{m_0} \ , \tag{2.282}$$

where $m_0 = 1 \, \mathrm{mol \, kg^{-1}}$, simultaneously indicating that pH is defined in terms of the molality scale. (The parallel definition of pH in terms of molarities as given in [2.257] will not be recommended in the future edition of IUPAC recommendations because two simultaneous recommendations, on the molality and the molarity scale, would violate the fundamental principle of traceability of measured quantities to a single definition). For simplicity, the hydrogen ion activity is simply written as a_H, omitting the hydration water molecule and the positive charge of the ion in the subscript, compare (2.282) with (2.277)–(2.280). Other ions are treated accordingly in solution theory as long as errors are excluded.

The corresponding quantity pOH is the negative logarithm of the hydroxyl ion activity,

$$pOH = - \log a_{OH-} \ , \tag{2.283}$$

which, according to (2.279), is connected to pH by

$$\mathrm{pH} + \mathrm{pOH} = \mathrm{p}K_\mathrm{W} \ . \tag{2.284}$$

Neutral solutions are characterized by an equal hydrogen and hydroxyl ion activity and thus by

$$\mathrm{pH} = \mathrm{pOH} = \frac{1}{2}\mathrm{p}K_\mathrm{W} \ , \tag{2.285}$$

which implies that the pH (and pOH) of neutral solutions is a function of temperature because of the (rather strong) temperature dependence of the thermodynamic ion product. For example, the pH of neutral solutions is $\mathrm{pH} = 0.5\,\mathrm{p}K_\mathrm{W} = 7.472$ at $0\,^\circ\mathrm{C}$, 6.998 at $25\,^\circ\mathrm{C}$, 6.631 at $50\,^\circ\mathrm{C}$, and 6.132 at $100\,^\circ\mathrm{C}$. Temperature-dependent values of the ion product can be found in textbooks and monographs, see for instance [2.2, 5, 258, 259], and can be interpolated by means of the polynomial [2.258]

$$\mathrm{p}K_\mathrm{W} = \frac{4471.99}{T} - 6.0846 + 0.017\,053\,T \ , \tag{2.286}$$

where T is the absolute temperature. Values of $\mathrm{p}K_\mathrm{W}$ up to $1000\,^\circ\mathrm{C}$ and up to $10\,000\,\mathrm{bar}$ "steam pressure" are given in [2.260].

2.10.2 Principles of pH Measurement

pH involves a single ion activity, which is immeasurable. As a way to determine pH, the method of comparing the solution with an unknown $\mathrm{pH(X)}$ with one or more standard buffer solutions with assigned standard $\mathrm{pH(S)}$ by means of electrochemical potentiometric cells was chosen. Measurement of pH, consequently, includes two fundamental problems. (1) The assignment of standard $\mathrm{pH(S)}$ values to certain standard buffer solutions involves necessarily a non-thermodynamic or semi-thermodynamic step, so that the assigned $\mathrm{pH(S)}$ is not a strictly thermodynamic-based quantity. However, $\mathrm{pH(S)}$ must be traceable to the definition of pH, (2.282), "through an unbroken chain of measurements, all measured intermediate quantities having stated uncertainties" [2.261]. (2) The cell for measuring $\mathrm{pH(X)}$ of unknown solutions unavoidably involves liquid junctions between the standard buffer solutions as well as the unknown solution and the solution of the reference electrode, generally a concentrated KCl solution. Each emf measured thus contains a liquid junction potential of a principally unknown magnitude. Consequently, differences of liquid junction potentials, which are called residual liquid junction potentials, are inherent in each calibration and in each measuring procedure and introduce a slight uncertainty into measured $\mathrm{pH(X)}$. Although thus lacking 100% exactness, measured $\mathrm{pH(X)}$ must be traceable to the assigned $\mathrm{pH(S)}$ and thus to the definition of pH.

A further requirement is that, as far as foreseeable, all standardized steps of pH measurement are open to future amendments, made necessary for instance by new insights into the physical chemistry of electrolyte solutions.

2.10.3 Primary Standard Buffer Solutions – Assignment of pH(PS)

A procedure of assigning primary pH(PS) values to certain qualified buffer solutions that meets the requirements mentioned is the application of the so-called Harned cell as reported by *Bates* [2.2, 262]. It is based on the Pt,H$_2$ electrode because its potential ε_{Pt,H_2} indicates the activity of hydrogen ions in a strictly thermodynamic way according to the Nernst equation,

$$\varepsilon_{Pt,H_2} = \varepsilon^0_{Pt,H_2} + \frac{2.303\,RT}{F} \log \frac{a_H}{\sqrt{p_{H_2}}} \ . \tag{2.287}$$

The standard potential ε^0_{Pt,H_2} of the hydrogen electrode is defined to be zero at all temperatures, a definition that is arbitrary and physically meaningless, and so are the consequences, zero standard energy and standard free energy at all temperatures and zero standard entropy of the reaction $H_2 = 2H^+ + 2e^-$ [2.263]. Nevertheless, the definition is indispensable on the grounds of the present knowledge of solution chemistry and must be kept in mind when standard pH values are assigned to buffer solutions at different temperatures.

   In order to preserve as much of the 100% traceability of the Pt,H$_2$ electrode potential to the hydrogen ion activity as possible, the assignment is carried out in a Harned cell, which is without transference and employs an electrode of the second kind,

$$\text{Pt} \big| \text{H}_2\, (p_{H_2} = 1\,\text{bar}) \big| \text{standard buffer, Cl}^-(m_{Cl}) \big| \text{AgCl} \big| \text{Ag} \ . \tag{2.XVII}$$

A liquid junction is thus avoided, and the emf consists merely of the potentials of the two electrodes involved,

$$E = \varepsilon_{Ag/AgCl} - \varepsilon_{Pt,H_2} \ . \tag{2.288}$$

The buffer solutions in cell (2.XVII) contain various concentrations m_{Cl} of chloride, which fix the potential of the silver/silver chloride electrode, so that the emf of cell (2.XVII) is given by

$$E = \varepsilon^0_{Ag/AgCl} - k \, \log(m_H\, \gamma_H\, m_{Cl}\, \gamma_{Cl}) \ , \tag{2.289}$$

for each solution with a certain chloride molality m_{Cl}. (The standard molality m^0 is omitted here for simplicity.) Rearrangement of (2.289) yields the so-called acidity function

$$p(m_H\, \gamma_H\, \gamma_{Cl}) = \frac{E - \varepsilon^0_{Ag/AgCl}}{k} + \log m_{Cl} \ . \tag{2.290}$$

Extrapolation of this quantity to zero chloride concentration, $m_{Cl} \to 0$, after the introduction of the standard potential of the silver/silver chloride electrode, results in the limiting value $p(a_H\gamma_{Cl})^0$ of the acidity function and to pa_H according to

$$pa_H = p(a_H \gamma_{Cl})^0 + \log \gamma_{Cl} , \tag{2.291}$$

which still contains the activity coefficient of chloride. This quantity is, however, not available thermodynamically, so that a non- or semi-thermodynamic convention must be chosen. Usually, the Bates–Guggenheim convention is applied [2.2, 264].

$$\log \gamma_{Cl} = -\frac{A\sqrt{I}}{1 + 1.5\sqrt{I}} . \tag{2.292}$$

It is based on the Debye–Hückel theory of electrolytes but has the disadvantage that it treats all buffer solutions equally by using the number 1.5 for the product $B\mathring{a}$ in the denominator. B is a constant, and $\mathring{a}$ is the so-called ion size parameter ("distance of closest approach"), which actually depends slightly on the buffer substance dissolved. I denotes ionic strength, and A is the (temperature-dependent) Debye–Hückel slope. The activity coefficient has been a matter of much and intense discussion, and it has been suggested to improve the evaluation of the activity coefficient of chloride [2.265], for instance on the basis of the so-called Pitzer theory [2.266]. Although such improvements would lead to a solution-dependent and thus individual activity coefficient of chloride for each primary buffer solution, it can be estimated that the resulting improvement of the traceability of the standard pH(PS) values will only be marginal.

The last step of the assignment finally is the definition of the quantity pa_H as the standard pH(PS) value of the primary standard buffer solution.

$$pa_H \equiv pH(PS) . \tag{2.293}$$

Standard pH(PS) are traceable to the definition of pH because each step of the assignment yields values with stated uncertainties.

- The Pt,H_2 electrode indicating the hydrogen ion activity in a thermodynamically correct way introduces no uncertainty.
- The standard potential of the silver/silver chloride electrode, which is measured in cell (2.XVII) employing the same electrodes as used for the assignment and containing 0.01 molal hydrochloric acid [2.2], the activity coefficient of chloride ions, and the ionic strengths of the standard buffer solutions with and without added chloride have stated uncertainties.
- The same is true for all steps of the procedure, for instance measuring temperatures, concentrations, hydrogen pressure of the Pt,H_2 electrode and its reduction to the standard pressure, emf measurements including meter accuracy, and the procedure of extrapolation to zero chloride concentration.

Besides, the assignment of pH(PS) is open to future changes, for example by the above-mentioned possible amendment by replacement of the Bates–Guggenheim convention.

The assignment of pH(PS) is an indirect procedure insofar as experimental data obtained from chloride-containing buffer solutions are extrapolated to zero chloride content. Because, consequently, the actual chloride-free standard solution, which is prepared at 25 °C, is not directly subjected to the assignment, its physical existence at all temperatures of application must be secured. That this is less trivial than it may sound was demonstrated by an example. We have found that 0.05 molal tetraoxalate standard buffer solution, whose pH(PS) were reported between 0 °C and 95 °C [2.2, 262] and certified by NBS [2.267], is a saturated solution at +4.8 °C. It thus does not exist physically at 0 °C, nor can it be supercooled to this temperature [2.268]. Obviously, its (non-existing) pH(PS) at 0 °C could nevertheless be determined at the NBS because the KCl-containing solutions yielding the data for the extrapolation to zero KCl concentration can be supercooled to 0 °C, as we found, even for rather long periods [2.268]. Another, rather trivial, explanation would be that the pH(PS) at 0 °C had not been measured but was obtained by extrapolation, which, however, is not quite conceivable. Similarly, 0.1 molal tetraoxalate buffer, to which pH(S) between 15 °C and 60 °C have been assigned, but which has never been certified, is a saturated solution at 22 °C and cannot be used below this temperature. Surprisingly, these errors have never been detected during the application of the solutions. Either they have never been used at temperatures below 4.8 °C and 22 °C, respectively, or scientists applying the solutions at the low temperatures have done so with great confidence and have thus not noticed the unexpected crystallization and, consequently, have used slightly faulty "standard" pH values.

Eight primary standard buffer materials with certified pH(PS) between pH = 1 and pH = 11 in the temperature range 0–50 °C are currently available from several standardization institutions, for instance the National Institute of Standardization and Technology (NIST) (formerly NBS), USA, and the Physikalisch-Technische Bundesanstalt (PTB) in Braunschweig, Germany. They meet certain criteria with respect to purity, handling, and stability of solutions [2.2]. Solution stability against mold growth can be considerably improved by heat treatment (sterilization) without noticeably changing the pH [2.269, 270]. Surprisingly, the solid standard material disodium tetraborate has been found to be stable only for limited periods because of structural changes connected with a loss of water, although the loss of water is not responsible for the observed pH(PS) changes [2.271].

For still unknown reasons, the pH(PS) values of all standard materials depend slightly on the particular lots of the materials. Therefore each quantity provided must be accompanied by a certificate (in German: "Kalibrierschein", see for instance [2.272]) that informs about the actual pH(PS) of the individual buffer material for which it is issued. Consequently, recommendations for measuring pH, for example DIN 19266 [2.273] and the new edition of *IUPAC Recommendations*, which is in preparation (see below), will merely give examples of pH(PS) values of more or less recently certified lots, indicating their approximate magnitude and temperature dependence, and will otherwise refer to the certificates to be used with each particular buffer material.

2.10.4 Secondary Standard Buffer Solutions – Assignment of pH(SS)

The assignment of pH(PS) to primary standard buffer solutions is rather laborious and is thus preferably carried out at national metrological institutes. Therefore, methods for deriving secondary standards with nominally the same composition as the primary standards to which they are referred, but with possibly minute pH differences ("quasi-identical" buffer solutions) have been developed and will be recommended in future *IUPAC Recommendations*, see Sects. 1.1.7 and 2.10.10, especially. Most frequently applied is the differential cell "quasi-without liquid junction" [2.257, 269, 274, 276],

$$\text{Pt,H}_2 | \text{primary buffer, pH(PS)} \| \text{secondary buffer, pH(SS)} | \text{Pt,H}_2 \ ,$$

$$(2.XVIII)$$

which employs a glass frit with porosity P 40 (according to ISO 4793) as a liquid junction device, through which the solutions are in direct contact. The hydrogen pressure of the Pt,H_2 electrodes is strictly identical (although not necessarily equal to the standard pressure), which is accomplished by an equal depth of the jets in the solutions of the half-cells through which the hydrogen is introduced and by a common outlet for the gas.

The potential difference of cell (2.XVIII) is given by the potentials of the Pt,H_2 electrodes,

$$E = \varepsilon(\text{SS}) - \varepsilon(\text{PS}) + \varepsilon_\text{j} \ ,$$

$$(2.294)$$

and involves a liquid junction potential ε_j. However, the measured emf represents the pH difference with a known maximum uncertainty. As long as the resulting pH difference is smaller than ± 0.02 and the pH of the solutions is between pH $= 3$ and pH $= 11$, the liquid junction potential is smaller than 10% of the measured potential difference, $\varepsilon_\text{j} \leq 0.1E$ [2.274]. This result was obtained by measurements employing standard buffer solutions with pH differences defined by the use of buffer values [2.2], by calculating the liquid junction potential by Henderson's equation, and by comparing the experimental and calculated data [2.274]. The pH(SS) of secondary standard buffer solutions is thus represented by

$$\text{pH(SS)} = \text{pH(PS)} - \frac{E \pm 0.1E}{k}$$

$$(2.295)$$

and therefore traceable with a stated uncertainty of $\pm 0.1E/k$ to the pH(PS) of the primary standard buffer to which it is referred and finally to the definition of pH.

Potential differences measured by cell (2.XVIII) are in the range of several ten μV to one mV, corresponding to ΔpH between 0.001 and 0.02 with an uncertainty of $\pm$0.0003. The uncertainty of pH(SS) values is thus only slightly larger than that of pH(PS) of the primary standard materials, which is $\pm$0.005 and thus constitutes the major contribution to ΔpH(SS). Uncertainty budgets will be found in the new edition of *IUPAC Recommendations*, see below. Measurements are carried out at accredited laboratories, for instance at Merck, Germany, or Radiometer, Denmark, which issue certificates (see for example [2.275]) containing the actual pH(SS) to be used with the secondary standard buffer materials in question. There are thus two kinds of standard material.

- Primary pH standard materials of highest metrological quality are certified by national metrological institutes. They are used by accredited laboratories to certify secondary pH standard materials but may also be used by laboratories for calibrating pH-measuring assemblies.
- Secondary pH standard materials of high metrological quality, but not fulfilling the highest requirement as primary standards, are certified by accredited laboratories and are used by laboratories for calibrating pH-measuring assemblies. They may also be applied to measure pH values of (tertiary) solutions which serve as derived standards, for instance for rough practical measurements at the industry if knowledge of an uncertainty is not required.

2.10.5 Calibration of pH Meter–Electrode Assemblies

Practical pH measurements, by which pH(X) of an unknown solution is referred to pH(S) of a standard buffer solution, are carried out in cells with transference according to the general cell scheme

$$\text{ref. el.} \left| \text{KCl}(3.5 \, \text{mol dm}^{-3}) \right\| \text{stand. buffer, pH(S), or sol'n, pH(X)} \left| \text{glass el.} \right.,$$
(2.XIX)

whose emf unavoidably contains a liquid junction potential ε_j between, respectively, the buffer and the unknown solution and the reference electrolyte, a 3.0 molal, 3.5 molal, or saturated KCl solution:

$$E = \varepsilon_{gl} - \varepsilon_{ref} + \varepsilon_j \, .$$
(2.296)

These cells must thus be calibrated. Besides, (a) they involve practical liquid junction devices, for instance ceramic plug, platinum, or glass sleeve junction devices causing individual, device-dependent liquid junction potentials that are larger than those arising at a free diffusion liquid junction, the most reproducible kind of liquid junction [2.106], see paragraphs on reference electrodes below. (b) Glass electrode cells exhibit slopes that are smaller than the ideal slope factor [2.106]. The calibration can be carried out by two procedures, both of which secure traceability of the unknown pH(X) to the standard pH(S) values.

2.10.6 Bracketing Procedure

The emfs $E(S_1)$, $E(S_2)$, and $E(X)$ of cell (2.XIX) containing standard buffer solutions with $pH(S_1)$ and $pH(S_2)$ and the unknown solution with $pH(X)$, respectively, are measured, and $pH(X)$ is obtained by

$$pH(X) = pH(S_1) + \frac{E(X) - E(S_1)}{k'} \ , \tag{2.297}$$

where k' is the practical slope,

$$k' = \frac{E(S_2) - E(S_1)}{pH(S_2) - pH(S_1)} \ , \tag{2.298}$$

which accounts for the deficiencies of cell (2.XIX), that is, liquid junction potentials [2.257, 276] and non-ideal slope of the glass electrode [2.2, 95]. In order to ensure that only one $pH(X)$ results from the measurements and any overdefinition is excluded, one of the standard buffer solutions must have a larger and the other a smaller $pH(S)$ value than the unknown $pH(X)$ ("$pH(S)$ bracket $pH(X)$"), and both $pH(S)$ must be the closest to the $pH(X)$ available. The expanded uncertainty of the bracketing procedure is 0.02–0.05, depending on the conditions chosen.

2.10.7 Multiple-Point Calibration with Linear Regression

Different from the bracketing procedure, the uncertainties of multiple-point calibration with linear regression can be numerically stated [2.106, 276, 277]. The procedure is recommended for pH measurements with a precision of typically $\Delta pH = \pm 0.02$ and when individual measuring cells are to be characterized with respect to their quantitative performance. In practice, a certain number of specified standard buffer solutions with $pH(S_1)$, $pH(S_2)$, and so forth, applied in cell (2.XIX) result in respective emfs $E(S_1)$, $E(S_2)$, etc. The data are subjected to linear regression, which yields the standard emf $E^{0'}$ and the practical slope k' of cell (2.XIX) according to the linear regression function

$$E(S) = E^{0'} - k'pH(S) \ , \tag{2.299}$$

see Fig. 2.69. Subsequently, $pH(X)$ of unknown solutions is obtained by measuring $E(X)$ and applying

$$pH(X) = \frac{E^{0'} - E(X)}{k'} \ . \tag{2.300}$$

The number and kind of the standard buffer solutions applied to the standardization must be agreed upon. From theoretical and practical considerations,

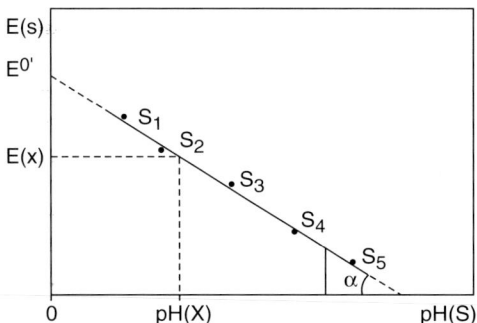

Fig. 2.69. Schematic presentation of multiple-point calibration with linear regression according to (2.296). $k' = \tan\alpha$ is the practical slope and $E^{0'}$ is the standard emf of practical cell (2.XIX)

five solutions with consecutive pH(S) values represent an optimum with respect to statistics [2.278, 279], handling of solutions, and design of modern pH meters. This agreement regards merely the precision of cell calibration and pH measurement and is in no way a convention with respect to the absolute magnitude of the pH(X) measured. Details of the statistical method including the calculation of uncertainties are given in [2.278]. The relatively large number of measurements is easily carried out automatically with modern pH-measuring equipment in a relatively short period of time. Such instruments are available [2.106] or under development at various companies. It is not surprising that multiple-point calibration was not proposed in the past, for instance by *Bates* [2.2] or *Schwabe* [2.4], because the time-consuming manual pH measurements in earlier times excluded any unnecessary additional measurements.

Standardization of the practical cell (2.XIX) at 100% pH selectivity of the glass electrode by multiple-point calibration involving certain sets of buffer solutions and a specified liquid junction device yields practical average slopes k' in (2.299) which are smaller than the ideal slope factor k [2.106]. Stated in a different way, the electromotive efficiency according to Bates' definition of the pH-dependent emf of cell (2.XIX) is smaller than unity, $\alpha_{\mathrm{cell}} = (k'/k) < 1$, under these conditions. This means that one or more of the potentials of which the emf consists must have an electromotive efficiency smaller than unity. Indeed, the following analysis shows that all three potentials do.

2.10.8 The Slope of Cells with Transference Containing Standard Buffer Solutions

pH(S) Dependence of the Liquid-Junction Potential

Surprisingly, cell (2.XIX), which contains standard buffer solutions and the Pt,H$_2$ instead of the glass electrode as in cell

$$\text{ref. el.} \left| \text{KCl} \,(3.5 \,\text{mol dm}^{-3}) \right\|_{\|} \text{standard buffer, pH(S)} \left| \text{Pt,H}_2 \right. \qquad (2.\text{XX})$$

and yields the linear regression equation

$$E'(\text{S}) = E^{0''} - k'' \text{pH}(\text{S}) \,, \qquad (2.301)$$

exhibits an average practical slope k'' which is not equal to the ideal slope factor, as would be expected for the average value of the slope, but slightly smaller, $k'' < k$, if the cell contains a specified liquid-junction device. The ratio $\beta = k''/k$, which denotes electromotive efficiency in analogy to the electromotive efficiency of cell (2.XIX) as proposed by *Bates* [2.2], is smaller than unity [2.106]. In addition, for each set of buffers, especially for 0.05 molal solutions, and for each liquid junction device used, the quantity $(1 - \beta)k = k - k''$ represents the average pH dependence of the liquid-junction potential of cell (2.XX),

$$\frac{\text{d}\varepsilon_\text{j}}{\text{dpH}(\text{S})} = (1 - \beta)k = \text{const.} \,, \qquad (2.302)$$

so that the linear regression equation, (2.301), of cell (2.XX) may be written as [2.106],

$$E'(\text{S}) = E^{0''} - \left(k - \frac{\text{d}\varepsilon_\text{j}}{\text{dpH}(\text{S})} \right) \text{pH}(\text{S}) \,, \qquad (2.303)$$

and the pH dependence of $E'(\text{S})$ is represented by

$$\frac{\text{d}E'(\text{S})}{\text{dpH}(\text{S})} = - \left(k - \frac{\text{d}\varepsilon_\text{j}}{\text{dpH}(\text{S})} \right) \,. \qquad (2.304)$$

The numerical values of the slope $(\text{d}\varepsilon_\text{j}/\text{dpH}(\text{S}))$ depend on the set of buffers and the type and individual property of the liquid junction device used with each of the buffer sets, all experimental values being positive and ranging from $+0.03$ to $+0.73\,\text{mV}$ [2.106]. This finding is supported by calculated liquid-junction potentials ε_j at the continuous-mixture junction sat'd $\text{KCl} \|$ buffer, Fig. 2.70, and by experimental residual liquid-junction potentials $\Delta\varepsilon_\text{j}$ referred to NIST phosphate 1:1, Fig. 2.71, both reported by *Bates* [2.2, 280]. The respective magnitudes of these slopes are $(\text{d}\varepsilon_\text{j}/\text{dpH}) = (+0.19 \pm 0.03)\,\text{mV}$ and $+0.16\,\text{mV}$. A quantitative explanation of this observation was given on the basis of ionic mobilities [2.106].

pH(S) Dependence of the Glass Electrode Potential

Theoretical Response. It was derived theoretically and verified experimentally by means of the sub-Nernstian slope [2.95], see Sect. 2.5.3, that the theoretically correct slope of the glass electrode potential at 100% pH selectivity is independent of pH and smaller than the ideal slope factor,

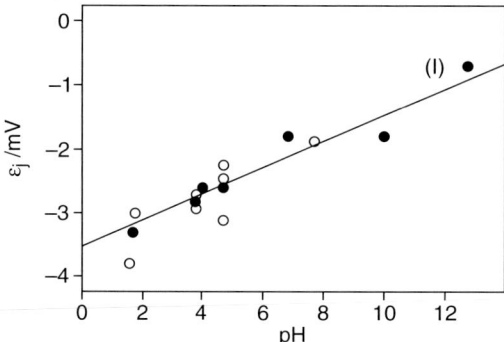

Fig. 2.70. Calculated liquid-junction potentials at the continuous-mixture junction KCl(sat'd)$\|$buffer solution [2.2] plotted as a function of pH [2.106]. Dots: total buffer concentration $0.05\,\mathrm{mol\,kg}^{-1}$; open circles: other concentrations; (1) $0.05\,\mathrm{mol\,kg}^{-1}$ NaOH

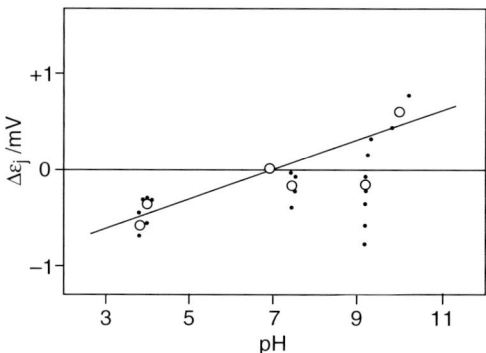

Fig. 2.71. Experimental residual liquid-junction potentials, referred to NIST phosphate 1:1 buffer, at the free-diffusion liquid junction KCl (3.0, $3.5\,\mathrm{mol\,kg}^{-1}$ or sat'd)$\|$buffer solution at several temperatures between $10\,^{\circ}\mathrm{C}$ and $40\,^{\circ}\mathrm{C}$ [2.2] plotted as a function of pH [2.106]. Dots: values as given; open circles: average values. For borax buffer $0.01\,\mathrm{mol\,kg}^{-1}$ only data below $25\,^{\circ}\mathrm{C}$ were used for calculating the average

$k_{\mathrm{gl}} = \alpha_{\mathrm{gl,th}}k =$ const $< k$. The electromotive efficiency of the glass electrode $\alpha_{\mathrm{gl,th}}$ in cell (2.XIX) can thus be written as

$$\alpha_{\mathrm{gl,th}} = \left(1 - \frac{\mathrm{d}\log a'_{\mathrm{SiO}^-}}{\mathrm{dpH(S)}}\right) , \tag{2.305}$$

the glass electrode potential as

$$\varepsilon_{\mathrm{gl}} = \varepsilon_{\mathrm{gl}}^0 - \left(1 - \frac{\mathrm{d}\log a'_{\mathrm{SiO}^-}}{\mathrm{dpH(S)}}\right) k\mathrm{pH(S)} , \tag{2.306}$$

and the slope of the glass electrode potential as

$$\frac{d\varepsilon_{gl}}{dpH(S)} = -\left(1 - \frac{d \log a'_{SiO^-}}{dpH(S)}\right) k \; , \tag{2.307}$$

where a'_{SiO^-} is the activity of siloxy groups at the membrane glass surface.

Response of Glass Electrode with an Electrode Shunt

Electrode shunts (i.e., partial short circuits of the opposite membrane potentials) are practically excluded with modern glass electrodes, even with glass electrodes having membranes with very high resistance [2.95]. However, because electrode shunts would cause pH-independent reductions of the glass electrode response, they are also treated in connection with cell (2.XIX). It can easily be shown that a glass electrode shunt causes a pH-independent reduction of the ideal electrode slope factor, $k_R = \alpha_R k$, where α_R is the electromotive efficiency of the pH-dependent glass electrode potential according to

$$\alpha_R = \frac{R_S}{R_S + R_m} \; , \tag{2.308}$$

where R_S and R_m are the resistances of shunt and glass electrode membrane, respectively. Because α_R always appears together with $\alpha_{gl,th}$, it is physically meaningful only to give an equation for the overall glass electrode potential, which is derived to be

$$\varepsilon_{gl} = \varepsilon_{gl}^0 - \alpha_R \alpha_{gl,th} k pH(S) = \varepsilon_{gl}^0 - \frac{R_S}{R_S + R_m}\left(1 - \frac{d \log a'_{SiO^-}}{dpH(S)}\right) k pH(S) \; . \tag{2.309}$$

Overall Slope dE(S)/dpH(S)

The overall emf $E(S)$ of cell (2.XIX) is obtained by combining (2.303) and (2.309),

$$E(S) = E^{0'} - \left[\frac{R_S}{R_S + R_m}\left(1 - \frac{d \log a'_{SiO^-}}{dpH(S)}\right) k - \frac{d\varepsilon_j}{dpH(S)}\right] pH(S) \; , \tag{2.310}$$

and yields the potential slope

$$\frac{dE(S)}{dpH(S)} = -\frac{R_S}{R_S + R_m}\left(1 - \frac{d \log a'_{SiO^-}}{dpH(S)}\right) k + \frac{d\varepsilon_j}{dpH(S)} \; . \tag{2.311}$$

The analysis thus shows that the ideal slope factor k is indeed reduced by a pH(S)-independent factor because the terms in the brackets on the right side of (2.310) and thus the slope, (2.311), are constant for each particular set of standard buffer solutions, in particular for 0.05 molal solutions, and for each liquid-junction device used with each buffer set, see Figs. 2.70 and 2.71 and [2.95, 106]. In principle, the slope of the emf of the practical glass electrode cell (2.XIX) can thus be calculated if the terms of (2.311), the theoretical (subideal) slope of the glass electrode, the pH(S) dependence of the liquid-junction potential, and the membrane and shunt resistance are known. As stated above, however, the fraction $R_S/(R_S + R_m)$ can generally be approximated to unity with modern glass electrodes because $R_S \gg R_m$.

2.10.9 Past Proposal of an Alternative pH Scale by BSI

When primary standard pH values pH(PS) assigned by means of the Harned cell (2.XVII) without transference are compared in cell (2.XX) with transference,

$$\text{ref. el.} | \text{KCl(conc'd)} \| \text{standard buffer, pH(PS)} | \text{Pt,H}_2 \ , \tag{2.XXI}$$

the resulting data points are not positioned exactly on a straight line. In other words, an evaluation of the data by means of an equation similar to (2.297) but involving the ideal instead of the practical slope,

$$\text{pH(PS}_n) = \text{pH(PS}_1) - \frac{E(\text{PS}_n) - E(\text{PS}_1)}{k} \ , \tag{2.312}$$

(where $k = (2.303RT)/F$, PS_n denotes any primary standard buffer, and PS_1 means standard buffer phosphate (1:1), to whose pH(PS$_1$) the other standard pH values are referred) is not possible. The exact comparison necessitates an individual practical slope k'_n in (2.312) for each buffer couple PS_n–PS_1. In other words, the internal consistency of the primary standard buffers is below 100%.

The reason for this deficiency is that the liquid-junction potentials ε_j between the standard buffer solutions PS_n and PS_1, respectively, and the KCl solution are not equal despite the low concentrations and similar ionic strengths of the buffers. This is expressed by the relationship

$$\text{pH(PS}_n) = \text{pH(PS}_1) - \frac{E(\text{PS}_n) - E(\text{PS}_1)}{k} - \frac{\varepsilon_j(\text{PS}_n) - \varepsilon_j(\text{PS}_1)}{k} \ , \tag{2.313}$$

where the difference of the liquid-junction potentials (third term on the right side of (2.313)) is called the residual liquid-junction potential $\Delta\varepsilon_j$. The misfit actually involves also a small contribution from the Bates–Guggenheim convention, (2.292), which is applied to each standard buffer solution independent of its individual ion size parameter. Residual liquid-junction potentials were, for instance, measured by *Bates, Pinching*, and *Smith* [2.281] and by *Paabo* and *Bates* [2.2], who used two different arrangements of the free diffusion liquid junction and 3 molar, 3.5 molar, and saturated KCl solution and obtained inconsistencies of the order of $\Delta pH(l.j.) = (\Delta\varepsilon_j/k) < 0.06$ within the pH range 3–9 at $25\,°C$ and slightly larger at higher temperatures and at higher and lower than medium pH values.

In order to circumvent these deficiencies of the primary pH standard buffer solutions, which were developed by the National Institute of Standards and Technology (NIST, USA) and have been adopted by the majority of national standardization institutions, the British Standards Institute (BSI) had developed a different philosophy. The history is sketched in [2.282]. It was argued that, for establishing a consistent series of standard buffer solutions, only one primary buffer solution with independently assigned pH(PS) would be needed, the other parameter being the ideal slope of the emf of cell (2.XX). The special kind of cell (2.XX) used for this purpose involved a free-diffusion liquid junction within a capillary with $1\,mm^2$ cross section, and $0.05\,mol\,kg^{-1}$ potassium hydrogen o-phthalate with assigned pH(PS) served as the so-called reference value standard (RVS). The buffers whose pH(OS) were determined were called operational standard buffers. Their number would be, in principle, unlimited.

At first sight, this proposal looked promising, and the series of the operational standard pH buffers was indeed consistent if the specified free-diffusion liquid-junction device was used. The trap, however, was that the liquid-junction potentials were in no way eliminated by the assignment of pH(OS) to the operational buffer solutions but were only hidden in the pH(OS) and reappeared in measured pH values, in particular when practical or commercial liquid-junction devices were used, as was the case in all practical measurements. The reappearance was confirmed experimentally [2.106]. In addition, unexpectedly large liquid-junction potentials, as for instance caused by linen fibre junctions [2.283], were not recognized and could result in unnoticed considerable errors of measured pH values. Moreover, the thermodynamic meaning of pH(OS) and thus of pH values measured on their basis, which is often required, for instance for acidity constants, was lost to a large extent. The most serious disadvantage, however, which has finally excluded the operational standard buffers from being adopted as the single basis of pH measurement, was that it offered no traceability of measured pH values to the definition of pH despite the high precision of the pH(OS). The uncertainties caused by the hidden liquid-junction potentials are unknown and thus cannot be stated.

2.10.10 International Recommendations of pH Measurement

Recommendations of electroanalytical methods are issued by several standardization organizations, for example by the International Electrotechnical Commission (IEC), International Organization for Standardization (ISO), Organization Internationale De Mètrologié Lègale (OIML), and by the International Union of Pure and Applied Chemistry (IUPAC). According to an unwritten agreement, IUPAC takes the lead in the recommendation of pH measurement. It is followed in this field by the other international organizations and by most national institutions; the Deutsche Institut für Normung (DIN), for instance, has issued nine recommendation documents on pH nomenclature, measurement, and measuring equipment [2.273, 284–291] which are basically in line with *IUPAC Recommendations*.

The latest *IUPAC Recommendations* on pH were issued in 1985 [2.257]. They are an excellent example for demonstrating the development of science even in the field of standardization. In 1985, when "pH experts were divided and IUPAC was unable to recommend, unreservedly, on scientific grounds, one or other of the two approaches to pH scale definition" [2.257], that is, primary or operational primary standard buffer solutions with pH(PS) or pH(OS), respectively, as the basis of pH measurement, IUPAC decided to recommend the simultaneous use of both protocols. It was also stated that "further research is needed to establish the respective merits of the two approaches and only then can a thermodynamically significant and metrologically sound pH scale be recommended". The *IUPAC Recommendations 1985* are thus clearly an interim document. Meanwhile, more information about, and especially better *metrological* insight into, the matter of pH measurement are available.

In 1997, an Interdivisionary Working Party on pH was established by IUPAC, see [2.292], which is to issue new recommendations. The amendments are also necessary because the *IUPAC Recommendations 1985*, in addition to recommending two parallel pH protocols, based on pH(PS) and pH(OS), contain at least four more metrologically intolerable shortcomings [2.293] whose significance obviously had not been noticed at the time the 1985 document was issued:

(1) pH is defined on the molal and on the molar scale. This contradicts metrological principles and can give rise to serious confusion because the numerical differences are not necessarily negligible (0.001 at $25\,°C$, 0.02 at $100\,°C$).

(2) Besides the so-called "*notional* definition", an *operational* definition of pH is given. It is based on cell (2.XIX), which may employ the Pt,H_2 electrode or a glass electrode. Evaluation is by

$$pH(X) = pH(S) - \frac{E(X) - E(S)}{k} \, , \tag{2.314}$$

where k denotes the ideal slope factor.

(3) The *operational* definition of pH according to (2.314) involves the ideal slope factor but, nevertheless, allows the use of different electrode types, the Pt,H_2 electrode and glass electrodes, with basically different and individually different slopes, respectively. This leads to a variety of pH values for each solution even if the measurements are performed strictly according to the *Recommendations*. The possible magnitude of the resulting pH differences is not limited because the *Recommendations* require no check

of the (ignored) practical slope of the applied glass electrodes prior to measurement.

(4) Two coexisting procedures of pH measurement are recommended, the *operational* procedure, (2.314), involving the ideal slope factor k, and the *bracketing* procedure, (2.297) and (2.298), involving the practical slope k'. Also this results in two pH values for each solution, even if only one electrode, the Pt,H_2 or a glass electrode, is applied.

As a member of the IUPAC Working Party on pH it is my conviction that the new *IUPAC Recommendations* will be free of major metrological inconsistencies because they follow three governing principles. (1) They will give only one definition of pH. (2) All measured quantities will be traceable to the definition with stated uncertainties of all intermediate values. (3) The new *Recommendations* will be open to future amendments, as far as foreseeable, in particular with respect to further developments of the theory of electrolyte solutions. These guidelines, which were recommended in detail in a critical paper in 1996 [2.293], served as the basis for the introduction and treatment of pH given here.

2.10.11 Measurement of Hydrogen Ion Concentrations

For certain, rare purposes, for instance for protonations and complex formation, the concentration of hydrogen ions, here either as molarity or molality(!), and the respective pH values are of interest:

$$pH_c = -\log\frac{c_H}{c^0} \tag{2.315}$$

and

$$pH_m = -\log\frac{m_H}{m^0} \ . \tag{2.316}$$

These quantities, and preferably pH_c, are obtained by means of the cell [2.2, 257],

$$\text{ref. el.}\left|KCl(\text{conc'd})\right\|MY\left\|(S + MY) \text{ or } (X + MY)\right|Pt,H_2 \ , \tag{2.XXII}$$

where MY is a 1,1-electrolyte, for instance KCl or $NaNO_3$, S is a strong (100% dissociated) acid, for instance perchloric acid, with known concentration, which serves as the standard, and X is an acid of unknown concentration and unknown degree of dissociation. pH_c is then obtained under the conditions that (a) the concentration of MY in the electrolyte bridge, the standard (S + MY), and the unknown solution (X + MY) is nearly the same, and (b) the concentrations of S and X are negligible with respect to the concentration of MY, which means that the pH of the standard and the unknown solution are between pH = 2 and pH = 12, the concentration of MY being 1 molar or larger at these extreme pH values.

Under these conditions, the ionic strength $I = 1/2 \sum_i m_i z_i^2$ (z_i = ionic charge) of the three solutions is nearly equal so that (a) the hydrogen ions have nearly the same activity coefficient in the standard and unknown solution and (b) liquid-junction potentials between bridge electrolyte and standard and unknown solution, respectively, are equal.

The unknown $pH_c(X)$ is then obtained by

$$pH_c(X) = pH_c(S) - \frac{E(X) - E(S)}{k} \, , \qquad (2.317)$$

where k is the ideal slope factor. If a glass electrode is used instead of the Pt,H_2 electrode in cell (2.XXII) and, in addition, high accuracy is required, the ideal slope factor k in (2.317) must be replaced by the practical slope k' of the glass electrode cell. It is measured by using two standard acid solutions (S + MY) whose pH_c values "bracket" that of the unknown solution.

2.10.12 Reference Electrodes

Reference electrodes used with practical pH measuring cells are not of primary interest but are "necessary complications" [2.294]. They add two potentials, the reference electrode and the liquid-junction potential, to the potential of interest, and this makes them as critical for the precision of the determined pH as the measuring glass or the Pt,H_2 electrode because an emf can only be as reproducible as the least reproducible of the potentials forming it. This significant fact has generated a great amount of work for developing new and improving known reference electrodes, see for instance [2.295]. Despite these efforts, only a small number of suitable reference electrodes have been found. Apart from a few systems with limited application [2.295], these are mainly the Ag/AgCl, the Hg/Hg_2Cl_2 or calomel, and the Hg,Tl (40 wt%)/TlCl or Thalamid® (*thal*lium *am*algam, thallium chlor*ide*) electrodes, and the calomel electrode even has only a limited temperature range. Their basis is briefly described in the following paragraphs.

Except for a few special cases [2.296], electrodes of the second kind are used as reference electrodes. They consist of the electrode metal, for instance silver, in contact with a solution that is saturated with one of its sparingly soluble salts, for example silver chloride, and contain a soluble salt with the same anion, for instance potassium chloride. The arrangement is demonstrated by the electrode scheme

$$Ag | AgCl(\text{sat'd}) | KCl(m_{KCl}) \, , \qquad (2.XXIII)$$

and the electrode reaction is represented by

$$Ag + Cl^-(\text{sol'n}) \rightleftarrows AgCl(s) + e^-(Ag) \, . \qquad (2.318)$$

The potential, which involves the solubility product $K_{0s} = a_{Ag^+} a_{Cl^-}$, is given by

$$\varepsilon_{\mathrm{Ag/Ag^+}} = \varepsilon^0_{\mathrm{Ag/Ag^+}} + \frac{RT}{F} \ln a_{\mathrm{Ag^+}} = \varepsilon^0_{\mathrm{Ag/Ag^+}} + \frac{RT}{F} \ln \frac{K_{0s}}{a_{\mathrm{Cl^-}}} \qquad (2.319)$$

or by the alternative equation

$$\varepsilon_{\mathrm{Ag/AgCl}} = \varepsilon^0_{\mathrm{Ag/AgCl}} - \frac{RT}{F} \ln a_{\mathrm{Cl^-}} \ , \qquad (2.320)$$

where $\varepsilon^0_{\mathrm{Ag/AgCl}} = \varepsilon^0_{\mathrm{Ag/Ag^+}} + (RT/F) \ln K_{0s}$ is the standard potential of the silver/silver chloride electrode. The condition that the silver is safely contacted by the silver chloride-saturated solution is best met if the metal is covered with the salt.

Standard potentials $\varepsilon^0_{\mathrm{Ag/AgCl}}$ have been determined not only by using electrodes of the second kind [2.297] but also by means of ion-sensitive AgCl membrane electrodes [2.298], and it has been pointed out that the extrapolation method for their determination may involve a serious pitfall [2.299]. For practical applications, for instance for practical pH measurements, the reference electrolyte is 3.5 molar or saturated with respect to KCl. The "standard potentials" of such practical "reference electrodes with fixed potential" are thus determined by the large, constant chloride activity and, for practical reasons, involve the liquid-junction potential between the concentrated KCl and the adjacent solution, usually a standard buffer solution as indicated by the dots in the electrode scheme

$$\mathrm{Ag \big| AgCl(sat'd) \big| KCl(3.5\,mol\,dm^{-3}\ or\ sat'd)} \|_{\|} \cdots . \qquad (2.\mathrm{XXIV})$$

Such standard potentials $(\varepsilon^{0'} + \varepsilon_{\mathrm{j}})$ of reference electrodes are usually not of interest to pH measurements because they are included in the standardization of the measuring cells. However, their temperature-dependent magnitudes must be correctly known in numerous other cases, for example in corrosion research [2.300], and have thus been measured for the silver/silver chloride [2.301–303], calomel [2.304], and Thalamid® electrodes [2.305–307] with 3.5 molar and saturated KCl solutions, and for the "Thalambid" electrode, a modified Thalamid® electrode with TlBr and KBr instead of TlCl and KCl [2.308]. For an overview, Fig. 2.72 presents these standard potentials as a function of temperature. The large temperature coefficients, which are of the order of 0.4–1.0 mV K^{-1} at 25 °C and slightly larger at higher temperatures, demonstrate the temperature uniformity which is required for isothermal pH measurements.

The electrochemically by far most satisfying reference electrode is the thallium amalgam/thallium chloride or Thalamid® electrode [2.309–311]. This applies also to its modification with thallous bromide [2.308] and to the application in non-aqueous solutions, for instance in dimethyl-sulphamide [2.312] and propylene carbonate [2.313]. Unfortunately, their poisonous character excludes them from a number of applications, for instance in food production and drinking water control. Due to their large exchange current densities [2.308], their potentials are extremely reversible and stable. In contrast to

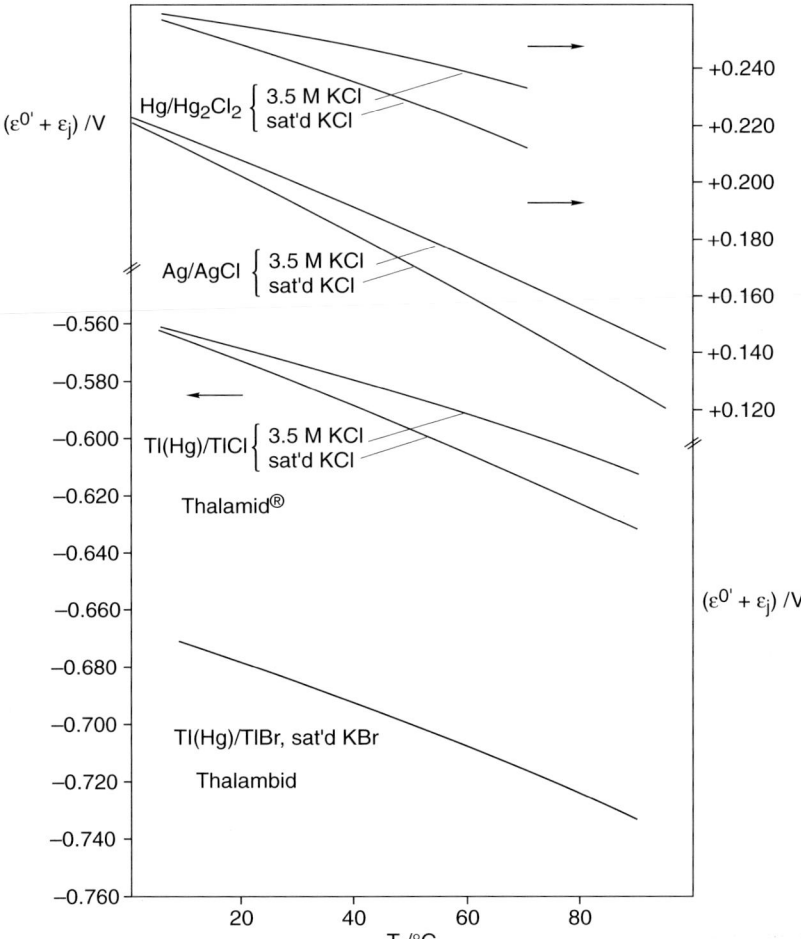

Fig. 2.72. Standard potentials $(\varepsilon^{0'} + \varepsilon_j)$ of reference electrodes with fixed potential as a function of temperature. Calomel electrode [2.304], silver/silver chloride, 10–40 °C [2.301], 0–95 °C [2.302, 303], Thalamid® [2.305], Thalambid [2.308]

silver chloride, thallous halides are subject to negligible complex formation, if any, which causes a fast potential response on temperature changes and excludes temperature hysteresis as observed with silver/silver chloride electrodes [2.309]. In contrast to the light sensitivity of the Ag/AgCl electrode, which can reduce the AgCl [2.314], Thalamid® electrodes are insensitive to light. They are applicable up to at least 135 °C [2.309], whereas silver/silver chloride electrodes suffer from the solubility of silver chloride as chloro-silver complexes, in particular at higher temperatures, and calomel disproportionates at temperatures above 60 °C in aqueous KCl solution and even spontaneously at room temperature in propylene carbonate [2.313]. The potential of thallium amalgam-based electrodes depends on the thallium content of the amalgam, which is not easy to prepare with exact concentrations. However, the most frequently used concentration, $40\,\mathrm{wt\%}$ thallium, exhibits only a small concentration dependence of the potential, which is $0.8\,\mathrm{mV}$ per weight percent at $25\,^\circ\mathrm{C}$ and only slightly larger at higher temperatures [2.305].

The standard potential of practical reference electrodes involves a liquid-junction potential, see cell scheme (2.XXIV), which, in principle, cannot be measured or calculated but can only be estimated [2.2]. This is obvious from the general expression

$$\varepsilon_\mathrm{j} = -\frac{RT}{F} \int_\mathrm{I}^\mathrm{II} \sum_\mathrm{i} \frac{t_\mathrm{i}}{z_\mathrm{i}} \mathrm{d}\ln a_\mathrm{i} \qquad (2.321)$$

for the liquid-junction potential ε_j between solutions I and II because single ion activities a_i and their concentration dependence, ionic transport numbers t_i and their concentration dependence, and the "geometry" of the boundary are unknown. Because the smallest liquid-junction potentials are generated between a concentrated solution of one of the rare 1,1-electrolytes with iso-mobile ions and relatively dilute solutions [2.2], reference electrodes with fixed potential contain high concentrations of such salts, for instance KNO_3, NH_4NO_3, and $CsCl$ [2.315] and preferably KCl because of the potential-determining chloride ion, and form "their own salt bridge" [2.305].

A very reproducible liquid junction is the so-called free-diffusion liquid junction [2.282]. It is formed between two adjacent solutions in a vertical tube, a device unsuited for practical, for instance industrial, purposes. Practical liquid-junction devices exclude convection between, but provide diffusional contact of, the different adjacent solutions. Several different junction devices have been developed. Here are some examples [2.294]. Platinum junction devices consisting of threaded thin platinum wires melted through a glass wall form defined thin channels and yield most reproducible and constant liquid-junction potentials [2.106, 310, 311] that are not disturbed by redox potential formation except for a few cases, for instance large concentrations of bromine/bromide. Ground glass sleeve junctions, actually glass joints, provide adjustable high leak rates, which exclude clogging by precipitates, can easily be cleaned, and also give rather reproducible liquid-junction potentials [2.310, 311]. Sintered ceramic and glass disks [2.310, 311] are also available with a large range of defined pore sizes [2.316], which can be chosen according to the measuring problem. Most frequently used are ceramic plugs, which are chemically inert but suffer from a slightly limited reproducibility of the liquid-junction potential due to the undefined channel structure of the ceramic. Linen fibre junctions, unlike asbestos fibres, generate unusually large liquid-junction potentials with tris buffer solutions [2.283] and are not generally recommended.

2.10.13 The Meaning of the Uncertainty of pH Values

According to the principle of traceability, primary and secondary standard and measured pH are reported with their uncertainties, for instance pH $\pm \Delta$pH. The uncertainty $\pm\Delta$pH is added/subtracted and, for example, has a certain magnitude for a certain calibration procedure, which is independent of the pH measured.

However, due to the logarithmic relation between pH and the hydrogen ion activity, which is actually the quantity of interest, the character of the uncertainty of a_H differs from that of the pH from which it is obtained, $a_H \pm \Delta a_H = 10^{-(pH\pm\Delta pH)}$. Thus the uncertainty range of the hydrogen activity is given by

$$\Delta a_H = \left[10^{-(pH-\Delta pH)} - 10^{-(pH+\Delta pH)} \right] , \tag{2.322}$$

from which an average uncertainty may be derived,

$$\pm\Delta \bar{a}_H = 0.5 \left[10^{-(pH-\Delta pH)} - 10^{(pH+\Delta pH)} \right] , \tag{2.323}$$

so that the average hydrogen ion activity with its average uncertainty is given by

$$\bar{a}_H \pm \Delta \bar{a}_H = 0.5 \left[10^{-(pH-\Delta pH)} + 10^{-(pH+\Delta pH)} \right]$$
$$\pm 0.5 \left[10^{-(pH-\Delta pH)} - 10^{-(pH+\Delta pH)} \right] . \tag{2.324}$$

These relationships show that neither the average hydrogen ion activity corresponds exactly to the average pH nor does its average uncertainty correspond to the uncertainty of pH. Nevertheless, this treatment seems to be justified as it eliminates positive and negative uncertainties of the hydrogen ion activity with different magnitudes, $+\Delta a_H \neq |-\Delta a_H|$, whereas $+\Delta \bar{a}_H = |-\Delta \bar{a}_H|$ as also $+\Delta$pH $= |-\Delta$pH$|$. For example, the pH value pH $\pm \Delta$pH $= 1.000 \pm 0.100$ yields the average hydrogen ion activity $\bar{a}_H \pm \Delta \bar{a}_H = 0.103 \pm 0.023$, whereas the hydronium ion activity $a_H(+\Delta a_{H1} - \Delta a_{H2}) = 0.1(+0.026 - 0.021)$ has an unsymmetric uncertainty, which is to be avoided as far as possible. (Note that the brackets denote uncertainties and not multiplication.)

2.11 Application of Glass Electrodes in Heavy Water

A full-scale treatment of the acidities of non-aqueous or partially aqueous solutions is beyond the scope of this book because, in Bates' words, this "is a problem of far greater complexity than the measurement of pH values in aqueous media" [2.2]. Detailed discussions of this topic are given for instance in [2.2, 4, 5, 280]. Discussing the application of glass electrodes in non-aqueous solutions in full is not feasible either because of the great number of solvents and their mixtures with water. However, the application of pH and pM glass electrodes in heavy water, where they function just as well as in ordinary water [2.255], will be covered here for three reasons.

(1) Heavy water is a special solvent in that it differs from ordinary water by its isotopic acidic component, the deuteron (D^+) or deutonium (D_3O^+) ion, whereas other non-aqueous solvents are characterized by the same acidic component as water, the proton or hydronium ion.

(2) It has been shown that the mechanism of the pD and pM response of glass electrodes in heavy water is the same as that of the pH and pM response in light water (dissociation mechanism) [2.108], which is of interest with respect to the foregoing sections.

(3) The mechanistic meaning of an empirical method of pD determination in D_2O, the deuteron effect, and of a related procedure for pNa determination, the deuterium oxide effect, have been elucidated [2.108], and this also should be presented in connection with the dissociation mechanism. In addition, heavy water has a technical significance as it is produced on the multitonne scale by electrolytic enrichment of normal water [2.317, 318] for the nuclear reactor industry so that pD measurements are also of considerable industrial interest.

Corresponding to the pH in light water (see Sect. 2.12), the acidity of heavy water is characterized by a pD value which is defined by the negative decadic logarithm of the deuterium ion activity a_D,

$$\text{pD} = -\log \frac{a_D}{m^0} , \tag{2.325}$$

where $m^0 = 1\,\text{mol}\,\text{kg}^{-1}$ [2.2]. pD is usually measured by means of electrochemical cells involving glass electrodes, whose functioning in heavy water has been safely established experimentally [2.255]. As in light water, standard pD(S) values which are traceable to the pD definition, see (2.325), have been assigned to appropriate buffer solutions in D_2O [2.2, 319], and serve as the traceable basis of pD measurements.

2.11.1 Mechanism of pD Response

It must be expected from the physical similarity of the isotopic solvents H_2O and D_2O and of the proton and deuteron, respectively, that the equal functioning of glass electrodes in light and heavy water is based on the same mechanism, which, for the pH response, has been shown to be the dissociation mechanism [2.72, 87, 95]. This expectation is supported by the well-known related homogeneous dissociation of proto- and deutero-acids in the isotopic solvents [2.319–323]. Accordingly, the pD response of glass electrode membranes in D_2O is expected to reflect the dissociation equilibrium

$$RD(D_2O) + D_2O = R^-(D_2O) + D_3O^+ \tag{2.326}$$

and the pM (or alkali) response the association equilibrium

$$R^-(D_2O) + M^+(D_2O)_n = RM(D_2O) + nD_2O , \tag{2.327}$$

and that these equilibria overlap in the transition range

$$RD(D_2O) + M^+(D_2O)_n + D_2O = RH(D_2O) + D_3O^+ + nD_2O , \tag{2.328}$$

where the transition pD is a characteristic quantity,

$$pD_{tr} = -\log(K_{D,D}\, K_{A,Na(D_2O)}\, a_{Na^+(D_2O)}) , \tag{2.329}$$

as is pH_{tr} in ordinary water (see Sect. 2.5.4) [2.72, 87, 95]. It can also easily be shown by a rigorous derivation of the electrode potential according to Sect. 2.5.3 [2.95] that the slopes of both the pD and the $pM(D_2O)$ response of glass electrodes in heavy water are subideal. The potential slope of glass electrodes, consequently, can be described by a single equation [2.95]:

$$\frac{d\varepsilon_m}{dpX} = -\left(1 - \frac{d(\log a'_{R^-})}{dpX}\right) k = -a_{gl} k , \tag{2.330}$$

where X is H, D, $M(H_2O)$, or $M(D_2O)$, R^- is any anionic glass surface group, and α_{gl} is the electromotive efficiency of the membrane glass [2.2], which, in principle, is below unity. Activity coefficients have probably little effect upon α_{gl} because of the extremely small concentration changes of R^- with pX [2.95] so that the subideal slope has a nearly equal magnitude in both isotopic solvents and, moreover, is practically independent of the surface group and thus of the glass composition and the ion to which the glass responds. Consequently it can also be predicted that, quite generally, all glass electrodes exhibit a sub-Nernstian slope as do pH-selective glass electrodes [2.2, 95].

The dissociation mechanism in heavy water was verified by the following experiments [2.108].

- Electrolyses in combination with concentration depth profiling (IBSCA and NRA) and infrared spectroscopy showed that membrane glass surfaces contain only RD groups at 100% pD response of the glass, RM groups at pure pM response, and both forms of groups at pD- and pM-determined concentrations within the transition range between pD and pM response.
- Corresponding to the findings in light water, 100% deuterated electrode membranes exhibited the same pD response in alkali-free D_2O solutions as before their deuteration. The response was even independent of whether the membranes were deuterated or protonated or contained alternatingly deuterated and protonated layers. This certainly contradicts the ion exchange theory also in heavy water solutions.

- Infrared spectra showed that D_2O-leached layers formed below RD group-covered glass surfaces in the same way as H_2O-leached layers did below RH group-covered surfaces (see Sect. 2.7.1). The slightly larger steepness of the D/Li concentration profiles compared to those of the H/Li profiles (obtained for lithium silicate glasses) indicated that the diffusion coefficients of deuterons in the inner, glass-like, leached layer range are slightly smaller than those of protons. An H/D isotope effect was also observed during leaching of a sodium calcium silicate glass [2.90]. The diffusion potentials caused by the interdiffusion of deuterons and lithium ions were not observable during pD measurements at 100% pD response, which corresponds to the same observation during pH measurements in light water [2.87].

2.11.2 Deuteron Effect

Before the assignment of pD(S) values to pD standard buffer solutions [2.2], an empirical method of pD measurement had been reported [2.253, 324], frequently verified [2.320, 321, 325–329], and thoroughly investigated [2.255]. This procedure consists of the measurement of the apparent (often termed operational) pH^D of a solution in heavy water by means of a glass electrode cell calibrated in standard buffers in light water and the addition of a correction term δ_{glass} which converts the pH^D into the pD value of the solution,

$$pD = pH^D + \delta_{glass} \ . \tag{2.331}$$

The numerical values of δ_{glass} are given as 0.41 on the molar and 0.45 on the molal scale [2.255]. This effect is called the deuteron effect in the following treatment. The first thorough report was based on 0.1 molal HCl and DCl solutions with pH and pD assumed equal [2.253], which, however, is only an approximation. The following mechanistic treatment is based on the meanwhile established pH and the pD scales because they provide an understanding of the effect [2.108].

Operational Equation

A glass electrode cell calibrated, for example, by a multiple-point calibration procedure with linear regression by using several standard buffer solutions in both light water [2.106] and heavy water, has the respective calibration functions

$$E_H = E_H^0 - k'_H pH \tag{2.332}$$

and

$$E_D = E_D^0 - k'_D pD \ , \tag{2.333}$$

where E_{H}^0 and E_{D}^0 are the respective standard potentials of the cell, and k_{H}' and k_{D}' are the respective average practical slopes of the cell emf. Applied in a solution in D$_2$O, the cell yields an emf that represents the emf $E_{\mathrm{H}}^{\mathrm{D}}$ and the related apparent pH^{D} on the light water calibration line (Fig. 2.73),

$$\mathrm{pH}^{\mathrm{D}} = \frac{E_{\mathrm{H}}^0 - E_{\mathrm{H}}^{\mathrm{D}}}{k_{\mathrm{H}}'} \; , \tag{2.334}$$

as well as the emf E_{D} and the pD on the heavy water calibration function

$$\mathrm{pD} = \frac{E_{\mathrm{D}}^0 - E_{\mathrm{D}}}{k_{\mathrm{D}}'} \; . \tag{2.335}$$

The difference of pD and pH^{D} is equal to the correction factor in (2.331), whose operational equation is thus given by

$$\mathrm{pD} - \mathrm{pH}^{\mathrm{D}} = \delta_{\mathrm{glass}} = \frac{E_{\mathrm{D}}^0 - E_{\mathrm{H}}^0}{k'} \; , \tag{2.336}$$

because $k_{\mathrm{H}}' = k_{\mathrm{D}}' = k'$ with good approximation and, according to the multiple-point calibration procedure, $E_{\mathrm{D}} = E_{\mathrm{H}}^{\mathrm{D}}$. The unknown liquid junction potential ε_{j}, which is actually contained in E, does not appear in the operational equation (2.336) because also $E - \varepsilon_{\mathrm{j}} = E_{\mathrm{D}} = E_{\mathrm{H}}^{\mathrm{D}}$. The correction term δ_{glass} is thus indeed independent of the liquid-junction potential if multiple-point calibration is applied.

The uncertainty of δ_{glass} is given by the difference of the average uncertainties of the calibration functions [2.106, 293]

$$\Delta\delta_{\mathrm{glass}} = \pm\frac{\overline{\varepsilon}_{\mathrm{j,H}} - \overline{\varepsilon}_{\mathrm{j,D}}}{k'} = \pm\frac{\Delta\overline{\varepsilon}_{\mathrm{j}}}{k'} \; , \tag{2.337}$$

which are both small and of comparable magnitude because the same liquid-junction device and similar types of standard buffers are applied in both

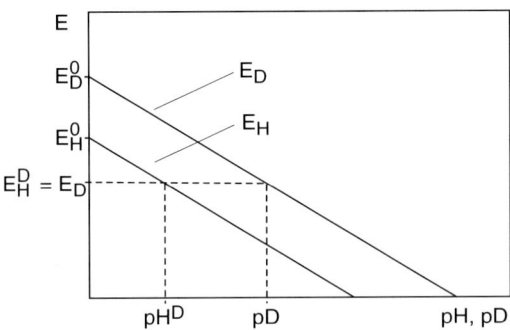

Fig. 2.73. Sketch of the calibration functions of a glass electrode cell in ordinary and heavy water, demonstrating the operational meaning of the deuteron effect

solvents [2.106]. $\Delta\delta_{glass}$ is estimated to be below 0.01, which corresponds to 2% of δ_{glass} and does not represent a significant contribution to the correction term. An exchange of the isotopic solvent of the reference electrode of the cell used for determining δ_{glass} does not appreciably change the results either [2.253, 255].

Mechanistic Meaning of δ_{glass}

For describing the mechanistic meaning of the operational correction term, the standard potentials of the cell contained in (2.336) must be explained by the glass electrode mechanism, and this is done by analysing the potentials of the cells involved [2.108]. Thus, the emf of cell (2.XXV) with ordinary water,

$$\text{ref. el.} | \text{KCl(s)}, H_2O \| \text{sol'n}, H_2O, \text{pH} | \text{glass membr.} | \text{int. buffer} | \text{int. ref. el.} \ , \tag{2.XXV}$$

consists of the sum of all potentials involved and can be represented by

$$E_H = \varepsilon_{m,H} + \varepsilon_{j,m,H} + \varepsilon_{j,H} + \sum \varepsilon_n \ , \tag{2.338}$$

where $\varepsilon_{j,H}$ is the liquid-junction potential between the reference and measuring solution, $\varepsilon_{j,m,H}$ is the diffusion potential in the leached layer below the external membrane surface, $\sum \varepsilon_n$ is the sum of all constant potentials of cell (2.XXV), and $\varepsilon_{m,H}$ is the phase-boundary potential at the interface membrane/measuring solution. It is determined by the dissociation equilibrium (see Sect. 2.5.2) and is given in its shortest version by

$$\varepsilon_{m,H} = -k \log K_{D,H} - k_{gl,H} \, \text{pH} \ , \tag{2.339}$$

where $K_{D,H}$ is the heterogeneous dissociation constant of the protonated form of the surface groups and $k_{gl,H}$ is the practical (thermodynamic) glass electrode slope, $k_{gl,H} = \alpha_H k$; α_H is the electromotive efficiency of the glass electrode (or glass membrane). Introducing the phase boundary potential, (2.339), into (2.338) yields the emf of cell (2.XXV),

$$E_H = -k \log K_{D,H} + \varepsilon_{j,m,H} + \sum \varepsilon_n - k_{gl,H} \, \text{pH} + \varepsilon_{j,H} \ , \tag{2.340}$$

whose first three terms on the right side represent the standard potential of the calibration function, (2.332), of cell (2.XXV) with light water,

$$E_H^0 = -k \log K_{D,H} + \varepsilon_{j,m,H} + \sum \varepsilon_n \ , \tag{2.341}$$

which is required for the mechanistic interpretation of the correction term (2.336).

An analogous derivation yields the emf of cell (2.XXV) with heavy instead of light water,

$$E_D = -k \log K_{D,D} + \varepsilon_{j,m,D} + \sum \varepsilon_n - k_{gl,D} \, pH + \varepsilon_{j,D} \,, \qquad (2.342)$$

where the sum of the first three terms on the right side constitutes the standard potential E_D^0 of the calibration function, (2.333), of cell (2.XXV) with heavy water:

$$E_D^0 = -k \log K_{D,D} + \varepsilon_{j,m,D} + \sum \varepsilon_n \,. \qquad (2.343)$$

Introducing the standard potentials (2.341) and (2.343) into the operational equation (2.336) finally yields the mechanistic expression of the correction term

$$\delta_{glass} = \log \frac{K_{D,H}}{K_{D,D}} + \frac{\Delta \varepsilon_{j,m}}{k} \qquad (2.344)$$

because, with good approximation, $k' = k$. The difference of the diffusion potentials in the isotopically leached layers of the membrane glass $\Delta \varepsilon_{j,m} = \varepsilon_{j,m,D} - \varepsilon_{j,m,H}$ may be called the "residual diffusion potential of the isotopically leached layer" [2.108] in analogy of the residual liquid-junction potential between solutions as used by *Bates* [2.2].

Equation (2.344) shows that, contrary to common belief, the correction term δ_{glass} is not a universal quantity but depends on two glass properties, the isotopic heterogeneous dissociation constants of the glass surface groups and the residual diffusion potential of the isotopically leached layers of the membrane glass. This is supported by the scatter of the reported and measured correction terms, in particular at 25 °C, Fig. 2.74, and the different temperature dependence of δ_{glass} for the different surface groups SiO and [AlOSi].

It will be shown below that the second term on the right side of (2.344) can be neglected as a first approximation so that reported correction terms yield rather accurate ratios of heterogeneous isotopic dissociation constants. They lie between 2.4 and 3.1, Table 2.11, and agree with corresponding ratios of isotopic acids in homogeneous solutions, which generally range from 2.0 to 3.5 [2.319–323]. Obviously, the surface charge of the glass has an equal effect upon both isotopic dissociation constants and does not affect their ratio although it reduces their absolute values by as much as several orders of magnitude, see Sect. 2.5.4 [2.72].

pH and pD values cannot strictly be compared because they are based on the concept of zero standard potentials of the Pt,H_2 and Pt,D_2 electrodes [2.2]. However, *Covington* et al. [2.255] have reported an empirical correction term δ_{gas} for determining pD by means of the Pt,gas electrodes. This term can be expressed operationally and mechanistically,

$$pD_g - pH_g^D = \delta_{gas} = \frac{E_{Pt,D_2}^0 - E_{Pt,H_2}^0}{k_g'} = \frac{\varepsilon_{Pt,D_2}^0 - \varepsilon_{Pt,H_2}^0}{k_g'} \,, \qquad (2.345)$$

Table 2.11. Average thermodynamic data of deuteron effect

Glass, Response	T °C	$\log \frac{K_{D,H}}{K_{D,D}}$	$\Delta(\Delta G_D^0)^a$ kJ mol^{-1}	$\Delta(\Delta H_D^0)_{av}$ [b] kJ mol^{-1}	$\Delta(\Delta S_D^0)_{av}$ [c] J K mol^{-1}	References
pH glass, pH resp.	25	0.45	−2.6	−3.8	−4.3	literature average and [2.108]
	50	0.40	−2.5			[2.108]
	78	0.33	−2.2			[2.327]
	100	0.29	−2.1			[2.327]
pNa glass, pH resp.	25	0.41	−2.3	−7.4	−16.9	[2.108]
	50	0.31	−1.9			[2.108]
	75	0.22	−1.5			[2.108]

[a] $\Delta(\Delta G_D^0) = \Delta G_{D,H}^0 - \Delta G_{D,D}^0$
[b] $\Delta(\Delta H_D^0)_{av} = (\Delta H_{D,H}^0 - \Delta H_{D,D}^0)_{average}$
[c] $\Delta(\Delta S_D^0)_{av} = (\Delta S_{D,H}^0 - \Delta S_{D,D}^0)_{average}$

where E^0 is the standard emf of the cell containing the Pt,gas electrode indicated and ε^0 is the standard potential of the Pt,gas electrode indicated. It is thus possible to obtain absolute ratios of the heterogeneous dissociation constants according to

$$\log \frac{K_{D,H}}{K_{D,D}} = \left(\delta_{glass} - \frac{\Delta\varepsilon_{j,m}}{k} \right) - \delta_{gas} , \qquad (2.346)$$

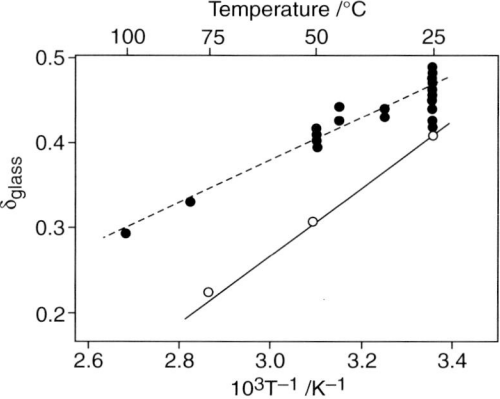

Fig. 2.74. Correction terms δ_{glass} of pH and pNa glass electrodes from various sources as a function of the reciprocal absolute temperature. *Filled circles:* pH electrodes (SiO surface group): 25 °C: *Covington* et al. [2.255], *Glasoe* and *Long* [2.253], *Forcé* and *Carr* [2.329], *Five* and *Bruice* [2.327], Schott electrodes [2.108]; 35 °C and 45 °C: *Forcé* and *Carr* [2.329]; 50 °C: Schott electrodes [2.108]; 78 °C and 100 °C: *Five* and *Bruice* [2.327]. *Open circles*: pH function of pNa electrodes ([AlOSi] surface group): 25 °C, 50 °C, 75 °C: Schott electrodes [2.108]

which are smaller than $(\delta_{\text{glass}} - \Delta\varepsilon_{j,m}/k)$ by $\delta_{\text{gas}} = 0.072$. This number was given by *Covington* [2.255] and is in agreement with a value obtained from measurements by *Gary* et al. [2.323]. Table 2.11, however, presents uncorrected reported data because of their better comparison.

As shown in Table 2.11, the difference of the standard entropies $\Delta(\Delta S_D^0)_{\text{av}}$ of the isotopic dissociation equilibria differs considerably for the pH functions of the different surface groups SiO (pH glass) and [AlOSi] (pNa glass). This seems to indicate that different forms of ions participate in the dissociation equilibria. The siloxy SiO^- group is known to combine with protons and deuterons to form SiOH and SiOD, respectively, whereas the $[AlOSi]^-$ group obviously prefers hydronium and deutonium ions generating $[AlOSi](H_3O)$ and $[AlOSi](D_3O)$ entities. The cause could be the dislocated electron of the [AlOSi] arrangement, which, unlike the electron located at the oxygen of the SiO^- group, does not favour the dissociation of, respectively, hydronium and deutonium ions into protons and deuterons and water and heavy water molecules. Besides, the spatial conditions of the [AlOSi] site, which favour its association with alkali ions over that with protons and deuterons [2.175], should be expected to bind hydronium and deutonium rather than protons and deuterons.

Lowe and *Smith* reported the occurrence of small potential drifts after transfers of glass electrodes between the isotopic solvents that depended on the square root of time [2.328], indicating diffusion control. Indeed, our infrared spectra measurements revealed that the phenomenon was caused by an exchange of the isotopes in the leached layers of the membrane glasses and that, obviously, the authors had measured the residual diffusion potential. The small magnitude of less than $1.5\,\text{mV}$ corresponding to $\delta_{\text{glass}} \leq 0.025$ justifies neglecting the residual diffusion potential for the calculation of the relative isotopic dissociation constants (see above). However, the drift should be observed when accurate pD are to be obtained by means of the deuteron effect [2.328].

2.11.3 Deuterium Oxide Effect

An empirical method of determining $pNa(D_2O)$ values of sodium-containing solutions in heavy water has also been reported [2.330]. It consists of the measurement of the apparent $pNa(H_2O)^{D_2O}$ in the heavy water solution by means of a glass electrode cell calibrated in standard sodium solutions in light water and the addition of a correction term $\delta_{\text{glass,Na}}$ (originally denoted Δ in [2.330]), which converts the apparent value into $pNa(D_2O)$,

$$pNa(D_2O) = pNa(D_2O)^{D_2O} + \delta_{\text{glass,Na}} . \tag{2.347}$$

The correction term $\delta_{\text{glass,Na}}$ has been estimated to be 0.09 ± 0.02 on the molar scale [2.330], which corresponds to 0.13 ± 0.03 on the molal scale. The effect is distinguished from the deuteron effect by the denotation deuterium oxide effect [2.108].

Operational Equation

The operational equation as well as the mechanistic meaning of the pNa
correction term are obtained in analogy to those of the pH correction term
and can be treated briefly in the following way. A cell involving a sodium-
selective glass electrode at 100% sodium ion response and calibrated by means
of standard sodium solutions in light and heavy water, when applied in a
sodium-containing solution in heavy water, yields an emf that characterizes
the apparent pNa in light water,

$$\mathrm{pNa(H_2O)^{D_2O}} = \frac{E^0_{\mathrm{Na(H_2O)}} - E^{D_2O}_{\mathrm{Na(H_2O)}}}{k'} \qquad (2.348)$$

as well as the pNa(D_2O) in the pNa scale in heavy water,

$$\mathrm{pNa(D_2O)} = \frac{E^0_{\mathrm{Na(D_2O)}} - E_{\mathrm{Na(D_2O)}}}{k'} \ , \qquad (2.349)$$

where the E^0_{Na} are the standard potentials of the calibration lines in the
isotopic solvent indicated, E is the measured emf signifying the indicated emfs
of the corresponding calibration function, and k' is the average practical slope
assumed equal in the solvents. Combination of (2.348) and (2.349) yields the
operational expression of the pNa correction term according to (2.347),

$$\mathrm{pNa(D_2O)} - \mathrm{pNa(H_2O)^{D_2O}} = \delta_{\mathrm{glass,Na}} = \frac{E^0_{\mathrm{Na(D_2O)}} - E^0_{\mathrm{Na(H_2O)}}}{k'} \ . \qquad (2.350)$$

Mechanistic Meaning of $\delta_{\mathrm{glass,Na}}$

For obtaining the mechanistic expression of (2.350), the emf of the cells containing the sodium solutions in light and heavy water are analysed in an
analogous way as those for δ_{glass} and yield the standard potentials in ordinary water,

$$E^0_{\mathrm{Na(H_2O)}} = k \log K_{\mathrm{A,Na(H_2O)}} + \varepsilon_{\mathrm{j},m,\mathrm{Na(H_2O)}} + \sum \varepsilon_n \ , \qquad (2.351)$$

and in heavy water,

$$E^0_{\mathrm{Na(D_2O)}} = k \log K_{\mathrm{A,Na(D_2O)}} + \varepsilon_{\mathrm{j},m,\mathrm{Na(D_2O)}} + \sum \varepsilon_n \ . \qquad (2.352)$$

Introducing (2.351) and (2.352) into (2.350) results in the mechanistic expression of the pNa correction term

$$\delta_{\mathrm{glass,Na}} = \log \frac{K_{\mathrm{A,Na(D_2O)}}}{K_{\mathrm{A,Na(H_2O)}}} \ , \qquad (2.353)$$

where the terms containing the diffusion potential in (2.351) and (2.352) have cancelled because the surface coverage with RNa groups at 100% pNa response of the glass and the resulting diffusion potential are independent of the solvent. However, two different situations must be distinguished. If the glass itself contains sodium ions, the diffusion potential is zero, whereas if it contains an ion other than sodium, the diffusion potential has a finite constant value. In both cases, the glass membrane must be covered with RNa groups by storing the electrodes in sodium-containing solutions with pH $>$ (pH$_{tr}$ + 2) or pD $>$ (pD$_{tr}$ + 2), respectively, to keep protons or deuterons and alkali ions other than sodium from penetrating the glass because interdiffusion of the sodium of the glass versus these ions would cause unpredictable diffusion potentials, in particular at frequently changing surface coverage [2.96].

The different magnitudes of the correction terms $\delta_{glass,Na} \approx 0.29\,\delta_{glass}$ is obviously caused by the different character of the interfacial equilibria. The dissociation equilibrium responsible for the deuteron effect involves the isotopic ions, the protons and deuterons, whereas the association equilibrium causing the deuterium oxide effect employs the same alkali ion in both solvents with the only modification that surface groups and alkali ions are differently solvated.

Selectivities in Light and Heavy Water

The different magnitudes of the pD and pNa correction terms result in different selectivity ranges of glass electrodes in the isotopic solvents, Fig. 2.75. This follows from a comparison of the transition pH in light water,

$$\mathrm{pH_{tr}} = -\log\left(K_{D,H}\, K_{A,Na(H_2O)}\, a_{Na^+(H_2O)}\right)\ , \tag{2.354}$$

with the transition pD in heavy water,

$$\mathrm{pD_{tr}} = -\log\left(K_{D,D}\, K_{A,Na(D_2O)}\, a_{Na^+(D_2O)}\right)\ , \tag{2.355}$$

for a given membrane glass [2.108]. Combining (2.354) and (2.355), assuming the same sodium ion activity in the isotopic solutions, and observing (2.344) and (2.353) for the correction terms yield the difference

$$\mathrm{pD_{tr} - pH_{tr}} = \delta_{glass} - \delta_{glass,Na} - \frac{\Delta\varepsilon_{j,m}}{k'}\ . \tag{2.356}$$

Because this difference is positive, pNa glass electrodes are expected to show a slightly wider sodium-selective range in H_2O than in D_2O, Fig. 2.75, and pH glass electrodes to exhibit a smaller sodium error in heavy than in light water solutions. From reported correction terms [2.255, 330] and residual diffusion potentials [2.328], it follows that (pD$_{tr}$ - pH$_{tr}$) = 0.30 ± 0.08, which should be measurable under good experimental conditions. As also concluded from Fig. 2.75, the effect is independent of the relative standard potentials, which merely shift the curves along the ordinate without affecting the position of the transition values.

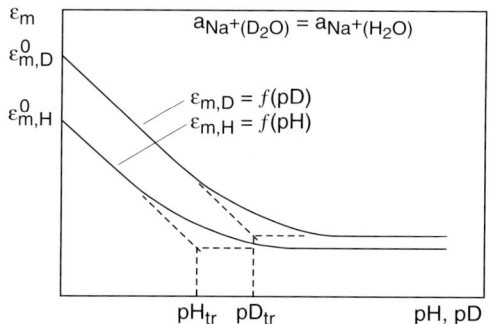

Fig. 2.75. Schematic diagram of glass electrode response in the transition range between pH and pNa(H$_2$O) in light water and between pD and pNa(D$_2$O) in heavy water at an equal sodium activity

References

2.1 G.D. Moody, J.D.R. Thomas: *Selective Ion Sensitive Electrodes* (Merrow, Watford 1971)

2.2 R.G. Bates: *Determination of pH. Theory and Practice*, 2nd ed. (Wiley, New York 1973)

2.3 U. Fritze: "Richtlinien für die pH Messung in industriellen Anlagen", Normenheft 22, ed. by Deutscher Normenausschuß (DNA) (Beuth, Berlin 1974)

2.4 K. Schwabe: *pH-Meßtechnik*, 4th ed. (Steinkopff, Dresden 1976)

2.5 H. Galster: *pH Measurement. Fundamentals, Methods, Applications, Instrumentation* (VCH, Weinheim 1991)

2.6 F. Honold, B. Honold: *Ionenselektive Elektroden. Grundlagen und Anwendungen in Biologie und Medizin* (Birkhäuser, Basel 1991)

2.7 R. Degner, S. Leibl: *pH messen. So wird's gemacht!* (VCH, Weinheim 1995)

2.8 K. Schwabe, H.D. Suschke: "Theorie der Glaselektrode", Angew. Chem. **76**, 39–49 (1964)

2.9 K. Schwabe, H. Dahms: "Untersuchung des Ionenaustauschs an Glaselektroden mit Radioindikatoren", Isotopentechnik **1**, 34–39 (1960)

2.10 G. Eisenman: "The origin of the glass electrode potential", in *Glass Electrodes for Hydrogen and Other Cations*, ed. by G. Eisenman (Dekker, New York 1967) pp. 133–173

2.11 L. Kratz: "Das Schrifttum über Glaselektroden", Z. Elektrochem. **46**, 259–264 (1940)

2.12 M. Dole: *The Glass Electrode*, 2nd. ed. (Wiley, New York 1947)

2.13 L. Kratz: *Die Glaselektrode und ihre Anwendungen* (Steinkopff, Frankfurt 1950)

2.14 H.T.S. Britton: *Hydrogen Ions. Their Determination and Importance in Pure and Industrial Chemistry*, Vol. 1 (Chapman and Hall, London 1955)

2.15 R.G. Bates: "The glass electrode", in *Reference Electrodes. Theory and Practice*, ed. by D.J.G. Ives, G.J. Janz (Academic Press, New York 1961) pp. 231–267

2.16 G. Eisenman (Ed.): *Glass Electrodes for Hydrogen and Other Cations* (Dekker, New York 1967)

2.17 R.P. Buck: "Transient electrical behavior of glass membranes", part I, J. Electroanal. Chem. **18**, 363–380; part II, ibid. 381–386; part III, ibid. 387–399 (1968)

2.18 M. Lavallee, O.F. Schanne, N.C. Hebert (Eds.): *Glass Microelectrodes* (Wiley, New York 1969)

2.19 B. v. Lengyel, B. Csakvari: "Zur Theorie der Glaselektrode", Chemiker Z., Chem. Apparatur, Verfahrenstechn. **93**, 763–770 (1969)

2.20 R.G. Bates: *Determination of pH. Theory and Practice*, 2nd ed. (Wiley, New York 1973) pp. 340–390

2.21 G. Johansson, B. Karlberg, A. Wikby: "The hydrogen-ion selective glass electrode", Talanta **22**, 953–966 (1975)

2.22 H. v. Helmholtz: "On the modern development of Faraday's conceptions of electricity" (Faraday Lecture, 1881), J. Chem. Soc. **39**, 277–304 (1881)

2.23 W. Giese: "Über den Verlauf der Rückstandsbildung in Leydener Flaschen bei constanter Potentialdifferenz der Belegungen", Ann. Phys. Chem. NF **9**, 161–208 (1880)

2.24 M. Cremer: "Über die Ursache der elektromotorischen Eigenschaften der Gewebe, zugleich ein Beitrag zur Lehre von den polyphasischen Elektrolytketten", Z. Biol. **47**, 562–608 (1906)

2.25 S.P.L. Sørensen: "Enzymstudien. II. Mitteilung. Über die Messung und die Bedeutung der Wasserstoffionenkonzentration bei enzymatischen Prozessen", Biochem. Z. **21**, 131–200, 201–304 (1909); in French: Compt. Rend. Lab. Carlsberg **8**, 1–168 (1909)

2.26 F. Haber, Z. Klemensiewicz: "Über elektrische Phasengrenzkräfte", Z. Phys. Chem. **67**, 385–431 (1909)

2.27 W. Nernst: "Über die Löslichkeit von Mischkrystallen", Z. Phys. Chem. **9**, 137–142 (1892)

2.28 O. Schott: "Über das Eindringen von Wasser in die Glasoberfläche", Z. Instrumentenkunde **9**, 86–90 (1889)

2.29 F. Förster: "Über das chemische Verhalten des Glases. Einwirkung der Lösungen von Alkalien und Salzen auf Glas", Ber. Deutsch. Chem. Ges. **25**, 2494–2518 (1892)

2.30 F. Förster: "Zur weiteren Kenntnis des chemischen Verhaltens des Glases", Ber. Deutsch. Chem. Ges. **26**, 2915–2922 (1893)

2.31 F. Förster: "Vergleichende Prüfung einiger Glassorten hinsichtlich ihres chemischen Verhaltens", Z. Anal. Chem. **34**, 381–396 (1894)

2.32 M. Dole: "The theory of the glass electrode", J. Am. Chem. Soc. **53**, 4260–4280 (1931)

2.33 W.S. Hughes: "On Haber's glass cell", J. Chem. Soc., 491–506 (1928)

2.34 P. Gross, O. Halpern: "Über Mischelektroden zweiter Art", Z. Phys. Chem. **115**, 54–60 (1925)

2.35 B. v. Lengyel: "Über das Phasengrenzpotential Quarz/Elektrolytlösungen", Z. Phys. Chem. A **153**, 425–442 (1931)

2.36 B. v. Lengyel: "Beiträge zum Verhalten der Quarzelektroden", Part I, Z. Phys. Chem. A **159**, 145–160 (1932); part II, ibid. 393–402

2.37 H. Freundlich, P. Rona: "Über die Beziehungen zwischen dem elektrokinetischen Potentialsprung und der elektrischen Phasenkraft", Sitzungsber. Preuss. Akad. Wiss. **20**, 397–402 (1920)

2.38 K.L. Cheng: "pH glass electrode and its mechanism", in *Electrochemistry, Past and Present*, ACS Symposium Series 390, ed. by J.T. Stock, M.V. Orna (Am. Chem. Soc., Washington, DC 1989) pp. 286–302

2.39 K.L. Cheng: "Capacitor theory for nonfaradaic potentiometry", J. Microchem. **42**, 5–24 (1990)

2.40 K.J. Vetter: *Elektrochemische Kinetik* (Springer, Berlin, Heidelberg 1961)

2.41 M. Cremer: Beitr. Physiol. **2**, 229 ff. (1924)

2.42 F.G. Donnan: "The theory of membrane equilibria", Chem. Rev. **1**, 73–90 (1924)

2.43 P. Henderson: "Zur Thermodynamik der Flüssigkeiten", Z. Phys. Chem. **59**, 118–127 (1907)

2.44 P. Henderson: "Zur Thermodynamik der Flüssigkeiten", Z. Phys. Chem. **63**, 325–345 (1908)

2.45 L. Michaelis: "Die Permeabilität von Membranen", Naturwissenschaft **14**, 33–42 (1926)

2.46 M. Dole: "The theory of the glass electrode. II. The glass as a water electrode", J. Am. Chem. Soc. **54**, 3095–3105 (1932)

2.47 K. Schwabe, H. Dahms: "Versuche zur Frage der Durchlässigkeit von Glaselektroden für Wasserstoffionen mit Hilfe von Tritiummarkierung", Monatsber. Deutsch. Akad. Wiss. Berlin **1**, 279–282 (1959)

2.48 P.R. Hammond: "Studies on the glass electrode in solutions containing hydrogen isotopes", Chem. Ind. **17**, 311–312 (1962)

2.49 Y. Abe, M. Maeda: "Do hydrogen ions really cross a glass membrane to enable pH measurement?", Phys. Chem. Glasses **37**, 176 (1996)

2.50 F.M. Ernsberger: "Comments on 'Do hydrogen ions really cross a glass membrane to enable pH measurement?' by Y. Abe, M. Maeda", Phys. Chem. Glasses **38**, 282–283 (1997)

2.51 Y. Abe, M. Nogami, M. Maeda: "Mobile hydrogen ion in glass to enable pH-measurement", J. Non-Cryst. Solids **209**, 204–206 (1997)

2.52 Y. Abe, H. Hosono, Y. Ohta, L.L. Hench: "Protonic conduction in oxide glasses – simple relations between electrical conductivity, activation energy, and OH bonding state", Phys. Rev. B **38**, 10166–10169 (1988)

2.53 Y. Abe, H. Hosonso, O. Akita, L.L. Hench: "Protonic conduction in phosphate glasses", J. Electrochem. Soc. **141**, L64–L65 (1994)

2.54 F.G.K. Baucke, H. Bach: "Investigation of glasses using surface profiling by spectrochemical analysis of sputter-induced radiation. II, Field-driven formation and electrochemical properties of protonated glasses containing various proton concentrations", J. Am. Ceram. Soc. **65**, 534–539 (1982)

2.55 F.G.K. Baucke: "Electrochemistry and glass structure", J. Non-Cryst. Solids **129**, 233–239 (1991)

2.56 F.G.K. Baucke: "Field-driven redistribution of 'guest protons' within protonated silicate glasses", J. Non-Cryst. Solids **40**, 159–169 (1980)

2.57 B.P. Nicolsky: "Theory of the glass electrode I", Acta Physicochim. USSR **7**, 597–610 (1937)

2.58 H. Schiller: "Über die elektromotorischen Eigenschaften der Gläser", Ann. Phys. **74**, 105–135 (1924)

2.59 K. Horovitz, J. Zimmermann: "Untersuchungen über Ionenaustausch an Gläsern", Sitzungsber. Akad. Wiss. Wien, Abt. IIa, **134**, 355–383 (1925)

2.60 B.P. Nicolsky, M.M. Shul'ts, A.A. Belyustin, A.A. Lev: "Recent developments in the ion-exchange theory of the glass electrode and its application in the chemistry of glass", in *Glass Electrodes for Hydrogen and Other Cations. Principles and Practice*, ed. by G. Eisenman (Dekker, New York 1967) pp. 274–222

2.61 G.A. Rechnitz: "Cation-sensitive glass electrodes in analytical chemistry", in *Glass Electrodes for Hydrogen and Other Cations. Principles and Practice*, ed. by G. Eisenman (Dekker, New York 1967) pp. 322–343

2.62 G. Eisenman: "Cation selective glass electrodes and their mode of operation", J. Biophys. **2**, 259–323 (1962)

2.63 J.O. Isard: "The dependence of glass-electrode properties on composition", in *Glass Electrodes for Hydrogen and Other Cations. Principles and Practice*, ed. by G. Eisenman (Dekker, New York 1967) pp. 51–100

2.64 Z. Boksay, B. Csákvári: "The corrected formula of the phase-boundary potential of the glass electrode consistent with the structure of glass", Acta Chim. Akad. Sci. Hung. **67**, 157–160 (1971)

2.65 F.G.K. Baucke: "Cation migration in electrode glasses", in *Mass Transport Phenomena in Ceramics*, ed. by A.R. Cooper, A.H. Heuer (Plenum, New York 1975) pp. 337–353

2.66 K. Schwabe, H. Dahms: "Vergleichende Untersuchungen der elektromotorischen Eigenschaften und des chemischen Verhaltens von Glaselektroden mit Hilfe radioaktiver Indikatoren, I. Untersuchungen im Gebiet des Alkalifehlers", Z. Elektrochem. **65**, 518–526 (1961)

2.67 B. v. Lengyel, B. Csákvári, J. Toperczer: "Über den Alkalifehler der Glaselektrode, III. Neuere Beiträge zur Deutung des Alkalifehlers", Acta Chim. Hung. **45**, 177–185 (1965)

2.68 K. Schwabe: "Über den Alkalifehler der Glaselektrode", Acta Chim. Hung. **51**, 1–2 (1967)

2.69 J.O. Isard: "The origin of the electric potential in the ion exchange theory of the glass electrode", Phys. Chem. Glasses **17**, 1–6 (1976)

2.70 A. Wikby: "The surface resistance of glass electrodes in alkaline solutions", J. Electroanal. Chem. **39**, 103–109 (1972)

2.71 F.G.K. Baucke: "The glass electrode – applied electrochemistry of glass surfaces", J. Non-Cryst. Solids **73**, 215–231 (1985)

2.72 F.G.K. Baucke: "Glass electrodes. Why and how they function", Ber. Bunsenges. Phys. Chem. **100**, 1466–1474 (1996)

2.73 J.O'M. Bockris, A.K.N. Reddy: *Modern Electrochemistry*, Vol. 1 (Plenum, New York 1970) pp. 16–17

2.74 G. Bouquet, S. Dobos, Z. Boksay: "Untersuchung der Oberflächenschicht des Glases", Ann. Univ. Sci. Budapest (Rolando Eötvös Nominatae) Sect. Chim. **6**, 5–13 (1964)

2.75 Z. Boksay, G. Bouquet, S. Dobos: "Diffusion processes in the surface layer of glass", Phys. Chem. Glasses **8**, 140–144 (1967)

2.76 A. Distèche, M. Dubuisson: "Transient response of the glass electrode to pH step variations", Rev. Sci. Instrum. **25**, 869–875 (1954)

2.77 G.A. Rechnitz, G.C. Kugler: "Transient phenomena at glass electrodes", Anal. Chem. **39**, 1682–1688 (1967)

2.78 A. Wikby, G. Johansson: "The resistance and intrinsic time constant of glass electrodes", J. Electroanal. Chem. **23**, 23–40 (1969)

2.79 A. Wikby: "The resistance of the surface layers of glass electrodes", Phys. Chem. Glasses **15**, 37–41 (1974)

2.80 A. Wikby, B. Karlberg: "Correlation between the gel layer properties and the electrochemical behaviour of hydrogen selective glass electrodes", Electrochim. Acta **19**, 323–328 (1974)

2.81 A. Wikby: "Chemical and electrical properties of the surface layers of some glass electrodes", Electrochim. Acta **19**, 329–336 (1974)

2.82 Z. Boksay, M. Varga, A. Wikby: "Surface conductivity of leached glass", J. Non-Cryst. Solids **17**, 349–358 (1975)

2.83 H. Bach: "Zur Bestimmung der Reichweiten von beschleunigten Ionen in dünnen Oxidschichten", Z. Angew. Phys. **28**, 239–244 (1970)

2.84 H. Bach: "Abtragraten und spezifische Energieverluste von 5,6 keV-Edelgasionen an Kieselglas", Z. Naturforsch. **27a**, 333–338 (1972)

2.85 W.A. Lanford: "Hydrogen profiling: scientific applications", Nucl. Instrum. Meth. **148**, 1–8 (1978)

2.86 F.G.K. Baucke: "Investigation of surface layers, formed on glass electrode membranes in aqueous solutions, by means of an ion sputtering method", J. Non-Cryst. Solids **14**, 13–31 (1974)

2.87 F.G.K. Baucke: "The modern understanding of the glass electrode response", Fresenius' J. Anal. Chem. **349**, 582–596 (1994)

2.88 H. Bach, F.G.K. Baucke: "Investigation of glasses using surface profiling by spectro-chemical analysis of sputter-induced radiation: I, Surface profiling technique with high in-depth resolution", J. Am. Ceram. Soc. **65**, 527–533 (1982)

2.89 W.A. Lanford, K. Davies, P. Lamarche, T. Laursen, R. Groleau, R.H. Doremus: "Hydration of soda-lime glass", J. Non-Cryst. Solids **33**, 249–266 (1979)

2.90 P. March, F. Rauch: "Leaching studies of soda-lime-silica glass using deuterium- and ^{18}O-enriched solutions", Glastechn. Ber. **63**, 154–162 (1990)

2.91 H. Scholze: "Der Einbau des Wassers in Gläsern", part I, Glastechn. Ber. **32**, 81–88 (1959); part II, ibid. 142–152; part III, ibid. 271–281; part IV, ibid. 314–320

2.92 J.F. Ziegler, C.P. Wu, P. Williams, et al.: "Profiling hydrogen in materials using ion beams", Nucl. Instrum. Meth. **149**, 19–39 (1978)

2.93 R.A. Robinson, R.H. Stokes: *Electrolyte Solutions*, 2nd ed. (Butterworths, London 1968) pp. 104–109

2.94 W. Vogel: *Chemistry of Glass*, ed. by N. Kreidl (Am. Ceram. Soc., Columbus, OH 1985)

2.95 F.G.K. Baucke: "Thermodynamic origin of the sub-Nernstian response of glass electrodes", Anal. Chem. **66**, 4519–4524 (1994)

2.96 F.G.K. Baucke: "The origin of the glass electrode response", in *Proc. Conf. on Glass Current Issues*, ed. by A.F. Wright, J. Dupuy, NATO ASI Series, Appl. Sci. Eng. (Nijhoff, Dordrecht 1985) pp. 481–505

2.97 F.G.K. Baucke: "Simultaneous transfer of different cations across anodic electrolyte solution–glass interfaces in electric fields", in *The Physics of Non-Crystalline Solids*, ed. by G.H. Frischat (Trans Tech Publications, Aedermannsdorf 1977) pp. 503–508

2.98 A.A. Belyustin: "Dynamics of the potential and mechanism of operation of ion-selective glass electrodes", Usp. Khim. **49**, 1880–1903 (1980)

2.99 G. Haugaard: "The mechanism of the glass electrode", J. Phys. Chem. **45**, 148–157 (1941)

2.100 E.A. Guggenheim: *Thermodynamics* (North Holland, Amsterdam; Wiley, New York 1967) pp. 298 ff.

2.101 G. Haugaard: "Studies on the glass electrode", Compt. Rend. Lab. Carlsberg, Ser. Chem. **22**, 199–204 (1938)

2.102 L. Kratz: *Die Glaselektrode und ihre Anwendungen*, Wissenschaftliche Forschungsberichte, Naturwissenschaftliche Reihe, Vol. 59, ed. by R. Jäger (Steinkopff, Frankfurt 1950) pp. 72 ff.

2.103 British Standard Specification BS 2586: "Glass electrodes" (British Standard Institution, London 1979)

2.104 W.H. Beck, A.E. Bottom, A.K. Covington: "Errors of glass electrodes in certain standard buffer solutions at high discrimination", Anal. Chem. **40**, 501–505 (1968)

2.105 T.S. Light, K.S. Fletcher III: "Accurate evaluation of glass electrodes in high ionic strength medium", Anal. Chem. **39**, 70–75 (1967)

2.106 F.G.K. Baucke, R. Naumann, C. Alexander-Weber: "Multiple-point calibration with linear regression as a proposed standardization procedure for high-precision pH measurements", Anal. Chem. **65**, 3244–3251 (1993)

2.107 G. Eisenman: "The electrochemistry of cation-sensitive glass electrodes", in *Advances in Analytical Chemistry and Instrumentation*, Vol. 4, ed. by C.N. Reilley (Wiley-Interscience, New York 1965) pp. 213–369

2.108 F.G.K. Baucke: "Further insight into the dissociation mechanism of glass electrodes. The response in heavy water", J. Phys. Chem. B **102**, 4835–4841 (1998)

2.109 L. Bousse, P. Bergveld: "The role of buried OH sites in the response mechanism of inorganic-gate pH-sensitive ISFETs", Sens. Actuators **6**, 65–78 (1984)

2.110 D.E. Yates, S. Levine, T.W. Healy: "Site-binding model of the electrical double layer at the oxide/water interface", J. Chem. Soc. Faraday Trans. **1** (70), 1807–1818 (1974)

2.111 U. Ösch, Z. Drzozka, A. Xu, B. Rusterholz, et al.: "Design of neutral hydrogen ion carriers for solvent polymeric membrane electrodes of selected pH range", Anal. Chem. **58**, 2285–2289 (1986)

2.112 R.K. Iler: *The Chemistry of Silica. Solubility, Polymerization, Colloid and Surface Properties, and Biochemistry* (Wiley, New York 1979) pp. 123 ff.

2.113 J.O'M. Bockris, A.K.N. Reddy: *Modern Electrochemistry*, Vol. 2 (Plenum, New York 1970)

2.114 F.G.K. Baucke: "The glass electrode – proposal of kinetic measurements for the development of improved membrane glasses", Glastechn. Ber. Sci. Technol. **70 C**, 369–381 (1997)

2.115 J.M. Hodgson: "Dependence upon work function of electrical conduction across the interface between a contact and a glass", Glass Technol. **26**, 208–211 (1985)

2.116 K.J. Vetter: *Elektrochemische Kinetik* (Springer, Berlin, Heidelberg 1961) pp. 8–9

2.117 B. Heinrich: *Experimentelle Untersuchungen zur Wanderung von Silberionen in oxidischen Gläsern*, Diploma Thesis (FH Fresenius, Wiesbaden 1988)

2.118 T. Kaneko: "The field-assisted penetration of a silver film into glass", J. Non-Cryst. Solids **120**, 188–198 (1990)

2.119 F.G.K. Baucke, J.A. Duffy: "Ion migration study in a sodium borate glass: proposal of a new oxide transport", J. Electrochem. Soc. **127**, 2230–2233 (1980)

2.120 H. Kahnt, Ch. Kaps, J. Offermann: "A new method of simultaneous measurement of tracer diffusion coefficient and mobility of alkali ions in glasses", Solid State Ionics **31**, 215–220 (1988)

2.121 F.G.K. Baucke: "Cation migration in electrode glasses", in *Mass Transport Phenomena in Ceramics*, ed. by A.R. Cooper, A.H. Heuer (Plenum, New York 1975) pp. 337–354

2.122 C.A. Kraus, E.H. Darby: "A study of the conduction process in ordinary soda-lime glass", J. Am. Chem. Soc. **44**, 2783–2797 (1922)

2.123 D.E. Carlson, C.E. Tracy: "Injection of ions into glass from a glow discharge", J. Appl. Phys. **46**, 1575–1580 (1975)

2.124 D.E. Carlson, C.E. Tracy: "Metallization of glass using ion injection", Ceram. Bull. **55**, 530–532 (1976)

2.125 D.L. Kinser, L.L. Hench: "Electrode polarization in alkali silicate glasses", J. Am. Ceram. Soc. **55**, 638–641 (1969)

2.126 T.M. Proctor, P.M. Sutton: "Static space charge distribution with a single mobile charge carrier", J. Chem. Phys. **30**, 212–220 (1959)

2.127   T.M. Proctor, P.M. Sutton: "Space-charge development in glass", J. Am. Ceram. Soc. **43**, 173–179 (1960)

2.128   F.M. Ernsberger: "Ion conduction in oxide glasses: blocking electrodes and space charge", Phys. Chem. Glasses **36**, 152–153 (1995)

2.129   D.E. Carlson, K.W. Hang, G.F. Stockdale: "Electrode 'polarization' in alkali-containing glasses", J. Am. Ceram. Soc. **55**, 337–341 (1972)

2.130   S.P. Mitoff, R.J. Charles: "Electrode polarization of ionic conductors", J. Appl. Phys. **43**, 927–934 (1972)

2.131   J.R. Macdonald: "Electrode polarization of ionic conductors", J. Appl. Phys. **44**, 3455–3458 (1973)

2.132   S.P. Mitoff, R.J. Charles: "Comments on 'Electrode polarization of ionic conductors'", J. Appl. Phys. **44**, 3786–3787 (1973)

2.133   F.G.K. Baucke: "Determination of cation mobilities in glasses by direct measurement of drift velocities (moving boundary)", Z. Naturforsch. **26a**, 1778 (1971)

2.134   C.C. Rüssel: *Polyvalente Elemente in oxidischen Glasschmelzen*, Habilitation Thesis (Erlangen-Nürnberg 1991) pp. 132–138

2.135   F.G.K. Baucke: "Transport properties of glasses containing mobile original and anodically introduced 'guest' ions", in *Proc. Int. Congr. on Glass* (CVTS Dum techniky Praha, Prague 1977) pp. 347–356

2.136   R.H. Doremus: "Interdiffusion of hydrogen and alkali ions in a glass surface", J. Non-Cryst. Solids **19**, 137–144 (1975)

2.137   N.N. Greenwood, A. Earnshaw: *Chemistry of the Elements* (Pergamon, Oxford 1984) p. 86

2.138   J. Bruinink: "Proton migration in solids", J. Appl. Electrochem. **2**, 239–249 (1972)

2.139   L. Glasser: "Proton conduction and injection in solids", Chem. Rev. **75**, 21–65 (1975)

2.140   A. Sendt: "Ion exchange and diffusion processes in glass", in *Advances in Glass Technology* (Plenum, New York 1962) pp. 307–332

2.141   W.A. Weyl, E.C. Marboe: "A new interpretation of the behavior of materials under stress", Silic. Ind. **38**, 5–19 (1973)

2.142   M. Abou-el-leil, A.R. Cooper: "Analysis of field-assisted binary ion exchange", J. Am. Ceram. Soc. **62**, 390–395 (1979)

2.143   Z. Boksay, B. Lengyel: "Vacancy type mechanism of the electrical relaxation processes in glass", J. Non-Cryst. Solids **14**, 79–87 (1974)

2.144   Z. Boksay: "Mass transport in non-crystalline solids", in *The Physics of Non-Crystalline Solids*, ed. by G.H. Frischat (Trans Tech Publications, Aedermannsdorf 1977) pp. 428–446

2.145   H. Scholze: "The build-in of water in glasses: I, Influence of water dissolved in glasses upon the infrared spectrum, and quantitative infrared-spectroscopic determination of water in glasses", Glastechn. Ber. **32**, 81–88 (1959)

2.146   H. Scholze: "The build-in of water in glasses: II, Infrared measurements on silicate glasses with systematically varied composition, and interpretation of OH bands in silicate glasses", Glastechn. Ber. **32**, 142–152 (1959)

2.147   N.N.: "Spectrometry nomenclature", Anal. Chem. **43**, 2038 (1971)

2.148   R.H. Doremus, A. Babinec, K. D'Angelo, M. Doody, W.A. Lanford, C. Burman: "Electrolysis of soda-lime silicate glass in water", J. Am. Ceram. Soc. **67**, 476–479 (1984)

2.149   M. Tomozawa: "Dielectric characteristics of glass", in *Glass I: Interaction with Electromagnetic Radiation,* Treatise on Materials Science and Technology, Vol. 12, ed. by M. Tomozawa, R.H. Doremus (Academic Press, New York 1977) pp. 283–342

2.150 J.O'M. Bockris, A.K.N. Reddy: *Modern Electrochemistry,* Vol. 2 (Plenum, New York 1970) pp. 883–888

2.151 K. Cammann: *Untersuchungen zur Wirkungsweise ionenselektiver Elektroden (Abschätzung von Standardaustauschstromdichten),* PhD Thesis (München 1975) pp. 105–118

2.152 F.G.K. Baucke: "The effect of the cooling rate on electrochemical surface and bulk properties of some lithium silicate glasses", in *Diffusion and Defect Data,* Vols. 53–54, ed. by R.A. Weeks, D.L. Kinser (Trans Tech Publications, Aedermannsdorf 1987) pp. 197–202

2.153 F. Quittner: "Einwanderung von Ionen aus wässriger Lösung in Glass", Ann. Phys. (IV. Folge) **85**, 745–769 (1928)

2.154 M.D. Ingram: "Ionic conductivity in glass", Phys. Chem. Glasses **28**, 215–234

2.155 U.K. Krieger, W.A. Lanford: "Field assisted transport of Na$^+$ ions, Ca^{2+} ions and electrons in commercial soda-lime glass I: Experimental", J. Non-Cryst. Solids **102**, 50–61 (1988)

2.156 J.O. Isard: "Mixed alkali effect in glasses", J. Non-Cryst. Solids **1**, 235–261 (1969)

2.157 R.H. Doremus: "Mixed alkali effect and interdiffusion of Na and K ions in glass", J. Am. Ceram. Soc. **57**, 478–480 (1974)

2.158 G. Tomandl, H.A. Schaeffer: "Relation between the mixed-alkali effect and the electrical conductivity of ion-exchanged glasses", in *The Physics of Non-Crystalline Solids,* ed. by G.H. Frischat (Trans Tech Publications, Aedermannsdorf 1977) pp. 480–485

2.159 K. Hughes, J.O. Isard: "Ionic transport in glasses", in *Transport Processes in Solid Electrolytes and in Electrodes, Physics of Electrolytes,* Vol. 1, ed. by J. Hladik (Academic Press, London 1972) pp. 351–400

2.160 J.-P. Lacharme: "Mobilités électriques du sodium et du potassium dans les verres à alcalis mixtes Na$_2$O–K$_2$O", Compt. Rend. Acad. Sci. Paris, Serie C, **t 275**, 993–996 (1972)

2.161 H. Jain, N.L. Peterson, H.L. Downing: "Tracer diffusion and electrical conductivity in sodium-cesium silicate glasses", J. Non-Cryst. Solids **55**, 283–305 (1983)

2.162 J.E. Davidson, M.D. Ingram, A. Bunde, K. Funke: "Ion hopping processes and structural relaxation in glassy materials", J. Non-Cryst. Solids **203**, 246–251 (1996)

2.163 Z. Boksay, G. Bouquet, E. Hári, J. Rohonczy: "Composition fluctuations in an alkali silicate glass", Glastechn. Ber. **66**, 9–14 (1993)

2.164 A.Y. Kuznetsov: "Electrical conductivity of glasses in the Li$_2$O–SiO$_2$ system", Russian J. Phys. Chem. **33**, 20–22 (1959) (Zh. Fiz. Khim. **33**, 1492–1494 (1959))

2.165 C. Wagner: "Oxidation of alloys involving noble metals", J. Electrochem. Soc. **103**, 571–580 (1956)

2.166 W.W. Mullins, R.F. Sekerka: "Stability of a planar interface during solidification of a dilute binary alloy", J. Appl. Phys. **35**, 444–451 (1964)

2.167 J.S. Kirkaldy, D.G. Fedak: "Nonplanar interfaces in two-phase ternary diffusion couples", Trans. Met. Soc. AIME **224**, 490–494 (1962)

2.168 D.P. Whittle: "Oxidation mechanisms for alloys in single oxidant glasses", in *High Temperature Corrosion,* ed. by R.A. Rapp (Nat. Ass. of Corrosion Eng., Houston 1983) pp. 171–183

2.169 S. Schimschal-Thölke, H. Schmalzried, M. Martin: "Instability of moving interfaces between ionic crystals KCl/AgCl", Ber. Bunsenges. Phys. Chem. **99**, 1–6 (1995)

2.170 M. Chemla: "Diffusion d'ions radioactifs dans des cristaux. Applications",
       Ann. Physique **13**, 959–1002 (1956)
2.171 H. Ohta, M. Hara: "Ion-exchange in sheet glass by electrolysis", J. Ceram.
       Assoc. Jpn. **78**, 158–164 (1970)
2.172 H. Ohta: "Migration of K^+ ions from the ion-exchanged layer under the
       reverse electric field", J. Ceram. Assoc. Jpn. **80**, 16–24 (1972)
2.173 R.H. Doremus: "Exchange and diffusion of ions in glass", J. Phys. Chem. **68**,
       2212–2218 (1964)
2.174 H. Franz, T. Kelen: "Conclusions on the structure of alkali silicate glasses
       and melts from the build-in of OH groups", Glastechn. Ber. **40**, 141–148
       (1967)
2.175 G. Eisenman: "The physical basis for the ionic specificity of the glass elec-
       trode", in *Glass Electrodes for Hydrogen and Other Cations, Principles and
       Practice*, ed. by G. Eisenman (Dekker, New York 1967) pp. 223–237
2.176 F.G.K. Baucke, J.A. Duffy: "Use of thallium(I) probe for identifying sites of
       mobile cations in glass during electrolysis", J. Chem. Soc., Faraday Trans.
       1, **79**, 661–667 (1983)
2.177 F.G.K. Baucke, J.A. Duffy: "Electrolysis of a sodium borate glass: a new
       mechanism of oxide ion transport", Glastechn. Ber. **56K**, 608–613 (1983)
2.178 K. Takizawa: "Ionic conduction of $Li_2O \cdot 2SiO_2$ glass under dc potential",
       J. Am. Ceram. Soc. **61**, 475–478 (1978)
2.179 F.G.K. Baucke: "Electrochromic applications", Mater. Sci. Eng. B **10**, 285–
       292 (1991)
2.180 J.A. Duffy, M.D. Ingram: "Establishment of an optical scale for Lewis basicity
       in inorganic oxyacids, molten salts, and glasses", J. Am. Chem. Soc. **93**,
       6448–6454 (1971)
2.181 J.A. Duffy, M.D. Ingram: "An interpretation of glass chemistry in terms of
       the optical basicity concept", J. Non-Cryst. Solids **21**, 373–410 (1976)
2.182 J.H. Binks, J.A. Duffy: "A molecular orbital treatment of the basicity of
       oxyanion units", J. Non-Cryst. Solids **37**, 387–400 (1980)
2.183 J.A. Duffy, E.I. Kamitsos, G.D. Chryssicos, A.P. Patsis: "Trends in local
       optical basicity in sodium borate glasses and relation to ionic mobility",
       Phys. Chem. Glasses **34**, 153–157 (1993)
2.184 F.G.K. Baucke: "The electrochemically structured interface between oxide
       glasses and solutions", in *Abstracts, Glass Meeting 89*, Lake Buena Vista,
       Sept. 17–19, 1989 (Am. Ceram. Soc., Columbus, OH 1989)
2.185 E. Pungor: "Working mechanisms of ion-selective electrodes", Pure Appl.
       Chem. **64**, 503–507 (1992)
2.186 T.M. Sullivan, A.J. Machiels: "Influence of the electric double layer on glass
       leaching", J. Non-Cryst. Solids **55**, 269–282 (1983)
2.187 H. Scholze, D. Helmreich, I. Bakardjiev: "Untersuchungen über das Verhalten
       von Kalk-Natrongläsern in verdünnten Säuren", Glastechn. Ber. **48**, 237–247
       (1975)
2.188 F.G.K. Baucke: "Equilibria of functional groups of glass surfaces with cations
       in contacting solutions", in *The Glassy State*, Proc.7th All-Union-Conf.,
       Leningrad, Oct. 13–15, 1981 (Acad. Sci. USSR, Leningrad 1983) pp. 96–108
2.189 R.H. Doremus: "Exchange and diffusion of ions in glass", J. Phys. Chem. **68**,
       2212–2213 (1964)
2.190 F.G.K. Baucke: "Contribution to the electrochemistry of pH glass electrode
       membranes", in *Ion-Selective Electrodes*, ed. by E. Pungor (Elsevier, Ams-
       terdam 1978) pp. 215–134

2.191 F.G.K. Baucke: "Reversible and irreversible reactions at the surface of glass electrode membranes", in *Extended Abstracts, Part I.*, 29th Meeting, Int. Soc. Electrochem., ed. by L. Sándor (MTA, Budapest 1978) pp. 50–52

2.192 F.G.K. Baucke: "Investigation of electrode glass membranes: proposal of a dissociation mechanism for pH glass electrodes", J. Non-Cryst. Solids **19**, 75–86 (1975)

2.193 H. Bach, F.G.K. Baucke: "Measurement of ion concentration profiles in surface layers of leached ('swollen') glass electrode membranes by means of luminescence exited by ion sputtering", Electrochim. Acta **16**, 1311–1319 (1971)

2.194 L.L. Hench, D.E. Clark: "Physical chemistry of glass surfaces", J. Non-Cryst. Solids **28**, 83–105 (1978)

2.195 K. Horovitz: "Der Ionenaustausch am Dielektrikum I. Die Elektrodenfunktion der Gläser", Z. Phys. **15**, 369–398 (1923)

2.196 H. Leng: "Adsorptionsversuche an Gläsern und Filtersubstanzen nach der Methode der radioaktiven Indikatoren", Sitzungsber. Akad. Wiss. Wien (IIa) **136**, 19–42 (1927)

2.197 J. Hensley, A. Long, J. Willard: "Reactions of ions in aqueous solution with glass and metal surfaces. Studies with radioactive tracers", Ind. Eng. Chem. **41**, 1415–1421 (1949)

2.198 J. Hensley: "Adsorption of tagged phosphate ions on glass surfaces as related to alkaline attack", J. Am. Ceram. Soc. **34**, 188–192 (1951)

2.199 A. Long, J. Willard: "Reactions of ions in aqueous solution with glass. Studies with radioactive tracers", Ind. Eng. Chem. **44**, 916–920 (1952)

2.200 E. Trebge, R. Fischer: "Spezielle Untersuchungen an Glaselektroden mit ^{24}Na unter besonderer Berücksichtigung des Alkalifehlers", Silikattechnik **10**, 385–389 (1959)

2.201 R.K. Iler: *The Colloid Chemistry of Silica and Silicates* (Cornell Univ. Press, Ithaca 1955) Chap. 8

2.202 Z. Boksay, G. Bouquet, S. Dobos: "The kinetics of the formation of leached layers on glass surfaces", Phys. Chem. Glasses **9**, 69–71 (1968)

2.203 A. Wikby: "The surface resistance of glass electrodes in neutral solutions", J. Electroanal. Chem. **38**, 429–443 (1972)

2.204 Th. Richter, G.H. Frischat, G. Borchardt, St. Scherrer: "Initial stages of glass corrosion in water", Glastechn. Ber. Glass Sci. Technol. **63**, 300–308 (1990)

2.205 F.G.K. Baucke: unpublished results

2.206 F.G.K. Baucke: "Corrosion of glasses and its significance for glass coating", Electrochim. Acta **39**, 1223–1228 (1994)

2.207 H. Bach, F.G.K. Baucke: "Investigations of reactions between glasses and gaseous phases by means of photon emission induced during ion beam etching", Phys. Chem. Glasses **13**, 123–129 (1974)

2.208 R.W. Douglas, J.O. Isard: "The action of water and of sulphur dioxide on glass surfaces", Glass Technology **33**, 289–335 (1949)

2.209 G. Karreman, G. Eisenman: "Electrical potentials and ion fluxes in ion exchangers. I. 'n-type' non-ideal systems with zero current", Bull. Math. Biophys. **24**, 413–427 (1962)

2.210 F. Conti, G. Eisenman: "The non-steady state membrane potential of ion exchangers with fixed sites", J. Biophys. **5**, 247–256 (1965)

2.211 R.H. Doremus: "Diffusion potentials in glass", in *Glass Electrodes for Hydrogen and Other Cation. Principles and Practice*, ed. by G. Eisenman (Dekker, New York 1967) pp. 101–132

2.212 H. Dutz: "Eigenschaften einer selektiven Natriumelektrode", Glastechn. Ber. **39**, 139–140 (1966)

2.213 K. Cammann: *Das Arbeiten mit ionenselektiven Elektroden*, 2nd ed. (Springer, Berlin, Heidelberg 1977) pp. 57/58

2.214 K. Sykut, A. Kusak: "Natriumselektive Glasmembran-Elektroden aus Lithium-Aluminium-Silikat-Gläsern mit Zusätzen von TiO_2 und ZrO_2", Ann. Univ. Mariae Curie-Sklodowska, Lublin - Polonia, Sectio AA **33**, 17–27 (1978)

2.215 Chung-Chang Young: "pH responsive glass compositions and electrodes", US Patent No. 4 028 196 (1977)

2.216 I.S. Ivanovskaja, A.A. Belyustin, I.D. Pozdnyakova: "The effect of treatment of cation-selective glass electrodes with $AgNO_3$ solution on electrode properties", Sens. Actuators B **24–25**, 304–308 (1995)

2.217 G.A. Perley: "pH-responsive glass electrode", US Patent No. 2 444 845 (1948)

2.218 G.A. Perley: "Glasses for measurement of pH", Anal. Chem. **21**, 394–401 (1949)

2.219 F.G.K. Baucke: "Phosphate and fluoride error of pH glass electrodes. Erroneous potentials caused by a component of some membrane glasses", J. Electroanal. Chem. **367**, 131–139 (1994)

2.220 A.E. Martell, R.M. Smith (Eds.): *Critical Stability Constants*, Vol. 4, *Inorganic Complexes* (Plenum, New York 1976) p. 56

2.221 I.V. Tananaev, I.A. Rozanov, E.N. Beresnev: "The solubility product in the method of residual concentrations", Izv. Akad. Nauk. SSSR, Neorg. Mater. **5**, 419–426 (English translation: 347–353) (1969)

2.222 R.C.L. Mooney: "Crystal structures of a series of rare earth phosphates", J. Chem. Phys. **16**, 1003 (1948)

2.223 R.C.L. Mooney: "X-ray diffraction study of cerous phosphate and related crystals. I. Hexagonal modification", Acta Cryst. **3**, 337–340 (1950)

2.224 F.M. Ernsberger: "Molecular water in glass", J. Am. Ceram. Soc. **60**, 91–92 (1977)

2.225 I.S. Ivanovskaja, A.A. Belyustin, M.M. Shul'ts, T.P. Vorob'eva: "Distribution of sodium in the surface layers of sodium silicate glasses after interaction with aqueous solutions", Sov. J. Glass Phys. Chem. **1**, 139–143 (1975)

2.226 M.M. Shul'ts, A.A. Belyustin, V.V. Mogileva, I.S. Ivanovskaja: "Concentration distribution and interdiffusion of ions in surface layers of a sodium-aluminosilicate glass treated with aqueous solutions", Proc. Acad. Sci. USSR **241**, 603–608 (1978)

2.227 F. Geotti-Bianchini, L. De Riu, G. Gagliardi, M. Guglielmi, C.G. Pantano: "New interpretation of the IR reflectance spectra of SiO_2-rich films on soda-lime glass", Glastechn. Ber. Glass Sci. Technol. **64**, 205–217 (1991)

2.228 B.C. Bunker: "Molecular mechanism for corrosion of silica and silicate glasses", J. Non-Cryst. Solids **179**, 300–308 (1994)

2.229 R.H. Doremus: *Glass Science*, 2nd ed. (Wiley, New York 1994) p. 260

2.230 J.P. Surman, I. Bosse: "Abfall des pH-Wertes wässriger Salzlösungen in Ampullen", Pharm. Ind. **54**, 66–68 (1992)

2.231 F.G.K. Baucke: "Konstante pH-Werte wässriger Lösungen in Ampullen aus Neutralglas", Pharm. Ind. **54**, 886–889 (1992)

2.232 P. Duffer: "How glass reacts with water and causes surface corrosion", Glass Industry **76**, 22–24, 27–28 (1995)

2.233 T.J. Hwang, W.T. Han: "Complex impedance analysis of the glass/solution interface in the glass electrode for pH measurement", J. Non-Cryst. Solid **203**, 345–352 (1996)

2.234 P.J. Jorgensen, F.J. Norton: "Proton transport during hydrogen permeation in vitreous silica", Phys. Chem. Glasses **10**, 23–27 (1969)

2.235 J.C. Bazán: "On silica glass (Suprasil) protonic conductor", Z. Phys. Chem. NF **110**, 285–288 (1978)

2.236 G.J. Hills, D.J.G. Ives: "The hydrogen electrode", in *Reference Electrodes. Theory and Practice*, ed. by D.J.G. Ives, G.J. Janz (Academic Press, New York 1961) pp. 71–126

2.237 R.G. Bates: *Determination of pH. Theory and Practice*, 2nd ed. (Wiley, New York 1973) pp. 279–294

2.238 J.A. Christiansen: "Manual of physico-chemical symbols and terminology", J. Am. Chem. Soc. **82**, 5517–5522 (1960)

2.239 F.G.K. Baucke, H. Bach: "Investigation of glasses using surface profiling by spectrochemical analysis of sputter-induced radiation: II, Field-driven formation and electrochemical properties of protonated glasses containing various proton concentrations", J. Am. Ceram. Soc. **65**, 524–539 (1982)

2.240 E. Pungor: "Ion-selective electrodes – analogies and conclusions", Electroanalysis **8**, 348–352 (1996)

2.241 F.G.K. Baucke: "Contribution to the electrochemistry of glass electrode membranes", in *Ion-Selective Electrodes*, ed. by E. Pungor (Elsevier, Amsterdam 1977) pp. 215–234

2.242 W. Schmickler: "Die Elektrochemie im Umbruch", Nachr. Chem. Tech. Lab. **10**, 872–877 (1985)

2.243 H.J.C. Tendeloo, A.J. Zwart Voorspuij: "Researches on adsorption electrodes. VI. Glass electrodes", Rec. Trav. Chim. **63**, 793–814 (1943)

2.244 L.R. Pederson, B.P. McGrail, G.L. McVay, D.A. Petersen-Villalobos, N.S. Settles: "Kinetics of alkali silicate and aluminosilicate glass reactions in alkali chloride solutions: influence of surface charge", Phys. Chem. Glasses **34**, 140–148 (1993)

2.245 G. Kortüm: *Lehrbuch der Elektrochemie*, 4th ed. (VCH, Weinheim 1966) pp. 288–292

2.246 R. Brdicka: *Grundlagen der Physikalischen Chemie*, 13th ed. (VEB Deutscher Verlag der Wissenschaften, Berlin 1976) pp. 675–686

2.247 P.M.S. Monk, R.J. Mortimer, D.R. Rosseinsky: *Electrochromism, Fundamentals and Applications* (VCH, Weinheim 1995) p. 26

2.248 R.G. Bates: *Determination of pH. Theory and Practice*, 2nd ed. (Wiley, New York 1973) pp. 341–344

2.249 F. Conti, G. Eisenman: "The steady state properties of ion exchange membranes with fixed sites", J. Biophys. **5**, 511–530 (1965)

2.250 D. Mackay, P. Meares: "Ion-exchange across a cationic membrane in dilute solutions", Kolloid Z. **171**, 139–149 (1960)

2.251 G. Eisenman, J.P. Sandblom, J.L. Walker, Jr.: "Membrane structure and ion permeation", Science **155**, 965–974 (1967)

2.252 R.H. Doremus: *Glass Science*, 2nd ed. (Wiley, New York 1994) pp. 260–262

2.253 P.K. Glasoe, F.A. Long: "Use of glass electrodes to measure acidities in deuterium oxide", J. Phys. Chem. **64**, 188–191 (1960)

2.254 R.G. Bates: *Determination of pH. Theory and Practice*, 2nd ed. (Wiley, New York 1973) pp. 251–253, 375–376

2.255 A.K. Covington, M. Paabo, R.A. Robinson, R.G. Bates: "Use of the glass electrode in deuterium oxide and the relation between the standardized pD (paD) scale and the operational pH in heavy water", Anal. Chem. **40**, 700–706 (1968)

2.256 S.P.L. Sørensen, K. Linderstrøm-Lang: "On the determination and value of π_0 in electrometric measurements of hydrogen ion concentrations", Compt. Rend. Lab. Carlsberg **15**, 1–40 (1924)

2.257 A.K. Covington, R.G. Bates, R.A. Durst: "Definition of pH scales, standard reference values, measurement of pH and related terminology", Pure Appl. Chem. **57**, 531–542 (1985)

2.258 G. Kortüm: *Lehrbuch der Elektrochemie*, 5th ed. (VCH, Weinheim 1972)

2.259 B.E. Conway: *Elektrochemische Tabellen* (Govi, Frankfurt 1957)

2.260 W.L. Marshal, E.U. Frank: "Ion product of water substance, 0–1000 °C, 1–10 000 Bars. New international formulation and its background", J. Phys. Chem. Ref. Data **10**, 295–304 (1981)

2.261 BIPM, IEC, ISO, OIML: *International Vocabulary of Basic and General Terms in Metrology*, 2nd ed. (ISO, Geneva 1994) pp. 45–47

2.262 R.G. Bates: "Revised standard values for pH measurement from 0 to 95 °C", J. Res. NBS – A. **66 A**, 179–184 (1962)

2.263 D.J.G. Ives, G.J. Janz: "General and theoretical introduction", in *Reference Electrodes. Theory and Practice*, ed. by D.J.G. Ives, G.J. Janz (Academic Press, New York 1961) pp. 1–67

2.264 R.G. Bates, E.A. Guggenheim: "Report on the standardization of pH and related terminology", Pure Appl. Chem. **1**, 163–168 (1960)

2.265 J.G.H.M. Lito, F.G.F.C. Camoes, I.A. Ferra, A.K. Covington: "Calculation of pH reference values for standard solutions from the corresponding acid dissociation constants", Anal. Chim Acta **239**, 129–137 (1990)

2.266 K.S. Pitzer (Ed.): *Activity Coefficients in Electrolyte* Solutions, 2nd ed. (CRC, Boca Raton, FL 1991)

2.267 NBS Certificate: "Standard sample 189, potassium tetroxalate (paH standard)" (NBS, Washington, DC 1964)

2.268 F.G.K. Baucke: "Lower temperature limit of NBS (DIN) pH standard buffer solution potassium tetroxalate", Electrochim. Acta **24**, 95–97 (1979)

2.269 F.G.K. Baucke: "Heißsterilisierte lagerfähige NBS-(DIN)-pH-Standardpufferlösungen. – Untersuchung ihrer thermischen Stabilität", Chem. Ing. Tech. **49**, 739–740

2.270 F.G.K. Baucke: "Stabile pH-Werte nach Heißsterilisation. – Untersuchung lagerfähiger NBS-(DIN)-pH-Standardpufferlösungen", Schott Information **4**, 11–13 (1977)

2.271 R. Naumann, C. Alexander-Weber, F.G.K. Baucke: "Limited stability of the pH reference material sodium tetraborate decahydrate ('borax')", Fresenius' J. Anal. Chem. **350**, 119–121 (1994)

2.272 PTB-Kalibrierschein: "PTB BO1, primäres pH-Wert-Referenzmaterial: di-Natriumtetraborat-Decahydrat" (Physikalisch-Technische Bundesanstalt, Braunschweig 1994)

2.273 DIN 19266: "pH-Messung, Standardpufferlösungen" (Beuth, Berlin 1999)

2.274 F.G.K. Baucke: "Differential-potentiometric cell for the restandardization of pH reference materials", J. Electroanal. Chem. **368**, 67–75 (1994)

2.275 DKD-Zertifikat: "DKD-K-14301, 9410, Referenzmaterial zur Kalibrierung von pH-Meßeinrichtungen: di-Natriumtetraborat-Decahydrat, sekundäres Referenzmaterial" (E. Merck, Darmstadt 1994)

2.276 R. Naumann, C. Alexander-Weber, F.G.K. Baucke: "The standardization of pH measurements", Fresenius' J. Anal Chem. **349**, 603–606 (1994)

2.277 R. Naumann, C. Alexander-Weber, F.G.K. Baucke: "High-precision pH measurements by means of electrochemical cells with transference", Fresenius' J. Anal. Chem. **349**, 639–642 (1994)

2.278 S. Ebel, W. Parzefall: *Experimentelle Einführung in die Potentiometrie* (VCH, Weinheim 1975)

2.279 S. Ebel, E. Glaser, H. Mohr: "Fehler und Fehlerfortpflanzung bei der Bestimmung von pH-Werten", Z. Anal. Chem. **293**, 33–35 (1978)

2.280 R.G. Bates: "The modern meaning of pH", Crit. Rev. Anal. Chem. **10**, 247–278 (1981)

2.281 R.G. Bates, G.D. Pinching, E.R. Smith: "pH standards of high acidity and high alkalinity and the practical scale of pH", J. Res. NBS **45**, 418–429 (1950)

2.282 A.K. Covington: "Recent developments in pH standardization and measurement for dilute aqueous solutions", Anal. Chim. Acta **127**, 1–21 (1981)

2.283 M.F. Ryan: "Unreliable results", Science **165**, 851 (1969)

2.284 DIN 19260: "pH-Messung, Allgemeine Begriffe" (Beuth, Berlin 1971)

2.285 DIN 19261: "pH-Messung, Begriffe für Meßverfahren mit Verwendung galvanischer Zellen" (Beuth, Berlin 1971)

2.286 DIN 19262: "Steckbuchse und Stecker geschirmt für pH Elektroden" (Beuth, Berlin 1959)

2.287 DIN 19263: "pH-Messung, Glaselektroden" (Beuth, Berlin 1989)

2.288 DIN 19264: "pH-Messung, Bezugselektroden" (Beuth, Berlin 1985)

2.289 DIN 19265: "pH-Messung, pH-Meßumformer, Anforderungen" (Beuth, Berlin 1994)

2.290 DIN 19267: "pH-Messung, Technische Pufferlösungen vorzugsweise zur Eichung von technischen pH-Meßanlagen" (Beuth, Berlin 1978)

2.291 DIN 19268: "pH-Messung von klaren, wäßrigen Lösungen" (Beuth, Berlin 1985)

2.292 F.G.K. Baucke, P. Spitzer, R. Naumann: "pH controversy revisited", Anal. Chem. News Features **70**(7), 226A (1998)

2.293 F.G.K. Baucke: "The definition of pH. Proposal of improved IUPAC recommendations", in *Traceability of pH Measurement, Lectures Delivered at the 126th PTB Seminar*, PTB-Bericht W-68, ed. by P. Spitzer (Physikalisch-Technische Bundesanstalt, Braunschweig 1997) pp. 10–20

2.294 F.G.K. Baucke: "Reference electrodes for measurements with ion-sensitive electrodes. The importance of the potential", in *Ion and Enzyme Electrodes in Biology and Medicine*, ed. by M. Kessler, L.C. Clark, D.W. Lübbers, I.A. Silver, W. Simon (Urban & Schwarzenberg, München 1976) pp. 200–204

2.295 D.J.G. Ives, G.J. Janz: *Reference Electrodes. Theory and Practice* (Academic Press, New York 1961)

2.296 F.G.K. Baucke, R. Bertram, K. Cruse: "The iodide–iodine system in acetonitrile", J. Electroanal. Chem. **32**, 247–256 (1971)

2.297 R.G. Bates, V.E. Bower: "Standard potential of the silver–silver-chloride electrode from 0° to 95 °C and the thermodynamic properties of dilute hydrochloric acid solutions", J. Res. NBS **53**, 283–290 (1954)

2.298 F.G.K. Baucke: "Thermodynamics of solid-state connected ion-sensitive membrane electrodes: the silver–silver chloride system. Part I. Standard potential E_M^0 at 25, 50, and 75 °C", J. Electroanal. Chem. **67**, 277–289 (1976)

2.299 F.G.K. Baucke: "Potentials of electrodes of the second kind at low concentrations of common ion electrolyte.– Part I. General discussion", Electrochim. Acta **17**, 845–849 (1972), Part II. "Quantitative treatment of electrodes with salts with negligible complex formation", ibid. 851–859

2.300 F.G.K. Baucke, G.H. Wagner: "Bezugselektroden für Korrosionsuntersuchungen bei höheren Temperaturen und Drücken", Mat.-Wiss. Werkstofftechnik **22**, 128–136 (1991)

2.301 R.G. Bates: "Inner reference electrodes and their characteristics", in *Glass Microelectrodes,* ed. by M. Lavallée, O.F. Schanne, N.C. Hébert (Wiley, New York 1969) pp. 1–24

2.302 F.G.K. Baucke: "Thermodynamics of solid-state connected ion-sensitive membrane electrodes: the silver–silver chloride system. Part II. Standard potentials E_{M}^{0} between 5 and 90 °C of the 2nd kind silver–silver chloride reference electrode with 3.5 M and sat'd KCl measured by means of membrane electrodes", J. Electroanal. Chem. **67**, 291–299 (1976)

2.303 F.G.K. Baucke: "Standardpotentiale $(\varepsilon^{0'}+\varepsilon_{\mathrm{j}})_{\mathrm{T}}$ der Silber/Silberchlorid-Elektrode in 3.5 m und in ges. KCl unter Verwendung entsprechender ('Cl^{-}-ionensensitiver') Membranelektroden (0–95 °C)", Chem.-Ing.-Techn. **47**, 565–566 (1975)

2.304 G.J. Hills, D.J.G. Ives: "The calomel electrode and other mercury-mercurous salt electrodes", in *Reference Electrodes. Theory and Practice*, ed. by D.J.G. Ives, G.J. Janz (Academic Press, New York 1961) pp. 127–178

2.305 F.G.K. Baucke: "Standard potentials $(\varepsilon^{0'} + \varepsilon_{\mathrm{j}})$ of the Thalamid® reference electrode, Hg,Tl(40 wt%)/TlCl(s)/KCl(s)// . . . , in aqueous solution between 5 and 90 °C", J. Electroanal. Chem. **33**, 135–144 (1971)

2.306 F.G.K. Baucke: "Standardpotentiale $(\varepsilon^{0'} + \varepsilon_{\mathrm{j}})$ und Polarisationsverhalten der Thalamid®-Bezugselektrode (3.5 mol/L und ges. KCl) zwischen 5 und 90 °C", Chem.-Ing.-Techn. **46**(71) (1974)

2.307 F.G.K. Baucke: "Potenziali standard $(\varepsilon^{0'} + \varepsilon_{\mathrm{j}})$ dell' elettrodo di riferimento al Talamide® Hg,Tl(40 wt%)/TlCl(s)|KCl(s)// . . . , in soluzione acquosa tra 5 e 90 °C", Italglas-Riv. trimestrale d'informazione **8**, 3–8 (1972)

2.308 F.G.K. Baucke: "The electrode Hg,Tl(40 wt%)/TlBr(s), KBr(s)// . . . as reference electrode in aqueous solution: Reversibility of the system and standard potentials $(\varepsilon^{0'}+\varepsilon_{\mathrm{j}})$ of the half-cell between 5 and 90 °C", J. Electroanal. Chem. **39**, 263–273 (1972)

2.309 H.K. Fricke: "Eine neue Bezugs- und Ableitelektrode für Glaselektroden-Meßketten", DECHEMA-Monographie, Vol. 43, *Meß- und Regeltechnik* (VCH, Weinheim 1962) pp. 161–172

2.310 Schott Glas: *Elektroden für Labor und Umwelt*, Catalogue 3105 (Schott-Geräte, Hofheim 1992)

2.311 Schott Glas: *Elektroden für Prozeßchemie, Biotechnologie und Wasserwirtschaft*, Catalogue 3106 (Schott-Geräte, Hofheim 1991)

2.312 W.C. Smyrl, C.W. Tobias: "Thermodynamic properties of LiCl in dimethylsulfoxide", J. Electrochem. Soc. **115**, 33–36 (1968)

2.313 F.G.K. Baucke, C.W. Tobias: "Thallium-thallous halide reference electrodes in propylene carbonate", J. Electrochem. Soc. **116**, 34–37 (1969)

2.314 W. Ingold: "Silber-/Silberhalogenid-Ableitelektrode für Meßketten", DE 1 168 120 (1957) ICP:G01n

2.315 P.R. Mussini, F. D'Andrea, A. Galli, P. Longhi, S. Rondinini: "Characterization and use of aqueous caesium chloride as an ultra-concentrated salt bridge", J. Appl. Chem. Electrochem. **20**, 651–655 (1990)

2.316 Schott Glas: *Duran Laboratory Glassware*, Catalogue 50020, 1st ed. (Schott Glas, Mainz 1991) pp. 75 ff.

2.317 G. Vasaru, D. Ursu, A. Mihala, P. Szentgyörgyi: *Deuterium and Heavy Water* (Elsevier, Amsterdam 1975)

2.318 H.K. Rae (Ed.): *Separation of Hydrogen Isotopes*, ACS Symposium Series, Vol. 68 (Am. Chem. Soc., Washington, DC 1978)

2.319 R. Gary, R.G. Bates, R.A. Robinson: "Dissociation constant of acetic acid in deuterium oxide from 5 to 50 °C. Reference points for a pD scale", J. Phys. Chem. **69**, 2750–2753 (1965)

2.320 N.C. Li, P. Tang, R. Mathur: "Deuterium isotope effects on dissociation constants and formation constants", J. Phys. Chem. **65**, 1074–1076 (1961)

2.321 P. Salomaa, L.L. Schaleger F.A. Long: "Solvent deuterium isotope effects on acid-base equilibria", J. Am. Chem. Soc. **86**, 1–7 (1964)

2.322 C.K. Rule, V.K. La Mer: "Dissociation constants of deutero acids by emf measurements", J. Am. Chem. Soc. **60**, 1974–1981 (1974)

2.323 R. Gary, R.G. Bates, R.A. Robinson: "Second dissociation constant of deuteriophosphoric acid in deuterium oxide from 5 to 50 °C", J. Phys. Chem. **68**, 3806–3809 (1964)

2.324 R. Lumry, E.L. Smith, R.R. Glantz: "Kinetics of carboxypeptidase action. I. Effect of various extrinsic factors on kinetic parameters", J. Am. Chem. Soc. **73**, 4330–4340 (1951)

2.325 H.H. Hyman, A. Kaganove, J.J. Katz: "The basicity of amino acids in D_2O", J. Phys. Chem. **64**, 1653–1655 (1960)

2.326 K. Mikkelsen, S.O. Nielsen: "Acidity measurements with the glass electrode in H_2O–D_2O mixtures", J. Phys. Chem. **64**, 632–637 (1960)

2.327 T.H. Fife, T.C. Bruice: "The temperature dependence of the pD correction for the use of the glass electrode in D_2O", J. Phys. Chem. **69**, 1079–1080 (1961)

2.328 B.M. Lowe, D.G. Smith: "Glass electrode measurements in deuterium oxide", Anal. Lett. **6**, 903–907 (1973)

2.329 R.K. Forcé, J.D. Carr: "Temperature-dependent response of the glass electrode in deuteriumoxide", Anal. Chem. **46**, 2049–2052 (1974)

2.330 B.M. Lowe, D.G. Smith: "The behaviour of cation selective glass electrodes in deuterium oxide", Electroanal. Chem. Interfacial Electrochem. **51**, 295–303 (1974)

3. Electrochemistry of Glass-Forming Melts

3.1 Introduction

Friedrich G.K. Baucke

This chapter is concerned with the electrochemistry of oxidic glass-forming melts, and it may be mentioned at the outset that, in a strict sense, the term glass-forming melts is a more accurate name for this class of materials than just glass melts, at least with regard to glass production, because they are produced to become, but have never before been, glasses. Because, however, both terms are frequently used also for glasses that have been remelted, for instance in laboratories, melts treated in this chapter will not only be called glass-forming melts but will occasionally be termed glass melts.

Unlike solid glasses, oxidic glass-forming melts are good electrolytes. Despite their usually high viscosities, their ionic conductivity at temperatures of about 800–1600 °C, which are of interest here, is comparable to that of aqueous salt solutions at environmental temperatures. Glass-forming melts are highly reactive, not only chemically but also electrochemically, because kinetic hindrances of electrode reactions are mostly excluded at the high temperatures. Moreover, glass melts are characterized by a reduction-oxidation or, briefly, redox state caused by dissolved oxygen and by differently charged ions of polyvalent oxides, which are generally dissolved in the melt. They form so-called redox equilibria, which make the melts more or less oxidizing or reducing materials.

These properties, *inter alia*, make glass-forming melts ideal media for electrochemical research and its application in technical glass melting. Thus, crucibles and containers for batch-wise production, which contain up to tens of tonnes of melt, as well as continuously working tanks, which produce up to several hundred tonnes of glass per day, must actually be viewed as large electrochemical cells. On the one hand, the various electrochemical processes proceeding in these cells are required as a part of the melting process; but on the other hand they are intolerable reactions which are caused by certain parameters of the melting tanks and disturb the production process and impair the quality of the glass product. Some examples may demonstrate the variety of electrochemical reactions. The melts are often subject to alternating currents of up to hundreds of kA for direct electric heating, their redox state is changed between the extremes of the equilibria by strong temperature variations, which are to generate millions of oxygen bubbles for fining the melt, electrolytic cells unintentionally set up by metal parts of the tanks and driven by stray voltages tend to electrolyse the melts, and concentration and thermoelectric cells formed with electron conductors would immediately cause deleterious reactions on short-circuiting. In addition, the refractories of the containers are frequently corroded electrochemically, and the electrodes supplying the electric energy for heating are subject to corrosion as a direct or sometimes indirect consequence of the high current densities of up to several hundred amperes per square centimetre.

In view of this situation, the electrochemical research done in the field of glass melts by the Electrochemical Laboratory of Schott Glas has necessarily been practice- and production-oriented. We endeavoured to deepen our understanding of the electrochemical phenomena with the ultimate goal of improving the economy and ecology of the melting process as well as the glass quality on the basis of a better understanding of the electrochemical phenomena. The electrochemical characterization of oxidic glass-forming melts given in Sect. 3.2 provides a base for understanding the information given in the following. Section 3.3 is on the basicity of glasses. It is concerned with "oxide", the reduced compound of the intrinsic redox couple "oxide"/oxygen of oxidic melts, the development of basicity concepts and especially on the meaning of optical basicity and its application to redox reactions in melts. This section was written by one of the workers who developed optical basicity. Section 3.4 is on oxygen in glass-forming melts, which represents the oxidized counterpart of "oxide". It is directly connected to the production at Schott Glas, as it reports on the development of oxygen sensors for laboratory and technical application. It also describes how basic problems of the platinum measuring and the zirconia reference electrodes have been solved and how the thermodynamically correct functioning of the sensors was verified. Principle and application of alternative (metal) reference electrodes, which are mechanically more stable but electrochemically more demanding than zirconia electrodes, will also be described.

Section 3.5 is concerned with the application of oxygen sensors to control fining of glass-forming melts. It starts with a description of fining in general and redox fining in particular, including basic information about sulphur fining, which, although not practically applied at Schott Glas, is the most frequently applied fining process and is used for nearly all mass-produced glasses world-wide. Also reported is the determination of thermodynamic standard data of polyvalent elements dissolved in glass melts by indirect and *in situ* methods, which are increasingly needed for an improved evaluation of oxygen partial pressures measured in technical melting units. Subsequently, a comparative test of Schott sensors with sensors of different design and origin in a large-scale technical melter is reported. Finally we propose an alternative fining procedure, called electrolytic fining, which is based on electrolytically generated oxygen bubbles and thus needs less fining agents than conventional fining processes.

Section 3.6 focuses on non-isothermal melts, which have received little attention in the past. This is surprising in view of the fact that thermoelectric potentials cannot be excluded in melt containers at the high temperatures of glass melting. The lack of attention is especially astonishing because continuous melting units are designed to submit glass melts successively to locally different temperatures, $T(t) = f(x)$, the melt thus experiencing strong temperature gradients, $\mathrm{d}T/\mathrm{d}x = f(t)$, whereas in batch-wise production, the entire container is successively subjected to different temperatures, $T = f(t)$, and the melt is kept principally isothermal, $\mathrm{d}T/\mathrm{d}x = 0$. Thermoelectric reactions are a consequence when thermo-cells are short-circuited, and it is discussed whether or not, and if so, under which conditions, such reactions yield oxygen bubbles. The practical basis of this discussion is the knowledge of exact thermoelectric voltages, which could be measured by means of zirconia electrodes for the first time after these sensors had been developed.

Section 3.7 describes an arrangement for displacing electrochemical reactions locally, and Sect. 3.8 reports on the elucidation of a long-standing problem: bubble formation at the interface zirconium silicate (ZS) refractory/oxidic glass melt. Time-dependent potentiometric and electrolytic measurements and short-circuiting experiments showed that oxygen bubbles are generated by oxidation of oxide and that this oxidation is caused by an internal reduction of redox impurities of the refractory, for instance iron and titanium. The combination of the two redox reactions over large and growing distances in the ZS is enabled by (small) alkali ion and electron conductivities of the refractory. It is surprising at first sight that this process, in principle, is an electrochromic reaction. The work led to the (patented) proposal of a technical cell for pretreating refractories to exclude this kind of oxygen bubble formation.

Section 3.9, finally, is concerned with temperature-dependent conductivities of glass-forming melts. They are needed for electrically heated technical glass melters, which are of increasing significance for energy and environmental reasons. Because an analysis of reported conductivity cells was unsuccessful, a cell was developed for measuring exact temperature-dependent absolute conductivities at up to $1500\,^\circ\mathrm{C}$. The cell was applied to a series of seven mixed-alkali glasses, which confirmed that the mixed-alkali effect is not restricted to solid glasses but continues to exist in the melt range. Surprisingly, the conductivity loses the mixed-alkali effect at about $1200\,^\circ\mathrm{C}$, whereas the activation energy shows it up to the highest temperatures measured ($1500\,^\circ\mathrm{C}$).

3.2 Electrochemical Characterization

Friedrich G.K. Baucke

Oxidic glass-forming melts, mainly silicates, borates, phosphates, and their combinations, such as borosilicates and aluminosilicates, are not simply molten oxides or oxide mixtures, as could be concluded from their overall composition, but consist of polyoxyanions, long chains and networks of, for instance, connected $\equiv SiO_4$ tetrahedra, which are connected via oxygen atoms and are frequently interrupted by negatively charged, terminal oxides, for instance siloxy groups, $\equiv SiO^-$. The charge of these so-called non-bridging oxides is balanced by cations, mainly alkali and alkaline-earth ions. This structure, which differs basically from those of other molten salts, causes the characteristic properties of molten glasses, among which the wide temperature range of high viscosities, the lack of a defined melting point, and the connected strong tendency of glass formation are the most obvious and important. High temperatures, mainly above a thousand degrees Celsius, are required to obtain viscosities that are sufficiently small to allow reasonable melting processes. Because of these high temperatures and despite the high viscosities, the ionic conductivities of glass melts are rather large. They depend on the composition and specific structure of the melts. For example, that of sodium calcium silicate glass with (in mol%) $15.5\,Na_2O$, $10.8\,CaO$ and $73.7\,SiO_2$ is $0.8 \times 10^{-1}\,S\,cm^{-1}$ at $1000\,°C$, $1.8 \times 10^{-1}\,S\,cm^{-1}$ at $1200\,°C$, and 4.0×10^{-1} at $1500\,°C$ [3.1], and thus in the range of the conductance of aqueous 1 molal KCl solution, which is $1.1 \times 10^{-1}\,S\,cm^{-1}$ at $25\,°C$ [3.2]. Most glass-forming melts therefore are good electrolytes and can be processed by direct heating with electric currents, as mentioned in Sect. 3.1.

Another typical electrochemical property of glass-forming melts is their redox state, which can be understood in the following way. The terminal siloxy groups $\equiv SiO^-$ of the SiO_4 network are subject to a continuous exchange with the bridging oxygens $\equiv SiOSi \equiv$ because of thermal motion at the high temperatures [3.3]. On the average, this leads to an equilibrium like

$$2 \equiv SiO^-(m) \rightleftarrows\, \equiv SiOSi \equiv (m) + \left(O^{2-}(m)\right) \,, \tag{3.1}$$

which could be called basicity equilibrium. The brackets of the doubly charged oxide on the right side of (3.1) indicate small, perhaps negligible concentration. Not being zero, however, the oxides are in equilibrium with dissolved oxygen according to

$$\left(O^{2-}(m)\right) \rightleftarrows \frac{1}{2}O_2(m) + (2e^-) \,, \tag{3.2}$$

where the brackets of the second term on the right side indicate the tendency to provide (or take up) electrons. Combining (3.1) and (3.2) yields the equilibrium

$$2 \equiv SiO^-(m) \rightleftharpoons \equiv SiOSi(m) + \frac{1}{2}O_2 + (2e^-) \,, \tag{3.3}$$

which is inseparably connected with oxidic glass-forming melts and is therefore called the intrinsic redox equilibrium (or system) of oxidic glass forming melts [3.4]. The corresponding intrinsic redox equilibrium constant is given by

$$K_{\text{intr}} = \frac{a_{SiOSi}\sqrt{p_{O_2,m}}\,a_{e^-}^2}{a_{SiO^-}^2} \,. \tag{3.4}$$

Corresponding to (3.3), the activity of electrons in (3.4) merely indicates the tendency of the melt to provide (or take up) electrons, for instance to (or from) a polyvalent element that is dissolved in the melt. For example, an increase of the oxygen partial pressure of the melt has the tendency to depolymerize the network (to decrease a_{SiOSi}) by generating siloxy groups (to increase a_{SiO^-}) under uptake of electrons (a decrease of a_{e^-}). A higher oxygen content thus causes the melt to be both less reducing (more oxidizing) and more basic, and *vice versa*.

The oxygen partial pressure is the only free variable of the intrinsic redox equilibrium (3.3), (3.4). It is thus not possible to change the siloxy activity (except by a change of $p_{O_2,m}$) because changing the ratio a_{SiO^-}/a_{SiOSi} means changing the melt composition, and this would mean different thermodynamic standard data and thus also a different equilibrium constant K_{intr} and a different intrinsic redox equilibrium altogether.

Oxygen dissolved in a glass melt in equilibrium with the oxygen partial pressure according to (3.3) is dissolved purely physically (for detailed information see Sect. 3.5.1). It is present in the melt in uncharged form, O_2 or O, and the glass melt serves as an inert solvent. Its concentration $c_{O_2,m}$ is proportional to its partial pressure $p_{O_2,m}$ (Henry's law), and the temperature-dependent ratio

$$\alpha(T) = \frac{c_{O_2,m}}{p_{O_2,m}} \tag{3.5}$$

is called the physical solubility. The so-called internal $p_{O_2,m}$ generally differs from the (external) oxygen partial pressure $p_{O_2,g}$ of the atmosphere above the melt because of the high melt viscosity and connected small diffusion coefficients. Only if the internal and external oxygen partial pressure have eventually become equal after long equilibration times, $p_{O_2,m} = p_{O_2,g}$, is the physical solubility described also by Bunsen's law,

$$\alpha(T) = \frac{c_{O_2,m}}{p_{O_2,g}} = \frac{c_{O_2,m}}{p_{O_2,m}} \,. \tag{3.6}$$

When the oxide of a polyvalent element, for instance of iron, is dissolved in an oxidic glass-forming melt, the intrinsic redox system of the melt (3.3) competes with the dissolved redox couple

$$2Fe^{3+}(m) + (2e^-) \rightleftarrows 2Fe^{2+}(m) \tag{3.7}$$

for the electrons. The intrinsic and dissolved redox system form a reduction-oxidation or, briefly, redox equilibrium,

$$2 \equiv SiO^-(m) + 2Fe^{3+}(m) \rightleftarrows\equiv SiOSi \equiv (m) + 2Fe^{2+}(m) + \frac{1}{2}O_2(m) , \tag{3.8}$$

whose position is determined by the relative redox powers of the two systems. The corresponding redox equilibrium constant is given by

$$K(T) = \frac{a_{SiOSi}\, a_{Fe^{2+}}^2\, \sqrt{p_{O_2, m}}}{a_{SiO^-}^2\, a_{Fe^{3+}}^2} . \tag{3.9}$$

As in (3.4), the oxygen partial pressure is the only free variable also in (3.9). A change of the dependent variable siloxy activity would necessarily mean a different composition of the glass melt and a changed equilibrium constant, which excludes a thermodynamic treatment of the two glasses by the same equilibrium.

Different from (3.4), redox equilibrium (3.9) contains no quantities which merely indicate tendencies of changes. This is demonstrated by increasing the oxygen partial pressure of the melt, which results in a finite, although small, depolymerization of the network and a corresponding increase of the siloxy activity. Different from the intrinsic redox equilibrium (3.4), the electrons required for the generation of siloxy groups are actually provided by the lower valence state ions Fe^{2+} and their change to Fe^{3+}. An increase of the oxygen partial pressure thus actually makes the melt less reducing (more oxidizing) and more basic and does not merely cause tendencies for these changes as in the case of (3.4).

It is also noted that oxygen that dissolves in the glass melt to form siloxy groups according to (3.8) is distinguished from physically dissolved oxygen, see (3.5) and (3.6), by its bond to the silicon of the terminal siloxy group and its negative charge. It is therefore called chemically dissolved oxygen. Its concentration is equivalent to the concentration of the upper valence state redox ion, for instance $c_{O_2, m} = 4c_{Fe^{3+}}$, and can exceed the concentration of physically dissolved oxygen by orders of magnitude. Because they dissolve and release large quantities of the gas, glass melts containing polyvalent elements can also be understood as oxygen buffers whose capacity depends on the concentration of the dissolved polyvalent ions.

Two modes of treating equilibrium (3.9) are generally distinguished when redox reactions in oxidic glass-forming melts are investigated.

- At constant oxygen partial pressure and temperature, the lower to upper valence state ion activities or concentrations are investigated as a function of melt basicity and composition. This has resulted in the application of the optical basicity concept to redox problems. Although not directly connected to thermodynamics, optical basicity provides a common view of redox systems in melts and glasses with different basicities and compositions and even yields a connection of redox equilibria in oxidic melts and aqueous solutions (see Sect. 3.3 for details).

• At constant melt composition, the activities of lower to upper valence-state ions are investigated as a function of oxygen partial pressure and temperature. This work is of immediate significance to technical glass melting. It required the development of means to measure the oxygen partial pressure and of techniques to determine activities or concentrations of redox ions and thermodynamic standard data of redox couples. Results are reported *inter alia*, in Sects. 3.4 and 3.5.

3.3 Basicity of Glass-Forming Melts

John Duffy

3.3.1 Development of Basicity Concepts

The concept of acids and bases is one that pervades many aspects of chemistry and, as a result, several theories dealing with acid–base interaction have been proposed. The more important of these are described in chemistry textbooks (for a selection, see [3.5–8]).

For present purposes it is instructive to start by referring to acid–base interactions in aqueous solution. Here, both the acidic species (H^+) and basic species (OH^-) are heavily solvated, that is, they exist as hydrated species. In dilute solution, these species are isolated from each other, and it is possible to regard acidity or basicity in terms of their concentration or activity, usually with reference to the pH scale. Strong acids, examples of which include hydrochloric, sulphuric and nitric acids, are those which undergo almost complete dissociation into the hydrogen ion and the corresponding anion, for example:

$$HNO_3 = H^+ + NO_3^- \ . \tag{3.10}$$

Roughly speaking, a molar solution of nitric acid contains one mole of (hydrated) hydrogen ions per litre and therefore has a pH of approximately zero. Similarly, a strong base is one which undergoes almost complete dissociation into (hydrated) metal ions and hydroxide ions, for example:

$$NaOH = Na^+ + OH^- \ . \tag{3.11}$$

By contrast, ethanoic (acetic) acid is an example of a weak acid, since the equilibrium

$$CH_3CO_2H = H^+ + CH_3CO_2^- \qquad (3.12)$$

lies well over to the left-hand side with the constant, K_c, equal to 1.8×10^{-5} (at $25\,^\circ C$).

In media where there is little or no water, for example, pure sulphuric acid or its concentrated solutions, the hydrogen ion concentration is very low, and it is necessary to use a different concept of acids and bases in order to express sensibly the acidity of these media. Usually their acidic properties are regarded in terms of the tendency they have for donating protons. This tendency is probed by introducing into the medium a weak base, B, which then exists in equilibrium with the protonated species, BH^+ (the "conjugate acid"); simultaneously the acid is converted into its "conjugate base". Representing the formula of the acid as HA, the equilibrium can be written as

$$B + HA = BH^+ + A^- . \qquad (3.13)$$

When the acid is sulphuric acid, the conjugate base, A^-, is the HSO_4^- ion.

The concentration ratio of protonated to unprotonated base, $[BH^+]/[B]$, can usually be measured experimentally by spectrophotometry, and the tendency for the acid to protonate is given by the acidity function, H_0, which is defined as [3.9]

$$H_0 = pK_a^B + \log[BH^+]/[B] \qquad (3.14)$$

where pK_a^B is the negative logarithm of the dissociation constant, K_a^B, of the conjugate acid, BH^+, as measured under dilute aqueous conditions.

From the discussion so far, it is apparent that the acidity (or basicity) of a medium must be viewed in the context of the appropriate acid–base theory applicable to the conditions under consideration. For sulphuric acid, for example, the acidity of dilute aqueous solutions can be expressed in terms of hydrogen ion concentration (or activity), but this approach does not make sense when dealing with the concentrated acid because the hydrogen ion concentration is very low. Instead, its powerful acidity is best regarded in terms of the *tendency* to release protons (for example, in its reaction with a weak base), using the acidity function, H_0.

Oxide Systems

Many oxides are judged as basic or acidic by considering their behaviour with water. For example, CaO and Na_2O react with water to give alkaline solutions, whereas SiO_2 and P_2O_5 produce silicic and phosphoric acids. The chemistry of the Group II elements would indicate that MgO is also basic, but less so than CaO, while BaO is a stronger base. The amphoteric nature of aluminium hydroxide suggests that Al_2O_3 occupies an intermediate position between basic and acidic oxides, being less basic than MgO but less acidic than SiO_2. From the point of view of chemical bonding, basic oxides are essentially electrovalent, while acidic oxides are covalent. It is the existence of covalent bonding which provides scope for the atoms to join together in a "network" structure, for example, as in SiO_2 or B_2O_3.

In contact with each other, especially in the molten state, many of these oxides undergo chemical reactions which can be identified in general terms as acid–base reactions, for example, in the following

$$Na_2O + SiO_2 = Na_2SiO_3 \tag{3.15}$$

or

$$2Na_2O + SiO_2 = Na_4SiO_4 . \tag{3.16}$$

For the molten state, although such simple stoichiometric equations can be written, the reactions proceed in what might be described as a "non-stoichiometric" manner, and result in a breaking-up of the covalent network of the acidic oxide by the ionic, basic oxide. In this process, bridging oxygen atoms are converted into non-bridging oxygen atoms by the "free" oxide ions:

$$O^{2-} + \overset{|}{\underset{|}{Si}} - O - \overset{|}{\underset{|}{Si}} - = -\overset{|}{\underset{|}{Si}} - O^- + {}^-O - \overset{|}{\underset{|}{Si}} - . \tag{3.17}$$

(For convenience, the non-bridging oxygen atoms are represented carrying a charge of -1; the "real" charge is much less than this, however, owing to delocalisation through π-bonding, see below.) The extent of this breaking-up depends on the proportion of basic oxide present, and if, for the reaction between silica and sodium oxide, the Na_2O:SiO_2 ratio reaches 2:1, then the break-up is complete with the total conversion of SiO_2 into discrete SiO_4^{4-} tetrahedral units; the overall reaction is represented by (3.16). Silicate melts of interest to the glass scientist contain much smaller proportions of the basic oxide so that a substantial part of the covalent network is retained. For example, a glass with a Na_2O:SiO_2 composition which has a 1:2 mole ratio contains tetrahedral SiO_4 units in which, on average, one oxygen atom is non-bridging and three are bridging.

Although the chemical reaction between the ionic oxide and the covalent network oxide, for example Na_2O and SiO_2, is perceived as an acid-base reaction, it is not obvious at what stage the neutralisation "end-point" occurs. Indeed, the idea of end-points for reactions of this type is rarely considered. Nevertheless, since glass is composed of a mixture of basic and acidic oxides, some glasses are bound to be closer to a state of neutralisation than others.

In attempting to deal with the acidic or basic nature of silicate melts, early workers drew analogy between the dissociation of acids (to produce hydrogen ions) and what they perceived to be the dissociation of network oxyanions, such as silicates, to produce oxide, O^{2-}, ions [3.10, 11]. Equations were devised, such as

$$SiO_3^{2-} = SiO_2 + O^{2-} \qquad\qquad (3.18)$$

$$SiO_4^{2-} = SiO_3^{2-} + O^{2-} \qquad\qquad (3.19)$$

in an attempt to produce a p(oxide) scale analogous with the pH scale for aqueous solutions. Unfortunately, this approach precludes meaningful comparisons between, for example, a sodium silicate melt and a calcium silicate melt. This limitation arises owing to the impossibility of measuring single ion activities [3.12]. In spite of this defect, many workers dealing with molten oxides persisted in trying to obtain some measure of what they regarded as the "oxide ion activity". These approaches (see [3.13–20]) are empirical and a review by *Krämer* [3.21], dealing with the application of these approaches to problems in glass melts, indicates the very limited extent of their application.

The problem of basicity confronts not only the glass scientist but also the extraction metallurgist and the geochemist, and these workers, again, have thought of basicity in terms of the oxide ion activity. For example, in dealing with a calcium magnesium aluminosilicate slag, the oxide ion activity has been regarded as proportional to a "basicity ratio", R, [3.22]

$$R = \frac{[\text{CaO}] + \frac{1}{2}[\text{MgO}]}{[\text{SiO}_2] + \frac{1}{3}[\text{Al}_2\text{O}_3]}, \qquad\qquad (3.20)$$

where the fractions (1/2 and 1/3) were introduced arbitrarily to take account of the lesser basicity of MgO and the lesser acidity of Al_2O_3.

During the 1960s, Douglas and Paul noted the effect of glass composition on the stereochemistry and redox equilibria of transition metal ions. They were able to study these effects by measuring changes in the visible absorption spectra. For example, the Co^{2+} ion is octahedrally coordinated (and pink) in many glasses but becomes tetrahedrally coordinated (and deep blue) when the basic oxide content is increased. *Paul* has reported how these "colour indicators", as he called them, could be used for studying the acid–base nature of glass [3.23]. Again, the treatment was in terms of oxide ion activity.

3.3.2 Optical Basicity

Ordinary glass is the product of the chemical reaction between basic and acidic oxides. Its production therefore involves an acid–base neutralisation process, the extent of which depends on the glass composition. It should be possible to characterise a glass in terms of its basic or acidic nature, and there is a need for expressing this quantitatively. Since many important properties of a glass, both in the solid and molten state, depend very much on composition, it is desirable that there should be the facility for calculating this expression of basicity or acidity from the glass composition. Such a facility is provided by the optical basicity scale.

Classifying the oxygen atoms as bridging or non-bridging identifies them in two extreme states, bearing formal charges of zero or -1 (see (3.17)). However, owing to the effects of delocalisation, very few, if any, of the oxygen atoms exist in these extreme states. Instead, virtually all of the oxygen atoms in the silicate network exist over a range of states. The "average state" represents the basicity of the glass.

The optical basicity approach measures this average state by introducing into the glass, when it is molten, small quantities of probe metal ions [3.24–28]. These metal ions act as acids and become coordinated by oxygen atoms which donate a small degree of negative charge (shown schematically in Fig. 3.1). The magnitude of this negative charge donation relies on the glass basicity. All metal ions are affected by electron donation when they become coordinated and, as described presently, the extent of this effect is signalled in their optical (electronic) spectra. The p-block ions Tl^+, Pb^{2+} and Bi^{3+}, all with the $6s^2$ outer configuration, are particularly amenable to this measurement. An important feature of these ions is that usually they experience the electron donor power which is the average of the medium they are probing, see below however. Their behaviour might be described as "innocent" in the sense that they perform their role with little tendency to form specific coordination spheres or complexes.

The effect of coordination on any metal ion is primarily on its outer orbital energy levels. As far as transition metal ions are concerned, the effect is well understood from studies of their d-d spectra which indicate that coordination is always accompanied by orbital expansion. This expansion arises because of (i) the enhanced screening of the (positively charged) nucleus by the electron density of the coordinating atoms spilling into the metal ion, and (ii) the development of additional orbital volume which is a consequence of the conversion of an orbital from bonding mode to antibonding. These effects, (i) and (ii), are referred to as "central field" and "symmetry restricted" covalency, respectively [3.29]. Expansion of the outer orbitals leads to a decrease in the energy separation between the ground state and upper states owing to the decrease in electronic repulsion. Expressing this energy separation by the Racah parameter (B_f for the uncoordinated metal ion), it has been shown that the decrease, ΔB, can be related to a parameter, h, for the ligand and k for the metal ion such that

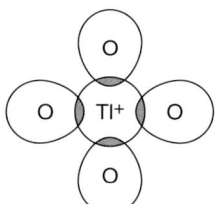

Fig. 3.1. Schematic (two-dimensional) diagram showing coordination of Tl^+ probe ion by oxygen atoms of a glass. Electron density in the valence shell of the oxygen atoms is donated, and this results in part of the oxygen negative charge being incorporated into the Tl^+ ion (shaded portion)

$$\Delta B / B_{\mathrm{f}} = h \times k \ . \tag{3.21}$$

The d-d spectra of aqua complexes have provided values of k for several metal ions by defining h for the H_2O ligand as unity. The k values have then been used in conjunction with spectra for chloro-, bromo-, etc., complexes for obtaining h values for these and other ligands. These data have indicated that the relationship (3.21) works extremely well, and h is found equal to 2.0 for Cl^-, 2.3 for Br^-, 2.7 for I^-.

Studies of this nature have contributed greatly to rationalising the optical spectroscopy of transition metal ions, and are a vital part of ligand field theory. By contrast, the ultraviolet and visible spectra of non-transition elements have been somewhat neglected. However, it has been shown for the p-block ions Tl^+, Pb^{2+} and Bi^{3+} (all with the outer $6s^2$ configuration) that a relationship analogous to (3.21) applies, but with the Racah parameter, B, replaced by the s-p energy separation (measured by the $^1S_0 \rightarrow {}^3P_1$ frequency, ν, [3.30] see Fig. 3.2). The plots of ν versus h for each of these ions (Fig. 3.3) have slopes which indicate that the ions are very sensitive to environmental change, for example for Pb^{2+} in going from a chloride environment to a bromide environment, which represents a change of 0.3 in h, the frequency shifts from $36\,800\,\mathrm{cm}^{-1}$ to $33\,200\,\mathrm{cm}^{-1}$. It is this sensitivity which makes these ions good probes for investigating sites in glass. Furthermore, the intensity of the absorption produced by these ions is high, and only small amounts are necessary for doping the glasses. A glass with thickness $1\,\mathrm{mm}$ needs a p-block ion concentration of approximately $10^{-3}\,\mathrm{mol\,dm}^{-3}$.

For chloride, bromide, and most other ligands, the value of h is more or less fixed, but for oxygen there is marked variation. For example, in the ionic state, as in crystalline CaO, the Pb^{2+} probe indicates $h = 2.56$, whereas for covalently bound oxygen in B_2O_3, h is found to be 1.08. When the Na_2O–B_2O_3 glass system is investigated, the spectral frequency shifts observed for Pb^{2+} (also Tl^+ and Bi^{3+}) show that h increases with increasing Na_2O content (Fig. 3.4a). This trend is general for glass systems (and other oxidic media) and it is possible to express the basicity of these media relative to the basicity of CaO by

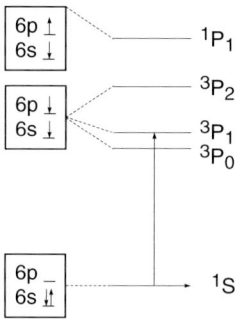

Fig. 3.2. Ground state, $^1S_0$, derived from the $6s^2$ configuration for the isoelectronic ions Tl^+, Pb^{2+} and Bi^{3+}, and the excited triplet and singlet states derived from the configuration $6s^1 6p^1$. The lowest energy Laporte-allowed transition, $^1S_0 \rightarrow {}^3P_1$, is indicated by the vertical arrow. Although this transition is formally spin-forbidden, it is intense (extinction coefficient approximately $5000\,\mathrm{L\,mol}^{-1}\,\mathrm{cm}^{-1}$) owing to the large spin–orbit coupling for these metal ions

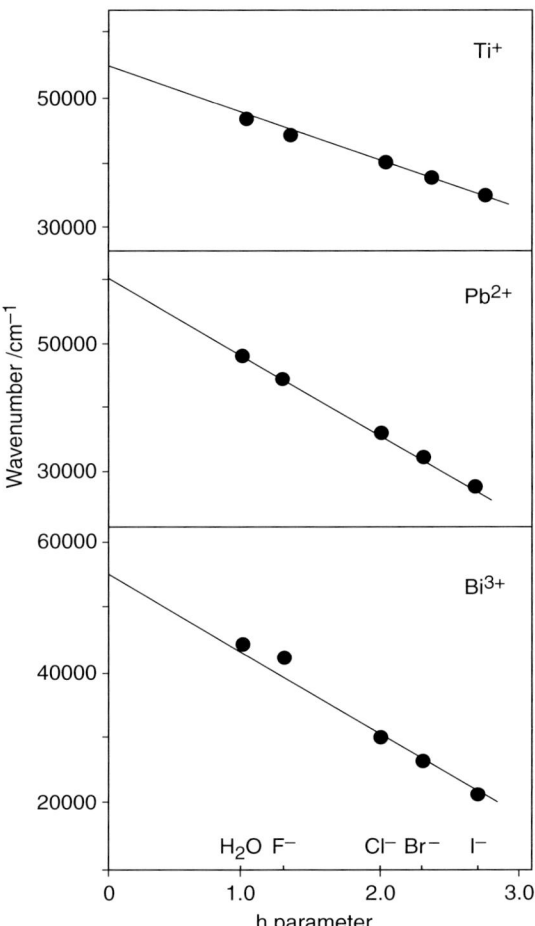

Fig. 3.3. Plot of the $^1S_0 \rightarrow \,^3P_1$ frequency maximum of complexed $6s^2$ ions (Tl^+, Pb^{2+}, Bi^{3+}) versus Jørgensen's orbital expansion parameter, h. The abscissa is marked for h values of the following ligands: water (1.00), fluoride (1.3), chloride (2.0), bromide (2.3), iodide (2.7)

$$\Lambda = h_{\text{glass}}/2.56 , \tag{3.22}$$

where h_{glass} is the h parameter as measured for the glass under investigation. Λ is the "optical basicity", so called because it is obtained from optical spectra. The relationship between h and ν allows the optical basicity to be expressed in terms of the $^1S_0 \rightarrow \,^3P_1$ frequency of the glass, ν_{glass},

$$\Lambda = \frac{\nu_{\text{f}} - \nu_{\text{glass}}}{\nu_{\text{f}} - \nu_{O^{2-}}} , \tag{3.23}$$

where $\nu_{O^{2-}}$ denotes the $^1S_0 \rightarrow \,^3P_1$ frequency in CaO, and ν_{f} the frequency for $h = 0$, that is, for the condition where the probe ion is in the free state

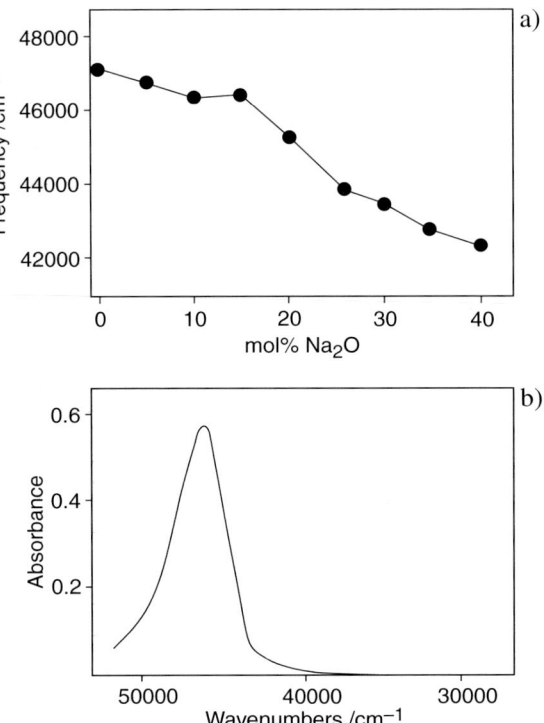

Fig. 3.4. (a) Fall in the $^1S_0 \to ^3P_1$ absorption frequency of Pb^{2+}, for the Na_2O–B_2O_3 glass system, with increasing Na_2O content. (b) Ultraviolet absorption spectrum of Pb^{2+} ion dissolved in $10\,mol\%$ Na_2O sodium borate glass showing the $^1S_0 \to ^3P_1$ frequency maximum at $46\,400\,cm^{-1}$

and unaffected by any surrounding electron density. Equation (3.23) indicates that the optical basicity of CaO is defined as unity. Specific expressions of (3.23) for Tl^+, Pb^{2+} and Bi^{3+} are in Table 3.1.

Figure 3.4b shows the ultraviolet absorption spectrum of Pb^{2+} in a sodium borate glass containing $10\,mol\%$ of Na_2O. The band maximum is at $46\,400\,cm^{-1}$, and substituting in the appropriate formula for the Pb^{2+} probe (Table 3.1) yields: $\Lambda = (60\,700 - 46\,400)/(60\,700 - 29\,700) = 0.461$. It should be noted that the absorption envelope exhibited by the probe ion is broadened not only by the usual (e.g. vibrational) effects but also because

Table 3.1. Relationship between optical basicity, Λ, and $^1S_0 \to ^3P_1$ frequency, $\nu(cm^{-1})$, for Tl^+, Pb^{2+}, and Bi^{3+}

Tl^+	Pb^{2+}	Bi^{3+}
$\Lambda = \dfrac{55\,300 - \nu}{18\,300}$	$\Lambda = \dfrac{60\,700 - \nu}{31\,000}$	$\Lambda = \dfrac{56\,000 - \nu}{28\,800}$

each probe ion has a tendency to distribute itself over a range of sites, each with its own basicity. Usually this range of sites is such that it focuses quite narrowly on the principal site. However, in certain glasses, more than one principal site is available, and this can lead to greater broadening. Sometimes the broadening leads to a splitting of the band so that two distinct maxima are observed, as occurs for the Tl^+ ion in the Na_2O–B_2O_3 glass system, for example [3.24, 31], (see Sect. 2.6.8).

Optical Basicity and Glass Composition

There have been many systematic studies of glasses and other materials relating optical basicity with chemical composition. Two main factors emerge which are best considered by adopting an initial viewpoint that the materials are composed entirely of monatomic ions. For example, in calcium metasilicate, $CaSiO_3$, there are present Ca^{2+}, Si^{4+} and O^{2-} ions. Before considering the introduction of covalency, it is seen that the Ca^{2+} ions neutralise one third and the Si^{4+} ions two thirds of the negative charge of the O^{2-} ions. This partitioning by charge is an important factor determining the basicity. Of course, the more realistic situation is that the calcium metasilicate consists of Ca^{2+} ions and a negatively charged silicate network with covalent Si–O bonds.

As well as neutralising the negative charge, the Ca^{2+} and Si^{4+} ions also polarise the O^{2-} ions. This is the second factor affecting basicity. The polarising effect of Si^{4+} is much greater than that of Ca^{2+} because it is much smaller and bears a greater positive charge (and also silicon is much more electronegative than calcium). Data from optical basicity studies have allowed the polarising power of the cations to be expressed numerically by so-called "basicity moderating parameters", γ, (Table 3.2) so that for calcium metasilicate, for example, the optical basicity can be expressed as [3.24–28]

$$\Lambda = \frac{1}{3} \times \frac{1}{\gamma_{Ca}} + \frac{2}{3} \times \frac{1}{\gamma_{Si}} \ . \tag{3.24}$$

Equation (3.24) shows that for a single oxide, Λ is simply $1/\gamma$. Therefore, in general terms (3.24) can be written as

$$\Lambda = X_A \times \Lambda(\text{oxide}(A)) + X_B \times \Lambda(\text{oxide}(B)) + \dots \ . \tag{3.25}$$

$\Lambda(\text{oxide}(A))$, $\Lambda(\text{oxide}(B))$, ... are the optical basicities of the constituent oxides, and X_A, X_B, ... are the proportions of oxygen atoms or ions which the oxides A, B, ... contribute, that is, their equivalent fractions. The value of Λ calculated from (3.25) is the average basicity of the glass. It represents the average electron donor power of the medium, taking into account the range and proportion of all the different sites existing in the glass. Wave mechanical calculations for various oxyanion species, with the general formula XO_4^{n-},

Table 3.2. Basicity moderating parameters for elements in oxidation states denoted, and values of Λ for individual oxides

Element	γ	Oxide	Λ
Caesium(I)	0.60	Cs_2O	1.7
Potassium(I)	0.73	K_2O	1.4
Sodium(I)	0.87	Na_2O	1.15
Lithium(I)	1.0	Li_2O	1.0
Barium(II)	0.87	BaO	1.15
Strontium(II)	0.91	SrO	1.1
Calcium(II)	1.00	CaO	1.00
Iron(II)	1.0	FeO	1.0
Manganese(II)	1.0	MnO	1.0
Magnesium(II)	1.3	MgO	0.78
Aluminium(III)	1.65	Al_2O_3	0.60
Silicon(IV)	2.1	SiO_2	0.48
Boron(III)	2.36	B_2O_3	0.42
Hydrogen(I)	2.5	H_2O	0.40
Phosphorus(V)	3.0	P_2O_5	0.33

Values of Λ are expressed to the nearest 0.05 for oxides of formula M_2O and MO (except CaO and MgO)

have indicated that the optical basicity is related to the negative charge borne by the oxygen atoms, q_O by [3.32]

$$\Lambda = q_O/1.15 \ . \tag{3.26}$$

In effect, the optical basicity model provides a means of estimating the extent of negative charge borne by the oxygen atoms in a glass. This cannot be calculated by wave mechanics for a glass owing to lack of stereochemical information. Many of the chemical and physical properties of a material arise owing to the manner in which the electronic charge is spread throughout the atomic/ionic array, and for oxidic glasses it is the value of q_O which is of prime importance. Attempting to correlate trends in a particular property with optical basicity is an attempt to make correlation with q_O. It therefore follows that when Λ is quantitatively related to a particular property, for example refractive index (see below) or redox equilibria (see Sect. 3.3.1), the correlation is of more than just empirical significance. Probably it is due to this reason, more than any other, that optical basicity has been successful over such a wide spread of application.

Optical basicity values for individual oxides are in Table 3.2, and it can be seen how these values provide a ranking of the acidic/basic properties. Whether an oxide is regarded as acidic or basic depends on relative comparison. Nevertheless, it is worth noting that Al_2O_3, with an optical basicity of 0.60, appears to be the dividing line between what are commonly thought of as basic oxides (Na_2O, CaO, ...) and acidic oxides (SiO_2, P_2O_5, ...).

Equation (3.25) can be used for obtaining the optical basicity of any glass from its chemical composition, provided the Λ values of the constituent oxides are available. An inspection of Table 3.2 indicates that calculation of optical basicity is possible for a very wide range of glass systems. For example, for a sodium silicate glass with $Na_2O:SiO_2$ ratio of 1:2, X for Na_2O is $1/5$ and for SiO_2 it is $4/5$; therefore, with $\Lambda(Na_2O) = 1.15$ and $\Lambda(SiO_2) = 0.48$ (Table 3.2), the optical basicity, Λ, is $(1.15 + 4 \times 0.48)/5$, that is, 0.61. For a glass of molar composition Na_2O (16%), CaO (12%), SiO_2 (72%), $\Lambda = (16 \times 1.15 + 12 \times 1.00 + 144 \times 0.48)/172 = 0.58$.

It should be noted that, since the basicity moderating parameters, γ, for the various cations, Na^+, Ca^{2+}, Si^{4+}, ..., are arrived at through the relationship, (3.23), they reflect the polarising power which the cations have on the oxide(-II) species as detected through the orbital expansion effect which the (polarised) oxygens have on the Pb^{2+}, etc., probe ions; γ is therefore an experimentally determined quantity, and in this respect it differs from *Dietzel's* (empirical) "cation field strength", F, [3.33] which has sometimes been invoked in attempting to handle glass basicity [3.21, 34]. F is simply the cation oxidation number divided by the square of the distance of the cation from the oxygen atom nucleus. In effect, it is an attempt to express the ionic-covalent nature of the bonding in glass in terms of *Fajans'* rules [3.35] (covalency favoured by small cations within a high oxidation state).

Optical Basicity and Electronegativity

The basicity moderating parameter, γ_M, represents the polarising power of the cation, M^{z+}, and is therefore related to the degree of covalency in the M–O bonding. It follows that there might be a relationship between γ_M and electronegativity. This possibility has been previously explored [3.25], and it was found that for a small number of elements (e.g. Si, Al, B, and the alkali and alkaline-earth metals) a simple relationship exists with Pauling electronegativity, x_M, such that

$$\gamma_M = (4x_M - 1)/3 . \tag{3.27}$$

It is emphasised that this relationship, (3.27), does not hold generally, and certainly not for the ions of the transition metals. For example, it has been found that Fe^{2+} has a γ-value of 1.0, whereas the Pauling electronegativity of iron (1.7) would indicate $\gamma = 1.9$.

The electronegativity of oxygen in oxidic compounds is not fixed, but varies depending on the degree of covalency or ionicity in the bonding. Usually a value of 3.5 is assigned for the electronegativity of oxygen, and this is the Pauling value in covalent compounds such as SiO_2 or P_2O_5. However, when oxygen bears a substantial negative charge, as in the alkali or alkaline-earth metal oxides, its electronegativity is much less; for example, in Na_2O, $x_O = 2.6$ [3.26]. This leads to the situation where the electronegativity difference in Na_2O and in SiO_2 are almost the same (approximately 1.65 in both cases), despite the much greater ionicity in the bonding for Na_2O compared with SiO_2. For all other elements (in a particular oxidation state), electronegativity is more or less fixed, and the exceptional behaviour of oxygen arises because its chemistry spans such a wide range of ionic-covalent character. In ionic compounds, the affinity of oxygen for electronic charge in the bonding (electronegativity) diminishes as its own negative charge increases.

Use of electronegativity for quantifying glass basicity has been made in the pB scale of *Baltã* [3.36]. Later modification required replacement of electronegativity by a combination of cation oxidation state and coordination number. For many purposes, the pB scale serves in a similar fashion to the optical basicity scale and, for certain glass compositions where electronegativity is a common link, the two scales parallel each other [3.37].

Refractivity and Optical Basicity

The electron charge clouds of atoms, whether neutral or charged (as cations and anions), are distorted by electric fields (Fig. 3.5). These fields can be static, for example, as occurs when the charge cloud of an anion is polarised by adjacent cations, or they can be oscillatory, as produced by electromagnetic radiation (light). The extent to which these oscillating electric fields induce corresponding oscillations of the orbital charge clouds depends on the refractivity of the cations and anions of the medium through which the light is travelling.

The refractivity of an ion can be pictured as its "floppiness" as it switches back and forth in response to the oscillating electric field. Cations with a high positive charge have their orbital charge clouds severely controlled by the nucleus (*Weyl* and *Marboe* [3.34] describe this as a "tightened" state). For anions, the situation is quite different and their charge clouds are much less controlled because the electrons outnumber the positive charge on the nucleus. The charge clouds of anions are therefore much more floppy than those of cations. This floppiness has two consequences for electromagnetic radiation passing through the medium. It causes a reduction in the velocity and, with increasing frequency, absorption of the radiation. Refractivity is therefore intimately related to the refractive index, both the real and imaginary parts.

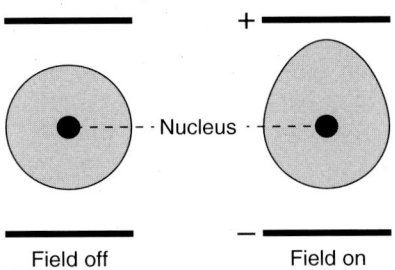

Field off   Field on

Fig. 3.5. Schematic diagram indicating distortion of electric charge cloud of an atom or ion by an external electric field

As far as the real part, n, is concerned, the refractivity of an isotropic medium is given by the Lorentz–Lorenz relationship [3.38]

$$R_{\mathrm{m}} = V_{\mathrm{m}}\frac{n^2 - 1}{n^2 + 2} \,, \tag{3.28}$$

where V_{m} is the molar volume (formula weight in grams divided by the density in grams per cm^3) and R_{m} is the molar refractivity in cm^3. For ionic materials, R_{m} is simply the sum of the ionic refractivities which, for most cations and anions, are more or less fixed (though arbitrary) quantities. Some values are given in Table 3.3. For the oxide ion, however, the refractivity can vary over a wide range depending on the extent of negative charge it bears, that is, depending on how electrovalently or covalently it is bound to other atoms or ions.

From what has been said about the relationship between refractivity and the floppiness of the charge clouds, it follows that increasing negative charge on oxygen should lead to increasing refractivity. Since optical basicity expresses this negative charge (see earlier), a simple relationship between optical basicity and oxide refractivity is expected [3.39, 40], and this is indeed illustrated in Fig. 3.6 for a set of alkali silicate glasses where the reciprocal of oxide refractivity is plotted against optical basicity for each glass. Correlations such as these between oxygen refractivity and optical basicity have been suggested as a means for expressing glass basicity in terms of a "refractivity basicity" scale [3.41].

Optical Basicity of Sites for Mobile Metal Ions

In recent years the technique of far-infrared spectroscopy has been used for elucidating the nature of the sites available for alkali metal ions in a number of glass systems, especially the alkali borates [3.31]. Since the vibrational frequency of the alkali metal ion depends on the electron density provided at the site, a straightforward relationship with optical basicity might be expected. Such a relationship has indeed been established, and it indicates how

Table 3.3. Molar refractivities, R_m, of selected ions

Ion	R_m cm^3	Ion	R_m cm^3
P^{5+}	negligible	Ca^{2+}	1.18
B^{3+}	0.008	K^+	2.09
Si^{4+}	0.042	Sr^{2+}	2.17
Li^+	0.073	Rb^+	3.53
Al^{3+}	0.131	Cs^+	3.58
Mg^{3+}	0.24	Ba^{2+}	3.91
Na^+	0.45	Pb^{2+}	9.1

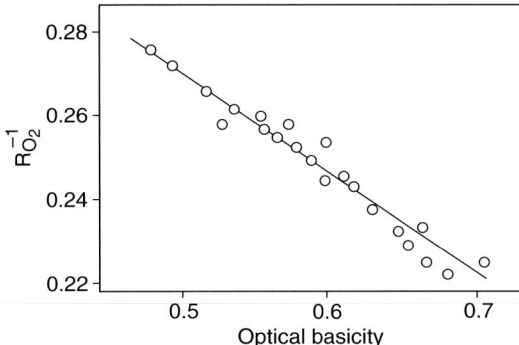

Fig. 3.6. Relationship between oxide refractivity and optical basicity for silicate glasses (based on data from [3.41])

frequency measurements in the far-infrared region can provide an alternative method for directly obtaining optical basicity. In the alkali borate glass systems, there is strong evidence that the alkali metal ions are divided between two types of site, one of lower than average basicity and the other of higher. The trend in optical basicity of these two sites in the Na_2O–B_2O_3 system, with increasing sodium oxide content, is shown in Fig. 3.7.

The existence of two types of site is of interest when considering the mobility of Na^+ ions during the electrolysis of the glass (see Sect. 2.6.8). When a 35 mol% Na_2O borate glass is electrolysed with a thallium amalgam anode and a platinum cathode, in-depth profiling (by ion sputtering techniques) shows depletion of sodium in the region of the glass just below the anode (Fig. 3.8) [3.42]. Spectroscopic examination of the glass indicates ultraviolet absorption which is typical of Tl^+ ions, with ν equal to $43\,300\,cm^{-1}$, which corresponds (using the appropriate formula in Table 3.1) to an optical basicity of 0.66. This is almost identical to the optical basicity of the higher basicity sites identified by far-infrared spectroscopy (Fig. 3.7). On the assumption that the Tl^+ ions which enter the glass (from the thallium amalgam) replace the Na^+ ions which have migrated from the anode region (Fig. 3.8), it appears that sites with higher than average electron density are the ones which favour

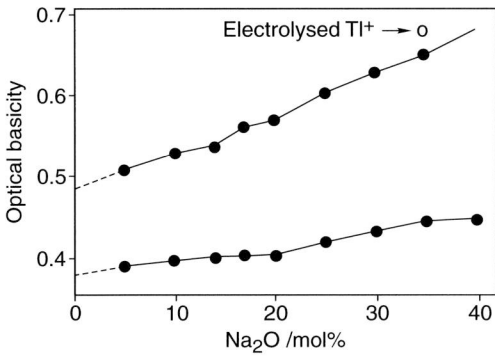

Fig. 3.7. Optical basicity trend for the two sites for cations in the Na_2O–B_2O_3 glass system as indicated by far-infrared "rattling" frequencies of the Na^+ ions

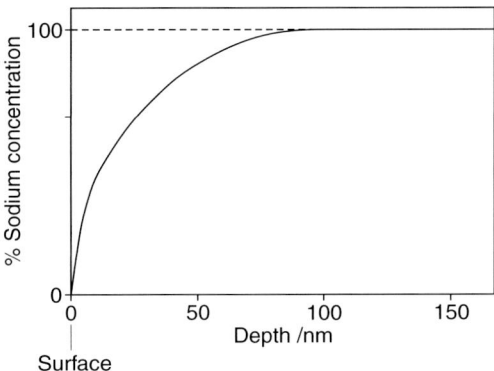

Fig. 3.8. Sodium concentration profile near anode surface of a sodium borate glass (35 mol% Na_2O) after electrolysis using a thallium amalgam anode

ion migration. A chemical bonding analysis has shown that these sodium ions bear a somewhat lower positive charge than the less mobile sodium ions in the lower basicity sites [3.43]. It has been pointed out [3.44] that if the Tl^+ ions do not replace the Na^+ ions (and size difference might be important here), then this conclusion is uncertain.

3.3.3 Redox Reactions in Melts

Glass can contain many impurities and additives. Some occur owing to contaminated starting materials, for example, iron and copper, or they enter the glass melt during surface contact with furnace linings. Some are deliberately introduced during manufacture in order to improve the quality of the glass, for example antimony for removing bubbles and dissolved gases (fining), see Sect. 3.5.1. Certain additives are present specifically for enhancing or developing a particular property, for example, lead and barium (for refractive index and dispersion) or neodymium (for laser action) or cobalt and chromium (for colour).

Many of these impurities, or additives, can exist in more than one oxidation state, and this gives rise to a redox equilibrium in the glass melt. For example, the presence of iron involves an equilibrium between Fe^{3+} and Fe^{2+}, while for copper, the equilibrium is between Cu^{2+} and Cu^+ (and sometimes also Cu^0 as metallic copper). Only a small number of equilibria have received a systematic study. Some investigations have been made directly on the melt using electrical measurements, but often the melt is rapidly quenched and the resulting glass then analysed (usually by spectrophotometry) to establish the proportion of oxidised and reduced states [3.25–27, 45–61], see Sect. 3.5.

Since the glass melt is usually exposed to air atmosphere, an equilibrium exists involving the oxygen/oxide half-reaction

$$O_2 + 4e^- = 2O^{2-} \ . \tag{3.29}$$

The O^{2-} ions are "stored" in the glass as non-bridging oxygen atoms (see (3.17), previously). The equilibrium (3.29) implies that the oxide ion activity, and therefore the glass composition, is an important factor for fixing the position of redox equilibrium for the impurity or additive. In the case of the Cu^{2+}/Cu^+ couple, for example, the overall reaction is represented by

$$4Cu^+ + O_2 = 4Cu^{2+} + 2O^{2-} . \tag{3.30}$$

Experiments show for alkali and alkaline-earth silicate glasses (quenched from $1400\,°C$) that with increasing glass basicity there is an increase in the Cu^+/Cu^{2+} ratio [3.54]. This would appear to make sense in terms of the application of Le Chatelier's principle and the law of mass action to (3.30). Furthermore, these principles again appear to operate in extraction metallurgy, [3.22] where it is found that the performance of slags in removing sulphur from molten iron, a reaction represented by

$$S_2(\text{metal}) + 2O^{2-}(\text{slag}) = 2S^{2-}(\text{slag}) + O_2 \tag{3.31}$$

is enhanced by increasing slag basicity. It is for this reason that much effort was expended in searching for a satisfactory expression of basicity in slags (see for example (3.20)). The optical basicity model has found considerable application in this area [3.63–70].

Many redox couples, however, do not behave in this apparently simple manner. Although it is important to know how changes in melt composition affect redox equilibria, the experimental difficulties have discouraged much work in this area, but data are available for the ion-couples Fe^{3+}/Fe^{2+}, Cr^{6+}/Cr^{3+}, Ce^{4+}/Ce^{3+}, Sn^{4+}/Sn^{2+}, As^{5+}/As^{3+}, and Cu^{2+}/Cu^+. All measurements are for $1400\,°C$ and air atmosphere. Apart from the Cu^{2+}/Cu^+ couple, increasing basicity is found to favour the upper oxidation state. The effect of oxide ion activity appears to be more than outweighed by another factor.

This factor arises from the stabilisation of the metal ions by the electron density at the sites provided by the oxygen atoms of the glass. In other words, it arises from the solvation of the ions. (It is important to note that the word "solvation" is used here in the general sense and does not imply the existence of solvating molecules.) In the free (gaseous) state, the positive ions of all elements are less stable than the neutral atom, and this instability (equivalent to the ionisation energy) is greater for the upper than the lower oxidation state (see Fig. 3.9). When hosted by a glass, both the upper and lower oxidation state ions become increasingly stabilised as the electron density at their sites is increased. In terms of *Pauling's* electroneutrality principle [3.71], the site will provide an optimum electron density corresponding to maximum stability of the metal ion. This electron density will be greater for the upper oxidation state ion because of its higher positive charge. Equating electron density with basicity (for example, see (3.24)) explains how increasing basicity favours the upper oxidation state at the expense of the lower. These ideas are shown schematically in Fig. 3.9.

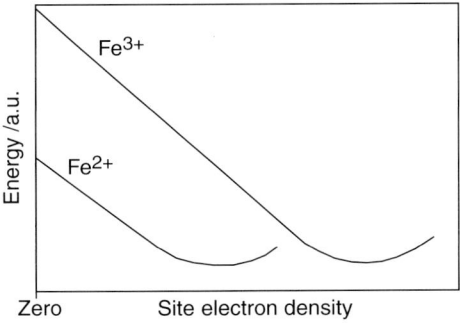

Fig. 3.9. Schematic diagram indicating how the Fe^{2+} and Fe^{3+} ions become stabilised by increasing negative charge of the environment provided by the oxygen atoms of the glass. Where this negative charge is zero, at the origin, is the condition for the ions in the free (gaseous) state, and the "destabilising energy" is the sum of the first and second ionisation energies for Fe^{2+} and of the first, second, and third for Fe^{3+}

When a metal ion exists in an environment of negative charge (here provided by the electrons of the surrounding oxygen atoms) it is said to be "coordinated". In this coordinated state, it experiences a reduction in the positive charge it bears. This is true generally: metal ions bear the exact charge of their oxidation numbers only when they exist in the gaseous state. The acceptance of electronic charge by the metal ion must be distinguished from the concept of reduction. Reduction (and oxidation) refer to a transference of charge which results in a change of electronic configuration. For example, reduction of Fe^{3+} to Fe^{2+} corresponds to a change in the 3d level from five electrons to six ($Fe^{3+}:3d^5 \rightarrow Fe^{2+}:3d^6$). Since these ions are uniquely identifiable by atomic spectroscopy, the numbers +3 and +2, for Fe^{3+} and Fe^{2+}, respectively, are referred to as "spectroscopic oxidation numbers" [3.72].

Redox Equilibrium and Optical Basicity

We take the example of iron in the alkali silicate system

$$4Fe^{2+} + O_2 = 4Fe^{3+} + 2O^{2-} \ . \tag{3.32}$$

Increasing stabilisation of the upper state (Fe^{3+}) with basicity has been demonstrated for a series of glasses of increasing alkali oxide molarity, and also as three distinct trends which become increasingly pronounced on going from the lithium to the sodium to the potassium silicate system (Fig. 3.10a). If the data are replotted so that the abscissa represents optical basicity instead of mol% alkali oxide, the three trends are united into a single trend (Fig. 3.10b) which is adequately expressed by the linear equation

$$\log([Fe^{2+}]/[Fe^{3+}]) = 3.2 - 6.5\Lambda \ . \tag{3.33}$$

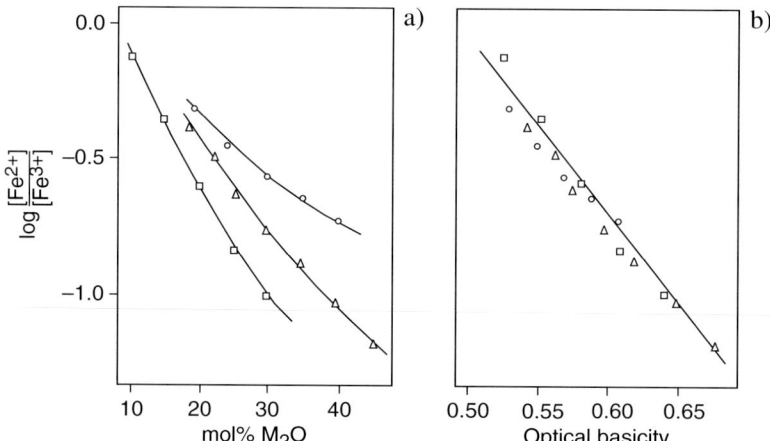

Fig. 3.10. Effect of glass basicity on the $[Fe^{2+}]/[Fe^{3+}]$ ratio in alkali silicate glasses:
(a) The upper oxidation state, Fe^{3+}, is enhanced either as the alkali oxide content
is increased, or by substituting Na_2O ($\triangle$) for Li_2O ($\bigcirc$) and K_2O ($\square$) for Na_2O.
(b) The distinct trends for the three glass systems are united into a single trend
when the plot is versus optical basicity instead of alkali oxide content

This relationship allows the $[Fe^{2+}]/[Fe^{3+}]$ ratio to be estimated for any alkali silicate glass composition simply by substituting the appropriate optical
basicity value of the glass. (The conditions would be for air atmosphere and
$1400\,^{\circ}C$ because the original data in Fig. 3.10a were obtained under these
conditions.) For example, a caesium silicate glass containing $20\,mol\%$ Cs_2O,
which has $\Lambda = (1.7 + 8 \times 0.48)/9 = 0.62$ (using (3.23)) and Λ values in
Table 3.2), would have $\log([Fe^{2+}]/[Fe^{3+}]) = 3.2 - 6.5 \times 0.62 = -0.8$.

Data for other redox couples have been treated similarly and in each
case the data points have been united into a single trend between the redox
ratio, R, ($R =$ lower/upper oxidation state) and the optical basicity [3.73, 74].
Linear expressions relating $\log R$ and Λ are in Table 3.4.

Redox Equilibria in Aqueous Solution

The optical basicity model is not restricted to glasses but applies to oxidic
media generally, and we now turn to metal ions in aqueous solution. Under
these conditions, the metal ion usually exists as the aqua complex, for example Fe^{3+} as $[Fe(H_2O)_6]^{3+}$, and the negatively charged environment at the
site for Fe^{3+} is provided by the electrons on the oxygen atoms of surrounding
water molecules. Such a site has been shown to have an optical basicity value
of 0.40 [3.75].

The redox behaviour of metal ion couples in aqueous solution, under acidic
and alkaline conditions, are characterised by their electrode potentials, E^0,
which refer to the equilibrium

Table 3.4. Relationship between the redox ratio, R, ($R =$ lower/upper oxidation state)[a] and the optical basicity, Λ. Upper part (from Fe to As) shows relationships obtained using experimental data for R; lower part contains expressions derived from E^0 aqueous solution data

Ion couple	Electrode potential E^0_{alk} V	E^0_{acid} V	Equilibrium	Redox relationship[b]
$Fe^{3+} \rightarrow Fe^{2+}$	−0.56	−0.036	$4Fe^{3+}+2O^{2-}=4Fe^{2+}+O_2$	$\log\{[Fe^{2+}]/[Fe^{3+}]\}=3.2-6.5\Lambda$
$Cr^{6+} \rightarrow Cr^{3+}$	−0.12	1.10	$4Cr^{6+}+6O^{2-}=4Cr^{3+}+3O_2$	$\log\{[Cr^{3+}]/[Cr^{6+}]\}$ $=8.2-13.7\Lambda$
$Ce^{4+} \rightarrow Ce^{3+}$	1.71	1.45	$4Ce^{4+}+2O^{2-}=4Ce^{3+}+O_2$	$\log\{[Ce^{3+}]/[Ce^{4+}]\}=5.4-8.3\Lambda$
$Sn^{4+} \rightarrow Sn^{2+}$	−0.96	0.15	$2Sn^{4+}+2O^{2-}=2Sn^{2+}+O_2$	$\log\{[Sn^{2+}]/[Sn^{4+}]\}=0.6-3.6\Lambda$
$As^{5+} \rightarrow As^{3+}$	−0.08	0.58	$2As^{5+}+2O^{2-}=2As^{3+}+O_2$	$\log\{[As^{3+}]/[As^{5+}]\}=5.2-8.9\Lambda$
$Mn^{3+} \rightarrow Mn^{2+}$	−0.40	1.51	$4Mn^{3+}+2O^{2-}=4Mn^{2+}+O_2$	$\log\{[Mn^{2+}]/[Mn^{3+}]\}=7-12\Lambda$
$Sb^{5+} \rightarrow Sb^{3+}$	−0.59	0.64	$2Sb^{5+}+2O^{2-}=2Sb^{3+}+O_2$	$\log\{[Sb^{3+}]/[Sb^{5+}]\}=6-11\Lambda$
$V^{5+} \rightarrow V^{4+}$	1.00	−0.74	$4V^{5+}+2O^{2-}=4V^{4+}+O_2$	$\log\{[V^{4+}]/[V^{5+}]\}=4-8\Lambda$
$Ni^{4+} \rightarrow Ni^{2+}$	1.93	0.49	$2Ni^{4+}+2O^{2-}=2Ni^{2+}+O_2$	$\log\{[Ni^{2+}]/[Ni^{4+}]\}=17-25\Lambda$
$Pb^{4+} \rightarrow Pb^{2+}$	1.46	0.28	$2Pb^{4+}+2O^{2-}=2Pb^{2+}+O_2$	$\log\{[Pb^{2+}]/[Pb^{4+}]\}=15-22\Lambda$
$Tl^{3+} \rightarrow Tl^{+}$	1.25	−0.05	$2Tl^{3+}+2O^{2-}=2Tl^{+}+O_2$	$\log\{[Tl^{+}]/[Tl^{3+}]\}=13-20\Lambda$
$Co^{3+} \rightarrow Co^{2+}$	1.84	0.20	$4Co^{3+}+2O^{2-}=4Co^{2+}+O_2$	$\log\{[Co^{2+}]/[Co^{3+}]\}=9-14\Lambda$

[a] for silicate melts at $1400\,°C$ in air atmosphere
[b] The ratios of concentrations are for equilibrium conditions.

$$M^{z+} = M^{(z+n)+} + ne^- \ . \tag{3.34}$$

Electrode potentials obtained under acidic conditions (E^0_{acid}) refer to the metal ion existing as the aqua ion (see above) for which (3.34) is more properly expressed as, for example

$$[Fe(H_2O)_6]^{2+} = [Fe(H_2O)_6]^{3+} + e^- \ . \tag{3.35}$$

When the metal ion exists as the hydroxo species as, for example, is supposed in the half-equation

$$Fe(OH)_2 + OH^- = Fe(OH)_3 + e^- \ , \tag{3.36}$$

the appropriate electrode potential is that obtained under alkaline conditions (E^0_{alk}). The environments provided for the metal ions in these species are by the oxygen atoms of hydroxide ions, and the optical basicity value has been shown to be 0.70 [3.75].

It is implicit in the optical basicity model that different media with the same optical basicity value provide (on average) sites of equivalent electron density for the hosted metal ions. For example, the electron density of sites in a hydroxo species is equivalent to those in a sodium silicate melt having $\Lambda = 0.70$ (i.e., by (3.23), containing 49% Na_2O). Thus, in spite of the enormous difference between aqueous conditions at $25\,°C$ and the molten silicate environment at $1400\,°C$, there is the possibility of a straightforward relationship existing between the electrode potential in alkaline solution and the redox ratio, R, obtained by substituting the value of 0.70 for Λ in the appropriate expression (Table 3.4) for the couples Fe^{3+}/Fe^{2+}, Cr^{6+}/Cr^{3+}, Ce^{4+}/Ce^{3+}, Sn^{4+}/Sn^{2+}, and As^{5+}/As^{3+}. It is found that plotting nE^0_{alk} versus $\log R$ (Fig. 3.11a) the points lie close to the expression [3.76–78]

$$\log R = 0.42nE^0_{\text{alk}} - 1.1 \ . \tag{3.37}$$

For acidic conditions (with $\Lambda = 0.40$) a similar relationship is found (see Fig. 3.11b):

$$\log R = 2.5nE^0_{\text{acid}} - 1.5 \ . \tag{3.38}$$

Figure 3.11b omits the single point for Cr^{6+}/Cr^{3+} which can be made to fit only by setting $\Lambda = 0.10$ (the condition for an environment provided by H_3O^+ ions). An explanation for this inconsistency has not yet been found.

Use of Aqueous E^0 Data for Predicting Melt Equilibria

The existence of the two relationships (3.37 and 3.38) offers the possibility for using electrode potential data in aqueous solution to derive equations relating $\log R$ with Λ which are applicable to ion couples in molten silicates (at 1400 °C and air atmosphere). Thus, (3.37) provides the equilibrium redox ratio, R, for $\Lambda = 0.70$ by substituting the value of E^0 for alkaline solution. Similarly, a value for R corresponding to $\Lambda = 0.40$ is obtained by substituting E^0 for acidic conditions in (3.35). The two values of R then yield a linear equation relating R with Λ.

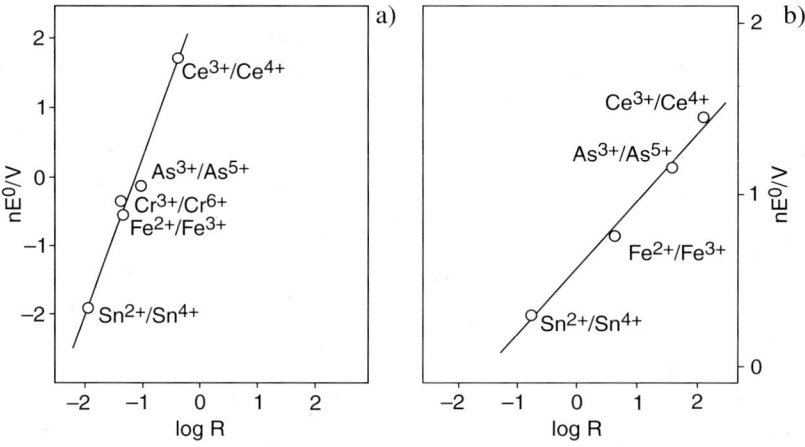

Fig. 3.11. Plot of (**a**) nE^0_{alk} and (**b**) nE^0_{acid} (aqueous solution data) versus logarithm of redox ratio, R, (obtained, Table 3.4, for optical basicity values of 0.70 and 0.40, respectively) for ion couples denoted

This is illustrated by the reaction

$$2Sb^{5+} + 2O^{2-} = 2Sb^{3+} + O_2 \ . \tag{3.39}$$

E^0 is known to be -0.59 V under alkaline conditions and 0.64 V under acidic, and (3.37) and (3.38) yield values of -1.6 and 1.7 for $\log R$. These values of $\log R$ are for $\Lambda = 0.70$ and 0.40 corresponding to the relationship $\log R = 6 - 11\Lambda$.

Relationships for other metal ion couples have been derived similarly (Table 3.4) and the plot of (the logarithm of) redox ratio against optical basicity is shown for them in Fig. 3.12. Ordinary glasses have optical basicities in the range approximately 0.55–0.65, and the redox ratios within this range are in accordance with the known behaviour of these metal ions in glass. For example, it is not possible to generate sensible quantities of Ni^{4+} under normal conditions. However, the plots for vanadium, manganese and antimony indicate that both oxidation states should coexist in glass, and this is what is observed experimentally. Cobalt usually exists as the Co^{2+} ion in glass but Co^{3+} can be obtained, although with more difficulty than for Mn^{3+}, and Fig. 3.12 indicates this. Previously published data [3.55] for the Mn^{2+}/Mn^{3+} couple in a glass melt of composition $3Na_2O\cdot 2CaO\cdot 15SiO_2$ (for which Λ is 0.567) indicate a $\log$(redox ratio) of 1.06 (though at $1450\,^\circ C$), which can be compared with the value of $(7 - 12 \times 0.567) = 0.2$ (see Table 3.4). The trend for lead is perhaps surprising since it does suggest that small amounts of Pb^{4+} can coexist with Pb^{2+}. Reports of Pb^{4+} in glass are unknown, but its possibility should be considered along with the trend for thallium where the data indicate the feasibility of Tl^{3+}. Although thallium has not been properly studied in silicate glasses, experiments in alkali borate glasses show that Tl^+

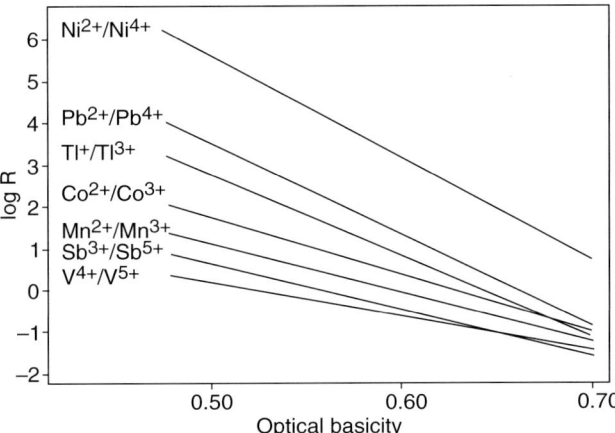

Fig. 3.12. Effect of glass basicity on the redox ratio of ion couples indicated by plotting $\log R$ (Table 3.4) against optical basicity

is the predominant species but that Tl^{3+} becomes increasingly important with increasing glass basicity [3.79].

It must be remembered for all of these relationships that the conditions are for $1400\,°C$ and air atmosphere. Also, there is the possibility that in quenching the melt to a glass at ambient temperature some equilibria might undergo change [3.80].

Two Ion-Couple Equilibria

In aqueous acidic solution Fe^{2+} is oxidised to Fe^{3+} when chromate ions, CrO_4^{2-}, are added.

$$CrO_4^{2-} + 3Fe^{2+} + 8H^+ = Cr^{3+} + 3Fe^{3+} + 4H_2O \ . \tag{3.40}$$

(Strictly, the oxidising species should be written as the dichromate ion, $Cr_2O_7^{2-}$, since this species predominates in acidic solution; however, this detail can be ignored for present purposes.) An analogous redox reaction occurs between Fe^{2+} ions and chromate ions in molten glass. The reaction can be written symbolically as

$$Cr^{6+} + 3Fe^{2+} = Cr^{3+} + 3Fe^{3+} \ . \tag{3.41}$$

This equilibrium has been shown experimentally to depend on the chemical composition of the glass melt [3.61]. In general, it is possible to obtain expressions for equilibria of this type by combining the expressions in Table 3.4, and it can be shown, as follows, that the equilibrium depends on optical basicity.

Writing these equilibria as the general equation

$$m M^{(z+n)+} + n M'^{y+} = m M^{z+} + n M'^{(y+m)+} \ , \tag{3.42}$$

it is seen for the equilibrium constant, K_c, that

$$\log K_c = \log \frac{[M^{z+}]^m [M'^{(y+m)+}]^n}{[M^{(z+n)+}]^m [M'^{y+}]^n}$$

$$= m \log \frac{[M^{z+}]}{[M^{(z+n)+}]} - n \log \frac{[M'^{y+}]}{[M'^{(y+m)+}]} \ . \tag{3.43}$$

Expressions for the two redox ratios (as their logarithms) in (3.40) are available from Table 3.4, enabling $\log K_c$ to be obtained [3.81]. For the iron/chromium equilibrium, for example, (3.43) becomes

$$\log K_c = \log \frac{[Cr^{3+}]}{[Cr^{6+}]} - 3 \log \frac{[Fe^{2+}]}{[Fe^{3+}]} \ . \tag{3.44}$$

Using expressions for $\log\{[Cr^{3+}]/[Cr^6]\}$ and $\log\{[Fe^{2+}]/[Fe^{3+}]\}$ in (3.44) (Table 3.4), it follows that $\log K_c$ is $(8.2 - 13.7\Lambda) - 3(3.2 - 6.5\Lambda) = 5.8\Lambda - 1.4$. Expressions of $\log K_c$ for further equilibria are in Table 3.5.

In principle, these expressions allow prediction of K_c, at $1400\,°C$, for any molten silicate for which Λ can be calculated.

Table 3.5. Relationship between K_c for equilibria, (3.43), and optical basicity

Equilibrium	log K_c
$Mn^{3+} + Ce^{3+} = Mn^{2+} + Ce^{4+}$	$1.6 - 3.7\Lambda$
$As^{5+} + Sb^{3+} = As^{3+} + Sb^{5+}$	$2.1\Lambda - 0.8$
$2Ce^{4+} + As^{3+} = 2Ce^{3+} + As^{5+}$	$5.6 - 7.7\Lambda$
$Ce^{4+} + Fe^{2+} = Ce^{3+} + Fe^{3+}$	$2.2 - 1.8\Lambda$
$2Mn^{3+} + As^{3+} = 2Mn^{2+} + As^{5+}$	$8.8 - 15.1\Lambda$
$2Ce^{4+} + Sb^{3+} = 2Ce^{3+} + Sb^{5+}$	$4.8 - 5.6\Lambda$
$2Mn^{3+} + Sb^{3+} = 2Mn^{2+} + Sb^{5+}$	$8 - 13\Lambda$
$Mn^{3+} + Fe^{2+} = Mn^{2+} + Fe^{3+}$	$3.8 - 5.5\Lambda$
$Sb^{5+} + 2Fe^{2+} = Sb^{3+} + 2Fe^{3+}$	$2\Lambda - 0.4$
$As^{5+} + 2Fe^{2+} = As^{3+} + 2Fe^{3+}$	$4.1\Lambda - 1.2$
$Cr^{6+} + 3Mn^{2+} = Cr^{3+} + 3Mn^{3+}$	$22.3\Lambda - 12.8$
$Cr^{6+} + 3Ce^{3+} = Cr^{3+} + 3Ce^{4+}$	$11.2\Lambda - 8.0$
$Cr^{6+} + 3Fe^{2+} = Cr^{3+} + 3Fe^{3+}$	$5.8\Lambda - 1.4$
$2Cr^{6+} + 3As^{3+} = 2Cr^{3+} + 3As^{5+}$	$0.8 - 0.7\Lambda$
$2Cr^{6+} + 3Sb^{3+} = 2Cr^{3+} + 3Sb^{5+}$	$5.6\Lambda - 1.6$
$Sn^{4+} + 2Fe^{2+} = Sn^{2+} + 2Fe^{3+}$	$9.4\Lambda - 5.8$
$As^{5+} + Sn^{4+} = As^{3+} + Sn^{4+}$	$4.6 - 5.3\Lambda$
$2Ce^{4+} + Sn^{2+} = 2Ce^{3+} + Sn^{4+}$	$10.2 - 13.0\Lambda$
$Sb^{5+} + Sn^{2+} = 2Sb^{3+} + Sn^{4+}$	$4.8 - 3.8\Lambda$
$2Cr^{6+} + 3Sn^{2+} = 2Cr^{3+} + 3Sn^{4+}$	$14.6 - 16.6\Lambda$
$V^{5+} + Fe^{2+} = V^{4+} + Fe^{3+}$	$0.8 - 1.5\Lambda$
$3V^{5+} + Cr^{3+} = 3V^{4+} + Cr^{6+}$	$3.8 - 10.3\Lambda$
$V^{5+} + Ce^{3+} = V^{4+} + Ce^{4+}$	$0.3\Lambda - 1.4$
$2V^{5+} + Sn^{2+} = 2V^{4+} + Sn^{4+}$	$7.4 - 12.4\Lambda$
$2V^{5+} + As^{3+} = 2V^{4+} + As^{5+}$	$2.8 - 7.1\Lambda$
$V^{5+} + Mn^{2+} = V^{5+} + Mn^{2+}$	$4\Lambda - 3$
$2V^{5+} + Sb^{3+} = 2V^{4+} + Sb^{5+}$	$2 - 5\Lambda$

Oxidation of Metallic Elements

Many metals undergo oxidation to the metal ion in silicate melts. Also, there are several instances where the reducing action of oxide ions is responsible for precipitation of the metal from the melt, for example, silver and copper. These redox reactions can be expressed generally in the form (again with the reduced species on the right-hand side)

$$4M^{n+} + 2nO^{2-} = 4M + nO_2 . \tag{3.45}$$

As previously, electrode potentials in aqueous solution are now used, with (3.37) and (3.38), to obtain expressions for the electrode potential in the melt (E_{melt}) for each metal. Also here, it must be remembered that the conditions are for $1400\,^\circ C$, for which temperature the expression $2.303RT/nF$ is $0.332/n$ (compared with $0.059/n$ for $25\,^\circ C$).

Choosing arsenic as an example, the E^0 value in alkaline solution for the half-reaction

$$AsO_2^- + 2H_2O + 3e^- = As + 4OH^- \tag{3.46}$$

is $-0.71\,\mathrm{V}$, and in acidic solution for

$$As_2O_3 + 6H^+ + 6e^- = 2As + 3H_2O\ , \tag{3.47}$$

E^0 is $0.234\,\mathrm{V}$. These yield values of -1.995, (3.34) and 0.255, (3.35), respectively, for $\log R$. Since the electrode potential for the melt at $1400\,^\circ\mathrm{C}$, E_{melt}, is related to the equilibrium redox ratio by (see above)

$$E_{\mathrm{melt}} = \frac{0.332}{n}\log R\ , \tag{3.48}$$

these values of $\log R$ yield $E_{\mathrm{melt}} = -0.22\,\mathrm{V}$ and $0.03\,\mathrm{V}$ corresponding to the conditions for $\Lambda = 0.70$ and 0.40, respectively, under which the E^0 values in aqueous solution were determined, that is, under alkaline conditions (where $\Lambda = 0.70$) and acidic conditions (where $\Lambda = 0.40$). These are also the Λ values for the melt. Therefore, for the melt, the linear equation fulfilling these values is

$$E_{\mathrm{melt}} = 0.36 - 0.83\Lambda\ . \tag{3.49}$$

It is possible to combine (3.37), (3.38), and (3.48) (with the conditions Λ of being 0.40 and 0.70) to yield

$$\begin{aligned}
E_{\mathrm{melt}} = {} & (0.46E_{\mathrm{alk}}^0 - 2.76E_{\mathrm{acid}}^0 + 0.44/n)\Lambda \\
& + (1.93E_{\mathrm{acid}}^0 - 0.185E_{\mathrm{alk}}^0 - 0.67/n)\ .
\end{aligned} \tag{3.50}$$

By making substitutions for E_{alk}^0, E_{acid}^0, and n, expressions are then obtained showing the dependence of E_{melt} on Λ (analogous to (3.50)). The results for a number of equilibria are in Table 3.6 (see right-hand column). These expressions can be used to obtain values of E_{melt} for any silicate composition simply by substituting the Λ value of the melt (calculated from (3.25)).

Voltammetric investigations in molten silicates have been made for a small number of metals, for example see [3.52, 55], and provide the opportunity to compare the electrode potentials derived from Table 3.6 with experimental electrode potentials. These experimental ones were obtained relative to the oxygen/oxide electrode, and it is therefore necessary to adjust the expressions for E_{melt} in Table 3.6 so that they are relative to the half reaction

$$O_2 + 4e^- = 2O^{2-}\ . \tag{3.51}$$

This gives a standard electrode potential (designated E_{melt}^0) by subtracting the E_{melt} expression for (3.51) from the other expressions in Table 3.6. For

Table 3.6. Electrode potential data[a]

Couple	$E^0_{\text{alk}}/\text{V}$ 25 °C	$E^0_{\text{acid}}/\text{V}$ 25 °C	E_{melt} relationship[b] 1400 °C
$Ag^+ \to Ag$	0.342	0.800	$-1.61\Lambda + 0.81$
$Pt^{2+} \to Pt$	0.16	1.2	$-3.02\Lambda + 1.95$
$Hg^{2+} \to Hg$	0.0984	0.854	$-2.09\Lambda + 1.30$
$Cu^{2+} \to Cu$	-0.224	0.3402	$-0.82\Lambda + 0.36$
$Tl^+ \to Tl$	-0.3445	-0.336	$1.21\Lambda - 1.25$
$Cu^+ \to Cu$	-0.361	0.522	$-1.17\Lambda + 0.40$
$Bi^{3+} \to Bi$	-0.46	0.32	$-0.95\Lambda + 0.48$
$Pb^{2+} \to Pb$	-0.578	-0.126	$0.30\Lambda - 0.47$
$Sb^{3+} \to Sb$	-0.66	0.212	$-0.74\Lambda + 0.31$
$Ni^{2+} \to Ni$	-0.66	-0.250	$0.61\Lambda - 0.70$
$As^{3+} \to As$	-0.71	0.234	$-0.83\Lambda + 0.36$
$Co^{2+} \to Co$	-0.73	-0.28	$0.66\Lambda - 0.74$
$Sn^{2+} \to Sn$	-0.79	-0.136	$0.23\Lambda - 0.45$
$Cd^{2+} \to Cd$	-0.815	-0.402	$0.95\Lambda - 0.96$
$Fe^{2+} \to Fe$	-0.877	-0.440	$1.03\Lambda - 1.02$
$Cr^{3+} \to Cr$	-1.2	-0.74	$1.64\Lambda - 1.43$
$Zn^{2+} \to Zn$	-1.216	-0.762	$1.76\Lambda - 1.58$
$Ga^{3+} \to Ga$	-1.22	-0.52	$1.02\Lambda - 1.00$
$Mn^{2+} \to Mn$	-1.47	-1.05	$2.44\Lambda - 2.09$
$Be^{2+} \to Be$	-2.28	-1.70	$3.86\Lambda - 3.19$
$Zr^{4+} \to Zr$	-2.33	-1.43	$2.99\Lambda - 2.50$
$Al^{3+} \to Al$	-2.35	-1.706	$3.77\Lambda - 3.08$
$Hf^{4+} \to Hf$	-2.60	-1.68	$3.55\Lambda - 2.93$
$Th^{4+} \to Th$	-2.64	-1.80	$3.86\Lambda - 3.15$
$Mg^{2+} \to Mg$	-2.67	-2.375	$5.55\Lambda - 4.42$
$O^{2-} \to O2$	0.401	1.229	$-2.99\Lambda + 1.96$

[a] values of E^0_{alk} and E^0_{acid} in aqueous solution are from [3.82]
[b] from (3.50)

example, E^0_{melt} for the reduction $Ag^+ \to Ag$ is $(-1.61\Lambda + 0.81) - (-2.99\Lambda + 1.96) = 1.38\Lambda - 1.15$. Figure 3.13 shows the plot of E^0_{melt} for a selection of these couples versus optical basicity (in effect, E_{melt} for (3.51) coincides with the abscissa). When the E^0_{melt} values calculated by this means (by substituting the appropriate Λ value of the melt) are compared with experimental data from [3.52, 55], good agreement has been found [3.77]. The comparison therefore provides support for using Fig. 3.13 at least as an approximate guide, in spite of the bold steps taken in this treatment.

It is necessary to comment on E_{melt} for the O^{2-}/O_2 couple, (3.51), which is obtained from the E^0 values for aqueous solution (Table 3.6) where the half reactions are

$$O_2 + 2H_2O + 4e^- = 4OH^- \tag{3.52}$$

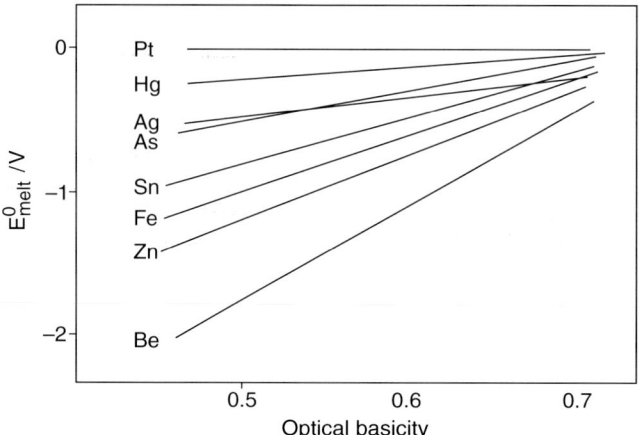

Fig. 3.13. Plot of E^0_{melt} (Table 3.6) for the couples designated versus optical basicity

and

$$O_2 + 4H^+ + 4e^- = 4H_2O \; . \tag{3.53}$$

Since O^{2-} ions are not present in these equations, it might be argued that the expression for E_{melt}, based on the E^0 data, does not correspond to the half reaction (3.51) for the melt (where oxide ions are involved). However, it must be remembered that (3.51) is schematic and that there are virtually no "raw" oxide ions present in the melt. In effect, the O^{2-} ions undergo a large degree of neutralisation by being incorporated into the silicate network as non-bridging oxygen atoms, see (3.1). The viewpoint of the optical basicity model is that there is severe polarization of the O^{2-} ion: by Si^{4+} cations in the silicate melt, and by H^+ ions in the H_2O molecule or OH^- ion. Thus, the application of the optical basicity concept provides not only the facility for comparing different oxidic media but the avoidance of such difficulties as those associated with handling oxide ion activities.

Soda-lime-silica glasses have optical basicities typically in the range approximately 0.55–0.65. For $\Lambda = 0.60$, Fig. 3.13 (or data in Table 3.6) indicates the electrochemical series Pt, Hg, Sb = Cu, Ag, Sn, Pb, Co, Ni, Cd, Fe, Tl, Zn. These metals are of interest to glass scientists from the point of view not only of corrosion, for example when used as container or electrode materials, but also with respect to the tendency for some of them to precipitate out from the glassy phase. The above series is rather different from that based on E^0 values in aqueous solution [3.78]. Since glass melts are regarded as basic media, the choice is for E^0 values in alkaline solution (Table 3.6) and this corresponds to the sequence Ag, Pt, Hg, Tl, Cu, Pb, Ni = Sb, Co, Sn, Cd, Fe, Zn.

When the expressions for E_{melt} in Table 3.6 are converted to express E^0_{melt} (see above), it is apparent that there is a trend towards convergence with increasing optical basicity. This can be seen in Fig. 3.13. At present there is no obvious explanation for this, but it might be significant that the optical basicity where the trends coincide is approximately 0.74, which is between that for Na_2SiO_3 and Na_4SiO_4 (Λ values of 0.70 and 0.82, respectively). The convergence therefore corresponds to the compositional range where the silicate network is fragmenting into single SiO_4^{4-} units. Under these conditions the metal ions are coordinated more and more by these units instead of being incorporated into the silicate network.

It is important to emphasize that Fig. 3.13 applies to silicate melts at 1400 °C with air atmosphere. This is because the relationships in Table 3.6 are based on experimental data for ion couples (Table 3.4, upper part) under these conditions. Since the equilibria involving these ion couples are temperature-dependent, it follows that the relationships for E_{melt} will differ from those in Table 3.6 for different temperatures.

3.4 Oxygen in Glass-Forming Melts

Friedrich G.K. Baucke

The intrinsic redox couple oxygen/"oxide", which is inseparably connected to oxidic glass-forming melts due to their composition, is involved in all redox reactions taking place, and in all redox equilibria attained, in such melts [3.4]. This involvement is demonstrated formally by the general equilibria

$$a\,M^{m+}(\text{melt}) + O^{2-}(\text{melt}) \rightleftarrows a\,M^{n+}(\text{melt}) + 1/2\,O_2(\text{melt}) \qquad (3.54)$$

and

$$2 \equiv SiO^-(\text{melt}) \rightleftarrows \equiv SiOSi \equiv (\text{melt}) + O^{2-}(\text{melt}) , \qquad (3.55)$$

where the ionic charges $m+$ and $n+$ and the number a of ions are connected by the relationship $m = n + 2/a$. In (3.54), either M is a solid or liquid metal when $n = 0$ with M^{m+} being the corresponding metal ion, or M^{n+} and M^{m+} are a polyvalent ion couple. The oxygen dissolved in the melt, (3.54), is nearly never in equilibrium with the gaseous oxygen of the atmosphere above the melt, i.e.,

$$O_2(\text{melt}) \rightleftarrows O_2(\text{gas}) , \qquad (3.56)$$

is virtually never attained during practical glass melting because of the high melt viscosities. Equations (3.54) and (3.55) demonstrate the significance of the intrinsic redox couple oxygen/"oxide" also for industrial glass melts, and this is demonstrated by the following examples.

M *is a solid metal* ($n = 0$). Platinum and platinum alloyed with other platinum metals are frequently applied as materials for melting tanks or parts of melting units because of their relative inertness. However, under the conditions of glass melting, these metals are chemically not as inert as one might think but undergo oxidation at high oxygen partial pressures, which results in dissolved platinum ions and intolerable colouration of the glasses produced (in particular optical glasses), serious economic losses, and a waste of platinum resources [3.83]. For example, there are large industrial glass melting units for which a content of as low as 1 ppm platinum of the produced glass would mean a loss of several tens of kilograms of platinum per year. These effects can be avoided by maintaining a sufficiently low oxygen partial pressures of the melt. If, however, ionic platinum has accidentally entered a glass melt, for example by oxidation of tank material, the oxygen partial pressure must be kept above a certain minimum value in order to avoid platinum reduction and recrystallization because metal crystallites are not tolerable in most glasses, for example, in optical and particularly in laser glasses, where platinum crystals even in the hundred micrometre range can lead to the damage of large glass blocks.

M *is a molten metal* ($n = 0$). Melts of lead-containing optical and silver-containing photochromic glasses [3.84] are not extremely stable with respect to redox conditions. The ions of these metals are reduced at rather moderate oxygen partial pressures. The reduction can even be brought about by pumping off the oxygen. The reduced metals precipitate and form a liquid phase at the bottom of the melting tank, which is called a "silver lake" in the case of silver-containing glasses. This not only changes the composition of the melt and renders the produced glasses useless due to connected property changes but also leads to the damage of melting tanks by the reducing, corroding, and alloying action of the molten metals upon the metal or refractory materials of the melters.

M^{n+}, M^{m+} *are polyvalent ions.* Even small concentrations of impurities often cause undesirable colouration, for instance of optical glasses [3.85]. In some cases, however, the transparency of the glass produced can be adjusted to its application by controlling the redox state of the melt. An example is iron, which is a frequent impurity of raw materials, as iron(II) absorbs at low and iron(III) at high energies of the spectrum [3.86]. This characteristic difference offers the possibility to eliminate effects of iron traces in the particular spectral range for which, for instance, optical glasses are produced by selecting appropriate oxygen partial pressures of the melts during and especially at the end of the melting process.

M^{n+}, M^{m+} *also are polyvalent ions.* However, different from the foregoing cases, certain steps of this example involve an exchange of oxygen between the melt and the atmosphere above the melt. The redox equilibria, (3.54) and (3.55), are employed as the basis of a technical process during glass melting and not in order to avoid or correct damage by faulty redox conditions. For this reason, the redox equilibrium involving, for instance, the polyvalent element antimony, of which small contents of the order of tenths of a percent are added to the base material, is applied to remove gaseous impurities such as nitrogen, nitrogen oxides, carbon dioxide, and water from the melt, which are unavoidably introduced by the raw materials and by reactions during the melting process [3.87]. (Arsenic, which has been used for the same reason in the past, has mostly been banished for environmental reasons). This so-called "redox fining" of glass-forming melts is based on the strong temperature-dependence of the redox equilibrium constant

$$K(T) = \frac{a_{Sb^{3+}}^2 \, p_{O_2}}{a_{Sb^{5+}}^2 \, a_{O^{2-}}^2} \ , \tag{3.57}$$

which is larger than unity, $(dK(T)/dT) > 1$, and thus generates high oxygen partial pressures of the melt at high temperatures. The result can be twofold. (1) If, as is often the case, the melt contains small bubbles, so-called "blisters", which are generated by the reactions of the raw materials in the melt, have radii of the order of tens of a millimetre, and are extremely difficult to remove because of their small rise velocity, the high oxygen partial pressure at high temperatures causes oxygen diffusion into these blisters, an increase of their volume and buoyancy, and thus their removal from the melt. An additional effect is the perturbation of the equilibrium between melt and blisters due to the dilution of the impurity gas by the oxygen, which leads to a diffusion also of the gaseous impurity from the melt into the blisters and thus to a faster removal of the gas. (2) If, in contrast, the melt contains dissolved impurity gases but neither blisters or bubbles, nor any phases that serve as bubble nuclei, the oxygen partial pressure is increased to above 1 bar by increasing the temperature; this leads to spontaneous formation of oxygen bubbles, into which the gaseous impurities diffuse. The resulting increase of bubble size and buoyancy also removes the impurities from the melt. The additional effect of dilution of the original gas of the bubbles is now reversed: the oxygen of the bubbles is diluted by the impurity gases, which causes further oxygen diffusion into the bubbles and an additional bubble growth. After these fining steps, the oxygen partial pressure of the melt is decreased by lowering the temperature (and shifting the equilibrium (3.54) to the left) so that remaining pure oxygen bubbles are eliminated by their redissolution in the melt. This "reversed" process is usually called the "resorption step" of glass fining.

As these examples demonstrate qualitatively, both components of the intrinsic redox couple of the oxidic glass melt, oxygen and "oxide", are intimately involved in all redox equilibria and their partial pressure and activity, respectively, should actually be known if the redox state of technical glass melting is to be controlled. Fortunately, however, most industrial glasses are produced with highly constant compositions and properties over long periods, often for years. In addition, the majority contains sufficiently small concentrations of polyvalent ions so that the "oxide activity" or basicity of the melts forms a relatively large reservoir and remains practically unchanged when the redox equilibria are shifted by changes of the oxygen partial pressure. This offers the possibility to control the redox state of production glass melts by merely one quantity, i.e., the oxygen partial pressure. It was thus decided to develop oxygen sensors on the basis of constant melt basicities, for which electrochemical cells are particularly suited because they yield on-line oxygen partial pressures with high, concentration-independent accuracies and can be applied at almost any location in glass melting tanks. They thus allow the measurement of the relative redox state of the melt at any phase of the melting process. The development and characteristics of these sensors are described in this section.

If required, for example for the production of melts with large amounts of various polyvalent ions for atomic waste glasses, the basicity must additionally be determined by separate experiments or by calculations, for instance, on the basis of the optical basicity concept. This is also often done with slags for steelmaking, where the Optical Basicity Databank for technological and scientific applications has been set up [3.70]. If, on the other hand, the parameters of a technical glass production are subject to frequent changes but, nevertheless, the fining process is to be controlled accurately, the activity (or concentration) ratio of the polyvalent ions must be known as a function of oxygen partial pressure and temperature. An example of such measurements on the basis of combined oxygen partial pressure and Mössbauer measurements [3.88, 89] will be given in Sect. 3.5.2, where also other methods, for instance square-wave voltammetry, will be described [3.55, 90–95].

3.4.1 Electrochemical Cells for On-Line Measurement of Oxygen Partial Pressures

The oxygen sensors developed are principally simple electrochemical cells without transference employing solid electrolytes, as introduced by *Wagner* [3.96, 97], tested with respect to their applicability in glass melts by *Besson* [3.98], and first applied in laboratory glass melts by *Plumat* [3.99]. They are represented by the isothermal cell scheme

$$\mathrm{Pt,O_2(r)|ZrO_2(Y_2O_3)|melt,O^{2-},O_2|Pt} \ . \tag{3.I}$$

The platinum measuring electrode is immersed in the melt, the platinum reference electrode denoted by (r) is maintained at a defined constant oxygen partial pressure and separated from the melt by an yttria-doped zirconia solid electrolyte, which has unit oxide transport number within the entire ranges of oxygen partial pressure and temperature of interest [3.100]. Partial stabilization of the ceramic by yttria with an approximate composition, $(\mathrm{ZrO_2})_{0.955}(\mathrm{Y_2O_3})_{0.045}$, was chosen because of its stability and relatively high corrosion resistance in oxidic glass melts [3.101]. Zirconias doped with twofold positive cations, containing, for instance, calcium or magnesium oxide [3.100], which are economically more attractive, were found to exchange their "stabilizing" ions readily for ions of contacting glass melts at the high temperatures and had to be excluded from the application.

In cell (3.I), the measuring electrode forming a separate unit can be applied in melts at any distance from the reference electrode and thus at any location in glass melting tanks as long as the melt around the two electrodes has the same composition. This distinguishes the cell from the arrangement as applied for oxygen measurements in gases [3.102], where direct contact of measuring electrode and solid electrolyte is unavoidable and the construction represents a combination or one-rod electrode. Such combination electrodes, which have the cell scheme

$$\mathrm{Pt,O_2|ZrO_2(Y_2O_3)|Pt|melt,O^{2-},O_2} \ , \tag{3.II}$$

were also tested in glass melts because they promised the development of highly corrosion-proof sensors. However, the tightness and the adherence of sufficiently thin, sputtered-on or vapour-deposited, oxygen-permeable platinum layers to the zirconia solid electrolyte were found to be insufficient under the conditions of glass melting, which excluded the further development of this electrode construction. This shows the high demands arising from an application in glass melts, compared to those of so-called Lambda Sensors for automobile engines [3.102], whose metal layers merely contact atmospheres, and of corresponding oxygen sensors as applied in molten steel and other metals [3.100, 102], where the measuring electrode is formed by the liquid metal whose oxygen content is to be measured [3.4, 103].

The cell reaction of both cells (3.I) and (3.II) is simply an electrochemical oxygen transport between melt and reference compartment, $\mathrm{O_2(m) \rightleftarrows O_2(r)}$, and the cell emf is given by

$$E(T) = \frac{RT}{4F} \ln \frac{p_{\mathrm{O_2}}(\mathrm{m})}{p_{\mathrm{O_2}}(\mathrm{r})} \ , \tag{3.58}$$

whose rearrangement yields the oxygen partial pressure of the melt,

$$p_{\mathrm{O_2}}(\mathrm{m}) = \exp\left(\frac{4FE}{RT} + \ln p_{\mathrm{O_2}}(\mathrm{r})\right) \ , \tag{3.59}$$

where (m) denotes melt, E is the measured isothermal emf, and F, R, and T have their usual meaning. Since the oxygen partial pressure of the melt is referred to the partial pressure $p_{\mathrm{O_2}}(\mathrm{r})$ of gaseous oxygen and thus to its standard state, the measured $p_{\mathrm{O_2}}(\mathrm{m})$ is the oxygen partial pressure in equilibrium with the activity of dissolved oxygen, independent of whether or not the gas phase is actually present [3.104]. If required, the oxygen activity of the glass melt can be obtained by an additional measurement of the solubility coefficient of oxygen, which, however, is rarely of concern in industrial glass melting.

3.4.2 Measuring Electrodes

Platinum is used as the metal of the measuring electrodes because of its relatively high chemical stability. The properties of glass melts, however, impose certain conditions on the metal and necessitate special constructions of the electrodes if accurate results are to be obtained [3.4, 104, 105]. Thus, the electrodes must consist of pure platinum although platinum alloyed with, for instance, rhodium or iridium is usually preferred in glass melting because of their higher mechanical stability. Owing to their less noble character, these metals are oxidized more readily than platinum, also in the alloyed state. The oxidation for instance of rhodium leads to the formation of rhodium oxide at the surface or to rhodium loss from surface ranges of the alloy, formation of mixed potentials, and to irreproducible and inconstant, faulty potentials. Figure 3.14 demonstrates this effect by depicting simultaneously measured, temperature-dependent emfs of cells employing platinum electrodes alloyed with 1% iridium and 5% and 10% rhodium in a sodium calcium silicate glass melt after different periods of contact between metals and melt. Pure platinum electrodes served as reference electrodes. The reactions between the alloys and the melt are not known in detail, and it seems that also diffusion of rhodium and iridium from the bulk alloys to the surface is involved. At any rate, the measurements demonstrate that platinum alloys are useless metals for the construction of electrodes which are to contact glass melts.

A further point of concern is the local extension of platinum measuring electrodes in glass melts. Because they are usually introduced into the glass melt through the melt surface, which is in contact with the atmosphere above, they contact melt volumes with different oxygen partial pressures be-

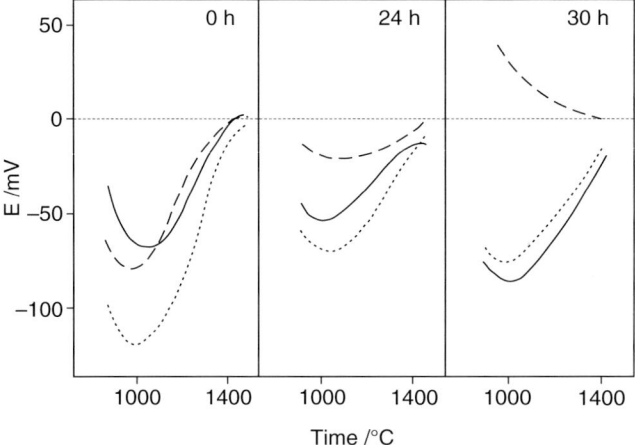

Fig. 3.14. Time- and temperature-dependent isothermal emfs of the cell Pt/Na,Ca silicate melt/Pt,M with M = 1% iridium (dashed curves), 5% rhodium (solid curves), and 10% rhodium (dotted curves)

cause suitable insulation materials to separate the upper electrode part from the melt are not available. Consequently, platinum electrodes in glass melts form complicated, short-circuited oxygen concentration cells that give rise to changes of the oxygen partial pressure along the metal (e.g., the platinum shaft), and, even more serious, are subject to the formation of mixed potentials, which can differ considerably from the equilibrium potentials to be measured. Although these deficiencies cannot completely be eliminated, they can be reduced to a negligible level by an electrode design which is based on electrode-kinetic considerations. Figure 3.15 demonstrates the principle by depicting a platinum electrode with two parts, an upper and a lower one, with different surface areas, A_1 and A_2, respectively, which contact melt volumes with respective oxygen partial pressures $p_{O_2,1}$ and $p_{O_2,2}$, the remaining centre part of the electrode being assumed inert for convenience. Because the absolute currents through the interfaces between the active electrode parts and the melt are equal, $I_1 = I_2$, the current densities are inversely proportional to the corresponding surface areas, $|i_1| / |i_2| = A_2/A_1$, i.e., the larger the surface area is, the smaller is the current density through that surface. The schematic current density-polarization curve in Fig. 3.15 shows that the relative current densities determine the relative polarization, i.e., the absolute difference between mixed and equilibrium potential of the electrode parts. The mixed potential $\varepsilon_{\mathrm{mixed}}$ of the platinum electrode is closer to the equilibrium potential ε_2^0 of the larger lower electrode part than to that of the smaller upper one. Thus, in order to measure a mixed potential that represents approximately the oxygen partial pressure of the melt, the electrode part in the bulk of the melt must have a larger surface area than the perturbing part near the melt surface, as indicated by the model electrode in Fig. 3.15. It was found that a ratio of approximately $A_2/A_1 = 25$ generally suffices to neglect the effect of the short circuit and to make $\varepsilon_{\mathrm{mixed}} \approx \varepsilon_2^0$ within a millivolt. The design also reduces the specific rates of oxygen formation and degeneration at the large lower electrode surface and minimizes corresponding errors, which, however,

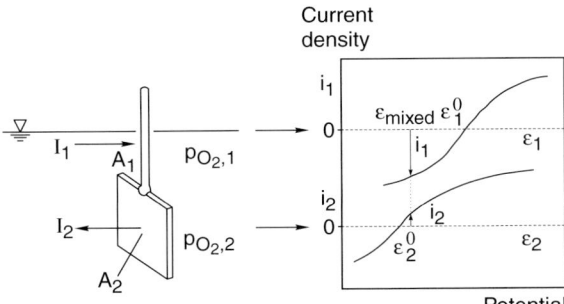

Fig. 3.15. Principle of reducing errors of the potential of platinum measuring electrodes caused by the contact of short-circuited electrode parts with melt volumes having different oxygen partial pressures, here $p_{O_2,1} > p_{O_2,2}$

are mostly eliminated by convection if the electrode is applied in streaming melts.

Depending on the surface tension of glass melts, bubbles tend to stick to platinum surfaces more or less easily. If the oxygen partial pressure of such adhering bubbles is not in equilibrium with the surrounding melt, as is often the case, this contact also causes mixed potentials resulting in faulty measurements. This deficiency cannot basically be excluded, but its probability is drastically reduced by installing platinum electrodes in glass melting tanks in a vertical position and by choosing locations where the melt contains as few bubbles as possible, for instance by applying constructions that reduce the contact of electrodes with bubbles in streaming melts by hydrodynamic means. Besides, experienced personnel often knows when bubbles are caught by an electrode because the time-dependent emf usually indicates these events by sudden steps, which are particularly indicative when two platinum electrodes are applied in parallel in the same location of the melting unit.

3.4.3 Reference Electrodes

The reference electrode of the oxygen sensors consists of a platinum electrode in a reference gas with defined oxygen partial pressure, which is separated from the melt by a doped-zirconia solid electrolyte with unit oxide transport number. Practical units applied in laboratory and industrial glass melts consist of a combination of the essential reference electrode and the zirconia electrolyte for easy handling and are called "reference electrodes", "zirconia reference electrodes", or briefly "zirconia electrodes" . The latter term can certainly be used as long as the underlying basis is clear to the researcher or user of these electrodes, and thus this terminology will also be used in this text. It is emphasized, however, that this is not merely a matter of semantics but concerns also the potentials of these "zirconia electrodes". Thus, the "potential" of a zirconia electrode, that is, the Galvani voltage "ε_{ZrO_2}" between the platinum electrode and the glass melt is actually the sum of two Galvani voltages, i.e., the potential difference between the platinum electrode and the zirconia and that between the zirconia and the glass melt, as seen by

$$\text{``}\varepsilon_{ZrO_2}\text{''} = (\varphi_{Pt} - \varphi_{ZrO_2}) + (\varphi_{ZrO_2} - \varphi_m) = (\varphi_{Pt} - \varphi_m) , \qquad (3.60)$$

which, under isothermal conditions, reduces to the potential difference between platinum and melt because of the uniform potential of the solid electrolyte [3.4]. However, it does not if the zirconia is subject to temperature gradients. In addition, (3.60) shows that a zirconia electrode can be looked at as a platinum electrode in a melt whose oxygen partial pressure can momentarily and arbitrarily be changed by outside means.

Several types of zirconia electrodes have been developed [3.4, 103–108]. The simplest form is the zirconia tube electrode, whose prototype was already tested by *Besson* [3.98] and applied in laboratory glass melts by *Plumat*

[3.99]. Figure 3.16 presents a cross section. Inside a closed-end zirconia tube, a four-bore alumina tube insulates the leads of the platinum reference electrode and of a thermocouple and supplies the reference gas to the lower section at the contact of platinum electrode and zirconia wall, which is simply and efficiently provided by zirconia grit [3.109]. The gas leaves the lower end of the unit through the space between the two concentric tubes. An electrode head made of stainless steel provides mechanical stability of the arrangement and tightness of the electrode interior. The electrochemically active part of zirconia tube electrodes is restricted to the lower end of the tube which is in contact with the platinum electrode via the zirconia grit. They are thus strictly isothermal and yield the most accurate results. However, they are not well suited for technical applications because of their sensitivity to thermal shock and the limited lifetime due to the corrosion of the thin zirconia wall. A similar arrangement consists of a zirconia disk sintered to the lower end of a platinum tube, which contains the same internal parts as the zirconia tube electrode [3.110]. This arrangement is extremely insensitive to thermal shock but suffers also from short lifetimes due to the thin zirconia tablet. In addition, the horizontal zirconia slab is particularly susceptible to a phenomenon called "bubble drilling", which is often observed when gas bubbles stick to a horizontal lower surface of a ceramic material [3.107]. Such bubbles can drill holes with their own diameter through the ceramic, sometimes through cm-thick plates within periods of days, depending on the melt composition. The cause of this phenomenon is not clear. Zirconia parts of reference electrodes must therefore be positioned as vertically as possible, or should be protected

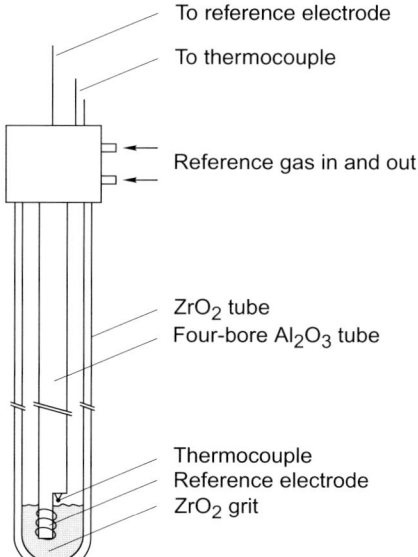

To reference electrode

To thermocouple

Reference gas in and out

ZrO_2 tube
Four-bore Al_2O_3 tube

Thermocouple
Reference electrode
ZrO_2 grit

Fig. 3.16. Cross section of a zirconia tube electrode

from contact with gas bubbles by hydrodynamic means like those applied to platinum measuring electrodes.

The dissolving zirconia electrode, schematically shown in Fig. 3.17, was specifically designed for technical applications [3.4, 105, 111, 112]. Their lifetime is considerably extended, compared to zirconia tube electrodes, by an up to 12 cm long zirconia electrolyte bridge arranged between the reference electrode and the melt surface and providing sufficient ionic contact at the high melting temperatures. The rod-shaped zirconia bridge is inserted into the lower end of an alumina electrode shaft and only secured by an alumina bolt because the slightly different coefficients of expansion of the two ceramic materials exclude sintering and application of ceramic cements. The reference gas must thus be kept at a slight overpressure in order to protect the reference compartment from contamination by gaseous impurities. The electrode is continuously or stepwise lowered at a rate given by the dissolution of the lower end of the zirconia bridge due to the corrosion by the glass melt. The arrangement is particularly suited for streaming melts because they remove traces of dissolved zirconia and thus eliminate errors caused by diffusion potentials, which can result from the slight changes of the melt compositions.

Because, however, the temperatures of glass melt and reference electrode in the alumina shaft generally differ considerably, measuring cells employing dissolving zirconia electrodes are non-isothermal electrochemical cells according to the cell scheme [3.4, 105]

$$\text{Pt,O}_2(\text{r})(T_\text{r})|\text{ZrO}_2|(T_\text{m})\text{melt,O}^{2-}, \text{O}_2(T_\text{m})|\text{Pt} . \tag{3.III}$$

The standard thermoelectric emf, $E_{\text{Th,ZrO}_2}(T_\text{r}, T_\text{m})$, along the zirconia bridge between the reference and melt temperatures, T_r and T_m, respectively, must be known for correcting the measured thermoelectric emf, $E_{\Delta T}(T_\text{r}, T_\text{m})$, in order to obtain the oxygen partial pressure of the melt,

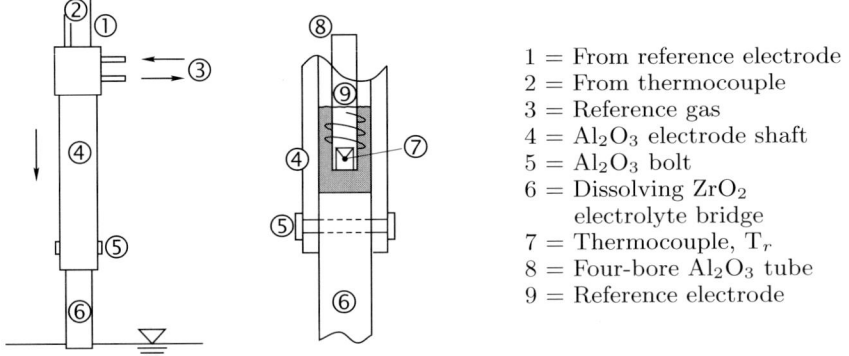

1 = From reference electrode
2 = From thermocouple
3 = Reference gas
4 = Al_2O_3 electrode shaft
5 = Al_2O_3 bolt
6 = Dissolving ZrO_2
    electrolyte bridge
7 = Thermocouple, T_r
8 = Four-bore Al_2O_3 tube
9 = Reference electrode

Fig. 3.17. Scheme and cross section of a dissolving zirconia electrode

$$p_{O_2,m}(T_m) = \exp\left\{\frac{4F}{RT_m}\left[E_{\Delta T}(T_r, T_m)\right.\right.$$
$$\left.\left. - E_{Th,ZrO_2}(T_r, T_m)\right] + \frac{T_r}{T_m}\ln p_{O_2,r}\right\}, \tag{3.61}$$

which is assumed isothermal in this equation.

Thermoelectric powers of yttria-doped zirconia, which are needed for the evaluation of (3.61), had been measured before [3.113, 114]. However, the standard thermoelectric emfs applied to (3.61) must be known with high accuracy in order to obtain sufficiently accurate oxygen partial pressures, which could not be guaranteed by the available data. Besides, the exact magnitude of the thermoelectric power depends on the yttria concentration [3.113–115] and on the properties of the individual zirconia charges, such as impurities and sintering conditions [3.113]. The measurement of thermoelectric data of the applied zirconias was thus unavoidable, and it was necessary to devise a method for measuring the exact value of each zirconia material to be applied as electrode material [3.116]. As indicated in Fig. 3.18, the arrangement consisted of two zirconia electrodes with 1 bar oxygen partial pressure, which were pressed against the ends of the specimen under investigation. Variable temperature profiles were supplied by a specially designed gradient furnace, which also provided either nearly constant temperatures or nearly equal temperature gradients to the ends of the furnace tube containing the zirconia electrodes in order to eliminate errors due to the material of the test electrodes. Also, the temperature gradients could be reversed in order to eliminate any asymmetries [3.116].

The measuring cell is represented by the cell scheme

$$Pt,O_2(T_1)|ZrO_2|(T_2)Pt,O_2 \tag{3.IV}$$

and the thermoelectric emf is given by [3.4, 105],

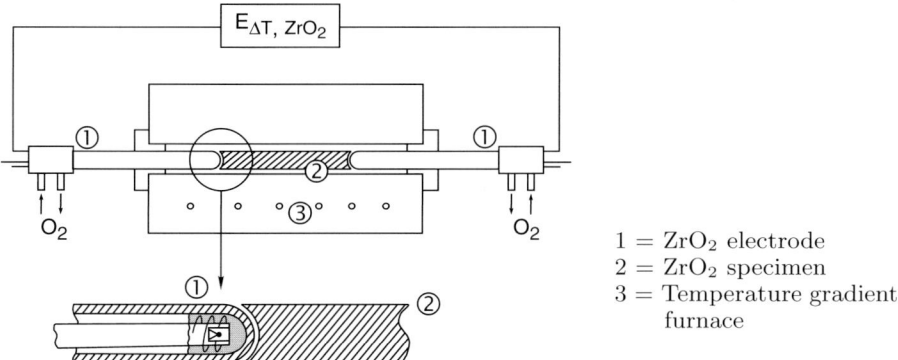

1 = ZrO$_2$ electrode
2 = ZrO$_2$ specimen
3 = Temperature gradient
     furnace

Fig. 3.18. Scheme of an arrangement for measuring thermoelectric emfs of rod-shaped zirconia specimens

$$
\begin{aligned}
E_{\Delta T, \mathrm{ZrO_2}} = {} & \frac{RT_2}{4F} \ln p_{\mathrm{O_2}}(T_2) - \frac{RT_1}{4F} \ln p_{\mathrm{O_2}}(T_1) \\
& + \left[\varepsilon_{\mathrm{ZrO_2}}^0(T_2) - \varepsilon_{\mathrm{ZrO_2}}^0(T_1) + \frac{RT_1}{2F} \ln a_{\mathrm{O^{2-}},\mathrm{ZrO_2}}(T_1) \right. \\
& \left. - \frac{RT_2}{2F} \ln a_{\mathrm{O^{2-}},\mathrm{ZrO_2}}(T_2) + \varepsilon_{\mathrm{ZrO_2}}(T_1,T_2) + \varepsilon_{\mathrm{Pt}}(T_2,T_1) \right]
\end{aligned}
\tag{3.62}
$$

(see also the detailed derivation of a corresponding equation for glass melts in Sect. 3.6) or by

$$
E_{\Delta T, \mathrm{ZrO_2}} = \frac{RT_2}{4F} \ln p_{\mathrm{O_2}}(T_2) - \frac{RT_1}{4F} \ln p_{\mathrm{O_2}}(T_1) + E_{\mathrm{Th},\mathrm{ZrO_2}}(T_1, T_2) \tag{3.63}
$$

if the square bracket on the right side of (3.62) containing only unattainable quantities is expressed by $E_{\Delta T, \mathrm{ZrO_2}}(T_1, T_2)$, which is called the standard thermoelectric emf because it is directly measured under the standard condition of 1 bar oxygen partial pressure in both zirconia electrodes [3.4, 105, 116],

$$
E_{\Delta T, \mathrm{ZrO_2}}(T_1, T_2) = E_{\mathrm{Th},\mathrm{ZrO_2}}(T_1, T_2) . \tag{3.64}
$$

With its temperature coefficient, $\mathrm{d}E_{\mathrm{Th},\mathrm{ZrO_2}}(T)/\mathrm{d}T$, defined standard Seebeck coefficient [3.4], the standard thermoelectric emfs necessary for the evaluation of (3.61) can be obtained by integration.

$$
E_{\mathrm{Th},\mathrm{ZrO_2}}(T_1, T_2) = \int_{T_1}^{T_2} \frac{\mathrm{d}E_{\mathrm{Th},\mathrm{ZrO_2}}(T)}{\mathrm{d}T} \mathrm{d}T . \tag{3.65}
$$

In agreement with the literature, standard Seebeck coefficients of yttria-doped zirconias were found to be independent of temperature [3.113–115] and proportional to the molar yttria concentration of the ceramic [3.114, 115]. For the materials investigated, they can be represented by the linear relationship

$$
\frac{\mathrm{d}E_{\mathrm{Th},\mathrm{ZrO_2}}(T)}{\mathrm{d}T} = a\, c_{\mathrm{Y_2O_3}} + b , \tag{3.66}
$$

where the constants are $a = 4.643 \times 10^{-3}\,\mathrm{mV/(K\,mol\%)}$ and $b = -0.4949$ $\mathrm{mV\,K^{-1}}$ [3.116]. For example, the standard Seebeck coefficient of the most frequently applied zirconia, $(\mathrm{ZrO_2})_{0.9547}(\mathrm{Y_2O_3})_{0.0453}$, is (-0.4739 ± 0.0015) $\mathrm{mV\,K^{-1}}$ within the temperature range 700–1550 °C.

Four modifications of dissolving zirconia electrodes were developed.

(1) The "corrosion-protected zirconia electrode" contains a platinum sleeve around the alumina shaft, which protects the ceramic from condensing vapours of melt components with high vapour pressures [3.117]. An additional gutter at the lower end of the sleeve ensures also that the condensed compounds do not contact or trickle down the zirconia bridge.

(2) The "isothermal zirconia electrode" involves a heating device which keeps the temperature of the reference electrode equal to that of the measuring electrode by means of an additional electronic control unit and thus eliminates thermoelectric voltages in the first place [3.118].

(3) The "contamination-protected zirconia electrode" incorporates an additional closed-end zirconia tube inside the alumina shaft, which is ionically contacted by the zirconia electrolyte bridge through zirconia grit and contains the platinum reference electrode, which is thus protected from melt vapours when the overpressure of the reference gas accidentally drops to the pressure of the tank atmosphere [3.119].

(4) Various designs of the "hook-shaped zirconia electrode" with horizontal electrode shaft and perpendicular zirconia electrolyte bridge allow the sensors to be applied through the side wall of glass melting tanks. "Top application" through the melt surface becomes unnecessary, and the introduction and removal of such reference electrodes is made considerably more practicable [3.120].

All types of zirconia electrodes can be checked for correct functioning by defined changes of the oxygen reference partial pressure during their application because such changes result in corresponding changes of the emfs or thermoelectric emfs only if the electrodes are unimpaired [3.105]. Due to the high response rates of the oxygen sensors at the melt temperatures, the check can even be applied while the measured oxygen partial pressure and thus the measured emf are slowly changing with time.

3.4.4 Verification of the Thermodynamically Correct Response of Oxygen Sensors

The described oxygen sensors have been developed for measuring oxygen partial pressures of oxidic glass melts with constant basicity and are to characterize the redox state of such melts independent of any other redox couples present. Their reversible response to the intrinsic redox couple oxygen/"oxide" of these melts also in the presence of polyvalent ions is thus essential and had necessarily to be verified before their application for laboratory and technical purposes. However, because glass-forming melts, as all liquids, are open systems with respect to dissolved gases and melts with defined oxygen contents and partial pressures, which could serve as a series of "standard melts", are not available, this had to be done in a different way, for which a thermodynamic method was chosen [3.4, 105].

For this reason, the oxygen sensors were applied in several glass melts containing up to 3 wt% of the polyvalent ions antimonic and antimonous or arsenic and arsenous. The respective relative concentrations of the ions were fixed by adjusting the redox state of the glass melts by "electrochemical pumping", i.e., by electrolysing oxygen into or out of the melt. As sketched in Fig. 3.19, this was done by means of a cylindrical rotating platinum electrode

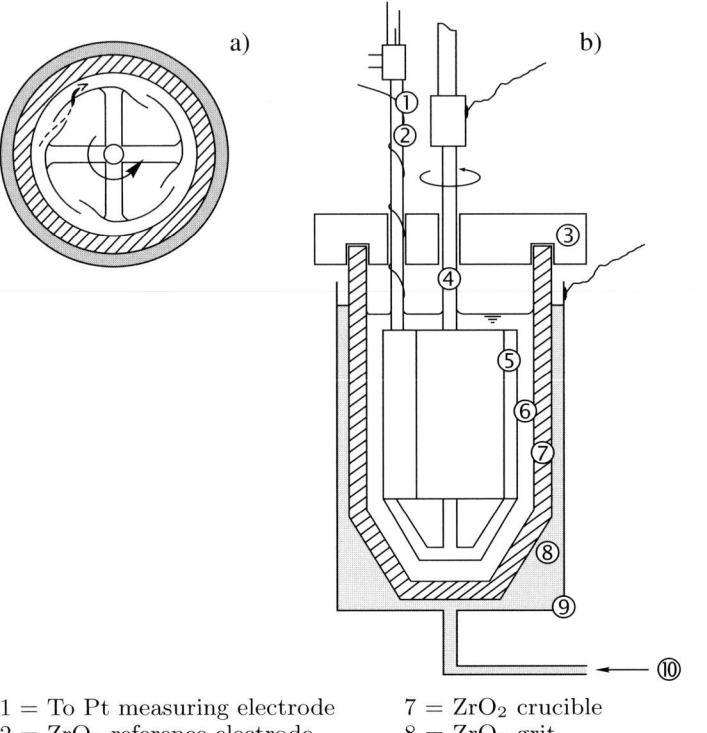

1 = To Pt measuring electrode
2 = ZrO$_2$ reference electrode
3 = Ceramic lid
4 = Axle
5 = Pt "stirring electrode"
6 = Glass melt

7 = ZrO$_2$ crucible
8 = ZrO$_2$ grit
9 = Pt crucible: counter-electrode
       and induction heating
10 = Gas supply (O$_2$ or Ar)

Fig. 3.19. Sketch of the arrangement for adjusting the redox state of a glass melt by electrochemical pumping of oxygen, (**a**) horizontal, (**b**) vertical cross section

with a large surface area which sheared and simultaneously stirred the melt ("stirring electrode"), thus keeping oxygen concentration gradients at the electrode small and suppressing oxygen bubble formation also at relatively high current densities during the electrolyses [3.88, 89]. A zirconia crucible externally contacted by the crucible-shaped platinum counter-electrode via zirconia grit contained the melt and served as an oxide ion-conducting wall between melt and external electrode [3.88]. The oxygen partial pressure of the melt was continuously monitored during the electrolyses by means of the oxygen sensors, which were subsequently used for the purpose of the main experiments.

The thermodynamic basis is derived in the following way [3.4]. The isothermal measuring cell containing the melt with, for instance, the polyvalent ions Sb(III) and Sb(V) is represented by the cell scheme

$$Pt, O_2(r)|ZrO_2|melt, Sb_2O_3(m), Sb_2O_5(m)|Pt \tag{3.V}$$

and is characterized by a cell reaction consisting of a redox process between the dissolved polyvalent ions in the melt and oxygen in the reference compartment:

$$Sb_2O_5(m) \rightleftarrows Sb_2O_3(m) + O_2(r) . \tag{3.67}$$

The entropy of the cell reaction is given by

$$\Delta S_{\text{cell}} = S^0_{\text{Sb}_2\text{O}_3,\text{m}} - S^0_{\text{Sb}_2\text{O}_5,\text{m}} + S^0_{\text{O}_2,\text{r}} - R \ln \frac{a_{\text{Sb}_2\text{O}_3,\text{m}}}{a_{\text{Sb}_2\text{O}_5,\text{m}}} - R \ln p_{\text{O}_2,\text{r}}$$
$$- RT \frac{\mathrm{d} \ln (a_{\text{Sb}_2\text{O}_3,\text{m}}/a_{\text{Sb}_2\text{O}_5,\text{m}})}{\mathrm{d}T} - RT \frac{\mathrm{d} \ln p_{\text{O}_2,\text{r}}}{\mathrm{d}T} , \tag{3.68}$$

where S^0 and a are standard entropies and the activities, respectively, of the species indicated. Because the reference oxygen partial pressure is independent of temperature and the activity ratio of the polyvalent ions can be assumed approximately equal to the ratio of their concentrations, which is made unity by electrochemical pumping, (3.68) reduces to

$$\Delta S_{\text{cell}} = \left(S^0_{\text{Sb}_2\text{O}_3,\text{m}} - S^0_{\text{Sb}_2\text{O}_5,\text{m}} \right) + S^0_{\text{O}_2,\text{r}} - R \ln p_{\text{O}_2,\text{r}} . \tag{3.69}$$

In addition, the (absolute) difference of the standard entropies of the redox species in (3.69) [3.121] is considerably smaller than the standard entropy of oxygen [3.122]

$$\left| S^0_{\text{Sb}_2\text{O}_3,\text{m}} - S^0_{\text{Sb}_2\text{O}_5,\text{m}} \right| < S^0_{\text{O}_2,\text{r}} , \tag{3.70}$$

so that (3.69) reduces further to

$$\Delta S_{\text{cell}} \cong S^0_{\text{O}_2,\text{r}} - R \ln p_{\text{O}_2,\text{r}} , \tag{3.71}$$

which yields the temperature dependence of the emf of cell (3.V),

$$\frac{\mathrm{d}E_{\text{cell}}}{\mathrm{d}T} = \frac{\Delta S_{\text{cell}}}{4F} \cong \frac{1}{4F} \left(S^0_{\text{O}_2,\text{r}} - R \ln p_{\text{O}_2,\text{r}} \right) \tag{3.72}$$

and its dependence on the logarithm of the reference partial pressure of oxygen

$$\frac{\mathrm{d} \left(\mathrm{d}E_{\text{cell}}/\mathrm{d}T \right)}{\mathrm{d} \log p_{\text{O}_2,\text{r}}} = -\frac{2.303 R}{4F} = -0.0499 \, \mathrm{mV \, K^{-1}} . \tag{3.73}$$

Because the standard entropy of oxygen is constant within several percent within the temperature range of interest, i.e., between $800 \, °C$ and $1500 \, °C$ [3.122], the temperature coefficient of the cell emf, (3.72), is expected to be approximately temperature-independent and equal to $0.59 \, \mathrm{mV \, K^{-1}}$ at $1000 \, °C$, $0.60 \, \mathrm{mV \, K^{-1}}$ at $1250 \, °C$, and $0.61 \, \mathrm{mV \, K^{-1}}$ at $1500 \, °C$ if the oxygen partial pressure of the reference electrode is kept at 1 bar. This expectation was met within a few percent by many different model and industrial glass melts containing various contents of antimony(III,V) and arsenic(III,V) present in respective equal concentrations as required for the derivation of (3.72). This is demonstrated by the example in Fig. 3.20a for a borate glass melt containing antimony(III,V) and a temperature change rate $dT/dt = 450\,\mathrm{K\,h^{-1}}$, which yielded an approximately linear temperature dependence with an average value $dE/dT = 0.62\,\mathrm{mV\,K^{-1}}$ of the emf between at least $800\,^\circ\mathrm{C}$ and $1200\,^\circ\mathrm{C}$. In addition, Fig. 3.20b shows for a flint glass with antimony(III,V) and the temperature range 700–$1400\,^\circ\mathrm{C}$ that also (3.73) is experimentally verified as, for instance, a change of the reference oxygen partial pressure between 0.01 bar and 1 bar and back resulted in a change of the temperature dependence of the emf of $-0.05\,\mathrm{mV\,K^{-1}}$ per decade oxygen partial pressure. Although conducted in the closed cell containing the electrochemical oxygen

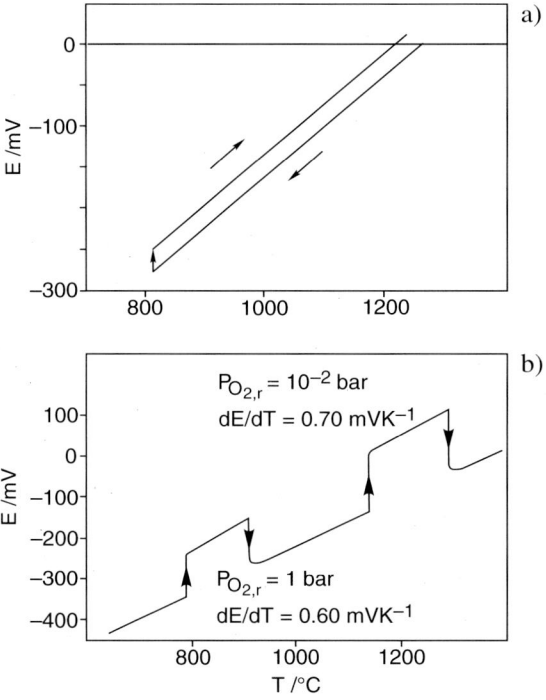

Fig. 3.20. Verification of the thermodynamically correct response of the developed oxygen sensors. (**a**) The nearly linear dependence of the emf of cell (3.I) on temperature with an average slope of $0.62\,\mathrm{mV\,K^{-1}}$ agrees with the theoretical (3.72). (The emf increase at $800\,^\circ\mathrm{C}$ was caused by the internal short circuit of the platinum electrode during a ten hour interruption of the measurements.) (**b**) Changes of the reference oxygen partial pressure during a temperature run result in the theoretical change $d(dE/dT)/d\log p_{O_2} = -0.05\,\mathrm{mV\,K^{-1}}$ according to (3.73)

pumping device, the experiments were carried out dynamically at temperature change rates which were small enough to guarantee isothermal conditions throughout the cell ("quasi-isothermal measurements") but sufficiently large to ensure negligible oxygen exchange of the melt with the atmosphere in the device ("quasi-closed system") [3.4, 105]. Because, in addition, the oxygen concentrations of the melts were considerably smaller than the various equilibrium concentrations of the polyvalent ions, $c_{O_2,m} \ll (c_{Sb_2O_3,m} + c_{Sb_2O_5,m})$, it must be concluded that the concentrations of the polyvalent ions were essentially constant during the temperature changes and that the oxygen sensors, consequently, respond reversibly to the intrinsic redox couple oxygen/"oxide" of the melts and not to the added redox ion couples. At first sight, this result is surprising because of the large concentrations of polyvalent ions, which should generate much larger exchange current densities at the platinum electrodes than the small oxygen concentrations. The reason for this unexpected behaviour might be found in the overwhelming concentration of "oxides", which comprises all charged oxygen species of the glass melt and whose sum represents the "electron donor power" of the melt. Obviously, the high total concentration causes exchange current densities that are larger than those of the polyvalent ions despite the much smaller concentration of uncharged oxygen. Further systematic investigation of this problem will certainly be rewarding, as it may open up the field of electrode kinetics in glass melts, which has been almost completely neglected in the past [3.123].

3.4.5 Alternative Reference Electrodes

Although dissolving zirconia reference electrodes provide the maximum lifetime possible for zirconia electrodes and offer a relatively high mechanical stability because of the rod-shaped zirconia electrolyte bridge and the alumina electrode shaft, the remaining possibility of breakage of the ceramic, in particular at the rough service in industrial melters, the corrosion of zirconia in melts with, for instance, high lead contents, and the nearly exclusive application of the sensors as top electrodes, i.e., through the melt surface, made an alternative reference electrode without these disadvantages highly desirable. Such electrodes are represented by metal/metal oxide electrodes, whose satisfactory functioning, for instance, in aqueous solutions is well known [3.124]. They were developed for application in glass melts on the basis of an early observation during measurements on industrial melters. It was then found that molybdenum electrodes, which had originally been introduced into the tanks as heating electrodes, but were tried out as "electrical reference" instead, provided extremely constant (albeit at that time still unidentified) potentials for electrochemical measurements. Today, after detailed work, the reason for the stability of their potential seems to be clear [3.125].

The molybdenum rods in those melting tanks consisted actually of the metal with a thin, tight layer of one of their oxides, which were in equilibrium with each other when they were applied for reference purposes. Such electrodes function on the basis of their oxide formation,

$$xM + yO \rightleftarrows M_xO_y , \tag{3.74}$$

and are thus represented by the scheme $M|M_xO_y$. If the oxide forms a tight layer at the metal surface, which, after some time, assumes a steady state with a relatively small formation and dissolution rate, (3.74) represents approximately an equilibrium as indicated, whose equilibrium constant can be written as

$$K(T) = \frac{a_{M_xO_y}}{(a_M)^x(a_O)^y} , \tag{3.75}$$

where a is the activity of the species indicated and a_O is the oxygen activity at the metal/metal oxide phase boundary. Because both metal and metal oxide are pure condensed phases and their activities are constant and defined to be unity, the oxygen activity at the phase boundary is fixed by the temperature according to the phase law or, expressed in a different way, its temperature dependence is given by that of the equilibrium constant, $a_O(T) = f(K(T))$, which can be calculated according to

$$\frac{d \ln p_{O_2(M/M_xO_y)}}{d(1/T)} = -\frac{2\Delta H^0}{yR} , \tag{3.76}$$

where ΔH^0 is the standard energy of formation of the pure oxide, (3.74). Melts contacting the oxide can lead to several disturbing effects, for example, chemical reactions with the oxide, formation of additional solid phases, and alloying of the metal with reduced metals from the melt. It is thus unavoidable for practical quantitative applications of the electrodes to conduct measurements which either confirm the absence of disturbing reactions or serve as standardizing measurements and yield the "practical $\Delta H^{0'}$" for the particular glass melt studied.

Several requirements must be met for a satisfactory functioning of metal/metal oxide electrodes. The metal must have a sufficiently high melting point and must not be too inert to allow the oxide to be formed at a sufficient rate, yet inert enough not to form the oxide at too high a rate so that the lifetime of the electrodes is long enough in the particular melt. The oxide also must have a high melting point, and it must be extremely sparingly soluble in the melt so that, instead of being dissolved on its formation, it forms a steady-state surface layer with a thickness in the range of from several atomic dimensions to several 10 nm. For this reason, it must be tight to ensure that the reaction, see (3.74), takes place entirely at the phase boundary between the metal and the oxide after the oxygen has diffused through the layer. In addition, it should be a pure oxide ion or a pure cation conductor for an ideal functioning of the electrode, although the effect of mixed conduction can be taken into account by measurements in the steady state of the electrode [3.125].

The main advantages of $M|M_xO_y$ electrodes are the absence of ceramic parts so that high corrosion rates and breakage are excluded, high thermal stability, and the possibility of their application, besides as top electrodes, through the wall or the bottom of technical melters, where they can easily be introduced also during the melting process.

A minor disadvantage is the above-mentioned necessity to standardize the electrodes in each envisaged melt before their application. This is done by means of three electrodes, a platinum, a zirconia, and the metal/metal oxide electrode, which form three interconnected electrochemical cells in the particular melt.

(1) The "standardizing cell" containing the zirconia as the reference electrode, cell scheme

$$Pt,O_2(r)|ZrO_2|melt,O_2|Pt , \tag{3.VI}$$

and exhibiting the emf

$$E_1(T) = \varepsilon_{Pt}(T) - \varepsilon_{ZrO_2}(T) , \tag{3.77}$$

yields the temperature-dependent oxygen partial pressure of the melt according to (3.58).

(2) The "reference cell" according to the cell scheme

$$Pt,O_2(r)|ZrO_2|melt, O_2|M_xO_y|M(Pt) \tag{3.VII}$$

refers the potential of the metal/metal oxide electrode to that of the zirconia electrode,

$$E_2(T) = \varepsilon_{M|M_xO_y}(T) - \varepsilon_{ZrO_2}(T) , \tag{3.78}$$

which thus serves as the ultimate reference electrode.

(3) The "measuring cell" containing the metal/metal oxide as the reference electrode, cell scheme

$$M|M_xO_y|melt, O_2|Pt , \tag{3.VIII}$$

is the cell actually to be applied during the practical measurements envisaged and thus yields the emf of interest,

$$E_3(T) = \varepsilon_{Pt}(T) - \varepsilon_{M|M_xO_y}(T) . \tag{3.79}$$

Combination of (3.77)–(3.79) gives

$$E_3(T) = E_1(T) - E_2(T) , \tag{3.80}$$

which shows that either the directly measured emf $E_3(T)$ of the "measuring cell" and the oxygen partial pressure $p_{O_2}(T)$ obtained from $E_1(T)$, or the measured emf $E_2(T)$ of the "reference cell", the measured emf $E_1(T)$

of the "standardizing cell", and the oxygen partial pressure $p_{O_2}(T)$ (from $E_1(T)$) yield the isothermal standardization curve $p_{O_2}(T_m) = f(E_3(T_m))$ at any measuring temperature T_m, which is required for the evaluation of the measurements. For application in non-isothermal melts, the temperatures of measuring and reference electrode and the possibly temperature-dependent Seebeck coefficients of the glass melts must be known and taken into account, see Sect. 3.6.

Several metal/metal oxide electrodes were tested in various glass melts, and it was found that particularly the electrodes $Mo|MoO_2$, $W|WO_2$, and $Ta|Ta_2O_5$, whose metals and oxides have the required sufficiently high melting points [3.116], Table 3.7, show good agreement of practical and theoretical thermodynamic data, that is standard energies and standard entropies of oxide formation. These results were obtained in a sodium calcium silicate glass melt containing antimony as the fining agent as well as in technical glass melts, for example in borosilicate glass melts fined with sodium chloride. Of these metals, molybdenum exhibited the least corrosion, which appeared as thin, yellow surface layers. Besides, they did not cause any technical problems during several weeks of application. The logarithm of the equilibrium oxygen partial pressure at the metal/metal oxide interfaces is a linear function of the reciprocal absolute temperature. It ranges from approximately $\log p_{O_2} = -15$ at $1000\,°C$ to $\log p_{O_2} = -8$ at $1600\,°C$ in the cases of molybdenum/molybdenum oxide and tungsten/tungsten oxide, and from $\log p_{O_2} = -22$ at $1000\,°C$ to $\log p_{O_2} = -12$ at $1600\,°C$ with tantalum/tantalum oxide. These results promise a general application of metal/metal oxide electrodes in technical glass melters in the future, in parallel to the well-established use of zirconia electrodes for research purposes in laboratories and for the standardization of the alternative electrodes.

It is interesting to note the essential steps of the development of reference electrodes for use with oxygen-indicating electrodes. Very early, electrodes employing calcia- and magnesia-stabilized zirconia tubes were developed for

Table 3.7. Melting points of some metals and their oxides and temperature-dependent standard free energies of formation of the oxides [3.126], which are of interest for application as alternative reference electrodes in glass-forming melts

	Melting point °C	$-\Delta G^0$ kJ mol^{-1}			
		1000 K	1200 K	1400 K	1600 K
Ta	2996				
Ta$_2$O$_5$	1877	1607	1523	1443	1364
Mo	2610				
MoO$_2$	1927	389			318
W	3410				
WO$_2$	1570	401	367	333	

short-time oxygen activity measurements in molten steel and slags in the
steelmaking industry [3.127, 128]. As with the alternative electrodes for glass
melts just described, the reference oxygen partial pressure of these zirconia
electrodes was fixed by a metal/metal oxide system, for instance iron/iron
oxide, nickel/nickel oxide, cobalt/cobalt oxide, chromium/chromium oxide,
and others, which were contained in the zirconia tubes. These first electrodes
thus actually represented a combination of the zirconia tube electrodes described in Sect. 3.4.3 and the alternative metal/metal oxide reference electrodes we just discussed. Their reference oxygen partial pressure was fixed
by a solid-state reaction (metal/metal oxide), but the lack of suitable properties of these oxides, i.e., chemical inertness with respect to glass melts, and
tightness and a sufficient ionic conductivity of thin layers of the oxides, neces-
sitated an additional protecting and oxide ion-conducting solid electrolyte,
for instance zirconia, between the reference system and the glass melt. Incidentally, at the early stages of our development, zirconia electrodes with
nickel/nickel oxide and with cobalt/cobalt oxide reference systems were also
tested for applicability in glass melts. However, they were found unsuited for
long-term application in glass melts because of the rather limited lifetime of
the metal/metal oxide reference systems, which is probably caused by the
relatively high oxygen partial pressures of the oxides at high temperatures,
resulting in the fast decomposition of the fine-grain oxides, followed by fast
exhaustion of their oxygen content.

As a consequence, three alternative reference electrodes for glass melts
are presently in use.

(1) Zirconia tube electrodes (Sect. 3.4.3) employ a reference gas and thus
    have the advantage of having a temperature-independent reference oxy-
    gen partial pressure and of being strictly isothermal. On the other hand,
    they are subject to temperature-shock-induced and mechanically induced
    breakage. Zirconia tube electrodes are thus particularly suited for appli-
    cation in research and development, where they are even the only choice
    when standard Seebeck coefficients of glass melts are to be measured
    (Sect. 3.6).
(2) Dissolving zirconia reference electrodes (Sect. 3.4.3) also employ a ref-
    erence gas with temperature-independent oxygen partial pressure. If the
    standard Seebeck coefficient of the particularly doped zirconia is known,
    the emfs can automatically be corrected for thermoelectric voltages of
    the electrolyte bridge. Because dissolving zirconia electrodes are less
    temperature-shock-sensitive than zirconia tube electrodes and, in addi-
    tion, have rather long lifetimes, they are well suited as technical reference
    electrodes in industrial melting units.
(3) Metal/metal (in particular Mo/Mo) oxide electrodes offer the advantage
    of the high mechanical stability of metal rods, which can be inserted into
    glass melting furnaces without any danger of breakage. Despite their
    isothermal character, the reference oxygen pressure depends on temper-

ature, which necessitates their standardization with respect to zirconia tube electrodes and a temperature-dependent evaluation of the emfs measured (see above).

3.5 Application of Oxygen Sensors

Friedrich G.K. Baucke

The oxygen sensors whose development has been described in detail in the preceding section, are used for laboratory, i.e., scientific and development, purposes, but their main application is in technical melting units. Many glass properties depend on the redox state, for example colour, heat absorption in the molten and solid state, brittleness, and the tendency to form gas bubbles on remelting and thus during further processing. The ability of a melt to form oxygen bubbles under defined conditions is the basis of a main production step called fining, during which gaseous impurities of melts are removed. All of these properties are redox-related or redox-based and involve dissolved oxygen, whose partial pressure must thus be known and, if possible, continuously monitored.

As a basis, in Sect. 3.5.1 we describe fining in a general way and subsequently treat redox fining, which, being the main fining method for melting of special glasses, is the fining process most frequently applied at Schott Glas. It will become obvious that the application of the oxygen sensors described in Sect. 3.4 is a necessary requirement for developing and conducting an economically and ecologically optimal fining process that results in the highest glass quality and yield achievable.

In certain cases of constant production parameters, for instance, of mass-produced glasses over long periods of time, the continuous measurement of the oxygen partial pressure suffices to control the fining process. However, critical melts and the development of new fining procedures require the additional knowledge of thermodynamic standard data of the redox couple or couples involved. These data inform about the concentrations of the redox ions when the oxygen partial pressure and the temperature are known, which by themselves provide only minimum information about the redox state. The determination of redox standard data is thus the subject of Sect. 3.5.2, which first treats the conventional "indirect" way of analysing the quenched glasses by physical or wet-chemical means after the oxygen partial pressure of the melt has been measured before the quench. Subsequently, several modern *in situ* electrochemical methods are described, which allow the thermodynamic data to be directly determined in the melts at melt temperatures and which are more versatile, less subject to errors, and less time-consuming and costly than the conventional methods.

The main fining procedure for mass-produced glasses is sulphur fining, a special type of redox fining, which is a rather economic process. Since the sulphur redox system is fairly complicated mechanistically and because of its wide-scale application, it will be treated in Sect. 3.5.3 although it is practically not applied at Schott Glas.

In Sect. 3.5.4 we report a long-time comparative test application of technical oxygen-measuring cells. The performance of an industrial oxygen sensor of Schott Glas, constructed for highly accurate measurements in special glasses, was compared with that of a sensor of different origin, designed preferably for mass-produced glasses, at the same locus of a melting unit. The different sensor types yielded very much the same results, with accuracies that corresponded to their different purposes of application. Besides, the test resulted in some new knowledge on specific properties of industrial melters.

In Sect. 3.5.5, finally, an alternative fining process called electrolytic fining, which has been developed by the Electrochemical Laboratory at Schott Glas and was patented in 1993, is briefly described. Electrolytic fining has the advantage that the fining bubbles are provided by an electrolytic process and not by a redox reaction and that, consequently, the amount of necessary fining agents, for instance arsenic or antimony, can be drastically reduced to make glass production a more ecological process than in the past.

3.5.1 (Redox) Fining of Oxidic Glass-Forming Melts

Friedrich G.K. Baucke

As stated above, fining is the removal of the main quantity of (ideally all) gaseous impurities, such as Ar, N_2, O_2, CO, CO_2, H_2O, SO_2 and others, which have remained in a glass melt from the raw materials and their reactions. These gases are either dissolved in the melt physically as atoms or molecules, for instance Ar, N_2, O_2, and SO_2, or chemically, as, for instance, O_2, CO_2, SO_2, and H_2O, by reacting with glass components. In either case, their removal is difficult because either the formation of bubbles is hindered kinetically due to the lack of bubble nuclei, or bubbles, if present, are very small, forming so-called "blisters", whose rate of rising to the melt surface is negligibly small.

In order for gases to form a stable bubble, the power balance

$$\sum_i p_i = p_{\text{atm}} + p_{\text{hydr}} + p_\sigma \tag{3.81}$$

must be met, where $\sum_i p_i$ is the sum of partial pressures of all gases in the bubble, p_{atm} is the atmospheric pressure, $p_{\text{hydr}} = \rho \times g \times h$ is the hydrostatic pressure (ρ = density of the glass melt, g = gravitational constant, h = distance between the bubble and the melt surface), and $p_\sigma = 2\sigma/r$ is the pressure within a bubble with radius r due to the surface tension σ of the melt. Thus, in principle, fining of a melt is assisted by a decrease of the sum of the pressures on the right side of (3.81) and/or an increase of the sum of the partial pressures of the gases on the left side of (3.81), both leading to an increase of bubble radius and volume and thus to an increase of buoyancy and rate of bubble rise. Various mathematical models describing the behaviour of gas bubbles in glass melts have been given. Some of them [3.140–144] take into account chemical fining reactions, but most do not [3.129–139]. The evolution of gases (H_2O, N_2, NO, NO_2, CO_2) during melting of the raw materials of a television glass has been investigated by means of gas chromatography and mass spectrometry [3.145]. For details, the reader is referred to these publications.

In practice, a decrease of the pressures on the right side of (3.81) can be accomplished by the following steps.

(1) The atmospheric pressure p_{atm} above the melt is reduced by applying a vacuum. This so-called "vacuum fining" is frequently applied in laboratories and has also been applied technically [3.146].

(2) The hydrostatic pressure p_{hydr} of the melt is decreased by flowing the melt in a thin layer over a table-like design in the furnace. This so-called "thin-layer fining", which is sometimes applied in practical melting units, also reduces the distance of the bubbles to the melt surface above the table and thus the rise time necessary for the bubbles to reach the melt surface. However, the increased flow rate of the thin melt layer and the short residence time of the melt above the "table" may partly reduce the desired effect.

(3) The pressure p_σ due to the surface tension of the melt is reduced when the bubble radius increases during steps (1) and (2). Besides, the formation of bubble nuclei is supported by the presence of melt/solid interfaces, for instance of tank walls and bottom and still unreacted raw materials. Bubble nucleation can also be triggered by ultrasonic means without [3.147–150] and with [3.151] additional stationary magnetic field. Frequently, the fining process is also assisted by introducing gas bubbles into the melt by so-called "forced bubbling" [3.152], which, in addition, supports melt homogenization.

An increase of the pressure within bubbles (left side of (3.81)) can be effected by decreasing the solubility of the gases in the melt by temperature changes and by adding substances with high vapour pressures, so-called fining agents, to the melt. These materials, for instance $NaCl$ and NaF, evaporate, the vapour forming bubbles at sufficiently high temperatures, and the bubbles represent sinks for dissolved gases, which diffuse into the vapour bubbles and increase their volume, buoyancy and rise rate. These fining agents operate on the basis of a physical process (evaporation of the physically dissolved material) and require a modification of the power balance (3.81)

$$\sum_i p_i + p_{fa} = p_{atm} + p_{hydr} + p_\sigma \,, \tag{3.82}$$

where p_{fa} is the partial pressure of the (physical) fining agent, which often also acts as a nucleating agent. In most cases, p_{fa} is much larger than $\sum_i p_i$ at the fining temperature.

In the past, also solid materials expelling water vapour and/or generating carbon dioxide and thus causing H_2O and CO_2 bubbles in the melt at high temperatures were applied as fining agents. For example, glass melts were agitated by means of wet wooden rods (a process called "bülwern" in German), and potatoes (!) attached to metal bars were pushed into the melt for short times. However, this kind of fining was restricted to glasses whose contents in dissolved water and carbon dioxide were not critical. Besides, it is completely uncontrollable and was thus given up some time ago.

Most significant today are fining agents based on temperature-dependent chemical equilibria involving a gas phase. These substances, for instance the polyvalent elements arsenic [3.153], antimony [3.89] and cerium, and sulphur in the form of sulphate [3.154], form temperature-dependent dissociation or redox equilibria involving a gas, whose partial pressure depends on temperature and can thus be increased sufficiently to cause spontaneous formation of bubbles by increasing the temperature. The most important of these fining processes for Schott Glass and other special glass producers is redox fining, which will thus be treated in some detail in the following paragraphs of this section.

Redox Fining

Redox Equilibria of Fining Agents. The ions of polyvalent elements, whose oxides are usually dissolved by a few tenths of a percent in oxidic glass melts for the purpose of fining, are in equilibrium with the intrinsic redox couple oxygen/"oxide" of the melts [3.4]. An example with high practical significance is antimony, which is particularly used as a fining agent of special glasses and will serve as the example in the following description. The redox equilibrium is described most realistically by

$$2Sb^{5+}(m) + 4 \equiv SiO^-(m) \rightleftarrows 2Sb^{3+}(m) + 2 \equiv SiOSi \equiv (m) + O_2(m) , \tag{3.83}$$

where $\equiv SiO^-(m)$ and $\equiv SiOSi \equiv (m)$ characterize the respective tendency of electron-donation by the oxygens of the melt [3.4, 155]. The redox equilibrium can formally also be represented by the equation

$$2Sb^{5+}(m) + 2O^{2-}(m) \rightleftarrows 2Sb^{3+}(m) + O_2(m) , \tag{3.84}$$

where now O^{2-}, which is generally denoted as the basicity of the melt, characterizes the sum of the electron-donating oxygens.

Both equations are strictly valid only at constant basicity (and thus melt composition) [3.74], and their application to different melts (and basicities) leads to what has been called the "thermodynamic paradox" [3.58, 59]. This term refers to the experimental observation that, in most cases, high basicity (large $a_{O^{2-},m}$ in (3.85)) favours the upper oxidation state ion, here Sb^{5+}, and thus contradicts Le Chatelier's principle [3.74]. Several explanations of this contradiction have been put forward. *Budd* proposed that the paradox was caused by neglecting the formation of complexes of the polyvalent ions with oxide ions [3.156]. *Jeddeloh* explained it by the fact that the different sites of the differently charged ions in the network of the melt had not been taken into account [3.73]. *Baucke* and *Duffy*, finally, traced it back to the different but unknown standard free energies of the equilibrium of the different glass compositions, which basically forbid the application of equilibria like (3.83) and (3.84) to glass melts with different compositions [3.74]. They give a quantitative explanation of the experimental properties of glass melts on the basis of optical basicity, see Sect. 3.3.3.

Equilibrium (3.84) is characterized by the temperature-dependent thermodynamic equilibrium constant

$$K(T) = \frac{a_{Sb^{3+}}^2 \, p_{O_2,m}}{a_{Sb^{5+}}^2 \, a_{O^{2-},m}^2} \,, \tag{3.85}$$

where a is the activity of the species indicated, and $p_{O_2,m}$ is the equilibrium partial pressure of oxygen. Because, in principle, activities of single ions, here antimonous and antimonic ions, are not measurable and because the basicity $a_{O^{2-},m}$ is not attainable either, (3.85) is usually rearranged by incorporating the activity coefficients of the redox ions and the oxide activity into a practical equilibrium constant $K'(T)$,

$$K'(T) = K(T) \frac{\gamma_{Sb^{5+}}^2 \, a_{O^{2-},m}^2}{\gamma_{Sb^{3+}}^2} = \frac{c_{Sb^{3+}}^2 p_{O_2,m}}{c_{Sb^{5+}}^2} \,, \tag{3.86}$$

where γ denotes the activity coefficient of the redox species indicated. Unlike the thermodynamic redox equilibrium constant $K(T)$, (3.85), the practical equilibrium constant $K'(T)$ cannot be expected to be strictly constant at constant temperature, which must be taken into account when experimental results are evaluated and discussed on the basis of (3.86). Using the oxygen partial pressure $p_{O_2,m}$ instead of the oxygen fugacity $\dot{p}_{O_2,m}$ in (3.85) and (3.86) is well justified because the fugacity coefficient $f_{O_2,m}$ of oxygen is known to be close to unity,

$$\dot{p}_{O_2,m} = f_{O_2,m} \, p_{O_2,m} \cong p_{O_2,m} \,, \tag{3.87}$$

also at the high temperatures of glass melting.

Because of the high viscosity of glass melts and the connected small diffusion coefficients, the (external) oxygen $O_2(g)$ of the atmosphere above the melt is practically never in equilibrium with the (internal) oxygen $O_2(m)$ which is physically dissolved in the melt, at least in technical melting units.

Thus, (external) Henry's law between the oxygen concentration $c_{O_2,m}$ in the melt and the oxygen partial pressure $p_{O_2,g}$ of the atmosphere above the melt, known as Bunsen's law,

$$c_{O_2,m} = \alpha(T)p_{O_2,g} , \tag{3.88}$$

is not met in industrial furnaces and is reached in laboratories only after extended equilibration periods [3.106]. In contrast, the oxygen partial pressure $p_{O_2,m}$, which is in equilibrium with the concentrations of the redox ions according to (3.85) and (3.86), is the internal oxygen partial pressure of the melt. It is thus in equilibrium with the concentration of the physically dissolved oxygen molecules (or atoms) according to (internal) Henry's law,

$$c_{O_2,m} = \alpha(T)p_{O_2,m} , \tag{3.89}$$

where, as in (3.88), $\alpha(T)$ is the temperature-dependent solubility coefficient or the temperature- and partial-pressure-dependent physical solubility of oxygen,

$$\alpha(T) = \frac{c_{O_2,m}}{p_{O_2,m}} . \tag{3.90}$$

(The dimension of $\alpha(T)$ is thus mole per volume and pressure.) The internal oxygen partial pressure $p_{O_2,m}$, although sometimes called fictive because of the absence of a gas phase, is a real quantity because it is measurable, for example by oxygen sensors (see Sect. 3.4), and takes part in equilibria like (3.85).

Combining the oxygen solubility (3.90) with the practical equilibrium constant (3.86) yields the expression

$$K''(T) = K'(T)\alpha(T) = \frac{c_{Sb^{3+}}^2 \, c_{O_2,m}}{c_{Sb^{5+}}^2} , \tag{3.91}$$

which shows that the concentration $c_{O_2,m}$ of the "physically dissolved" oxygen $O_2(m)$ is in equilibrium with the redox ions of the polyvalent element and, consequently, also with the concentration $c_{O_2,chem}$ of the "chemically dissolved" oxygen $O_2(chem)$, which is present in the melt as an oxide concentration equivalent to the concentration of the upper oxidation state (Sb^{5+}) ions,

$$c_{O_2,chem} = 2c_{Sb^{5+}} = c_{Sb_2O_5} . \tag{3.92}$$

Consequently, a concentration change of the physically dissolved oxygen, for instance by a change according to (3.89), changes not only the concentration ratio of the polyvalent ions according to (3.91) but also the concentration of the oxygen which is chemically dissolved as the charge-balancing equivalent of the antimonic ions according to (3.92). It is clear that the sum of the concentrations of chemically and physically dissolved oxygen is the total concentration of oxygen dissolved,

$$c^0_{O_2,m} = c_{O_2,m} + c_{O_2,chem} = c_{O_2,m} + 2c_{Sb^{5+}} = c_{O_2,m} + c_{Sb_2O_5} . \qquad (3.93)$$

The concentration of chemically dissolved oxygen is generally much larger than that of physically dissolved oxygen and, depending on the content of polyvalent elements, can be up to 10^3–10^4 times larger than the latter [3.157]. Polyvalent elements dissolved in oxidic glass melts are thus often termed oxygen buffers; their buffer capacity is determined by the total amount of the polyvalent element(s) dissolved [3.4].

It is noted that the reality in glass melting furnaces is generally more complicated, as the conditions described may differ quantitatively from volume element to volume element of the melt because of the high viscosity of the melt and may cause additional cross diffusion and diffusion potentials between the volume elements. These complications, however, are probably averaged out to a large extent by the convection of the melt, at least in large melting units, which are usually designed for rather strong natural convection, except for the quiet zone of fining, where the melt has mainly attained its final homogeneity. The conditions in practical melting tanks are also complicated by the furnace atmosphere and by contaminants of the raw materials, for instance of recycled cullet, as has been discusses in detail by *Beerkens* et al. [3.158].

The Redox Fining Procedure. Redox fining is based on the temperature dependence of the redox equilibrium constant, which can be described by some function of the type

$$K'(T) = Ae^{-B/T} , \qquad (3.94)$$

where A and B are melt-dependent characteristic constants. High temperatures increase the equilibrium constant, shifting the redox equilibrium (3.84) to the right side and increasing the internal oxygen partial pressure of the melt according to (3.85) and (3.86), and *vice versa*. To make use of this dependence for the fining of glass melts, high internal oxygen partial pressures are generated at high temperatures of the melt. This causes oxygen diffusion into blisters, if present, and an increase of blister volume, buoyancy and rise rate, and/or spontaneous formation of oxygen bubbles, which, acting as sinks for dissolved impurity gases, cause a diffusion of the impurities into the bubbles, and an increase of their volume and of their rate of rising to the melt surface. Pure oxygen bubbles remaining in the melt after the fining process are subsequently resorbed due to small internal oxygen partial pressures, which are generated by lowering the temperature. This general scheme has lead to the following characteristic steps of the redox fining process.

(1) During the *heating period*, the temperature of the melt is increased. Up to somewhat below the fining temperature, two processes are possible. (i)

In the absence of blisters, the sum of the partial pressures of impurity gases and oxygen are below 1 bar. Because bubbles are not formed and practically no gas is given off, the melt is a closed system during this period. (ii) In the presence of blisters, however, oxygen diffusion into the small bubbles starts at relatively moderate temperatures and increases with increasing temperature so that the melt is not a completely closed system but loses some of its oxygen to the growing blisters.

(2) During the *fining period*, the temperature is raised above the fining temperature, where the partial pressure of oxygen exceeds the sum of the atmospheric, hydrostatic, and surface tension pressures,

$$p_{O_2,m} \geq (p_{atm} + p_{hydr} + p_\sigma) \,, \tag{3.95}$$

and is kept at a high value for a time sufficient for the removal of the majority (ideally of all) gas bubbles. During the fining period, the melt is an open system, losing large parts of its chemically dissolved oxygen (and most of the impurity gases). The oxygen loss leads to an increase of the concentration ratio of antimonous to antimonic ions, $(c_{Sb^{3+}}/c_{Sb^{5+}})$, until the equilibrium values of (3.91) corresponding to the high temperature are reached, and, in turn, cause a decrease of the oxygen partial pressure according to

$$p_{O_2,m} = \frac{c^2_{Sb^{5+}} \, K'(T)}{c^2_{Sb^{3+}}} \,. \tag{3.96}$$

This decrease of the oxygen partial pressure requires that either a sufficiently high fining temperature is initially chosen or, if not so, that the temperature is subsequently corrected. The expression (3.91) also shows that the fining temperature is not merely determined by the melt composition and the kind of fining agent but depends also on the initial concentration ratio of the redox ions present before the melt is heated up. The fining efficiency is the greater and the fining temperature can be the lower, the larger the ratio $(c_{Sb^{5+}}/c_{Sb^{3+}})$. Besides, the fining efficiency depends on the total concentration of fining agents, which determines the volume of oxygen released within a certain temperature range during heating, and on the rate of temperature increase, which determines the ratio of the oxygen to impurity content of the bubbles leaving the melt.

(3) During the *resorption period*, finally, oxygen bubbles remaining from the preceding fining period are resorbed by the melt since the oxygen partial pressure is drastically reduced when the melt temperature is decreased, (3.86) and (3.94). The melt, taking up oxygen, is thus an open system during this period. However, the resorption stops at a lower temperature limit because – despite the small oxygen partial pressures attained by the melt – diffusion kinetics prevents further sufficiently fast dissolution of the gas owing to the high melt viscosity at lower temperatures. Kinetics thus determines the rate of temperature decrease during the resorption period: In order to produce a bubble-free glass, remaining oxygen bubbles must have

been resorbed when the melt returns to a closed state at low temperatures. As an example, Fig. 3.21 presents the time dependence of the oxygen partial pressure of a lead silicate glass melt during oxygen resorption at various temperatures [3.4]. The melt contained arsenic but was free of impurity gases, and was only briefly heated and not fined. As all melts were subjected to the same heating and cooling rates, the plots demonstrate that the oxygen partial pressure (logarithmic scale!) becomes constant after the longer resorption times, the lower the temperature is. Also, the initial resorption rate, $(dc_{O_2,m}/dt) = \alpha(T)(dp_{O_2,m}/dt)$, decreases with decreasing temperature and is nearly zero at 800 °C. When the experiments were terminated after 25 h, the glass specimens treated at 850 °C and 800 °C still contained some small oxygen blisters.

3.5.2 Thermodynamic Standard Data of Redox Equilibria

Friedrich G.K. Baucke

In many cases, on-line measurements of the oxygen partial pressure and its evaluation on the basis of gathered experience suffice to control industrial glass melting and fining. This is especially true with continuously working glass melting tanks with a large output of, for instance, household or appliance glass products with nearly unchanged composition, containing negligible redox impurities and small concentrations of redox fining agents under conditions that are constant for long periods. However, for initially optimizing

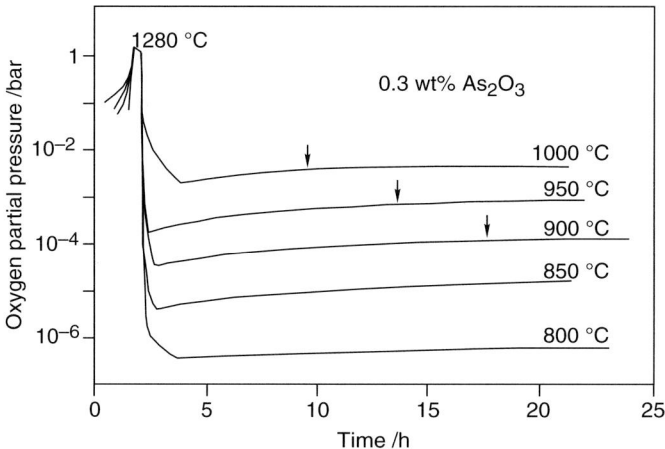

Fig. 3.21. Time dependence of the oxygen partial pressure (logarithmic scale) of a lead silicate glass melt containing arsenic after defined short heating to 1280 °C and fast cooling to various temperatures for oxygen resorption. The resorption period increases (see arrows), and the initial resorption rate decreases, with decreasing temperature

the fining conditions of such melts, for instance for determining the minimum concentration of fining agent needed and the minimum starting concentration ratio of upper to lower oxidation state ions, knowledge of the thermodynamics of the redox couple (or couples) is indispensable. This is also true when the fining of critical glass melts and of melts produced under variable conditions are to be continuously controlled. In these cases, thermodynamic standard data must be available to determine the concentration ratio of upper to lower oxidation state ions from the measured oxygen partial pressure and temperature.

The determination of such standard data has experienced a fast development during the past twenty years. Two basically different types of measurement must be distinguished. Between 1981 and 1985, one of the first attempts conducted was an "indirect" method [3.88, 89], "indirect", in this case meaning that some of the interrelated data needed (oxygen partial pressure) are measured in the melt at melt temperatures and others (redox ion concentrations) in the solid glass after quenching the melt. Nearly at the same time, the development of "in-situ" methods started, which made use of electrochemical analytical techniques [3.159–161], in particular those based on cyclovoltammetry [3.55, 90, 91], which had been developed for analysis in aqueous and non-aqueous solutions in the early 1950s [3.162–166]. These methods have the great advantage of being applicable in the melt at melt temperatures, thus avoiding errors arising from glass quenching and analysis at room temperature, and yield directly the standard emf and thus the standard free enthalpy without prior determination of oxygen partial pressure and redox concentrations. In addition, these techniques are far less time-consuming than the "indirect" techniques, although, due to the high temperatures and other melt-connected causes, their reproducibility and particularly their sensitivity are lower than those of the original methods in solution [3.91, 162]. The present state of the "competition" between the two kinds of measurement is that the *in-situ* methods are applied much more frequently than the "indirect" techniques, at least in simple melts such as sodium calcium silicates. The evaluation of data measured in multi-component technical melts, however, presents serious difficulties and necessitates the development of special deconvolution techniques for the curves obtained, which is still not completed.

In the following, both the "indirect" and the *in situ* methods will be briefly described in order to demonstrate their theoretical bases, modes of application, and essential differences. This presentation, moreover, is to point out that changes in the history of science can happen unexpectedly fast. Within less than ten years, two competing physico-chemical techniques have been developed with the goal to determine thermodynamic data in glass melts, and it seems already decided which of them will survive. The situation brings to mind the development of surface-analytical techniques in the sixties. The first ("indirect") method for measuring (discontinuous) concentration profiles in glass surfaces consisted of layerwise glass dissolution and subsequent analysis of the dissolved glass fractions [3.167] and was published in 1964. Only seven years later, it was overtaken by the (*in situ*) IBSCA technique, which yields continuous, highly depth-resolved concentration profiles by ion ablation within less than an hour [3.168], see Sect. 2.3. Such rapid changes repeatedly occur in the history of science. They always happen when the sudden and urgent need for a certain kind of experimental information coincides accidentally with the appearance of a new *in situ* technique which is elegantly applicable to a problem that had already been tackled by applying established but time-consuming and less practical "indirect" methods.

"Indirect" Methods

In "indirect" methods, thermodynamic standard data of redox reactions, for instance of the antimony equilibrium, see (3.83) and (3.84),

$$2Sb^{5+}(m) + 2O^{2-}(m) \rightleftarrows 2Sb^{3+}(m) + O_2(m) \tag{3.97}$$

are obtained by determining the temperature-dependent practical equilibrium constant

$$K'(T) = \frac{c_{Sb^{3+}}^2 \, p_{O_2,m}}{c_{Sb^{5+}}^2} \ , \tag{3.98}$$

which then yields the temperature-dependent standard free energy

$$\Delta G'^0(T) = -RT \ln K'(T) \ , \tag{3.99}$$

the standard energy

$$\Delta H'^0(T) = RT^2 \frac{\mathrm{d} \ln K'(T)}{\mathrm{d}T} \ , \tag{3.100}$$

and the standard entropy

$$\Delta S'^0(T) = \frac{\Delta H'^0(T) - \Delta G'^0(T)}{T} \tag{3.101}$$

of the redox reaction. The primed symbols indicate that the data obtained experimentally contain the same approximations as the practical equilibrium constant $K'(T)$ from which they are calculated, see (3.86). Besides, the standard energy and entropy may be found to depend slightly on temperature because of these approximations and also because of the wide temperature range normally applied to the investigation; they are therefore given as temperature-dependent quantities in (3.99)–(3.101).

Principle of Indirect Measurements. Strictly speaking, the temperature-dependent equilibrium constant (3.98) is determined experimentally by measuring the oxygen partial pressure and the concentration ratio of the polyvalent ions as a function of temperature. Fortunately, however, this cumbersome method is avoided in that the concentration ratio of the polyvalent ions must only be known at one of the temperatures at which the oxygen partial pressure is determined [3.88]. This treatment is justified provided the concentrations of both polyvalent ions are much greater than the concentration of the physically dissolved oxygen,

$$(c_{Sb^{5+}} \text{ and } c_{Sb^{3+}}) \gg c_{O_2,m} , \qquad (3.102)$$

so that the change of the oxygen concentration with temperature does not change the concentration ratio of the polyvalent ions. This precondition is generally met by polyvalent elements used as fining agents because they are added to melts at least in amounts of tenths of a percent and thus at concentrations exceeding that of the physically dissolved oxygen by at least two orders of magnitude, see below. Besides, the experimental conditions must exclude any exchange of oxygen between atmosphere and melt so that the total concentration of oxygen is constant, $c^0_{O_2} =$ const. Under these conditions, the measurement of the temperature-dependent oxygen partial pressure, for instance by means of an oxygen sensor, yields the temperature function of the equilibrium constant,

$$\frac{dK'(T)}{dT} = \left(\frac{c_{Sb^{3+}}}{c_{Sb^{5+}}} \right)^2 \frac{d\,p_{O_2,m}}{dT} , \qquad (3.103)$$

and the concentration ratio of the polyvalent ions subsequently obtained by an analysis of the quenched melt, in combination with the oxygen partial pressures, gives the corresponding temperature-dependent equilibrium constants according to (3.98).

In order to secure correct experimental results over a wide range of data and to exclude errors due to extrapolation, it is necessary to conduct the described measurements at several different total oxygen contents of the melt [3.88, 89]. This is favourably done by adjusting the oxygen content by electrochemical pumping, that is, by an electrolysis of the melt between two platinum electrodes, one of which is an oxygen-generating and -degenerating electrode positioned in the melt, and the other a counter-electrode separated from the melt by an oxide ion-conducting solid electrolyte, for instance stabilized zirconia. The practical arrangement described below employs zirconia as the crucible containing the melt [3.88, 89][3.88, 3.89].

Two modes of measurement are possible.

(a) The melt with oxygen sensor, pumping electrode, and stirrer is contained in a hermetically closed crucible, which allows the temperature cycles to be conducted and excludes an exchange of oxygen. Samples are withdrawn by means of a special trap system at appropriate temperatures.

(b) In a less accurate but more economical arrangement, the melt and the necessary equipment are contained in an open platinum or zirconia crucible. An exchange of oxygen between atmosphere and melt is nevertheless nearly excluded by selecting an appropriate temperature change rate [3.105]. The change in temperature must happen so fast as to make the melt a quasi-closed system with constant total oxygen content, and so slow as to guarantee a quasi-isothermal measurement of the temperature-dependent oxygen partial pressure. Samples for redox ion analysis must be withdrawn quickly from the melt and quenched immediately.

Use of Optical Basicity Data. As described in some detail in Sect. 3.3.3, equations exist for a number of polyvalent elements which give the concentration ratio of the lower to upper oxidation state ions as a function of the optical basicity at a certain oxygen partial pressure $p_{O_2(m)}$, usually for air atmosphere, and a certain temperature T_m, mostly at 1400 °C, see Table 3.4. An example is the expression (3.33) for iron,

$$\log\left(\frac{c_{Fe^{2+}}}{c_{Fe^{3+}}}\right) = a\,\Lambda + b\,, \tag{3.104}$$

where Λ is the optical basicity and, for the example iron, $a = -6.5$ and $b = 3.2$. These relationships are of great help when thermodynamic standard data of the listed elements are to be determined in glass melts whose average optical basicity can be calculated from the optical basicity increments of the single oxides forming the glass, see (3.25) and Table 3.2. The practical equilibrium constant at T_m is obtained by

$$K'(T_m) = \left(\frac{c_{Fe^{2+}}}{c_{Fe^{3+}}}\right)^4 p_{O_2(m)}\,, \tag{3.105}$$

which yields the standard free energy at T_m according to

$$\Delta G'^0(T_m) = -RT_m \ln K'(T_m) \tag{3.106}$$

or, in terms of optical basicity,

$$\Delta G'^0(T_m) = -2.303 RT_m \left[4\,(a\,\Lambda + b) + \log p_{O_2(m)}\right]\,. \tag{3.107}$$

The determination of the temperature-dependent standard free energy, the standard energy, and the standard entropy requires only the additional measurement of the temperature-dependent oxygen partial pressure of the melt containing the same total concentration of the polyvalent element and equilibrated with $p_{O_2(m)}$ at T_m. This should not be too difficult because, in nearly all cases, the measurements for establishing the expression (3.104) were carried out with melt equilibrated with air atmosphere at T_m so that even the simpler mode (b) using a quasi-isothermal measurement in a quasi-closed melt (yet saturated with air atmosphere at T_m) can be applied. Further treatment of the data is as described above.

Limitation of Indirect Measurements to One Polyvalent Element. These indirect methods of measuring thermodynamic standard data are basically restricted to melts containing only one polyvalent element, which is easily shown by a melt containing, for example, iron and cerium. Their redox equilibria are given by

$$4Fe^{3+} + 2O^{2-} \rightleftarrows 4Fe^{2+} + O_2 \qquad (3.108)$$

and

$$4Ce^{4+} + 2O^{2-} \rightleftarrows 4Ce^{3+} + O_2 \;, \qquad (3.109)$$

and the respective practical equilibrium constants are represented by

$$K'_{Fe}(T) = \frac{c^4_{Fe^{2+}} \, p_{O_2,m}}{c^4_{Fe^{3+}}} \qquad (3.110)$$

and by

$$K'_{Ce}(T) = \frac{c^4_{Ce^{3+}} \, p_{O_2,m}}{c^4_{Ce^{4+}}} \;, \qquad (3.111)$$

so that their combination results in the expression

$$\left(\frac{K'_{Fe}}{K'_{Ce}} \right)^{1/4} = \frac{c_{Fe^{2+}} \, c_{Ce^{4+}}}{c_{Fe^{3+}} \, c_{Ce^{3+}}} = \frac{c_{Fe^{2+}}/c_{Fe^{3+}}}{c_{Ce^{3+}}/c_{Ce^{4+}}} \;. \qquad (3.112)$$

If thus the equilibrium constants of the different redox ion couples exhibit different temperature functions,

$$\frac{dK'_{Fe}(T)}{dT} \neq \frac{dK'_{Ce}(T)}{dT} \;, \qquad (3.113)$$

which is true with nearly all polyvalent elements, the relative concentration ratios on the right side of (3.112) will also change with temperature. Consequently, knowledge of the ionic concentrations at only one temperature, in contrast to melts containing only one polyvalent element, does not suffice to determine the temperature-dependent equilibrium constants.

Analysis of the Quenched Glass. The quenched glass is analysed for polyvalent ions either by wet-chemical methods or physical techniques. For example, a study of the redox system As^{3+}/As^{5+} in a barium borosilicate glass melt has been reported [3.169], where the total and the fivefold positively charged arsenic were determined by wet-chemical analyses. Similarly, antimony in unspecified glasses [3.170] was investigated by ion chromatography interfaced with inductively coupled plasma atomic emission spectrometry to determine Sb^{3+} and Sb^{5+}, which exhibit well separated chromatographic peaks. Nevertheless, Sb^{5+}, found less reproducible, was determined as the difference between total antimony (after reduction) and antimonous contents. *Schreiber* reported standard data of more than thirty polyvalent elements in glass melts with simple compositions [3.171, 172] and studied the dependence of some data on the melt composition [3.173]. In this work, many of the redox ions were analysed by wet-chemical analyses [3.174], and others by physical means, mainly spectroscopy, see for instance [3.48]. Other authors also report on wet-chemical and spectroscopic analyses of redox ions in quenched glasses, see for instance [3.175, 176].

Despite the number of investigations employing wet-chemical analysis, a critical work by *Stahlberg* [3.88] comes to the conclusion that analyses of dissolved quenched glasses are more critical than generally assumed because of a possible uptake of gaseous or dissolved oxygen during dissolution, solution transfers, and titration. These sources of error are excluded when the quenched glass is analysed by physical techniques, of which Mössbauer spectroscopy was chosen for determining antimonous and antimonic species, both of which are Mössbauer-active. Being practically the first attempt to determine redox standard data of antimony in glass melts by employing a physical technique [3.88, 89], this work is briefly reported in the following paragraphs.

Example: Analysis of the Quenched Glass by Mössbauer Spectroscopy. The design for preparing and treating the melts is depicted in Fig. 3.19 and explained in Sect. 3.4.4. Mössbauer spectroscopy, which is based on repulsion-free γ emission and absorption of ^{121}Sb nuclei, has been treated extensively in the literature, see for example [3.177–191]. It was ensured by applying antimony tetroxide, which contains unit ratio of antimonic and antimonous species, that the measured Mössbauer absorption correctly yielded the concentrations of the respective antimony species, and this was the case if the glass samples containing the γ-absorbing ^{121}Sb atoms were analysed at or below 4.2 K (liquid helium). The investigated glass had the approximate composition (in mol%): 72SiO$_2$, 1BaO, 12 Na$_2$O, 4K$_2$O, 6CaO, 4Al$_2$O$_3$, 1ZrO$_2$, 0.22Sb$_2$O$_3$. Several different concentrations of oxygen were electrolysed into the melts. After each electrolysis, the oxygen partial pressure of the melt was measured as a function of temperature, and a sample was withdrawn for Mössbauer analysis at an appropriate temperature.

Figure 3.22 presents the temperature-dependent redox equilibrium constant thus obtained, and Fig. 3.23 shows standard energy, free energy, and entropy of the redox reaction. As expected (see above), standard energy and entropy indeed depend slightly on temperature. Figure 3.24, finally, shows the "diagram of redox state", which presents the concentration ratio of the antimony species as functions of oxygen partial pressure and temperature and can thus be applied to control the fining process as described above.

Several runs performed with antimony-free melts yielded the physical oxygen solubility of the order of 10^{-4} mol dm^{-3} bar^{-1} and thus confirmed that the concentration ratio of the antimony species was indeed independent of temperature as presumed for the evaluation. The temperature dependence of

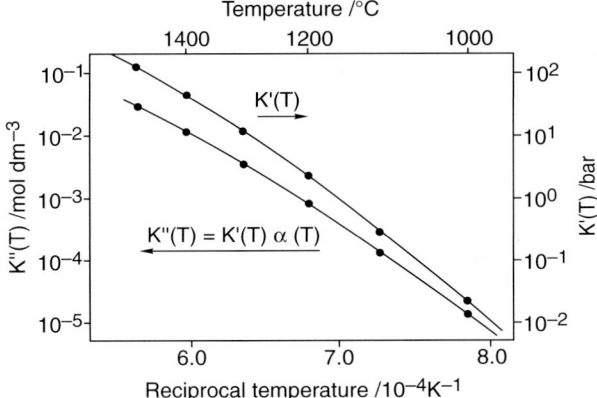

Fig. 3.22. Practical equilibrium constant $K'(T)$ according to (3.98) and modified equilibrium constant $K''(T) = K'(T)\alpha(T)$ of the redox reaction of antimony in a multi-component silicate glass melt as a function of reciprocal temperature

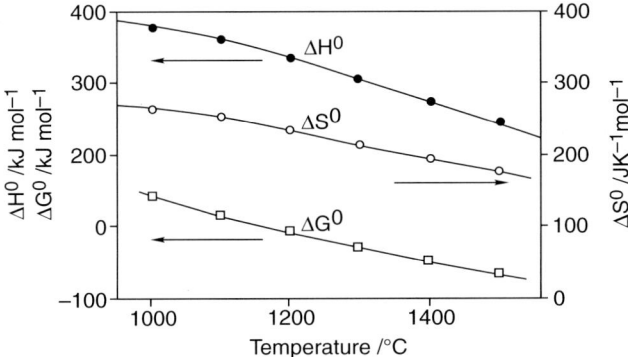

Fig. 3.23. Standard thermodynamic data of the redox equilibrium of antimony in a multi-component silicate glass melt as a function of temperature

the physical oxygen solubility obtained from the "best runs" is depicted in Fig. 3.25. Different from a statement in the literature [3.157], but in agreement with reported experimental data [3.193], the oxygen solubility in this melt decreases slightly with increasing temperature. Besides, the order of magnitude measured agrees with that reported for a glass melt with comparable composition but determined by a different method [3.193]. In addition to the practical equilibrium constant $K'(T)$, Figure 3.22 presents the modified redox equilibrium constant $K''(T) = K'(T)\alpha(T)$, which is referred to the oxygen concentration instead of the oxygen partial pressure, see (3.91). To give a feeling of the amount of oxygen that dissolves physically in a melt in equilibrium with air atmosphere, Table 3.8 presents the oxygen solubility in mass and in volume per unit melt volume and per 0.21 bar oxygen partial pressure at two melt temperatures.

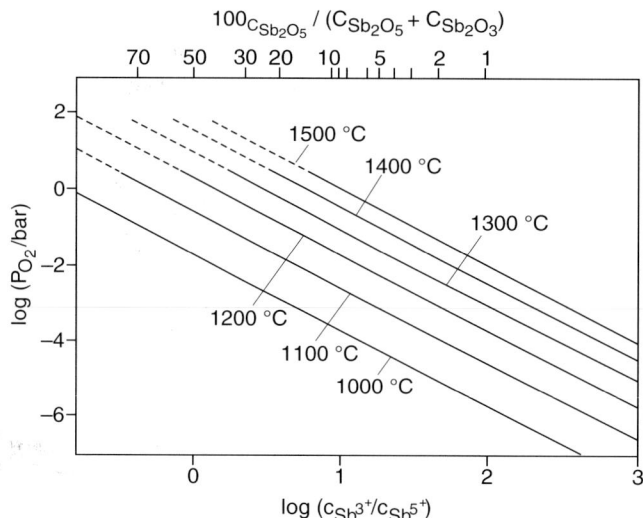

Fig. 3.24. "Diagram of redox state" of antimony in a multi-component silicate glass melt

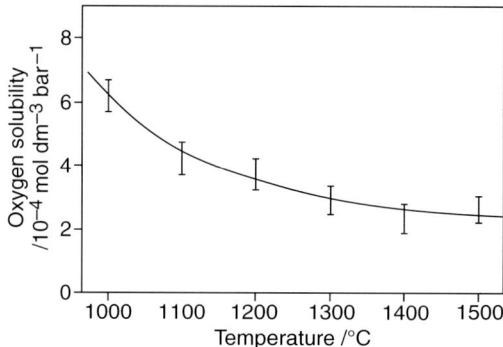

Fig. 3.25. Physical solubility $\alpha(T) = c_{O_2,m}/p_{O_2,m}(T)$ of oxygen in a redox ion-free multi-component silicate glass melt as a function of temperature

Table 3.8. Physical solubility of oxygen in a multi-component silicate glass melt under 1 bar oxygen partial pressure and in air atmosphere (0.21 bar oxygen) expressed in mole per volume and pressure, mass per volume and pressure, and volume (at 25 °C and at T_m) per volume and pressure. T_m = temperature of melt

Solubility	$T_m = 1000\,°C$	$T_m = 1500\,°C$
$\alpha(T_m)/\mathrm{mol\,dm^{-3}\,bar^{-1}}$	6.2×10^{-4}	2.4×10^{-4}
$\alpha(T_m)/\mathrm{mol\,dm^{-3}\,(0.21\,bar)^{-1}}$	1.3×10^{-4}	0.5×10^{-4}
$c_{O_2}/\mathrm{mg\,dm^{-3}\,(0.21\,bar)^{-1}}$	4.2	1.6
$c'_{O_2,25\,°C}/\mathrm{cm^3\,dm^{-3}\,(0.21\,bar)^{-1}}$	2.9	1.1
$c'_{O_2,T_m}/\mathrm{cm^3\,dm^{-3}\,(0.21\,bar)^{-1}}$	12.4	6.7

In Situ Electrochemical Methods

Simultaneously with the development of the described method involving Mössbauer spectroscopy, *in situ* electrochemical techniques were tried elsewhere and were applied in aqueous and nonaqueous solutions, also in glass melts. These methods offer the following significant advantages:

- They are carried out directly in the melt at melt temperatures, which eliminates difficulties arising from quenching the melt, avoids related errors, and is less time-consuming and costly.
- The electrochemical measurements themselves are fast techniques and, depending on the method chosen, yield results within the minute, second, and even millisecond range.
- In contrast to indirect methods, electrochemical *in situ* measurements allow the investigation of mutual equilibria of several redox elements in melts.
- Particularly, methods based on cyclic voltammetry directly yield standard potentials of the polyvalent elements of interest so that measurements at various temperatures result in standard data including equilibrium constants of redox reactions and, consequently, diagrams of redox state.

While electrochemical measurements, by themselves, are not particularly useful for the identification of dissolved species, the *in situ* methods adapted for the investigation of glass melts involve the factor time, which makes the techniques powerful tools for qualitative and quantitative investigations. They are concerned with the current–voltage–time relationship at an inert, mostly platinum, electrode in contact with the medium of interest, here the glass melt, or, in other words, they make use of the response of an inert electrode during an electrolysis conducted under controlled conditions.

Measurements are carried out in three-electrode cells. In addition to the analysing (or working) electrode, a counter (or auxiliary) electrode is used for applying the electrolysing voltage, which usually has a large surface so that small current densities are guaranteed and connected complications are avoided. The potential of the working electrode is measured against that of a reference electrode, preferably a zirconia electrode with oxygen or air as the reference gas, so that the measured data are automatically referred to the intrinsic redox system oxygen/"oxide" of the particular melt at the measuring temperature [3.4, 105]. Platinum wires in contact with melt and atmosphere frequently reported to be used as reference electrodes [3.159, 194, 195] are not recommended because of insufficient reproducibility due to the badly defined temperature of the only partly immersed platinum and the formation of mixed potentials between near-surface and more deeply submerged parts of the platinum wire.

Independent of the method chosen, the current through the working electrode consists of three parts. (a) The Faradaic current i_F through the melt/electrode interface represents the rate by which the redox material is reduced or oxidized at the electrode, which is the quantity of interest. It is necessarily accompanied by (b) the charging current i_C which charges the condenser formed by the layerwise arrangement of ions at the electrode surface ("double layer capacity"). It can be largely reduced or eliminated by certain modes of current measurement (see below). (c) A further possible complication is a Faradaic current due to the electrolysis of impurities or melt constituents discharged at the potentials of interest. This so-called background or residual current i_B can either be neglected or measured separately on the "blank" melt [3.159] and subtracted from the total current measured. Also, background currents limit the useful potential range called the "electrolytic window" of melts, whose span is given by the melt composition [3.159]. They thus also limit basically the applicability of electrochemical *in situ* measurements in melts.

Chronopotentiometry. In chronopotentiometry, the potential of the working electrode is measured as a function of time while it is subjected to a constant cathodic current (density), which is maintained by an appropriate amperostat, $E = f(t)_{i=\text{const}}$. The measurements are carried out in the millisecond to hundred-millisecond range. As indicated in the schematic diagram given in Fig. 3.26, the potential shows an initially steep increase due to the charging of the double layer. This step-like change is followed by a slow negative increase of the potential, which drives the electrolysis of appropriate cations, say cadmium ions, and the connected deposition of metallic cadmium at the working electrode due to the condition of constant current. When the concentration of the cadmium ions at the electrode surface approaches zero, the potential increases rapidly to more negative values determined by the "next" cathodic process, because then the current cannot be maintained by diffusion of cadmium ions from deeper parts of the electrolyte. The "next" electrolytic reduction may be the discharge of another kind of reducible ion dissolved in

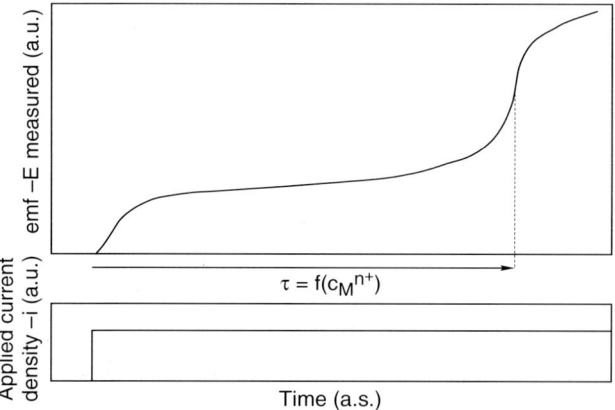

Fig. 3.26. Schematic chronopotentiogram of a reversible metal deposition from a glass melt

the melt, or the reduction of a melt constituent if the potential has reached the potential boundary given by the electrolytic window. The time necessary for discharging all cadmium ions at the melt/electrode interface is called the transition time t_{tr}. It depends on the concentration of the cadmium ions in the melt, which makes the method applicable for quantitative analysis if appropriate conditions are chosen.

Chronopotentiometry was first applied to glass melts as "solvents" by *Takahashi* and *Miura* in 1979 [3.196, 197] and again six years later by *Strzelbicka* and *Bogacz* [3.161]. However, it soon became apparent that an exact evaluation of the curves measured becomes increasingly difficult when the electrolytic windows of the melts investigated become more narrow. The first mentioned authors thus turned to a different *in situ* method and applied cyclic voltammetry more successfully to the problems at hand [3.52, 159, 160].

Cyclic Voltammetry or Potential Sweep Voltammetry. Different from chronopotentiometry, the working electrode in cyclic voltammetry is subjected to a controlled potential, which usually changes linearly with time, $E = E_i \pm (\mathrm{d}E/\mathrm{d}t)t$, where E_i is the initial potential and $(\mathrm{d}E/\mathrm{d}t)$ is the sweep rate. It thus experiences a triangular wave form with an initial cathodic and a following anodic sweep. The current through the melt/electrode interface is measured as a function of the changing potential (or time), $i = f(E)_{\mathrm{d}E/\mathrm{d}t=\mathrm{const}}$. When, during the cathodic sweep, the potential approaches the equilibrium potential of a redox couple dissolved in the melt, the oxidized form of this element is reduced, which is indicated by a cathodic peak in the current/potential curve. During the following anodic sweep, the particles thus reduced are re-oxidized near the equilibrium potential, and this causes an anodic peak in the voltammetric curve, which appears at more anodic potentials than the initial cathodic peak [3.55, 3.90, 3.159, 3.55, 90, 159], Fig. 3.27. The peak potentials, consequently, yield the standard potential E^0 of the polyvalent element in the glass melt either according to the expression for the cathodic sweep potential,

$$E_{\mathrm{p,cathodic}} = E^0 - 1.11\frac{RT}{nF} , \qquad (3.114)$$

or for the anodic sweep potential [3.159],

$$E_{\mathrm{p,anodic}} = E^0 + 1.09\frac{RT}{nF} , \qquad (3.115)$$

both of which were derived theoretically [3.198, 199]. The potential difference between the peaks amounts to

$$E_{\mathrm{p,anodic}} - E_{\mathrm{p,cathodic}} = +2.19\frac{RT}{nF} , \qquad (3.116)$$

which reveals that the separation of the cathodic and the anodic peak of the voltammogram is expected to be proportional to the absolute temperature

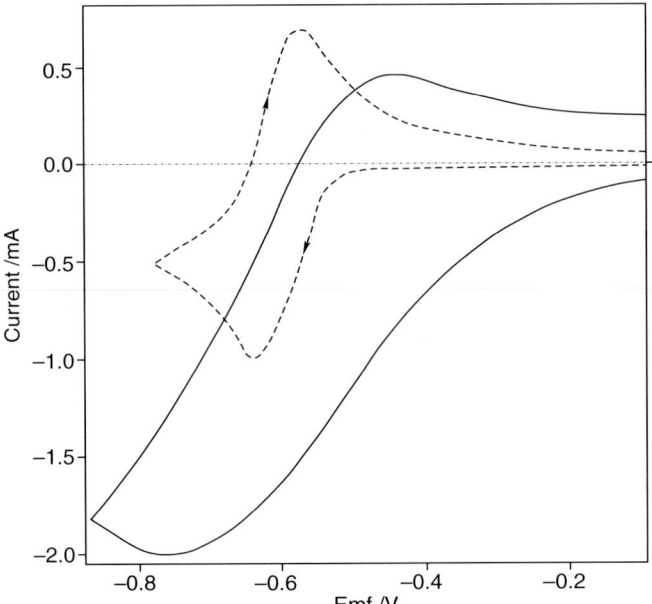

Fig. 3.27. Typical cyclic voltammograms for reversible one-electron reduction at different temperatures; (—) measured in a melt at $1100\,^{\circ}$C ($1\,$mol%), (- - -) calculated for $25\,^{\circ}$C ($1\,$mol% Fe(III)). Other relevant data comparable. After [3.55]

and inversely proportional to the charge n exchanged at the electrode. In addition, the peak current i_p (the current at the peak potential) is given by the theoretical equation [3.159]

$$i_\mathrm{p} = 0.447 n F A c_\mathrm{O}^2\, D_\mathrm{O}^{1/2} \left(\frac{zFv}{RT}\right)^{1/2}, \tag{3.117}$$

where c_O is the bulk concentration and D_O the diffusion coefficient of the oxidized form of the redox ions, A is the electrode area, and $v = \mathrm{d}E/\mathrm{d}t$ is the sweep rate. According to (3.117), consequently, the diffusion coefficient of the oxidized species is attainable from the measured peak potential since the other quantities on the right side of (3.117) are known (see, however, below).

In addition to the given equations (3.114)–(3.117), which are valid for a reversible redox equilibrium of soluble reduced and oxidized species, *Takahashi* and *Miura* also presented sets of equations for irreversible processes and for reversible metal depositions [3.159]. Moreover, Takahashi and Miura were probably the first to measure electrochemical series in glass melts (two silicate melts, $1250\,^{\circ}$C) by means of an *in situ* electrochemical method and to combine data of different redox couples to obtain the mutual equilibrium constants [3.159]. Incidentally, they later added more data obtained by cyclic voltammetry in silicate and borate melts at various temperatures (1100–

1400 °C) [3.52]. Meanwhile other authors have also reported on the successful application of this analytical tool [3.200–206].

However, two critical points concerning equations (3.114)–(3.117), are worth mentioning.

(a) The equations are based on the assumption of 100% initial concentration of the oxidized species (and thus the initial absence of the reduced form) of the redox couple [3.159]. Because this condition is only rarely given, *Sasahira* and *Yokogama* presented extended equations that take into account an initial finite concentration of the reduced form of the polyvalent element [3.200]. These rather complicated extended expressions should actually be applied to the evaluation of experimental results, which, however, would necessitate knowledge of the relative initial concentrations of both redox species. Even if this makes the application of the extended equations too complicated for routine applications, the expressions, together with estimated relative initial concentrations, at least give an estimate of the accuracy of the results obtained, which should be stated together with the results.

(b) Some doubt may arise concerning the physical meaning of the quantity D_O in (3.117) even if c_O is assumed to be correct. Although surrounded by charged species, the oxidized ion of the redox couple moves towards the electrode not only due to its concentration gradient caused by the electrolytic depletion of this species near the electrode but also because of its charge, which causes an ionic migration towards the electrode due to the electric field. The initial movement of the oxidized ion is even solely caused by the electric field! This is obviously of no concern for the determination of standard potentials. But it can be questioned whether the quantity D_O is really the diffusion coefficient of the oxidized ion, or whether it rather represents some mixed quantity consisting of diffusion coefficient and ionic mobility. Since diffusion coefficient and mobility differ quantitatively according to the Einstein equation, $D/u_{abs} = kT$, where u_{abs} is the absolute mobility, it will probably be difficult to obtain a detailed expression of D_O with a physical meaning. However, it seems to be misleading qualitatively and quantitatively to call the D_O "diffusion coefficients of the oxidized ion of a redox couple" [3.159]. That this quantity has even been called the "self-diffusion coefficient" of dissolved ions in general [3.207], specifically of Fe^{3+} [3.46], and also without specifying the self-diffusing species [3.45], shows that its meaning has obviously not been understood yet.

Square-Wave Voltammetry. One of the major drawbacks of cyclic voltammetry is that each cathodic and anodic potential sweep involves an uninterrupted, continuous sequence of potentials within the entire voltage range applied. Besides, the separation of the cathodic and anodic current peak increases with increasing temperature, as indicated by (3.116), and the increase of this separation is accompanied by a broadening of the peaks. These effects result in a reduction of sensitivity and resolution of the current peaks obtained [3.90], which can even exclude the application of cyclic voltammetry in certain instances.

These shortcomings are reduced with so-called pulse voltammetric techniques. They are based on a modification of the voltage sweep such that the momentary, electrolysing voltage is applied, and the resulting current is measured, only during short pulses. Each cathodic pulse thus generated is directly followed by an anodic pulse which reverses the electrolysis so that each cathodic pulse starts with a "new" cathodically clean electrode surface. Besides, the time during which the current is measured is chosen shorter than, and at the end of, each pulse of the electrolysing voltage. This nearly eliminates the charging current because charging currents decay much more rapidly than Faradaic currents [3.163]. In addition, some of the pulse voltammetric techniques are designed as differential methods: the difference between cathodic and anodic currents for each pulse is measured and yields a pulsating direct current with high frequency, which is plotted as a function of the changing voltage. This increases the sensitivity and resolution of the voltammetric method because the difference of a (positive) anodic and a (negative) cathodic current represents the sum of their absolute values, $\Delta i = i_+ - i_- = i_+ + |i_-|$, which exceeds either of them [3.162–166].

Various forms of pulse voltammetric techniques have been applied to the investigation of redox equilibria in glass melts [3.208–213], and two voltammetric sensors for use in technical glass melting tanks have been constructed [3.95, 214]. Obviously, the most successful modification is square-wave voltammetry (SWV), which has been introduced to, and extensively applied in, glass melts by *Rüssel* and coworkers [3.45, 46, 55, 90, 91, 208, 215–221]. SWV, based on the principles of pulse voltammetric methods, applies the sum of a synchronized square-wave and staircase potential to a stationary (platinum) electrode. Figure 3.28 demonstrates schematically the potential-time waveform applied, and in Fig. 3.29 is plotted the (normalized) response of the cathodic, anodic, and differential currents on the potential normalized with respect to the peak or standard potential for a soluble reversible redox couple (for the definition of the normalized current i_n, see [3.163]). An overview comparing the various pulse voltammetric techniques in glass melts has been presented [3.208], and a comparison of redox analyses by square-wave voltammetry of melts and by spectroscopy of quenched glasses is reported in [3.209].

For the evaluation of the curves measured it is important that the peak potential equals the standard potential of the redox couple, $E_p = E^0$ [3.90], and that the peak current in melts is represented by the expression [3.91]

$$i_p = \frac{0.31 A n^2 F^2 D_O^{1/2} E_{SW} c_O}{RT \tau^{1/2}} \,, \tag{3.118}$$

where A is the electrode surface area, n is the number of electrons exchanged, D_O and c_O are the diffusion coefficient and the bulk concentration, respectively, of the oxidized species (see, however, the comment above), E_{SW} is the

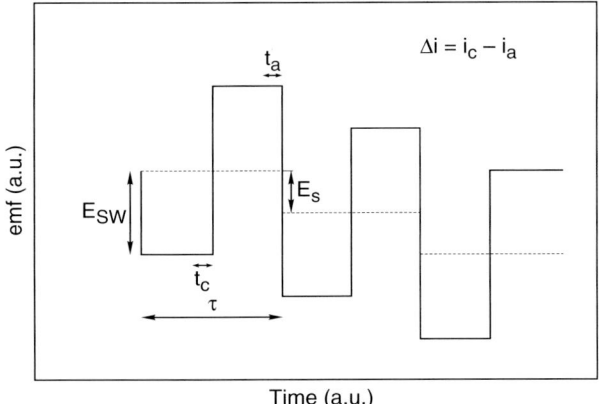

Fig. 3.28. Schematic of cathodic square-wave voltammetric potential sweep. t_c, t_a = measuring times of cathodic and anodic current, E_s = potential step

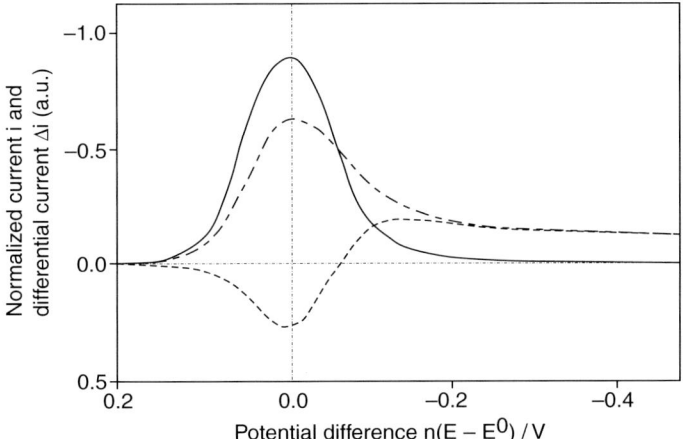

Fig. 3.29. Response to a cathodic SWV potential sweep for a reversible electrode process (after [3.163]). $i_{n,c}$ (– – – –), $i_{n,a}$ (- - - -), and Δi_n (———) = normalized cathodic, anodic, and resulting differential current, respectively, E^0 = standard potential [3.90]

square-wave amplitude, and τ is the pulse time. Typically, a pulse time of 1–100 ms and a square-wave amplitude of 50–250 mV have been applied in melts [3.91, 217]. These numbers may be the reason for using (3.118) in melts, which is actually the equation describing a square-wave polarographic peak current rather than an SWV peak current. According to (3.118), the peak current is proportional to the square of the number of electrons exchanged and inversely proportional to the temperature, and also in SWV, the peaks broaden with increasing temperature. This is particularly obvious from reported half-width peaks at the different temperatures, which are proportional to the absolute temperature according to $E_{P/2} = T b/n$, where b is $0.30 \, \mathrm{mV \, K^{-1}}$ and n is the number of exchanged electrons. The (identical) calculated and measured half peak width in aqueous $10^{-3} \, \mathrm{M \, Fe(III)}$ solution at $25 \, ^\circ\mathrm{C}$ is $0.09 \, \mathrm{V}$ (E_{SW} = 5–50 mV) [3.164], whereas in a melt containing $1 \, \mathrm{mol}\%$ $\mathrm{Fe(III)}$ at $1100 \, ^\circ\mathrm{C}$ it is calculated to be $0.41 \, \mathrm{V}$ ($E_{\mathrm{SW}} = 100 \, \mathrm{mV}$) [3.55], all other relevant data being comparable. The broadening of the peaks is particularly serious in melts of multi-component production glasses containing small concentrations of various reducible ions, which thus necessitate the development of special deconvolution methods for the evaluation of experimental curves.

3.5.3 Sulphate Fining

Detlef Köpsel

Sulphate is one of the most widespread fining agents, particularly for float and container glasses. Often it is used in combination with carbon or any similar reducing agent.

Redox States of Sulphur in Glass Melts

Several oxidation states of sulphur are known in glass melts. Which of it predominates depends on the oxygen partial pressure.

S^{+6}	chemically dissolved as sulphate SO_4^{2-} under oxidizing conditions	[3.222, 223]
S^{+4}	physically dissolved as SO_2 and possibly as chemically dissolved sulphite SO_3^{2-}, but in very low concentrations	[3.224]
S^0	only physical dissolution possible	[3.223, 224]
S_x^{2-}, S_y^-	chemically dissolved sulphides and poly-sulphides under reducing conditions	[3.224, 225]

The physical solubility is very low compared with chemical solubility [3.226]. Therefore it may be neglected in the discussion of the role of sulphur during fining. Two conditions under which sulphur exists in glass melts must be distinguished: oxidizing and reducing states.

Solubility Under Oxidizing Conditions

Under oxidizing conditions sulphur is dissolved in glass melts as sulphate:

$$SO_2 + 1/2 \, O_2 + O_{(glass)}^{2-} \rightleftarrows SO_4^{2-} \, . \tag{3.119}$$

The well-known equilibrium between the sulphur oxides SO_2 and SO_3 in the gas phase is shifted to the SO_2 side at temperatures above $1000 \, ^\circ\mathrm{C}$:

$$SO_3 \rightleftarrows SO_2 + 1/2 \, O_2 \, , \tag{3.120}$$

and the equilibrium constant for reaction (3.119) may be written as

$$K = \frac{a_{SO_4^{2-}}}{a_{O^{2-}} \, f_{SO_2} \, f_{O_2}^{1/2}} \, , \tag{3.121}$$

where f_i denotes the fugacity of the gases and a_i the activity of the components in the glass melt. Assuming ideal behaviour of the gas phase (i.e., fugacity is equal to the partial pressure $f_i = p_i$), constant activity of free oxygen ions in the glass melt ($a_{O^{2-}} = \mathrm{const}$), and constant activity coefficient for the dissolved sulphate ($a_{SO_4^{2-}} = \gamma_{SO_4^{2-}} \, x_{SO_4^{2-}}$ with $\gamma_{SO_4^{2-}} = \mathrm{const}$), (3.121) changes to

$$K' = \frac{x_{SO_4^{2-}}}{p_{SO_2} \, p_{O_2}^{1/2}} \, , \tag{3.122}$$

where x_i means the mole fraction of components in the glass melt. If sulphur were dissolved as SO_3^{2-} under oxidizing conditions,

$$SO_2 + O_{(glass)}^{2-} \rightleftarrows SO_3^{2-} \, , \tag{3.123}$$

the concentration of sulphur would not depend on the oxygen partial pressure. If sulphate SO_4^{2-} and sulphite SO_3^{2-} are coexisting, the plot of sulphur solubility against oxygen partial pressure should have a slope less then $1/2$ as (3.122) predicts. In many glass systems (3.122) is fulfilled [3.15, 154, 222, 227] which confirms the incorporation of sulphur mainly as sulphate. The solubility of sulphate in a soda-lime-silica glass (16% Na_2O 10% CaO 74% SiO_2) is shown for different partial pressures of SO_2 and O_2 as a function of temperature in Fig. 3.30.

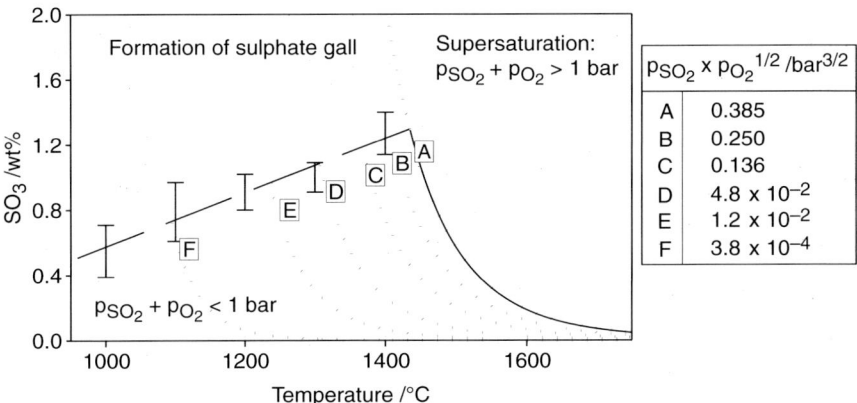

Fig. 3.30. Miscibility gap between soda-lime-silica glass (16% Na_2O 10% CaO 74% SiO_2) and sodium sulphate and equilibrium concentrations of sulphate for various O_2 and SO_2 partial pressures under oxidizing conditions [3.222]

Besides the chemical equilibrium a second phase equilibrium determines the solubility of sulphate at relatively high sulphate additions [3.222, 228–230]:

$$2M^+_{(glass)} + SO^{2-}_{4(glass)} \rightleftarrows M_2SO_{4(sulphate\ gall)} . \tag{3.124}$$

For the binary sodium silicate system this miscibility gap is well known (Fig. 3.31). The sulphate gall is formed as a separate phase on the glass melt, because of its lower density. For soda-lime-silica glass it consists mainly of Na_2SO_4. If only $CaSO_4$ is added in excess to the batch the following reaction takes place [3.231]:

$$CaSO_{4(excess)} + 2Na^+_{(glass)} \rightleftarrows Na_2SO_{4(gall)} + Ca^{2+}_{(glass)} . \tag{3.125}$$

The extension of the miscibility gap depends on temperature, i.e., with increasing temperature the solubility of sulphate added in excess increases if the decomposition of sulphate is low (Fig. 3.30). The equilibrium constant of reaction (3.121) is estimated in the following way [3.222]:

- The sulphate gall consists mainly of Na_2SO_4 ($x_{Na_2SO_4} \geq 99\,mol\%$), i.e., the activity of sulphate is close to unity and the activity coefficient may be expressed as

$$\gamma_{Na_2SO_4} = \frac{a_{Na_2SO_4}}{x_{Na_2SO_4}} \approx \frac{1}{x^{sat}_{Na_2SO_4}} = f(T) . \tag{3.126}$$

- The equilibrium constant in reaction (3.121) is calculated by using the thermodynamic data for the pure substances [3.232].
- It is further assumed that the activity of the free oxygen ions is mainly determined by the activity of Na_2O. The activity of sodium oxide is calculated from compiled literature data [3.233–236].

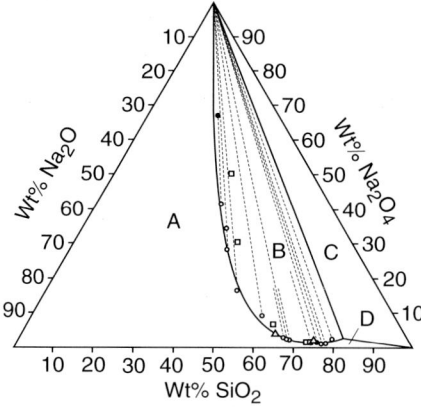

Fig. 3.31. Miscibility gap in the Na_2O–SiO_2–Na_2SO_4 system at 1200 °C [3.229] (A) 1 liquid, (B) 2 liquids, (C) 2 liquids + tridymite, (D) 1 liquid + tridymite

If in addition K' in (3.122) is obtained from K in (3.121) as explained above, the following thermodynamic data are derived: $\Delta H \approx 345\,\mathrm{KJ\,mol^{-1}}$, $\Delta S \approx 231\,\mathrm{J\,mol^{-1}\,K^{-1}}$.

Solubility of Sulphur Under Reducing Conditions

The solubility of sulphur in oxidation states less than $+6$ predominates at oxygen partial pressures $p_{O_2} \leq 10^{-4}\,\mathrm{bar}$, depending on temperature and glass composition [3.237–239]. For a slag composition (37% CaO 27% Al$_2$O$_3$ 36% SiO$_2$) the solubility of sulphur under both reducing and oxidizing conditions is shown in Fig. 3.32 [3.237]. With increasing temperature the minimum of sulphur solubility is shifted to higher oxygen partial pressures. The solubility of sulphate is decreased and the sulphide solubility is increased. Because several polysulphide chains are found besides the single sulphide ion S^2, the reaction of sulphide solubility should be generally expressed in the following way:

$$z\,SO_2 + O^{2-}_{(glass)} \rightleftharpoons S_z^{2-} + (z + 1/2)O_2 \ . \tag{3.127}$$

The corresponding equilibrium constant should be

$$K' = x_{S_x^{2-}}\,\frac{p_{O_2}^{(z+1/2)}}{p_{SO_2}^z} \ . \tag{3.128}$$

For example, in a colourless alkali borosilicate glass the factor z was determined to be unity [3.154]. Proportionally high amounts of polysulphides cause a colouration of the glass from blue and green over yellow to brown and

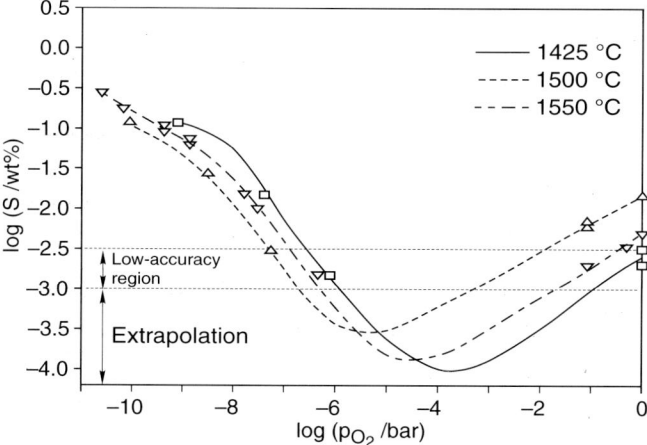

Fig. 3.32. Solubility of sulphide and sulphate in a 37% CaO–27% Al$_2$O$_3$–36% SiO$_2$ melt at $p_{SO_2} = 0.01\,\mathrm{bar}$ as a function of the oxygen partial pressure [3.237]

red, depending on the length of the polysulphide chains. Fining under both oxidizing and reducing conditions is possible and will be briefly described in the following paragraphs.

Fining With Sulphate Under Oxidizing Conditions

The vapor pressure of alkali sulphates in the temperature range below $1600\,^{\circ}C$ is very low [3.239–242] and does not considerably support fining as is sometimes assumed. The usable amount of fining agent is limited because of sulphate gall formation (Fig. 3.30). Depending on the chosen amount of sulphate the total sulphate decomposition pressure according to reaction (3.119) attains the value of 1 bar in a temperature range above $1450\,^{\circ}C$ for soda-lime-silica glass (Fig. 3.30). The gas release due to sulphate decomposition is shown in Fig. 3.33. During a temperature increase of the melt the first amount of both SO_2 and O_2 is already released in the rough melt within the temperature range around $1100\,^{\circ}C$. This is the result of the reaction between sulphate and sand. Although the impact of this first reaction of sulphate on fining is low, it accelerates the dissolution of sand. The release of the actual fining gas is observed at higher temperatures ($1400\,^{\circ}C \leq T \leq 1500\,^{\circ}C$) and leads to considerable bubble growth [3.244–247]. The remaining sulphate concentration in the glass after fining is usually less than the finally chosen amount by a factor of more than 2.

Fining With Sulphate in Combination With Reducing Agents. Usually carbon is used as reducing agent. The reactions between sulphate and carbon during

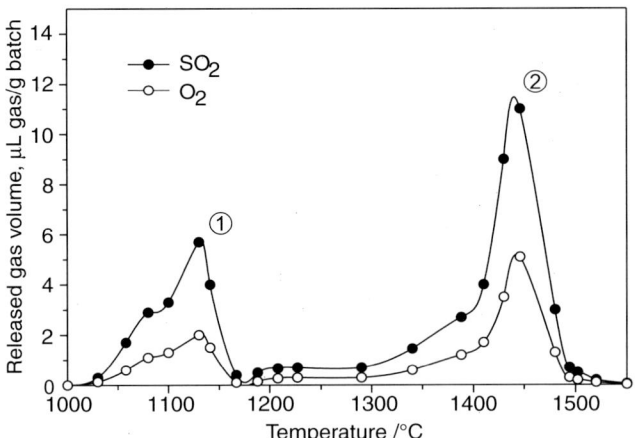

Fig. 3.33. Gas evolution during fining of a TV-glass melt containing sulphate under oxidizing conditions as a function of temperature (redox number $\approx +45$). (1) Sand dissolution (rough melt): $Na_2SO_4 + SiO_2 \rightarrow Na_2SiO_3 + SO_2 + 1/2\,O_2$, (2) fining: $SO_4^{2-} \rightarrow O^{2-} + SO_2 + 1/2\,O_2$ [3.243]

fining of glass under neutral or reducing conditions are very complex. There are some reactions necessary for fining and some side reactions which do not contribute to the fining process. In the first melting stage the carbon is oxidized and according to Boudouard's equilibrium ($T < 1000\,^{\circ}\mathrm{C}$) carbon monoxide is formed as the major product [3.249]:

$$2C + O_2 \rightleftarrows 2CO \tag{3.129}$$

In the next step carbon monoxide reacts with sulphate:

$$SO_4^{2-} + 4CO \rightleftarrows S^{2-} + 4CO_2 \ . \tag{3.130}$$

In the rather rare case of contact between carbon and sulphate also a direct reaction may take place:

$$SO_4^{2-} + 2C \rightleftarrows S^{2-} + 2CO_2 \ . \tag{3.131}$$

The sulphate must not completely be transformed to sulphides according to (3.130) or (3.131). This is because a part of the sulphate should be left for the essential fining reaction

$$SO_4^{2-} + S^{2-} \rightleftarrows SO_2 \uparrow + O_{(\mathrm{glass})}^{2-} \ . \tag{3.132}$$

Under reducing conditions the release of fining gas according to (3.132) takes place at lower temperatures than during fining under oxidizing conditions (Fig. 3.34). In the example in Fig. 3.34 sulphate is left in excess because of the low carbon additions. Therefore a second peak for gas release is observed

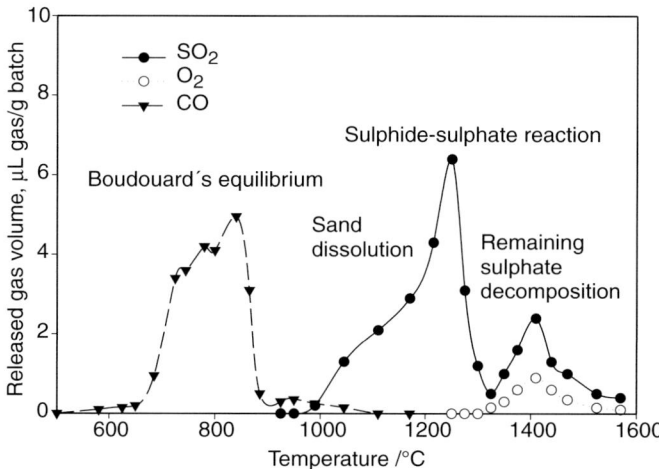

Fig. 3.34. Gas evolution during fining of a TV-glass melt containing sulphate and carbon under low reducing conditions as a function of temperature (redox number $\approx +2$) [3.243]

at a higher temperature due to the excess sulphate decomposition. The combination of (3.129), (3.130), and (3.132) yields a total fining reaction as given by

$$SO_4^{2-} + C + 1/2O \rightleftarrows SO_2 + CO_2 + O^{2-}_{(glass)} \ . \tag{3.133}$$

A minimum of remaining sulphur (as sulphate and sulphide) should actually be found in the glass after fining at a stoichiometric relation of $m_{Na_2SO_4}/m_c = 12$ according to (3.133), if all useless side reactions are neglected. In reality the minimum of total sulphur is found at more reducing conditions. The reason for that unexpected result is that on the one hand some carbon is oxidized and thus lost without any impact on the sulphate. On the other hand all raw materials and cullets already contain a certain amount of sulphur and reducing impurities. It should be taken into account, for instance, by determining the chemical oxygen demand (COD) of each batch component. On the basis of these COD values an empirical redox number can be calculated [3.250, 251]. The relation between redox number and remaining sulphur concentration in the glass is shown in Fig. 3.35. The most effective fining is achieved for soda-lime-silica glass for redox numbers within the range of –5 to –12 [3.248, 252], corresponding to an $[Fe^{2+}]/[Fe_{tot}]$ ratio of 45% to 58%.

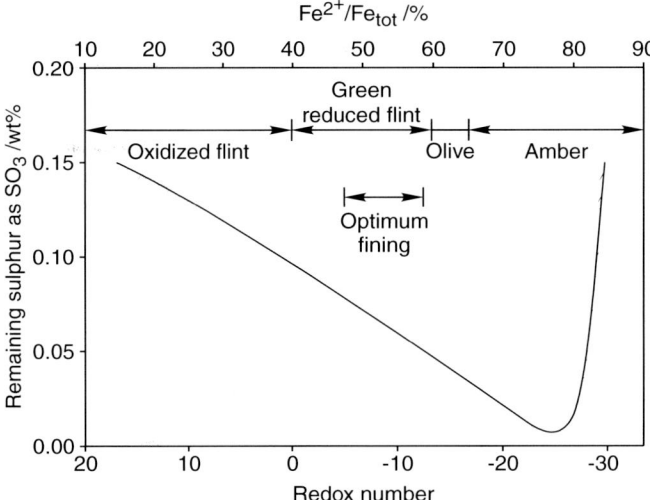

Fig. 3.35. Sulphur concentration remaining in the glass melt after fining as a function of the redox number [3.248]

3.5.4 Test of Technical Oxygen Sensors

Friedrich G.K. Baucke

As reported in Sect. 3.4, one type of oxygen sensor was developed especially for applications in industrial melters. It is characterized by a dissolving reference electrode with a zirconia electrolyte bridge between Pt,O_2 reference electrode and melt surface and a measuring electrode which carries a flag-like lower end for suppressing mixed potentials. Although these highly accurate units are frequently applied in technical melting tanks, where they yield correct data according to several tests, it was nevertheless of interest to compare their performance and the accuracy of the results with those obtained from sensors developed and constructed at a different laboratory. This opportunity was given by an oxygen sensor that had been particularly developed for melts of mass-produced articles at the Hüttentechnische Vereinigung der Deutschen Glasindustrie (HVG) in Frankfurt [3.253, 254] in parallel with the Schott sensors and is meanwhile commercially available under license of HVG and Schott Glas [3.255]. Comparative measurements were conducted in a container glass tank of the Glashütte Budenheim, Germany [3.256], where the parallel installation of sensors is less complicated and hence less costly than in tanks for special glass production as, for instance, at Schott Glas.

Sensor Characterization

The two sensors are constructed according to different philosophies. The Schott sensor is to yield highest accuracy as required for the production of optical and special glasses, whereas the HVG sensor, which is based on sensors developed earlier [3.106, 257], is mainly applied in the container and flat glass industry, where the use of recycled cullet and filter dust with a considerably varying redox state must be controlled and easy handling and low cost are the main objectives. As the schematic of the HVG sensor given in Fig. 3.36 shows, the reference electrode of this combined unit is also ionically connected to the melt via a zirconia bridge, which makes the sensor a nonisothermal cell as in the case of the Schott electrodes. The zirconia bridge of the HVG reference electrode consists of zirconia fully stabilized with yttria, that of the Schott reference electrode of partially stabilized material. The rod-shaped platinum measuring electrode of the HVG sensor is dipped into the melt for only 3–4 cm so that measurements are restricted to near-surface ranges of the melt. This eliminates the necessity of applying means for suppressing mixed potentials of different parts of the metal but also excludes any measurements in greater depths of the melt. Thermocouples are contained in the upper part of the reference electrode and at the lower end of the measuring electrode so that merely an approximate average value of the Seebeck coefficients of melt and zirconia can be applied to the evaluation of the measurements [3.257, 258]. Besides, the reference gas is supplied to the

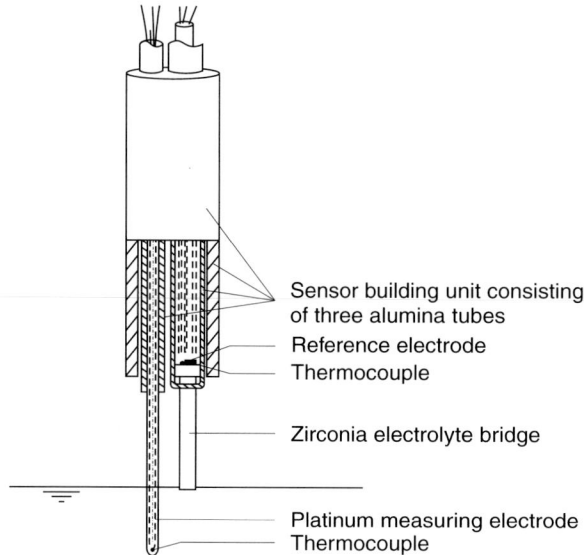

Sensor building unit consisting
of three alumina tubes
Reference electrode
Thermocouple

Zirconia electrolyte bridge

Platinum measuring electrode
Thermocouple

Fig. 3.36. Oxygen sensor as developed by HVG for application in the container and flat glass industry

HVG sensor through PVC tubing during the measurements, to the Schott reference electrode through stainless steel equipment (Swage-Lock®).

Location of the Oxygen Sensors

The sensors were installed in the feeder channel of a melting tank containing green glass melt (soda–lime glass with small concentrations of iron and chromium) for container glass production. This melt is distinguished by two constant standard Seebeck coefficients, $-0.575\,\mathrm{mV\,K^{-1}}$ at temperatures below $1020\,^\circ\mathrm{C}$ and $-0.453\,\mathrm{mV\,K^{-1}}$ above $1020\,^\circ\mathrm{C}$ (see Sect. 3.6). Measuring and reference electrodes were positioned in the flow direction, the measuring electrode being placed upstream. The different sensor types had a distance of about 20 cm perpendicular to the flow direction. The emfs of each sensor as well as the potential differences between either electrode of the different sensors were measured via separate contacts to reference and measuring electrodes. Measurements were carried out for three months; the HVG sensor was kept in the feeder channel for an additional three months.

The feeder channel was chosen as the location of the sensors for easy installation of and access to the electrodes and because the melt in this tank range was expected to be sufficiently homogeneous, which was essential for a meaningful comparison of the electrodes. However, the feeder channel is by no means a standard location of oxygen sensors in melting tanks. On the contrary, the actual place of oxygen-measuring electrodes depends on the objective of the measurements. Although a location near the drain of the melter, as in the case at hand, informs about the final quality of the melt and the glass articles produced, it causes long dead times for correcting errors resulting from raw materials or fining. If fining of a glass melting tank is to be controlled, the sensor or at least the measuring electrode should be positioned within or at the end of the fining chamber. The reaction zone (before the fining chamber), where the melt, still subject to chemical reactions, is not fully homogenized yet, is distinguished by the shortest dead time with respect to the raw materials. But it necessitates the measuring and reference electrode to be positioned very closely to each other because a uniform melt basicity of the melt, which is the basis of the measurements (see Sect. 3.4.1), has not yet been reached. Except for special investigations, oxygen sensors will thus generally not be positioned appreciably before the fining range of the melter.

The discussion shows that the positioning of oxygen sensors is a critical subject and that each melting tank and each glass melt produced actually requires specific locations of the sensors. Moreover, it must closely be examined whether or not several measuring electrodes positioned at different locations may be combined with only one reference electrode for the measurements, even if the standard Seebeck coefficient of the final melt is known. A "several measuring-one reference electrode arrangement", although more economical than a one-to-one application, can certainly not be recommended for the rough melt. According to the experience gained so far, the best compromise for normal production control is probably to install four oxygen sensors, one within the fining chamber, one at the entrance and one at the exit of the resorption zone, and a fourth sensor near the drain of the tank, for example in the feeder channel. This will allow one (a) to control the fining process, that is, to measure whether the oxygen partial pressure in the fining chamber reaches values sufficiently above 1 bar (see Sect. 3.5.1), (b) to measure the effect of the fining, which is indicated by the relative oxygen partial pressures before and after the resorption step, and (c) to characterize the final melt before it leaves the tank for the forming machines. However, sensor application for quality control implies that the melt near the drain is homogeneous also with respect to its oxygen content over the entire cross section of the feeder channel so that the oxygen partial pressure in this range is merely a function of temperature (see below).

Zirconia Reference Electrodes

After three months of continued contact with flowing melt, both the partially stabilized zirconia of the Schott reference electrode and the fully stabilized zirconia of the HVG sensor showed only traces of corrosion. In addition, no effect of the extended use upon the measurements could be detected. This was verified by comparing the oxygen partial pressures measured at the end of this period with those of newly installed sensors, which were equal within a few percent. Because the lifetime of oxygen sensors with a zirconia bridge is determined by the corrosion rate of the ceramic (see Sect. 3.4.3), this result means that the service life of both sensors in green glass melt at temperatures between $1100\,^{\circ}\mathrm{C}$ and $1200\,^{\circ}\mathrm{C}$ amounts to at least one year, independent of the yttria content of the zirconia. It is noted, however, that the corrosion rate of stabilized zirconia depends strongly on melt composition and must be determined before a sensor application.

The electrochemical performance of the different zirconia reference electrodes was compared by measuring their potential difference for extended periods, during which the reference gases were repeatedly changed between oxygen and synthetic air (argon with 20 vol% oxygen) in the case of the Schott sensor and between oxygen and air (containing 21 vol% oxygen) in the case of the HVG sensor. Each combination was measured for at least 40 min. Typical results are given in Table 3.9, where the upper part shows the conditions of the measurements, sensor temperatures and reference gases, and the lower part contains the potential differences measured, $\Delta\varphi_{\mathrm{meas'd}} = \varphi_{\mathrm{HVG}} - \varphi_{\mathrm{Schott}}$, and calculated, $\Delta\varphi_{\mathrm{calc'd}}$. For the calculation, the different temperatures of the $\mathrm{Pt,O_2}$ references of the zirconia electrodes ($1118\,^{\circ}\mathrm{C}$ of HVG and $1132\,^{\circ}\mathrm{C}$ of Schott electrode) and the different oxygen contents of the respective reference gases air and synthetic air were taken into account. Besides, the bottom line of Table 3.9 gives the difference of measured and calculated potential differences. They show good agreement of the sensor performance, the maximum difference being $\Delta(\Delta\varphi) = +1.1\,\mathrm{mV}$ in all cases, of which only a few are presented as examples in Table 3.9. However, measurements with the HVG sensor operated with pure oxygen are an exception, see columns 5 and 6 of Table 3.9. This surprising disagreement was traced back to the oxygen used as the reference gas for the HVG zirconia electrode, which was continuously diluted by nitrogen (by up to 9 vol%), which diffused through the rather long PVC tubing between the oxygen container and the electrodes. Obviously, PVC is not sufficiently impervious at temperatures as high as those prevailing near the melting tank (up to $100\,^{\circ}\mathrm{C}$ by heat radiation). The ideal solution of this problem is to apply stainless steel tubing as with the Schott sensors, which is highly recommended for such critical measurements as carried out here. However, the instant solution urgently needed for a continuation of the

Table 3.9. Electrochemical performance of Schott and HVG zirconia reference electrodes

Sensor	$T\ /^{\circ}\mathrm{C}$		Reference gas			
HVG	1118	air	air	O_2	O_2	air
Schott Glas	1132	O_2	synth. air	synth. air	O_2	O_2
$\Delta\varphi_{\mathrm{meas'd}}/\mathrm{mV}$ [a]		-39.0	$+8.0$	$+52.0$	$+4.0$	-39.0
$\Delta\varphi_{\mathrm{calc'd}}/\mathrm{mV}$ [b]		-40.1	$+8.6$	$+55.4$	$+6.6$	-40.1
$\Delta(\Delta\varphi)/\mathrm{mV}$ [c]		$+1.1$	-0.6	-3.4	-2.6	$+1.1$

[a] $\Delta\varphi_{\mathrm{meas'd}} = \varphi_{\mathrm{HVG}} - \varphi_{\mathrm{Schott}} =$ potential difference measured
[b] $\Delta\varphi_{\mathrm{calc'd}} =$ potential difference calculated
[c] $\Delta(\Delta\varphi) = \Delta\varphi_{\mathrm{meas'd}} - \Delta\varphi_{\mathrm{calc'd}}$

comparative measurements was to further apply only air as the reference gas of the HVG sensors, which removed the problem by eliminating oxygen diffusion into the tubing.

Sensors and Platinum Measuring Electrodes

As the reference electrodes had been found to show an identical behaviour, comparing the performance of the different oxygen sensors actually meant comparing the performance of their platinum measuring electrodes. As an example, Fig. 3.37a shows the logarithmic plot of the oxygen partial pressures of the melt as measured by the different sensors over a period of 34 days. Both units exhibit small oxygen partial pressures between 2×10^{-4} bar and 7×10^{-3} bar, which correspond to the reducing character of the green glass melt. Surprisingly, however, the two traces, although showing nearly the same contours, differ by $\Delta \log p_{O_2} = 0.24$, Fig. 3.37a. The difference does not indicate different oxygen partial pressures in the bulk and in the surface range of the melt since this would certainly cause different potential traces. Besides, the oxygen partial pressure at the melt surface in contact with the atmosphere would be expected to be higher than that in the bulk of the melt and not lower, as observed. The difference was traced to different temperatures, which are caused by the strong cooling of the melt in this feeder section. Figure 3.38 sketches two vertical temperature profiles of the melt measured at the sensor location on two different days and the relative position of the measuring electrodes and shows that the temperature of the HVG electrode is considerably lower than the average temperature of the (flag-like) measuring part of the Schott electrode. As shown in Fig. 3.37b, the potential traces become nearly identical when the oxygen partial pressures measured by the HVG sensor are adjusted to 1250 °C, corresponding to the appropriate thermoelectric voltage of the melt. The measurements thus not only demonstrate that the oxygen sensors are reliable analytical tools but also emphasize the significance of correct temperature measurements when absolute and not merely relative oxygen partial pressures of glass melts are to be measured.

Additional Effects

The measurements described yielded two additional results [3.256], which may be of general interest.

(a) As has been frequently observed in laboratory and in technical melters, emfs of newly installed oxygen sensors show considerable drifts for various periods after their immersion. During our experiments, this "induction period" was shown to be a viscosity and equilibration effect of the platinum measuring electrode. For this purpose, one of two platinum electrodes of the same type was withdrawn from the melt, kept in the furnace atmosphere above the melt, and was reinserted after several hours. After its re-installation, the electrode

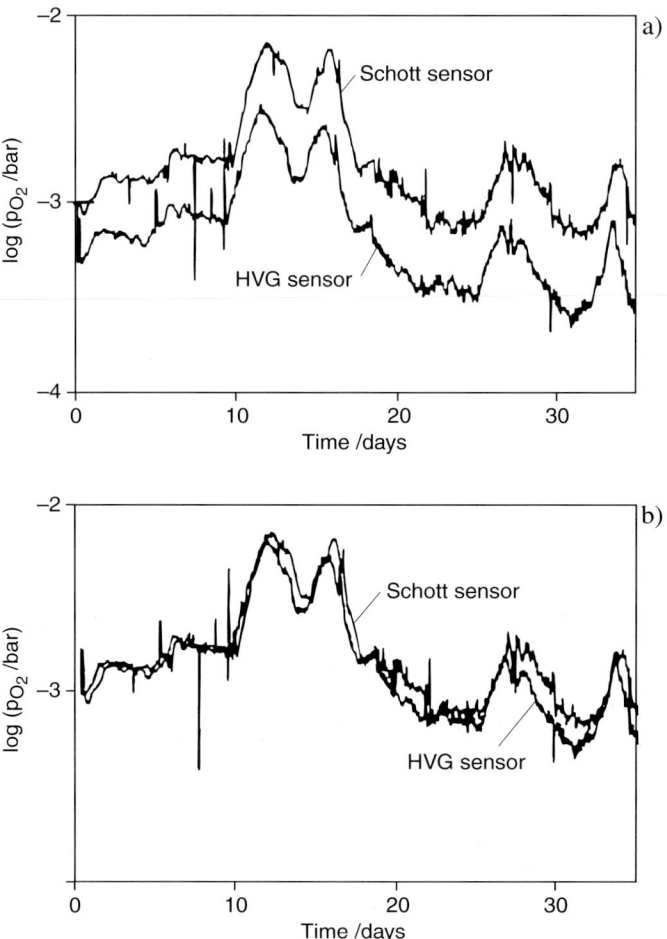

Fig. 3.37. Time-dependent oxygen partial pressure of green glass melt. (**a**) As measured by Schott and HVG sensors, (**b**) after temperature correction to 1250 °C of the HVG trace

showed a strong potential drift relative to the electrode that had remained in the melt. Obviously, the adhering melt took up oxygen from the atmosphere, whose subsequent re-equilibration with the oxygen of the melt, which has a viscosity-dependent rate, caused the observed drift. The exchange lasted about 15–20 min at 1200 °C, after which the electrodes exhibited an identical potential. This explanation is also valid for the newly installed platinum electrodes. A thin surface range of the melt acts like an extensible sheet of rubber that is pushed into the melt, covering the new electrode during its immersion. The layer of melt thus formed at the electrode surface has the oxygen partial pressure of the melt surface, which necessarily differs from that of the bulk

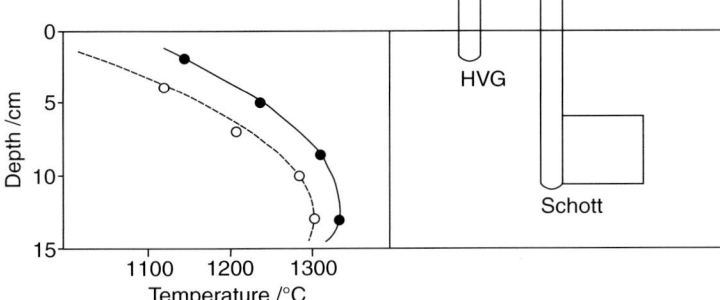

Fig. 3.38. Temperature–depth profiles of the melt at the location of the measuring electrodes (on two different days) (left), and relative immersion depths of the platinum electrodes in the melt (right)

of the melt. The equilibration of the different partial pressures is then measured as the potential drift, whose extension is called the "induction period". Although trivial, the effect should be known when platinum electrodes are applied in glass melts in order to avoid erroneous measurements and faulty explanations.

(b) After each increase of the tank load, the potential difference of two simultaneously and continuously applied platinum measuring electrodes indicated the appearance of oxygen inhomogeneities in the streaming melt. They were obviously caused by volume elements of highly viscous melt which were held back in cooler parts of the working chamber during periods of low load to be subsequently released at higher temperatures during high-load periods. Obviously, these melt volumes had small extensions because they were detected by either one or the other of the measuring electrodes despite an electrode distance of only 20 cm. The effect demonstrates that an oxygen sensor applied for quality control can well miss inhomogeneities of the melt, and it is also conceivable that locally restricted inhomogeneities are not detected even if several sensors are applied. The observation is of interest for glass melting because it not only demonstrates the importance of efficient mixing, even in the final production state of melts, but also emphasizes the significance of sampling, which obviously requires great attention, at least in the case of critical and sensitive glasses.

3.5.5 Electrolytic Fining

Friedrich G.K. Baucke

The preceding subsections have shown that fining is one of the most important processes in industrial glass melting. Without fining, no bubble-free glass and thus no useful optical, technical, appliance and household glass products could be produced, and no glass part containing a vacuum, for instance television tubes, would be safe from implosion. Correspondingly, several different fining processes, in particular redox and sulphate fining, have been developed. They have been briefly described, and it has been emphasized that these fining procedures are critical processes, which must be carefully controlled. It has also been mentioned that the most commonly used fining agents are not completely harmless with respect to the environment and are about to be replaced (arsenic oxide) by economically more compatible fining agents (antimony oxide) or should be replaced as soon as (and if) other compounds are available. Even a reduced concentration of these components is ecologically and economically desirable.

A discussion of possible solutions to these problems resulted in the proposal of a meanwhile patented alternative fining procedure [3.259, 260], which is based on oxygen as the fining gas, as is redox fining. Different from redox fining, however, the oxygen bubbles are not generated by a chemical reaction, the temperature-dependent dissociation of oxides, but by a physical process, the electrolysis of the glass melt. The use of fining agents is eliminated from the fining step and drastically reduced for the subsequent resorption period, during which stray oxygen bubbles are removed. The necessary amount of fining agent, consequently, is only a part of the quantity needed for redox fining and, in addition, depends on how many surplus oxygen bubbles are generated and how effectively they are removed during the fining period. Their quantity can thus be minimized. The principle of electrolytic fining is simple. Because, however, the electrolysis is characterized by strict but automatically controllable boundary conditions, the proposed procedure is described in some detail in the following.

Principle

Electrolytic fining is based on the anodic generation of oxygen bubbles at a platinum working electrode or "fining electrode", melt$|$Pt$(+)$, which is positioned at an appropriate location of the continuously working melter. The electrode reaction is described (for n moles of generated oxygen) by

$$2nO^{2-}(m) \rightarrow nO_2(g,b) + 4ne^-(Pt) , \qquad (3.134)$$

where the bracket (m) means melt and the bracket (g,b) denotes gas phase in the form of bubbles; oxygen traces necessary to saturate the melt physically are neglected. As indicated in Fig. 3.39, the fining anode is located at or near the bottom of the fining zone of the melter, from where the melt enters a quiet region where most of the bubbles (n_b moles), due to an uptake of impurity gases, have a diameter that increases with time, rise to the melt surface, and leave the melt:

$$n_bO_2(g,b) \rightarrow n_bO_2(g) . \qquad (3.135)$$

Some pure oxygen bubbles, however, which have smaller rise velocities because of their smaller radii, reach locations further downstream, see Fig. 3.39,

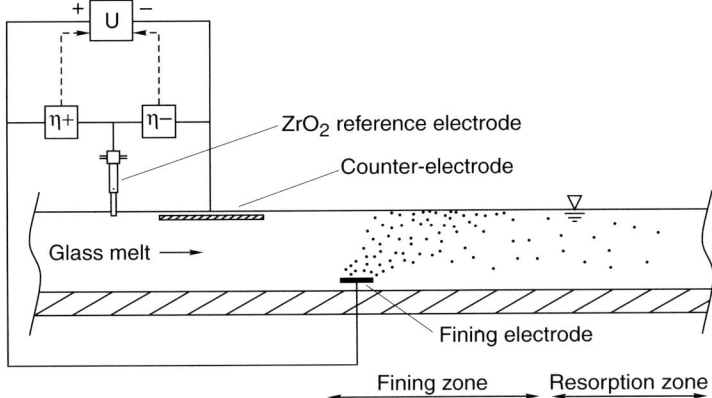

Fig. 3.39. Principle of electrolytic fining in a continuous melting furnace. The electrolysing voltage U is controlled by a zirconia reference electrode, which keeps the anodic, η_+, and cathodic polarization, η_-, and their sum U, within the "electrolytic window", see Fig. 3.40

and must consequently be eliminated before they enter into the glass product. Their removal is possible only by a resorption on the basis of polyvalent ions, for instance antimony, see Sect. 3.5.1,

$$(n - n_{\mathrm{b}})O_2(\mathrm{g,b}) + (n - n_{\mathrm{b}})Sb_2O_3(\mathrm{m}) \to (n - n_{\mathrm{b}})Sb_2O_5(\mathrm{m}) \ , \qquad (3.136)$$

so that the melt cannot be completely free of polyvalent elements. However, relatively small concentrations, compared to those usually applied for redox fining, will suffice. Thus, the melt, after leaving the fining zone, enters a resorption zone with a sufficiently low temperature and oxygen partial pressure, where the oxygen blisters are removed.

The galvanic cell is completed by a (cathodic) counter-electrode, which is located upstream from the fining electrode, Fig. 3.39, or in a by-pass and consists either of platinum or constitutes a zirconia electrode. The electrode reaction of a platinum counter-electrode, $(-)Pt,O_2$/melt, consists of the dissolution of oxygen in the melt,

$$nO_2(\mathrm{g}) \to nO_2(\mathrm{m}) \qquad (3.137)$$

and its reduction at the negative platinum electrode,

$$nO_2(\mathrm{m}) + 4ne^-(\mathrm{Pt}) \to 2nO^{2-}(\mathrm{m}) \ . \qquad (3.138)$$

The melt surrounding the counter-electrode, consequently, must continuously be saturated with oxygen. This necessitates that the electrode either has a relatively large surface area (see below) and is positioned near and parallel to the melt surface (if the melt contacts air atmosphere) or is continuously bubbled with oxygen.

The application of a zirconia counter-electrode, $(-)Pt,O_2|ZrO_2|$melt, eliminates the intermediate dissolution of oxygen in the melt because the oxygen is reduced at the platinum contacting the zirconia, through which the oxide ions generated are then transported into the melt,

$$nO_2(g) + 4ne^-(Pt) \rightarrow 2nO^{2-}(\text{external ZrO}_2 \text{ surface}) \tag{3.139}$$
$$\rightarrow 2nO^{2-}(\text{internal ZrO}_2\text{surface}) \rightarrow 2nO^{2-}(m) \ ,$$

where external and internal refer to the melt. The cross section of the zirconia wall, which is preferably tube-shaped, must be large enough to guarantee sufficiently small current densities and voltages which can be applied without danger. The application of large current densities causes a resistance decrease of the ceramic, as recently reported [3.261], and may even be advantageous for a long-time application of a zirconia cathode.

From the schemes of the proposed fining cells,

$$(-)O_2, Pt|\text{melt}|Pt(+) \tag{3.IX}$$

and

$$(-)O_2, Pt|ZrO_2|\text{melt}|Pt(+) \tag{3.X}$$

and the corresponding common cell reaction

$$nO_2(g) + (n - n_b)Sb_2O_3(m) \rightarrow n_bO_2(g,b) + (n - n_b)Sb_2O_5(m) \tag{3.140}$$

the approximate temperature profile of an electrolytically fined glass melt can be concluded. With both cells, the temperature of the fining electrode should be at or above the fining temperature of redox fining because the polyvalent oxide present should be completely dissociated so as not to compete with the electrolytic fining reaction. The temperature of the counter-electrodes is not critical and may be below the fining temperature. The temperature of the resorption zone, however, should be the same as that of the resorption zone of redox fining (for the same polyvalent element) in order to remove completely the remaining oxygen blisters.

Equations (3.134) and (3.138)–(3.140) involving the redox reaction of oxygen present only the essential particles, O_2 and O^{2-}, of the processes taking place. In detail, anodic and cathodic reactions will involve polymerization and depolymerization reactions, respectively, of the glass network. For example, the anodic generation of oxygen at the fining electrode is represented more completely by

$$4 \equiv SiO^-(m) \rightarrow 2 \equiv SiOSi \equiv (m) + O_2(g,b) + 4e^-(Pt) \tag{3.141}$$

than by (3.134). In both cases, the charge is balanced by an appropriate drift of ionic entities, for instance by alkali ions, between the electrodes. Participation of polyvalent ions in the electrode reactions is practically excluded because of the high temperatures of the melt at the fining electrode, which cause complete dissociation of the oxidized component and exclude an oxidation of the reduced ion.

Automatic Safety Control

The electrolysing voltage must be chosen such that the platinum electrodes are neither impaired anodically by oxidizing and dissolving the platinum nor cathodically by reducing melt components, for instance silicon, which react or alloy with platinum and thus also damage the electrodes. Figure 3.40 shows a schematic diagram of the (identical) current density-potential curves I, $i/A = f(\varphi)$, of the fining and the counter-electrode, which are assumed to have the same temperature. The equilibrium potential of the platinum electrodes is assumed to be φ_0, and it is seen that the application of a voltage U between the electrodes causes an anodic polarization $\eta_+ = (\varphi_+ - \varphi_0) > 0$ and a (positive) anodic current density $i_+/A_+ > 0$ as well as a cathodic polarization $\eta_- = (\varphi_0 - \varphi_-) < 0$ and a (negative) cathodic current density $i_-/A_- < 0$ (where A_+ and A_- are the surface areas of the anodic and cathodic electrode, respectively). The potentials are related by the requirement that anodic and (absolute) cathodic currents through the electrodes are equal, $i_+ = |i_-|$. Also given in Fig. 3.40 are the current density-potential curve II for the (anodic) platinum oxidation to platinum ions and curve III of the (cathodic) reduction of silicon. The potential difference between the beginning of these curves (at zero current density) may be called the "electrolytic window", which limits the anodic and cathodic potentials applied during the fining electrolysis.

The required rate of oxygen bubble formation and the equivalent electric current through the platinum anode, in combination with the maximum current density allowed according to Fig. 3.40, determine the maximum cur-

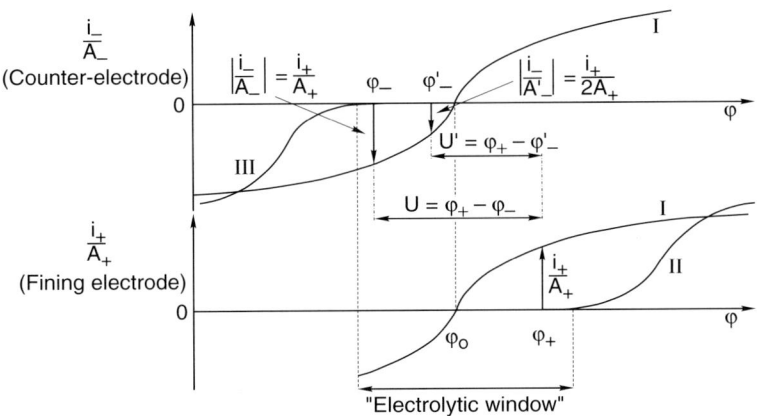

Fig. 3.40. Principle current density-potential curve of the reaction $O_2(m) + 4e^-(Pt) \rightleftarrows 2O^{2-}(m)$ at platinum in an oxidic glass melt (identical curves I) and its anodic and cathodic polarization by an outside voltage U as used for electrolytic fining. The current density-potential curves of platinum oxidation (curve II) and melt decomposition (curve III) limit the applicable polarizations, $\eta_+ = \varphi_+ - \varphi_0$ and $\eta_- = \varphi_0 - \varphi_-$, and the electrolysing voltage U to the "electrolytic window" of the melt

rent density and thus the minimum surface area of the fining electrode. Because of the condition $i_+ = |i_-|$, these quantities determine also the current and the maximum current density through the platinum counter-electrode, which, however, can be kept smaller and thus "on the safe side" by choosing a relatively large surface area of the counter-electrode. This is also shown in Fig. 3.40 by a potential change from φ_-, which corresponds to equal electrode surface areas, $A_- = A_+$, to φ'_-, corresponding to an assumed doubling of the surface area of the counter electrode, $A_- = 2A_+$.

Once the experimental curves I, II, and III in Fig. 3.40 are known, the fining electrolysis can be automatically controlled, for instance, by means of a zirconia reference electrode with an assumed equilibrium potential φ_0, Fig. 3.39. Both the anodic polarization η_+ and the cathodic polarization η_- can thus be restricted to the electrolytic window, and the electrolysing voltage $U = \eta_+ + \eta_- = \varphi_+ - \varphi_-$ can be kept below its critical value so that the arrangement in Fig. 3.39 represents an automatically controllable electrolytic fining method.

The potential of zirconia counter-electrodes, on the other hand, will mainly be determined by the resistance of the zirconia and thus by the voltage drop across the ceramic. In addition, zirconia cathodes are "safer" than platinum cathodes because their platinum electrode does not directly contact the melt. Zirconia electrodes have not been considered in Fig. 3.40 but must be treated separately.

Laboratory Tests

Electrolytic fining has been tested under stationary laboratory conditions. The arrangement is sketched in Fig. 3.41. The melt was contained in a platinum crucible and was divided into a concentric inner anodic and an outer cathodic part by an alumina tube, which had two horizontal, semicircular, narrow slits for an electrical connection of the anodic and cathodic compart-

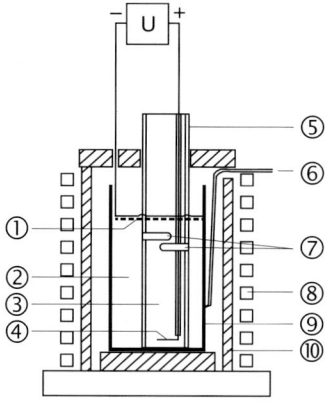

1 = Counter-electrode
2 = Cathodic half-cell
3 = Anodic half-cell
4 = Fining electrode
5 = Alumina tube
6 = From thermocouple
7 = Slits for electrolytic connection
8 = Induction coil
9 = Platinum crucible
10 = Ceramic protection tube

Fig. 3.41. Laboratory arrangement for testing the method of electrolytic fining

ments. A small circular anode was positioned at the bottom of the inner compartment so that bubbles generated rose inside the alumina tube and could be eliminated for gas analysis near the melt surface. A horizontal net-shaped cathode with a large surface area surrounded the alumina tube at the melt surface, where it was partly contacted by the atmosphere. The arrangement was heated by an induction coil.

Several test runs were conducted, and the results of two extreme experiments are reported in the next two paragraphs. The electrolyses were carried out at 1300 °C in sodium calcium silicate glass melts without fining agents. The first melt was strongly reducing; besides nitrogen, it contained carbon monoxide and hydrogen, which had been dissolved before the fining. Figure 3.42a presents the time-dependent contents of the bubbles, which were "caught" near the melt surface of the anode compartment during the electrol-

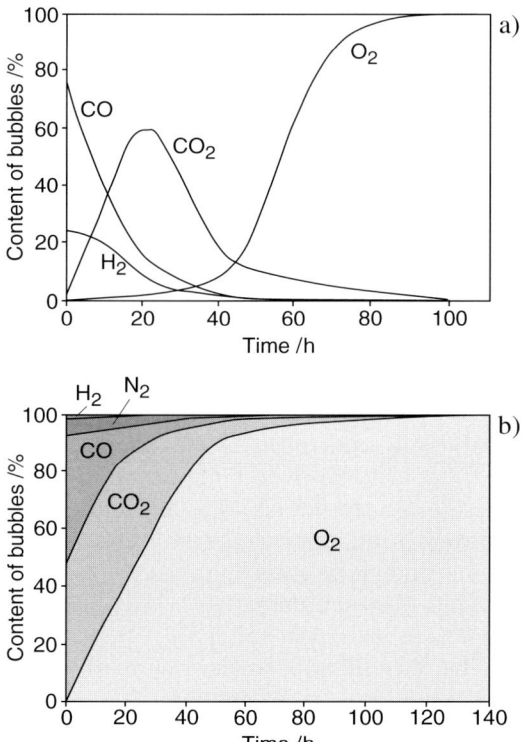

Fig. 3.42. Time-dependent composition of gas bubbles during electrolytic fining of a sodium calcium silicate melt without fining agent at 1300 °C. (**a**) Strongly reducing melt containing carbon monoxide and hydrogen. Only gas contents of special interest are plotted, the rest being nitrogen and water (not measured). (**b**) Weakly reducing melt containing equal amounts of carbon monoxide and carbon dioxide, less nitrogen and a trace of hydrogen (integral curves)

ysis. Only the contents of gases of special interest are given (in vol%), the rest being nitrogen and "defect gas", water, which was not analysed. The curves show that the generated oxygen oxidized the dissolved carbon monoxide (the original melt did not contain carbon dioxide) and the hydrogen during the first 40 h. During the about 50 h following the oxidation of CO and H_2, the rest of the carbon dioxide formed was eliminated, and the oxygen content of the bubbles increased continuously until only pure oxygen bubbles were detected during the remainder of the experiment, which indicated the removal of all impurity gases.

The second glass melt was much less reducing and contained about equal parts of carbon dioxide and carbon monoxide, about 8% nitrogen and a trace of hydrogen. Figure 3.42b shows time-dependent integral curves of the gas content of the bubbles found in the anode compartment. The result is similar to that of the first experiment, except that the carbon monoxide is not entirely oxidized and parts of the CO leave the melt unchanged. Nitrogen is removed from the melt simultaneously with the other gases, and after about 100 h, the fining bubbles consist of pure oxygen, indicating complete removal of the impurity gases as in the first experiment.

Significance and Future Work

As expected, these first experiments cannot solve all problems and leave many questions, particularly about details of the reactions. But the results show the scientific potential of these electrolyses and the technical usefulness of electrolytic fining, whose advantages may be summarized as follows.

- Electrolytic fining offers safe nucleation of bubbles even if bubble nucleation by other means is kinetically hindered or impossible.
- The fining process can be quantitatively controlled by electrical means, also with respect to the amount of impurity gases to be removed.
- No polyvalent elements are necessary for the fining step, and only small concentrations are needed for the resorption of stray oxygen bubbles. The process is thus ecologically favourable and yields glass qualities, for instance of optical glasses, that are higher than those of glasses with large contents of polyvalent ions.
- Addition of oxidizing agents, for instance nitrate, to the raw materials is not necessary.
- The location of the fining electrode in continuous melting units can be freely chosen.
- The location of the counter-electrode can also be chosen. For example, it may be positioned in a by-pass or overflow channel.
- Extreme requirements regarding temperature profiles in melters are not given. Only upper and lower temperature limits must be observed.
- The (average) size of the generated oxygen bubbles can be influenced by the application of alternating current which is superposed on the electrolysing

direct current. The particular electrical parameters must be found in practical experiments.

- Application of a (zirconia) reference electrode allows the electrolytic fining process to be conducted under automatic protection of the platinum electrodes from anodic and cathodic corrosion.

At present, however, electrolytic fining is but a method whose basic functioning has been experimentally verified in a laboratory crucible. The application in technical melters needs a large amount of development work. This task should be tackled open-mindedly because the possible result may be melting units with designs quite different from the familiar continuous melting furnaces of today. Obviously, the development is best carried out in semi-technical melters and will only be successful if the technical and scientific staff cooperate closely.

3.6 Non-Isothermal Glass-Forming Melts

Friedrich G.K. Baucke

In practical glass melting, one has often, if not always, to deal with non-isothermal melts for several reasons. On the one hand, it is nearly impossible to handle large industrial melting units – continuous melting tanks as well as crucibles for batch melting – without generating at least small temperature gradients. On the other hand, temperature gradients are often wanted and even required for certain steps of glass melting. For instance, the strong temperature dependence of the viscosity is used to control the flow rate of melts through electrically heated platinum tubes at the end of continuous melting processes, which allows the withdrawal of melts at defined rates for further treatment. This usually causes considerable temperature gradients within the platinum tubes and thus complicated non-isothermal cells, which are short-circuited by the platinum tubes. Another example is the application of the temperature-dependent equilibrium of the redox couple antimony(III)/antimony(V), for glass fining [3.87] as described in Sect. 3.4. Because in continuous melters spontaneous oxygen bubble formation at high temperatures is generated at a location different from that of the resorption of the remaining pure oxygen bubbles, glass melting tanks with redox fining are subject to continuous temperature gradients, which are deliberately and necessarily maintained. These examples demonstrate that the glass industry is normally concerned with non-isothermal glass melts and that isothermal melts are rare exceptions if they can be found at all.

Temperature gradients cause a "homogeneous polarization" of glass melts in that polar entities, for example dipoles and ion couples, attain a preferred direction along the temperature gradients. This directional order has two consequences [3.262].

- The local atomic electric fields along the temperature gradient add up and cause an electric potential difference between any two points of the melt with different temperatures, which has been measured in molten salts by appropriate electrodes [3.263]. This phenomenon is called the Seebeck effect.
- The gradient of ionic diffusion coefficients caused by the temperature gradient causes a thermodiffusion, which, after some time, is counter-balanced by backdiffusion of the ions due to the generated concentration gradients, which become constant when thermodiffusion and backdiffusion have attained equal rates. This so-called Soret effect is measurable by sensitive analytical methods but builds up at very small rates. For example, a build-up time of the concentration gradient of more than 80 hours has even been reported in 5% acetic acid and of more than 80 days in lead–tin melts at 360 °C and 600 °C [3.264]. This is in agreement with reports on a $40CaO–20Al_2O_3–40SiO_2$ melt, where concentration changes by thermodiffusion could not be detected [3.265], on several alkali silicate glasses, where the generation of concentration gradients due to thermodiffusion near T_g took 250–2100 h, depending on the glass composition [3.266], and with our own observation that thermoelectric potentials in glass melts were constant for several days. The Soret effect can thus be neglected when thermoelectric effects of glass-forming melts are studied and, particularly, is of no concern in industrial glass melting tanks, where, in addition, the melt is subject to continuous convection.

The Seebeck effect, however, is significant in glass-forming melts, and thermoelectric voltages must be known for several reasons.

- Emfs of oxygen sensors whose measuring and reference electrodes are positioned at different locations (and temperatures) in technical melting units must be corrected for thermoelectric potentials.
- Non-isothermal metal parts of melting tanks are often short-circuited either directly, as, for example, the different regions of electrically heated platinum tubes controlling the flow rate of the melt out of the melting tank and the heat-supplying metal electrodes connected to transformers, or indirectly, for instance, when metal parts are accidentally connected by badly designed or faulty circuitry or by unintentional grounding. The short circuit causes a non-isothermal cell reaction which can lead to metal deposition, alloying, and rapid corrosion of the involved parts or to the formation of oxygen and, under certain conditions, oxygen bubbles, which are intolerable in nearly all glass products. In addition, refractory materials in contact with non-isothermal melts can exhibit accelerated corrosion, depending on the composition of melt and refractory.

Due to this situation, it was decided to study properties and effects of non-isothermal glass melts [3.267–269]. This had to be done on a fundamental level because of the scarcity of available quantitative and even qualitative information on this subject, as only two papers had appeared on the measurement of thermoelectric voltages in glasses near T_g [3.270, 271] and one in glass melts [3.265], none of which, however, excluded effects of redox potentials and sufficed as the basis of the experimental investigations. This section will thus report on (1) the experimental determination of redox-free thermoelectric voltages and, different from the usual treatment, will derive the corresponding basic equations in a way that clearly shows the measurable and practically relevant as well as the unattainable quantities involved, (2) the application of the measured quantities to correct thermoelectric emfs of oxygen sensors obtained in non-isothermal glass melts, and (3) the derivation of thermoelectric potentials of platinum electrodes from the obtained quantities, which are not accurately measurable but serve to discuss thermoelectric reactions of short-circuited metal parts in non-isothermal melts.

3.6.1 Measurement of Thermoelectric Voltages – Standard Seebeck Coefficients

Different from the work reported by *Oldekop* [3.270], *Carlson* and *Trzeciak* [3.271], and *Ukyo* and *Goto* [3.265], who used platinum electrodes, thermoelectric voltages were measured by means of zirconia electrodes because they exclude the influence of disturbing redox potentials and yield pure thermoelectric emfs [3.267, 268]. Indeed, zirconia electrodes are applied as reference electrodes with redox potential measurements exactly for the reason of their inertness. The cell scheme is

$$\mathrm{Pt}'(T_i)\dots\mathrm{Pt}',\mathrm{O}_2(\mathrm{r})(T_1)|\mathrm{ZrO}_2(T_1)|\mathrm{melt}(T(x))$$
$$|\mathrm{ZrO}_2(T_2)|\mathrm{Pt},\mathrm{O}_2(\mathrm{r})(T_2)\dots\mathrm{Pt}(T_i)\ , \tag{3.XI}$$

where T_i is the temperature of the measuring instrument, and $T(x)$ the locally dependent temperature of the glass melt. Figure 3.43 gives cell components and related temperatures. The thermoelectric emf $E_{\Delta T,\mathrm{ZrO}_2}$ of cell (3.XI) is the sum of all relevant potential differences:

$$
\begin{aligned}
E_{\Delta T,\mathrm{ZrO}_2} &= [\varphi_{\mathrm{Pt}}(T_i) - \varphi_{\mathrm{Pt}}(T_2)] + [\varphi_{\mathrm{Pt}}(T_2) - \varphi_{\mathrm{m}}(T_2)] \\
&\quad + [\varphi_{\mathrm{m}}(T_2) - \varphi_{\mathrm{m}}(T_1)] + [\varphi_{\mathrm{m}}(T_1) - \varphi_{\mathrm{Pt}'}(T_1)] \\
&\quad + [\varphi_{\mathrm{Pt}'}(T_1) - \varphi_{\mathrm{Pt}'}(T_i)] \\
&= \varphi_{\mathrm{Pt}}(T_i) - \varphi_{\mathrm{Pt}'}(T_i)\ .
\end{aligned}
\tag{3.142}
$$

The second and the (negative) fourth square brackets of (3.142) denote the respective Galvani voltages of the isothermal zirconia electrodes at T_2 and T_1,

$$\varphi_{\mathrm{Pt}}(T_2) - \varphi_{\mathrm{m}}(T_2) = \varepsilon_{\mathrm{ZrO}_2}(T_2) = \varepsilon^0_{\mathrm{ZrO}_2}(T_2) + \frac{RT_2}{4F}\ln\frac{p_{\mathrm{O}_2,\mathrm{r}}(T_2)}{(a_{\mathrm{O}^{2-},\mathrm{m}}(T_2))^2} \tag{3.143}$$

Pt	ZrO$_2$	Melt	ZrO$_2$	Pt
T$_i$	T(x)	T$_1$	T(x)	T$_2$	T(x)	T$_i$

Fig. 3.43. Schematic of the temperature distribution in the electrochemical cell, consisting of two "zirconia electrodes", |Pt,O$_2$|ZrO$_2$|, in a non-isothermal glass melt, for measuring redox-free thermoelectric emfs

and

$$\varphi_{\text{Pt}'}(T_1) - \varphi_{\text{m}}(T_1) = \varepsilon_{\text{ZrO}_2}(T_1) = \varepsilon_{\text{ZrO}_2}^0(T_1) + \frac{RT_1}{4F} \ln \frac{p_{\text{O}_2,\text{r}}(T_1)}{(a_{\text{O}^{2-},\text{m}}(T_1))^2} \,,$$
(3.144)

because the potentials below both surfaces of each of the isothermal zirconia phases are equal and cancel (see Fig. 3.43). The third square bracket is the thermoelectric diffusion potential of the melt between temperatures T_1 and T_2, which can also be expressed by the entropy $\dot{S}$ transported by the ions,

$$\varphi_{\text{m}}(T_2) - \varphi_{\text{m}}(T_1) = \varepsilon_{\text{m}}(T_1, T_2) = \sum_i \frac{1}{z_i F} \int_{T_1}^{T_2} \dot{S}_i dT \,,$$
(3.145)

and the sum of the first and last square brackets is the thermoelectric potential of the platinum leads between temperatures T_2 and T_1, which can be expressed correspondingly by the entropy transported by the electrons,

$$\varphi_{\text{Pt}'}(T_1) - \varphi_{\text{Pt}}(T_2) = \varepsilon_{\text{Pt}}(T_2, T_1) = \frac{1}{F} \int_{T_2}^{T_1} \dot{S}_e dT \,.$$
(3.146)

This term is of the order of $0.025\,\mu\text{V K}^{-1}$ [3.272] and can numerically be neglected. The thermoelectric emf of cell (3.XI) is thus given by

$$E_{\Delta T, \text{ZrO}_2} = \frac{RT_2}{4F} \ln p_{\text{O}_2,\text{r}}(T_2) - \frac{RT_1}{4F} \ln p_{\text{O}_2,\text{r}}(T_1) + E_{\text{Th,m}}(T_1, T_2) \,,$$
(3.147)

if the unknown and unattainable quantities are combined as the standard thermoelectric emf,

$$E_{\text{Th,m}}(T_1, T_2) = \varepsilon_{\text{ZrO}_2}^0(T_2) - \varepsilon_{\text{ZrO}_2}^0(T_1) + \frac{RT_1}{2F} \ln a_{\text{O}^{2-},\text{m}}(T_1)$$
$$- \frac{RT_2}{2F} \ln a_{\text{O}^{2-},\text{m}}(T_2) + \varepsilon_{\text{m}}(T_1, T_2) + \varepsilon_{\text{Pt}}(T_2, T_1) \,,$$
(3.148)

where the term *standard* is used because, under the standard condition of 1 bar oxygen partial pressure in the zirconia electrodes, the thermoelectric emf is equal to the standard thermoelectric emf, $E_{\Delta T, ZrO_2} = E_{Th,m}(T_1, T_2)$, both between the same temperatures [3.268].

The standard thermoelectric emf, (3.148), between two temperatures is given by melt properties, which are principally unknown. However, it was made probable at least for certain melts that it is mainly given by the thermoelectric diffusion potential [3.267, 268]. Because it is not necessarily a linear function of the temperature nor of the temperature difference, it is useful to define and, in practice, to measure its derivative with respect to the temperature. We thus defined the standard Seebeck coefficient $dE_{Th,m}(T_1, T_2)/d(\Delta T)$ or $dE_{Th,m}(T_1, T_2 = \text{const})/dT_1$. It is independent of the oxygen partial pressures in the zirconia electrodes used as tools for their measurement as is also the standard thermoelectric emf [3.267, 268]. The temperature-dependent standard Seebeck coefficient is thus a specific quantity of glass melts and yields the standard thermoelectric emf, which is often used in practice, by integration, for instance by

$$E_{Th,m}(T_1, T_2) = \int_{T_1}^{T_2} \frac{dE_{Th,m}}{d(\Delta T)} dT \ , \tag{3.149}$$

or, if one temperature, for example T_2, is kept constant, by

$$E_{Th,m}(T_1, T_2 = \text{const}) = - \int_{T_2 = \text{const}}^{T_1} \frac{dE_{Th,m}}{dT_1} dT \ . \tag{3.150}$$

As shown in Fig. 3.44, the measurements were conducted by means of zirconia mini-electrodes with 6 mm outer diameter, which were immersed at different locations in the non-isothermal glass melt [3.4, 268]. The conductivity of the 12 cm long ceramic (e.g., zirconium silicate) boats containing the melt was shown to be negligible by applying boats of different shape and size, Fig. 3.44, which yielded identical results. Temperature profiles were provided by a specially designed gradient furnace [3.267, 273] and were directly plotted by pulling a thermocouple connected to the recorder through the melt [3.267]. Because of the finite extension of the zirconia electrodes, the region of the melt around the sensors was kept isothermal as shown schematically in Figs. 3.43 and 3.48. Besides, a reversal of the temperature profiles yielded equal results with reproducibilities of the standard Seebeck coefficients within a few hundredths of a mV K^{-1}.

The resulting standard Seebeck coefficients have magnitudes between -0.1 mV K^{-1} and -1.0 mV K^{-1} [3.269]. The negative sign denoting the sign of the potential of the hot electrode indicates that "interstitial" cations tend to transport energy from high to low temperature because anion transport in the reverse direction via cation "vacancies" is highly improbable. The

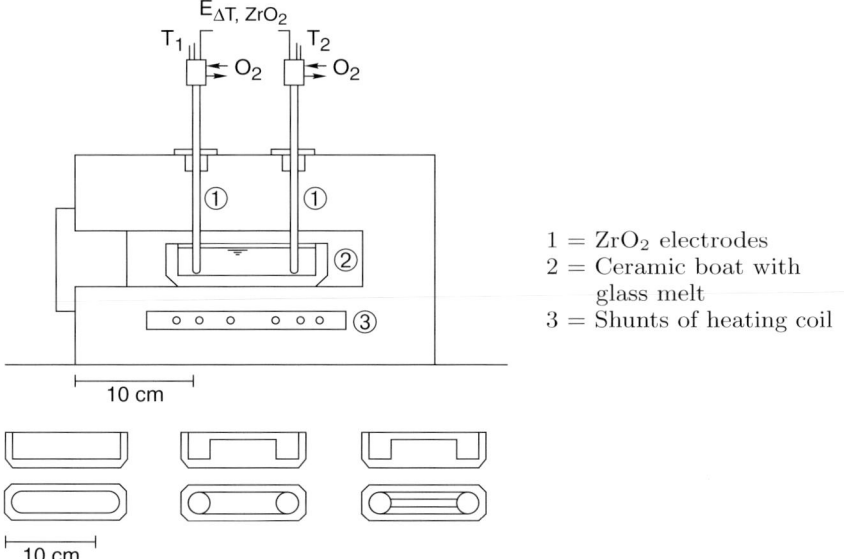

Fig. 3.44. Experimental arrangement for measuring thermoelectric emfs of glass melts by means of two zirconia electrodes. The temperature profiles are supplied by the gradient furnace containing a platinum heat coil with numerous shunts. Also shown are different shapes of the ceramic boats applied to verify negligible electric conductivity of the ceramic

measured data are thus in agreement with the general understanding of oxidic glass melts, which assumes that cations, preferably alkali ions, are the mobile species. Figure 3.45 presents some examples showing that standard Seebeck coefficients can have zero, positive, or negative temperature coefficients [3.269]. The most unexpected result was obtained by a green glass melt, which exhibited two temperature regions with constant standard Seebeck coefficients, $-0.575\,\mathrm{mV\,K^{-1}}$ below and $-0.453\,\mathrm{mV\,K^{-1}}$ above $1020\,^{\circ}\mathrm{C}$ [3.103].

Equation (3.147) shows that the thermoelectric emf of cell (3.XI) is determined by the temperature difference and the difference of the oxygen partial pressures of the zirconia electrodes, and it is of interest to compare the magnitudes of the different causes that result in the same potential difference of the electrodes and thus compensate each other. Thus, assuming zero thermoelectric emf and rearranging (3.147) results in

$$p_{\mathrm{O_2,r}}(T_1) = \exp\left[\frac{4FE_{\mathrm{Th,m}}(T_1,T_2)}{RT_1} + \frac{T_2}{T_1}\ln p_{\mathrm{O_2,r}}(T_2)\right] , \qquad (3.151)$$

which yields the oxygen partial pressure of the zirconia electrode at, for example, the lower temperature, T_1, referred to that at the higher temperature, T_2, which compensates the standard thermoelectric emf, $E_{\mathrm{Th,m}}(T_1,T_2)$, caused by

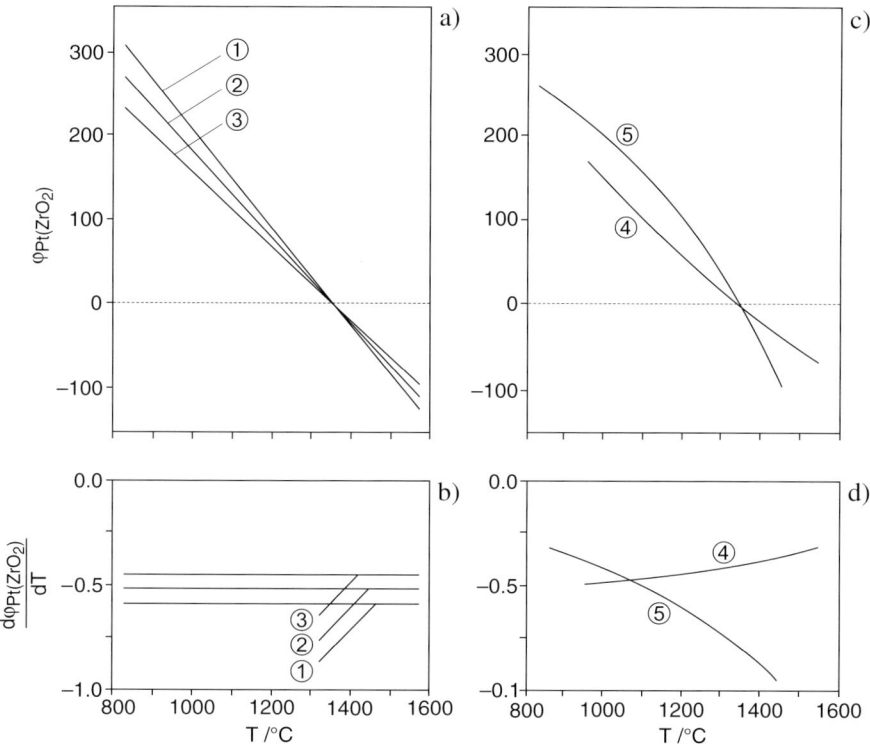

Fig. 3.45. Thermoelectric potentials of a zirconia electrode with 1 bar oxygen partial pressure as functions of temperature (**a, c**) and corresponding temperature-dependent standard Seebeck coefficients (**b, d**) of various glass melts. (1) Fiolax® klar, (2) $(Na_2O)_{0.078}(K_2O)_{0.078}(CaO)_{0.107}(SiO_2)_{0.737}$, (3) $(Na_2O)_{0.156}(CaO)_{0.107}(SiO_2)_{0.737}$, (4) BK 7, (5) phosphate-based optical glass

these temperatures. As an example, Fig. 3.46 presents the decadic logarithm of the oxygen partial pressure at T_1 as a function of the temperature difference $\Delta T = T_2 - T_1$ referred to $T_2 = 1673\,\mathrm{K}$ and 1 bar oxygen partial pressure, $p_{O_2,r}(T_2) = 1\,\mathrm{bar}$, of the hot zirconia electrode. A temperature-independent standard Seebeck coefficient $dE_{Th,m}(T_1, T_2)/dT = -0.5\,\mathrm{mV\,K^{-1}}$ was assumed for the calculations. An interesting feature of Fig. 3.46 is that a thermoelectric emf caused by only $150\,^\circ\mathrm{C}$ (between $1523\,\mathrm{K}$ and $1673\,\mathrm{K}$) is compensated by a ratio of the oxygen partial pressures in the zirconia electrodes that is as large as ten. In addition, the plot is not linear, so that this tendency increases with increasing temperature difference, being the ratio of the oxygen partial pressures of 10^5 required to compensate the thermoelectric emf caused by a temperature difference of $550\,\mathrm{K}$ (between $1123\,\mathrm{K}$ and $1673\,\mathrm{K}$). The example thus gives an idea of the significance of thermoelectric effects in oxidic glass melts and demonstrates that they can cause serious, impairing reactions in glass melting tanks, particularly because the

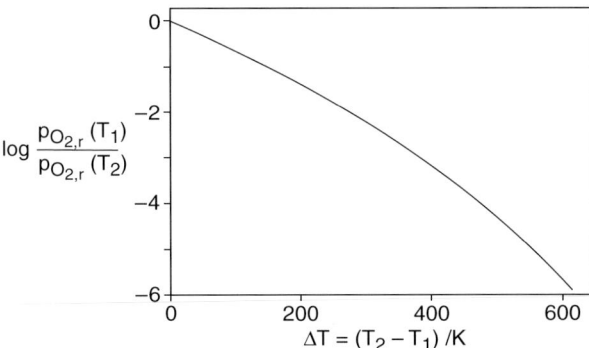

Fig. 3.46. Compensation of the thermoelectric emf of a cell consisting of two zirconia electrodes at temperatures T_1 and T_2 in a non-isothermal melt by the respective oxygen partial pressures $p_{O_2,r}(T_1)$ and $p_{O_2,r}(T_2)$ of the electrodes. Calculations are based on the higher temperature $T_2 = 1673\,\mathrm{K}$, the oxygen partial pressure of the hot electrode $p_{O_2,r}(T_2 = 1673\,\mathrm{K}) = 1\,\mathrm{bar}$, and an assumed constant standard Seebeck coefficient $\mathrm{d}E_{\mathrm{Th,m}}/\mathrm{d}T = -0.5\,\mathrm{mV\,K}^{-1}$

thermoelectric cells set up in these units are often obscured and difficult to analyse. As usual, the best "cure" is to avoid the generation of such damaging cells in the first place by devising a good scientific design for the melting tanks, to which specialists of all engineering and scientific fields, including electrochemists, contribute their share.

3.6.2 Application of Oxygen Sensors in Non-Isothermal Melts

Reference and measuring electrodes of oxygen sensors must often be applied at different locations in technical melting units and thus in non-isothermal glass melts. This becomes necessary either for economic reasons, when one reference electrode is used with several measuring electrodes (if this is possible and safe), or when prevailing technical conditions, for instance spatial problems, do not allow their application in one place. Such cells yield thermoelectric emfs, which must be corrected for their non-isothermal part(s). Because the derivation of the appropriate thermoelectric emfs corresponds to that of (3.147) followed by a simple rearrangement, the following is restricted to an accumulation of the expressions yielding the oxygen partial pressure in the different situations of technical melters and to the corresponding cell schemes for a clear identification of the quantities used; see also Fig. 3.47 [3.4].

(1) The basic arrangement is an isothermal cell represented by the cell scheme

$$\mathrm{Pt,O_2(r)|ZrO_2|melt|Pt}\ , \tag{3.XII}$$

where measuring and reference temperatures are equal, $T_m = T_r = T$, and the oxygen partial pressure of the melt results from the measured emf E and the reference oxygen partial pressure by [3.105]

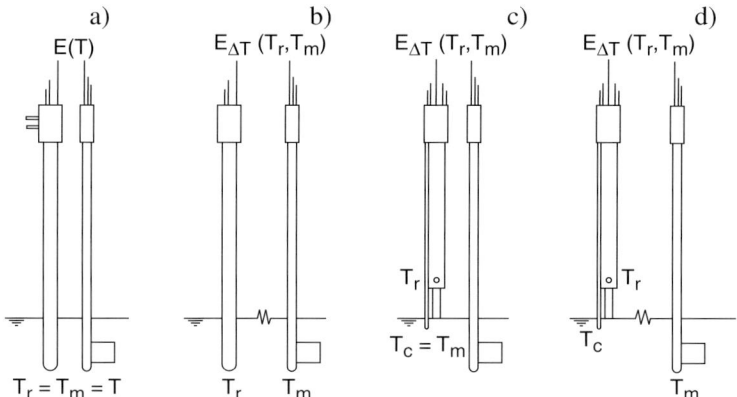

Fig. 3.47. Schematic of industrial applications of oxygen sensors in glass melts. (**a**) Isothermal measurement, (**b**) isothermal sensor in non-isothermal glass melt, (**c**) non-isothermal sensor containing a dissolving zirconia electrode in isothermal glass melt, (**d**) non-isothermal sensor in non-isothermal melt. T_r = temperature of reference electrode, T_m = temperature of measuring electrode, T_c = temperature at melt/zirconia electrolyte bridge contact

$$p_{O_2,m} = \exp\left(\frac{4F}{RT}E + \ln p_{O_2,r}\right) \ . \tag{3.152}$$

(2) A cell consisting of an (isothermal) zirconia tube electrode at reference temperature T_r and a platinum measuring electrode at measuring temperature T_m in a non-isothermal glass melt is represented by the cell scheme

$$\text{Pt,O}_2(\text{r})(T_r)|\text{ZrO}_2|(T_r)\text{melt}(T_m)|\text{Pt} \ , \tag{3.XIII}$$

whose thermoelectric emf $E_{\Delta T}(T_r, T_m)$ corrected by the standard thermoelectric emf of the melt, $E_{Th,m}(T_r, T_m)$, between the two temperatures yields the oxygen partial pressure of the melt at T_m according to

$$p_{O_2,m}(T_m) = \exp\left\{\frac{4F}{RT_m}\left[E_{\Delta T}(T_r, T_m) - E_{Th,m}(T_r, T_m)\right] + \frac{T_r}{T_m}\ln p_{O_2,r}\right\} \tag{3.153}$$

(3) The "reverse arrangement" of case (2) is a (non-isothermal) dissolving zirconia electrode in an isothermal melt, where the standard thermoelectric emf, $E_{Th,ZrO_2}(T_r, T_m)$, of the zirconia bridge replaces that of the melt in (3.153). The equation has been given as (3.61) in Sect. 3.4.3.

(4) The situation most interesting for industrial melters is a (non-isothermal) dissolving zirconia electrode in a non-isothermal glass melt. It is characterized by the cell scheme

$$\text{Pt,O}_2(\text{r})(T_r)|\text{ZrO}_2|(T_c)\text{melt}(T_m)|\text{Pt} \ , \tag{3.XIV}$$

which shows that the temperature T_c at the zirconia/melt contact must be known in addition to the reference and measuring temperatures. The measured thermoelectric emf must be corrected by the standard thermoelectric emfs of both the zirconia bridge and melt between the corresponding temperatures,

$$
p_{O_2,m}(T_m) = \exp\left\{\frac{4F}{RT_m}\left[E_{\Delta T}(T_r, T_m) - E_{Th,ZrO_2}(T_r, T_c)\right.\right.
$$
$$
\left.\left. - E_{Th,m}(T_c, T_m)\right] + \frac{T_r}{T_m}\ln p_{O_2,r}\right\} ,
$$

$$(3.154)$$

which necessitates exact knowledge of the standard Seebeck coefficients of both materials in the appropriate temperature ranges.

3.6.3 Short-Circuited Platinum Parts in Non-Isothermal Melts

A non-isothermal electrochemical cell with high practical significance is that involving two parts of identical metal as electrodes because their short circuiting can cause electrode reactions with possibly deleterious results. Thus, the metal can be dissolved. Elements, for example metals, can be formed from the melts and precipitate onto the metal parts of the glass melter, there to form corroding or soluble alloys. Also, the oxidation of oxide of the melt leading to the formation of elemental oxygen can, under certain conditions, give rise to the detrimental formation of oxygen bubbles, in particular when "inert" platinum parts are short-circuited [3.99, 274, 275] and other reactions are excluded. These cell and electrode reactions have been investigated in some detail [3.268, 269] and will be treated in the following by means of the non-isothermal cell

$$\mathrm{Pt}(T_1)|\mathrm{melt}(T(x))|\mathrm{Pt}(T_2) . \tag{3.XV}$$

Cell (3.XV) exhibits the thermoelectric emf,

$$
E_{\Delta T,Pt} = \frac{RT_2}{4F}\ln p_{O_2,m}(T_2) - \frac{RT_1}{4F}\ln p_{O_2,m}(T_1) + E_{Th,m}(T_1, T_2) ,
$$

$$(3.155)$$

which corresponds to that of cell (3.XI) with two zirconia electrodes but cannot be measured exactly because the oxygen partial pressures of the melt are unstable due to diffusion into and out of and between different volumes of the melt, particularly if it is not buffered by polyvalent ions. It can, however, be obtained by combining thermoelectric emfs of cell (3.XI) and temperature-dependent isothermal emfs measured by means of cell (3.XII) [3.268, 269], which are accurately measurable dynamically [3.105]. The combination even yields the relative, temperature-dependent, potentials of the platinum electrodes of cell (3.XV), which are more useful for the intended discussion. The following paragraph thus gives a brief conversion of the thermoelectric emf of a cell into the thermoelectric potentials of the corresponding electrodes, for which cell (3.XI) is taken as an example.

The thermoelectric emf of cell (3.XI) is equal to the potential difference of its "zirconia electrodes" or, stated more exactly, of the platinum reference electrodes of the zirconia electrodes,

$$E_{\Delta T, \mathrm{ZrO_2}} = \varphi_{\mathrm{Pt(ZrO_2)}}(T_2) - \varphi_{\mathrm{Pt(ZrO_2)}}(T_1) \, , \tag{3.156}$$

if the negligibly small thermoelectric voltage of platinum [3.272], (3.146), is again ignored. Equation (3.147) can thus be expressed in terms of the potential of, for instance, the zirconia electrode at the lower temperature, T_1,

$$\begin{aligned}
\varphi_{\mathrm{Pt(ZrO_2)}}(T_1) = {} & \frac{RT_1}{4F} \ln p_{\mathrm{O_2,r}}(T_1) - \frac{RT_2}{4F} \ln p_{\mathrm{O_2,r}}(T_2) \\
& - E_{\mathrm{Th,m}}(T_1, T_2) + \varphi_{\mathrm{Pt(ZrO_2)}}(T_2)
\end{aligned} \tag{3.157}$$

or

$$\varphi_{\mathrm{Pt(ZrO_2)}}(T_1) = -E_{\Delta T, \mathrm{ZrO_2}} + \varphi_{\mathrm{Pt(ZrO_2)}}(T_2) \, . \tag{3.158}$$

If an equal oxygen partial pressure is applied in the zirconia electrodes, as usual in experiments, $p_{\mathrm{O_2,r}}(T_1) = p_{\mathrm{O_2,r}}(T_2) = p_{\mathrm{O_2,r}}$, (3.157) changes to become

$$\varphi_{\mathrm{Pt(ZrO_2)}}(T_1) = \frac{R(T_1 - T_2)}{4F} \ln p_{\mathrm{O_2,r}} - E_{\mathrm{Th,m}}(T_1, T_2) + \varphi_{\mathrm{Pt(ZrO_2)}}(T_2) \, , \tag{3.159}$$

which is further reduced to (3.160) if 1 bar oxygen partial pressure is chosen, $p_{\mathrm{O_2,r}} = 1\,\mathrm{bar}$,

$$\varphi_{\mathrm{Pt(ZrO_2)}}(T_1) = -E_{\mathrm{Th,m}}(T_1, T_2) + \varphi_{\mathrm{Pt(ZrO_2)}}(T_2) \, . \tag{3.160}$$

Using, finally, the potential of the "hot" zirconia electrode as a reference potential, for instance $\varphi_{\mathrm{Pt(ZrO_2)}}(T_2) = 0$ at constant upper temperature T_2, yields

$$\begin{aligned}
\varphi_{\mathrm{Pt(ZrO_2)}}(T_1) &= -E_{\mathrm{Th,m}}(T_1, T_2 = \mathrm{const}) \\
&= \int_{T_2}^{T_1} \frac{\mathrm{d}E_{\mathrm{Th,m}}(T_1, T_2 = \mathrm{const})}{\mathrm{d}T_1} \mathrm{d}T \, ,
\end{aligned} \tag{3.161}$$

which describes the thermoelectric potential of the "cold" zirconia electrode with respect to zero constant potential of the hot zirconia electrode in a non-isothermal glass melt by the temperature-dependent standard Seebeck coefficient in the corresponding temperature range, see also (3.149) and (3.150).

In addition, the temperature coefficient of the thermoelectric potential of the cold zirconia electrode, (3.159), is given by

$$\frac{\mathrm{d}\varphi_{\mathrm{Pt(ZrO_2)}}(T_1)}{\mathrm{d}T_1} = \frac{R}{4F}\ln p_{\mathrm{O_2,r}} - \frac{\mathrm{d}E_{\mathrm{Th,m}}(T_1, T_2 = \mathrm{const})}{\mathrm{d}T_1} \tag{3.162}$$

if both zirconia electrodes contain an equal reference oxygen partial pressure, and at 1 bar oxygen partial pressure, it is equal to the standard Seebeck coefficient as shown by

$$\frac{\mathrm{d}\varphi_{\mathrm{Pt(ZrO_2)}}(T_1)}{\mathrm{d}T_1} = -\frac{\mathrm{d}E_{\mathrm{Th,m}}(T_1, T_2 = \mathrm{const})}{\mathrm{d}T_1}\,. \tag{3.163}$$

The thermoelectric potentials of a platinum electrode are obtained from those of the zirconia electrode by the principle sketched in Fig. 3.48. It shows a non-isothermal glass melt containing an isothermal cell according to cell scheme (3.XII) at the lower temperature T_1 and a non-isothermal cell according to cell scheme (3.XI) consisting of the zirconia electrode at T_1 and a zirconia electrode at the higher temperature T_2. The sum of the three potential differences between these electrodes, taken in one direction, is zero,

$$\left[\varphi_{\mathrm{Pt(m)}}(T_1) - \varphi_{\mathrm{Pt(ZrO_2)}}(T_1)\right] + \left[\varphi_{\mathrm{Pt(ZrO_2)}}(T_1) - \varphi_{\mathrm{Pt(ZrO_2)}}(T_2)\right]$$
$$+ \left[\varphi_{\mathrm{Pt(ZrO_2)}}(T_2) - \varphi_{\mathrm{Pt(m)}}(T_1)\right] = 0\,. \tag{3.164}$$

Replacing the first and second of the potential differences of (3.164) by the respective emf and thermoelectric emf as indicated in Fig. 3.48 and rearranging yields

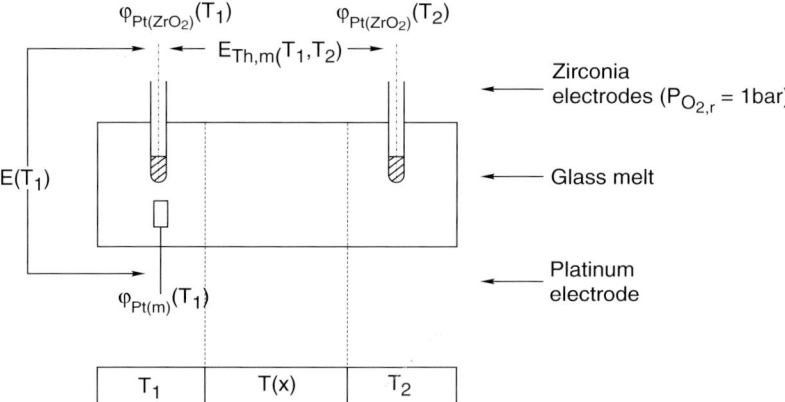

Fig. 3.48. Schematic for deriving temperature-dependent thermoelectric potentials of a platinum electrode in a non-isothermal glass melt with constant total oxygen content

$$\varphi_{\text{Pt,(m)}}(T_1) = \varphi_{\text{Pt(ZrO}_2)}(T_2) + E(T_1) - E_{\text{Th,m}}(T_1, T_2) \qquad (3.165)$$

or the equivalent

$$\varphi_{\text{Pt(m)}}(T_1) = \varphi_{\text{Pt(ZrO}_2)}(T_2) + E(T_1) + \int_{T_2}^{T_1} \frac{\mathrm{d}\varphi_{\text{Pt(ZrO}_2)}(T_1)}{\mathrm{d}T_1}\mathrm{d}T , \qquad (3.166)$$

and using the potential of the hot zirconia electrode (at T_2) as zero reference potential yields

$$\varphi_{\text{Pt(m)}}(T_1) = E(T_1) - E_{\text{Th,m}}(T_1, T_2 = \text{const}) . \qquad (3.167)$$

Equations (3.166) and (3.167) describe the thermoelectric potential of a platinum electrode by the isothermal emf of cell (3.XII) and the standard Seebeck coefficient of the melt. Figure 3.49 depicts the quantities as functions of temperature at a particular constant total oxygen content. The reference state was chosen to be zero emf and potentials at the reboil temperature of the glass melt, $E(T_{\text{reb}}) = \varphi_{\text{Pt(ZrO}_2)}(T_{\text{reb}}) = \varphi_{\text{Pt,(m)}}(T_{\text{reb}}) = 0$, where the melt has 1 bar oxygen partial pressure and oxygen bubble formation becomes possible (under 1 bar atmospheric pressure and if the effect of the hydrodynamic pressure of the melt is neglected). As the temperature-dependent isothermal emf of cell (3.XII) is zero at equal oxygen partial pressures of zirconia electrode and melt, this reference reduces simply to the assignment of

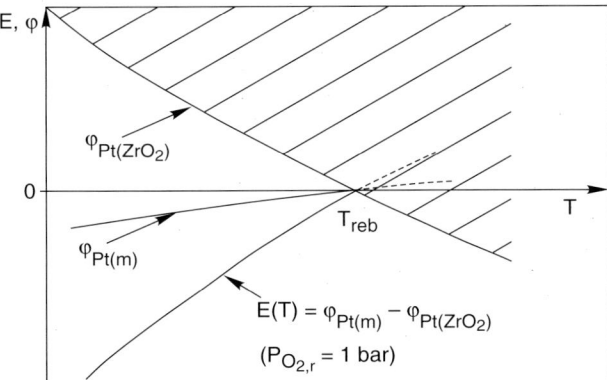

Fig. 3.49. Electrode potentials in a non-isothermal glass melt with constant total oxygen content. $E(T) =$ temperature-dependent isothermal emf of a cell consisting of a platinum electrode and a zirconia electrode with 1 bar oxygen partial pressure. $\varphi_{\text{Pt(m)}}$ and $\varphi_{\text{Pt(ZrO}_2)} =$ respective thermoelectric potentials of these electrodes. The reference state of the plot is zero potential at the reboil temperature T_{reb} of the melt. The potential of a platinum electrode in equilibrium with the melt is always equal to or below that of the zirconia electrode, independent of the total oxygen content of the melt, so that the shaded area is unattainable for platinum electrodes (1 bar atmospheric pressure, hydrodynamic pressure of melt neglected)

1 bar oxygen partial pressure to the zirconia electrode at the reboil temperature of the melt. The curve of the thermoelectric potentials of the zirconia electrode in Fig. 3.49 thus represents reboil temperatures at varying total oxygen contents of the melt and cannot be exceeded by the potentials of a platinum electrode in equilibrium with the melt, i.e., the shaded area of Fig. 3.49 denotes supersaturation of the melt with respect to oxygen. It is thus important for the following discussion to emphasize again that the plots $E(T)$ and $\varphi_{Pt(m)}(T)$ in Fig. 3.49 are for constant total oxygen content.

Depending on the relative temperature dependence of the isothermal cell emf, $E(T)$, and the thermoelectric potential of the zirconia electrode, $\varphi_{Pt(ZrO_2)}$, the temperature-dependent thermoelectric potential, $\varphi_{Pt(m)}$, of a platinum electrode can have positive, negative, or zero slope and can also exhibit extreme values. Figure 3.49, for example, shows conditions leading to a positive slope. In particular, melts containing polyvalent ions and thus buffered oxygen yield relatively small positive or negative slopes, whereas melts with unbuffered oxygen result in large, negative, slopes of the thermoelectric potentials of platinum electrodes, which are often comparable with those of the zirconia electrode. Consequently, oxygen generation, which always occurs at the negative electrode (anode) when non-isothermal platinum electrodes are short-circuited, is not restricted to the hot electrode as commonly assumed [3.99], but can also take place at the cold electrode. The formation of oxygen bubbles, however, depends on the steady-state oxygen partial pressure attained by the electrodes after the short circuit, which, besides by the basic electrode polarization, is given by the individual electrode properties such as shape and relative surface areas determining current densities and their gradients at the electrodes, kinetic factors such as transport of oxygen to and from the electrodes by convection and diffusion, and by the "quality" of the short circuit, i.e., the resistances of conductors and melt which complete the electrical circuit.

As an example, Fig. 3.50a shows the strongly temperature-dependent thermoelectric potential of a platinum electrode with a negative slope in a glass melt containing unbuffered oxygen. Also given for reference purposes is the temperature-dependent thermoelectric potential of a zirconia electrode with 1 bar oxygen partial pressure. Short-circuiting two platinum electrodes at T_1 and T_2 (which is assumed larger than T_1) equalizes their potentials so that both platinum electrodes attain the potential $\varphi'_{Pt(m)}$ (if small potential drops caused by resistances of the electric circuit and polarization are neglected) and leads to oxygen formation at the hot electrode and to oxygen degeneration at the cold electrode. Under the conditions of Fig. 3.50a, the steady-state potential $\varphi'_{Pt(m)}$ assumed by the hot, oxygen generating electrode, however, is smaller than the potential of the zirconia electrode at this temperature T_2, indicating that the steady-state oxygen partial pressure of the melt around the hot platinum electrode is below 1 bar. Correspondingly, the temperature dependence of the steady-state thermoelectric

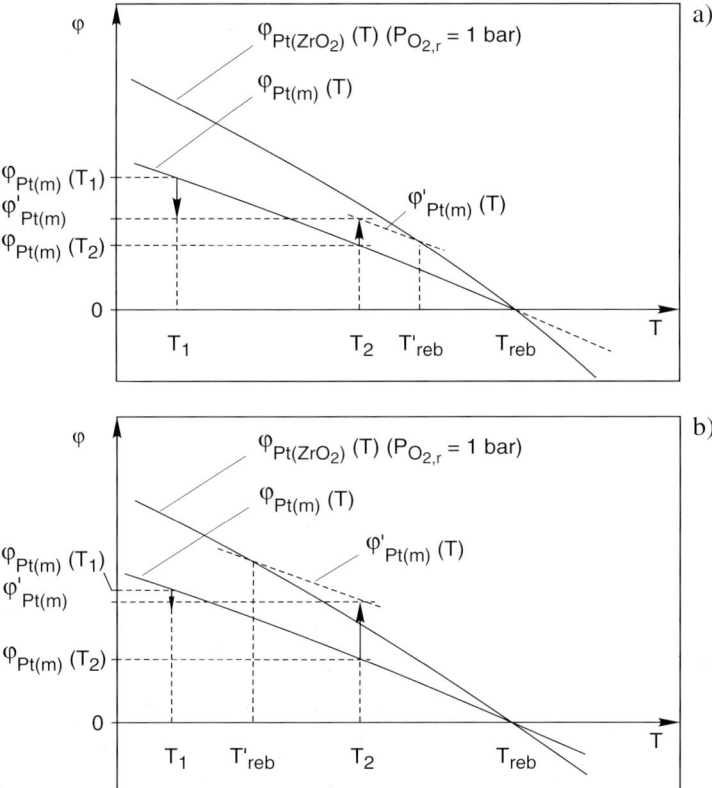

Fig. 3.50. Short circuit of two non-isothermal platinum electrodes with a negative temperature gradient of the thermoelectric potential in a glass melt with constant total content of unbuffered oxygen. (**a**) The steady-state potential $\varphi'_{Pt(m)}$ assumed by the hot electrode (T_2) due to oxygen formation is smaller than the potential of a zirconia electrode with 1 bar oxygen partial pressure at the same temperature indicating no tendency of the melt around the metal to form oxygen bubbles. (**b**) The steady-state oxygen partial pressure of the melt at the hot electrode is above 1 bar and can thus form oxygen bubbles. Correspondingly, the steady-state reboil temperature, T'_{reb}, is (a) above and (b) below the temperature T_2 of the hot electrode

potential, $\varphi'_{Pt(m)}(T)$, which, as a first approximation, is assumed as a curve parallel to that of the thermoelectric potential of the melt, $\varphi_{Pt(m)}(T)$, yields a steady-state reboil temperature, T'_{reb}, of the melt around the hot platinum electrode which is smaller than that of the melt, T_{reb}, but larger than the temperature T_2 of the hot electrode. Thus, short-circuiting non-isothermal platinum electrodes under the conditions assumed in Fig. 3.50a cannot result in the formation of oxygen bubbles.

Oxygen bubble formation, however, is not basically excluded under the thermodynamic conditions of Fig. 3.50. This is demonstrated by Fig. 3.50b, where, for example, a large ratio of the surface areas of the electrodes, $A(T_1) \gg A(T_2)$, causes a stronger polarization of the hot than of the cold electrode. This makes the steady-state potential $\varphi'_{Pt(m)}$ of the hot platinum electrode larger than that of the zirconia electrode at T_2, indicating supersaturation of the melt with respect to oxygen. Correspondingly, the steady-state reboil temperature, T'_{reb}, of the melt around the hot electrode, which is obtained in the same way as that in Fig. 3.50a, is smaller than T_2 and thus indicates that oxygen bubbles will form at this electrode if kinetic hindrances are excluded.

Figure 3.51 shows a potential–temperature plot explaining the short circuit of two non-isothermal platinum electrodes, whose thermoelectric potential $\varphi_{Pt(m)}$ exhibits a positive temperature dependence. The negative polarization of the hot electrode due to the short circuit causes a degeneration of oxygen at T_2 and thus an increase of the steady-state reboil temperature of the melt around this electrode. Oxygen bubble formation at the high-temperature electrode is thus excluded. Although the potential increase of the cold electrode leads to oxygen formation, the steady-state reboil temperature of the surrounding melt cannot attain values below or even close to the lower temperature, so that oxygen bubble formation is also excluded at T_1. Contrary to negative temperature gradients of thermoelectric potentials of platinum electrodes (Fig. 3.50), a positive temperature dependence (Fig. 3.51) thus does not present any danger with respect to oxygen bubble formation.

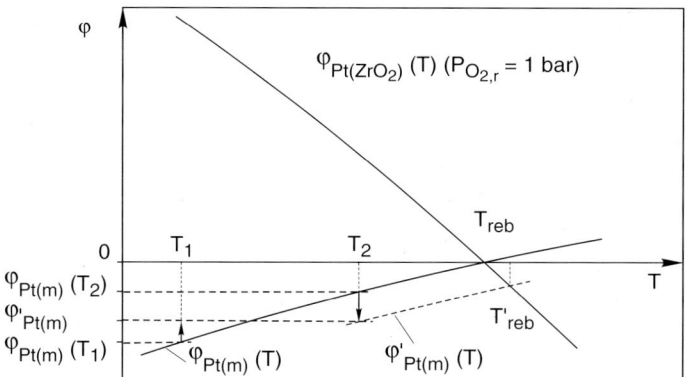

Fig. 3.51. Short circuit of two non-isothermal platinum electrodes with positive temperature gradient of thermoelectric potentials (see also Fig. 3.50). The steady state of neither electrode gives rise to oxygen bubble formation

3.6.4 Decomposition of Oxidic Glass Melts by Short-Circuited Non-Isothermal Metal Electrodes

The unavoidably non-isothermal character of technical glass melts causes a number of reactions in industrial glass melting, apart from oxygen generation. One of them is the formation of metals from the corresponding metal ions dissolved in the melt, as, for instance, silver from silver ion-containing glass melts and the frequent reduction of antimonous ions from melts fined by means of the redox couple antimony(III)/antimony(V) as described in Sect. 3.4. A possible thermoelectric reaction of this metal precipitation is described in the following.

In modern glass technology, especially in continuous melting, glass melts are increasingly heated by means of so-called primary energy, i.e., by electric currents, because of economic and ecological reasons and for energy conservation. The melt forms the heating resistor, and platinum, molybdenum, or other appropriate metal electrodes are applied as electrical contacts. Alternating currents of up to $1000\,\mathrm{kA}$ at about $6\,\mathrm{V}$ are supplied by transformers, which are directly connected to the electrodes. Energy loss is lower than with combustion heating, generation of carbon dioxide, carbon oxide, and nitrogen oxides is practically reduced to zero, and evaporation of melt components is much less than in the case of burning oil or gas, which stream over the melts with rather high rates thus causing considerable evaporation.

In the case of molybdenum, rod-shaped electrodes are applied so that the current density has its maximum at the electrode surfaces and decreases as a function of distance from the metals. As a consequence, the glass melt around the heating electrodes has maximum temperature. Small differences in electrode size, position, or wiring, i.e., an unbalanced connection of different electrodes, however, can cause considerably differing maximum temperatures of the melt around the electrodes involved. They thus form non-isothermal electrochemical cells, which are ideally short-circuited by the frequently applied, directly connected energy-supplying transformers. This causes thermo-electrical reactions at temperatures that are higher than the usual average temperatures of the melt and leads to reaction products or corrosion phenomena that are not expected under the usual "low-temperature" conditions.

One of these electrode reactions is favoured by the homogeneous equilibrium of the polyvalent ions antimony(III) and antimony(V), the basis of redox fining,

$$2\mathrm{Sb}^{5+} + 2\mathrm{O}^{2-} \rightleftarrows 2\mathrm{Sb}^{3+} + \mathrm{O}_2 \ , \tag{3.168}$$

and the connected equilibrium of the threefold positively charged antimony ion with metallic antimony,

$$4\mathrm{Sb}^{3+}(\mathrm{m}) + 6\mathrm{O}^{2-}(\mathrm{m}) \rightleftarrows 4\mathrm{Sb}^{\pm 0} + 3\mathrm{O}_2 \ , \tag{3.169}$$

which comes into play at higher temperatures. Both equilibria are shifted to the right side with increasing temperature but to a degree which ensures that antimony metal is not formed according to the equilibrium, (3.169), under ordinary melting conditions. However, (3.168) and (3.169) favour the electrode reactions of the non-isothermal cell, which is established by, for instance, two molybdenum heating electrodes at correspondingly high temperatures. In this cell, antimonous ions and oxygen are reduced at the positive hot electrode according to

$$Sb^{3+}(m) + 3e^-(electrode) \rightarrow Sb^{\pm 0}(metal) \tag{3.170}$$

and

$$O_2(m) + 4e^-(electrode) \rightarrow 2O^{2-}(m) \; , \tag{3.171}$$

and oxide is oxidized at the negative cold electrode,

$$2O^{2-}(m) \rightarrow O_2(m) + 4e^-(electrode) \; , \tag{3.172}$$

where the activity of antimony(III) and the partial pressure of oxygen are much lower. It can be estimated from the equilibrium constants that these reactions override the thermoelectric potential of most glass melts so that, indeed, antimony metal must be deposited thermoelectrically at the positive hot molybdenum electrode. This reaction mechanism is in agreement with observations on continuous melters, where antimony metal, so-called "antimony lakes", often alloyed with small amounts of other metals, is frequently found below molybdenum heating electrodes when the melting tanks are inspected after their service life. Whether the formation of metallic antimony also leads to corrosion of the molybdenum electrodes cannot be foreseen because of the wide spectrum of possible reactions and must be investigated experimentally in each particular case.

This section has demonstrated that the measurement of thermoelectric potentials, which became feasible by the availability and application of zirconia electrodes, has considerably added to the understanding of non-isothermal glass melts. The simple reason is that the temperature-dependent isothermal emf of cells of type (3.XII), which had been known for some time, could be subdivided into its components, the thermoelectric potentials of the platinum and of the zirconia electrode. This not only resulted in correction factors for oxygen sensors applied in non-isothermal glass melts but also revealed an unexpectedly large magnitude of thermoelectric effects in several cases, which are of high significance for industrial glass melting. The described study of oxygen bubble formation at short-circuited, non-isothermal platinum electrodes and the deposition of metals at heating electrodes with different temperatures are merely two examples, which will certainly be followed by further and possibly even more significant investigations.

3.7 Arrangement for Displacing Electrochemical Reactions

Friedrich G.K. Baucke

It is often difficult, if not impossible, to eliminate undesired electrochemical cell or electrode reactions in industrial glass melting units. For example, short circuits of non-isothermal metal parts are frequently caused by stray currents, whose complete elimination is extremely difficult in large melting tanks containing hundreds of tonnes of up to 1800 °C hot glass melt. It is also nearly impossible to balance current densities at heating electrodes (which are always ideally short-circuited by the energy-supplying transformers) so accurately as to exclude thermoelectric voltages between the electrode metals and a possibly connected decomposition of the melt, as described in Sect. 3.6. Electrically heated platinum tubes applied as means for controlling the flow rate of the melt out of the melting tank by making use of the strong temperature dependence of the melt viscosity are short-circuited in themselves. Their temperature and in particular their temperature gradients are dictated by the required flow rates and cannot be influenced at all. Similar conditions apply to platinum linings, which are often applied to protect refractories. Also, melts are sometimes electrolysed between metal parts of the melters and unidentified counter-electrodes by likewise unknown current sources, for instance, by direct currents caused by rectification of energy-supplying alternating currents.

In most of these cases, a basically simple technique developed at the Electrochemical Laboratory of Schott Glas between 1969 and 1972 can be applied to reduce or even exclude such undesired electrode reactions. The principle of this so-called auxiliary electrode arrangement is a shift of the electrical potential of the metal involved by means of an additional electrode, the auxiliary electrode, which serves as a counter-electrode and eliminates or at least reduces the disturbing electrode reaction. Two modifications of this arrangement have been developed, the ordinary auxiliary electrode consisting of an inert metal (e.g., platinum) and employing an outside voltage supply, and the autogenic auxiliary electrode, a reactive metal (e.g., molybdenum), which tends to supply electrons because of its tendency to dissolve in the melt. Although we never published this work, both arrangements have meanwhile been described in the literature. For example, application of the ordinary auxiliary electrode to prevent formation of oxygen blisters [3.276, 277] and to protect the refractory lining of melting furnaces [3.278] has been patented, and the principle and application of externally driven and autogenic auxiliary electrodes to avoid oxygen bubble formation at platinum parts of melting tanks have been reported [3.279], see also [3.280]. In the following we briefly describe and discuss the fundamental electrochemistry of these techniques.

Figure 3.52 explains the principle of this method by the application to a platinum tube used as a means to control the flow rate of the melt out

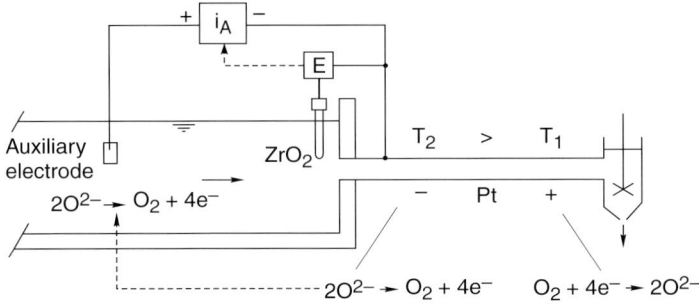

Fig. 3.52. Schematic of the auxiliary electrode arrangement as applied to short-circuited non-isothermal platinum electrodes in an industrial glass melter

of the melting tank as explained above. The melt, when entering the tube, has the temperature T_2, and it is assumed that, under the prevailing flow conditions, oxygen bubbles are formed at this high temperature T_2 and that oxygen is reduced at the lower temperature T_1 of the melt leaving the tube (see Sect. 3.6). The hot entrance part of the platinum tube thus forms the negative and the exit part the positive electrode of a non-isothermal electrochemical cell, which is ideally short-circuited by the tube metal. (Reactions at intermediate parts of the platinum need not be taken into account because they are automatically reduced when the extreme platinum parts are subjected to potential shifts.) In order to eliminate oxygen bubble formation, the hot part of the platinum tube is connected to the negative pole of the external current source i_A, whose positive pole, in turn, is hooked up to a platinum auxiliary electrode which is positioned at a location where bubble formation is of no concern, for instance in the fining region of the melting tank. A zirconia electrode controls the potential of the platinum tube and the auxiliary d.c. current supplied by the current source i_A.

Figure 3.53 demonstrates by means of the current–potential curves of the electrodes how this three-electrode assembly works. The equilibrium potentials $\varphi^0_{\mathrm{Pt}}(T_1)$ and $\varphi^0_{\mathrm{Pt}}(T_2)$ of the platinum at T_1 and T_2, respectively, differ due to the temperature difference. Their short-circuit by the platinum tube results in an anodic current i_S at T_2, which is the cause of the oxygen bubble formation, and in a cathodic current $-i_S$ at T_1 representing the reduction of oxygen. These currents have an equal absolute magnitude, $i_S(T_2) = |-i_S(T_1)|$, at the potential φ'_{Pt}. Connecting the auxiliary electrode and the current source to the platinum tube as shown in Fig. 3.52 and applying the auxiliary voltage U_A between platinum tube and auxiliary electrode has three effects.

(1) The potential of the platinum is shifted to the more negative equilibrium potential $\varphi^0_{\mathrm{Pt}}(T_2)$, if the correct auxiliary voltage is chosen, which causes zero current at the higher temperature T_2, $i_S \to 0$, and thus elimination of the oxygen bubble formation.

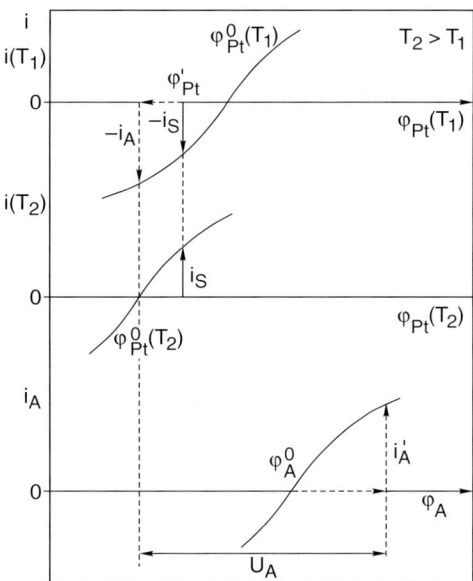

Fig. 3.53. Current–potential curves of the electrodes depicted in Fig. 3.52. The voltage U_A supplied by the external current source shifts the potential φ'_{Pt} of the platinum tube to the more negative equilibrium value $\varphi^0_{Pt}(T_2)$ by polarizing the auxiliary electrode in the positive direction. This reduces the anodic current i_S at T_2 to zero thus eliminating oxygen bubble formation and increases the cathodic current $-i_S$ at T_1 to become equal to the auxiliary current, $|-i_A(T_1)| = i'_A$ through the auxiliary electrode

(2) Due to this potential shift, the cathodic current at the lower temperature T_1 and thus the rate of oxygen reduction are increased, $-i_S \rightarrow -i_A$.

(3) The potential of the auxiliary electrode is shifted from its equilibrium value φ^0_A to a more positive value causing the anodic auxiliary current i'_A, which is equal to the cathodic current through the platinum at T_1, $i'_A = |-i_A(T_1)|$ and causes a rather high formation rate of oxygen bubbles at the auxiliary electrode.

Application of the correct auxiliary voltage U_A, consequently, changes the short-circuited non-isothermal cell

$$|Pt(T_1)|\text{melt}|Pt(T_2)| \,, \tag{3.XVI}$$

whose hot platinum electrode shows oxygen bubble formation according to the internal current i_S, to the electrolysing cell,

$$(-)Pt(T_1)|\text{melt}|Pt(\text{auxiliary})|(+) \,, \tag{3.XVII}$$

whose auxiliary electrode is subject to relatively strong oxygen bubble formation due to the externally driven current i'_A, which is larger than i_S. The platinum electrode at the higher temperature T_2 is eliminated and replaced by the auxiliary electrode, and the current caused by the short circuit is replaced by the electrolysing current. The practical effect is thus a displacement of the oxygen bubble formation from the platinum tube (at T_2) to the auxiliary electrode, where it is of no concern, and a corresponding increase of the oxygen formation rate. The correct magnitudes of the auxiliary voltage U_A and the auxiliary current i'_A are controlled by the zirconia electrode, which also prevents the reduction of melt components and thus protects the platinum tube from alloying and destruction. The auxiliary electrode being subject to an anodic polarization must be sufficiently large to secure a limited current density so that the electrode reaction at its surface is restricted to oxygen formation and the oxidation of platinum is excluded.

The application of the externally driven auxiliary electrode is basically not restricted to the elimination of anodic electrode reactions. Also cathodic processes, for instance, metal formation by reduction of metal ions at short-circuited non-isothermal heating electrodes and connected melt decomposition and electrode corrosion can be suppressed. However, appropriate electrical filters must be applied if the electrodes to be protected are subject to alternating currents. Moreover, means for indicating the degree of protection must be provided because, unlike the formation of oxygen bubbles, most reactions do not simply indicate themselves.

A modified version is the so-called autogenic auxiliary electrode, whose principle is shown in Fig. 3.54. The inert auxiliary (e.g., platinum) electrode and the external current source shown in Fig. 3.52 are replaced by a reactive metal (e.g., molybdenum) electrode, which has the tendency to dissolve anodically in glass melts. Figure 3.54 demonstrates its application to eliminate oxygen bubble formation which, in this particular case, is caused by an

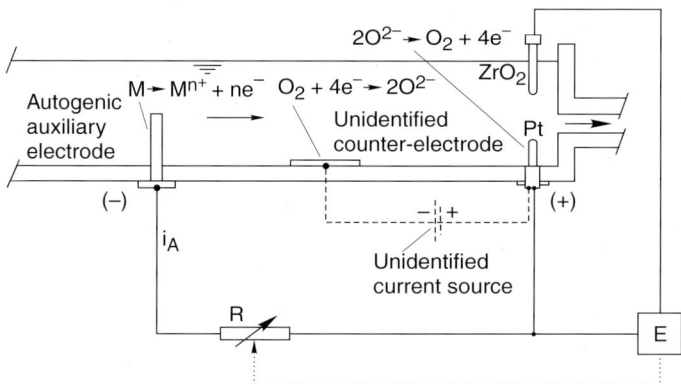

Fig. 3.54. Autogenic auxiliary electrode consisting of a reactive metal, for example molybdenum, as applied to suppress the electrolytic formation of oxygen bubbles at the platinum part of a glass melter. Counter-electrode as well as electrolysing current source are unidentified

electrolysis of unknown origin. Neither the counter-electrode and its location in the melting tank nor the electrolysing current source are identified. The autogenic auxiliary electrode is simply connected electrically to the platinum part of concern, and a zirconia electrode controls the auxiliary current i'_A between these electrodes via a variable resistor R.

Figure 3.55 shows the electrochemical operation. The voltage $U_{(?)} = \varphi'_{Pt} - \varphi'_{(?)}$ of the unidentified current source causes an anodic current i_E at the potential φ'_{Pt} through the platinum electrode, which is responsible for oxygen bubble formation, and an equally large cathodic current $-i_E$ at the potential $\varphi'_{(?)}$ through the unknown counter-electrode, which is connected with a likewise unknown electrode reaction. Connecting the autogenic auxiliary electrode, whose equilibrium potential is φ^0_A, to the platinum electrode shifts the platinum electrode potential to its more negative equilibrium value φ^0_{Pt}, thus reducing the anodic current (and oxygen formation) to zero, and also displaces the potential $\varphi'_{(?)}$ of the unknown counter-electrode to a likewise more negative value. Correspondingly, the potential of the auxiliary electrode is positively polarized, $\varphi^0_A \to \varphi_A$, thus causing the anodic auxiliary

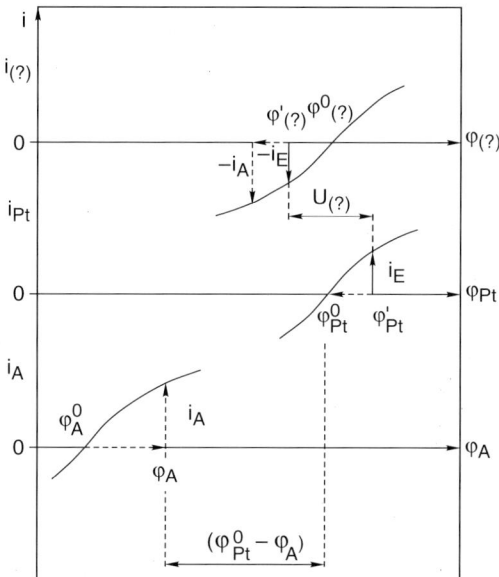

Fig. 3.55. Current–potential curves of the electrodes shown in Fig. 3.54. The short-circuited, slowly dissolving, negative, autogenic auxiliary electrode polarizes both the platinum and the unidentified counter-electrode to more negative values so that the anodic current i_E through the platinum is made zero at the equilibrium potential φ^0_{Pt} and the cathodic current $-i_E$ through the unknown counter-electrode is increased to become equal to the auxiliary current i_A. Both counter-electrode and electrolysing current source, which are responsible for oxygen bubble formation, need not be identified for the application of the auxiliary electrode

current i_A, whose absolute value is equal to the cathodic current through the unknown counter-electrode.

The application of the correct auxiliary potential $(\varphi_{Pt}^0 - \varphi_A)$ by means of the variable resistor R (Fig. 3.54) and of the corresponding auxiliary current i_A, consequently, replaces the electrolysed cell

$$(-)|\text{unidentified electrode}|\text{melt}|\text{Pt}|(+) \qquad\qquad (3.XVIII)$$

in which the electrolysis current i_E causes oxygen bubble formation at the platinum electrode, by the short-circuited cell

$$|\text{Mo}|\text{melt}|\text{unidentified electrode}| \,, \qquad\qquad (3.XIX)$$

which is subject to the auxiliary current i_A. Oxygen bubble formation at the platinum electrode is replaced by an anodic dissolution of the autogenic (e.g. molybdenum) auxiliary electrode, which may be located at any technically convenient place in the melting tank. An essential condition of the application is that the potential of the polarized autogenic auxiliary electrode must be more negative than those of the platinum as well as of the (unknown) counter-electrode, which is met by molybdenum because of its rather negative standard potential in all oxidic glass-forming melts.

Molybdenum ions generated during the application of the autogenic auxiliary electrode have a negligible effect upon the quality of the produced glass because of the small metal quantities dissolved. If, for example, an externally driven and an autogenic auxiliary electrode are compared and both assumed to have comparable current–potential curves, it can be estimated that 9300 oxygen bubbles with a diameter of 6 mm at 1200 °C, which are generated at an auxiliary electrode within approximately 5–10 h, are equivalent to the dissolution of 1 g (corresponding to 0.01 mol or 0.1 cm^3) molybdenum as fourfold positively charged molybdenum ions. Such dissolution rates, i.e., 0.1–0.2 g molybdenum ions per hour, have practically no effect upon the glass quality even if the molybdenum auxiliary electrode is applied in small melting units.

The application of external voltages in glass melts, as in the case of externally driven auxiliary electrodes, and of internal voltages, as with autogenic auxiliary electrodes, introduces the possibility of intolerable electrode reactions. Excessively negative potentials can cause reduction of melt components, whose reaction products, mostly elements, tend to alloy with the metals (e.g. platinum) at the high temperatures, leading to corrosion and, in inferior cases, to the melting of the metal part. High anodic current densities at positively polarized electrodes, on the other hand, can oxidize the electrode metal and, depending on the material applied, can cause oxide layers or dissolve the metal as ions, which may impair the glass products by inducing undesired colouring or by precipitating as crystallites under unfavourable redox conditions. The application of auxiliary electrodes thus necessitates considering all reactions that could possibly lead to intolerable effects and avoiding unnecessarily large positive and particularly negative potentials, in particular when unidentified counter-electrodes are involved. Despite these difficulties, auxiliary electrodes, applied with care, are helpful tools for the improvement of industrial glass melting.

3.8 Oxygen Bubble Formation at Melt/Refractory Interfaces

Friedrich G.K. Baucke

A major problem in glass melting is the formation of gas bubbles, which are intolerable in glass products not only for esthetic reasons, as in household glasses, and for functional reasons, as in optical glasses, but also because they can present considerable safety problems, for instance in television tubes and screens. Bubbles can be generated in glass melts by various causes. For example, insufficient fining of glass melts can lead to reboil at hot parts of the melters, yielding bubbles of variable gas contents. Water vapour from defective cooling arrangements of glass melters and sodium chloride, which is frequently used as a fining agent, induce the formation of vacuum bubbles in glass products because water and salt vapours condensate in the voids at ambient temperatures. Oxygen bubbles can be generated by electrolysis of melts and by short-circuiting electrochemical cells, for instance inert metal parts in non-isothermal glass melts, as described in Sect. 3.6. The danger that the bubbles reach the glass product is the greater, the closer the bubble source is to the drain of the technical melter and the higher the viscosity in this range of the melting unit.

It is of particular interest that bubbles can also be generated merely by the contact of refractory materials with oxidic glass melts. This effect is observed with new, untreated zirconium silicate (ZS), which is frequently used as building or lining material of melting tanks with long lifetime because of its extremely high chemical resistivity to oxidic melts particularly at high temperatures. Because of its often long-lasting duration and the resulting losses, the reaction is of high economic importance. As scientists quite early noticed, there are two possible causes for this bubble formation. (1) At the high melt temperatures, pores in the refractory expel their gas content, which consists mainly of air, and generate bubbles which also act as nuclei for other gases from the melts and thus lead to so-called flow-over bubbles of various composition. (2) An electrochemical reaction leads to the formation of oxygen and oxygen bubbles. The first kind of bubble generation corresponds to that frequently observed with various refractory materials, whereas the electrochemical formation of oxygen bubbles is of particular scientific interest, in addition to its economic importance, and had repeatedly been the object of speculations in the past. In 1966, *Bossard* and *Begley* suggested that the driving force of the reaction is given by a sodium concentration cell between ZS and the contacting sodium-containing melts [3.281]. Later, *Bedros* and *Fojtkovà* proposed that oxygen is formed by a redox reaction connected with sodium migration into the zirconium silicate [3.282]. In both papers, however, neither the cells employed for the investigations nor detailed reaction mechanisms are given. We have investigated this reaction because of its economic significance and will describe the resulting analysis [3.283–285] in detail in this section because of the unique character of the reaction, which, surprisingly, turned out to have an electrochromic mechanism [3.283]. Electrochromic reactions are generally observed in thin-layer systems at ambient temperatures [3.286] but have never been described for refractory materials and at high temperatures. The reaction of new zirconium silicate with oxidic glass melts above 1000 °C is thus of fundamental scientific interest.

The experiments were conducted on ZS materials of high density, for example on ZS 1300 with an apparent porosity of 0.5%, containing the impurities Fe_2O_3 (0.12 wt%) and TiO_2 (1.28 wt%) [3.287], and on ZS 835 dense with an apparent porosity of 2.9%–3.4%, containing 0.08 wt% Fe_2O_3 and 0.25 wt% TiO_2. Dissociation of these materials to ZrO_2 and SiO_2 is restricted to high-temperature high-alkaline environments [3.287] and was thus excluded during the experiments. The glass melts applied were mostly alkali calcium silicates with the composition (in mol%) 15.5 alkali oxide, 10.8 calcium oxide, 73.7 silicon dioxide, where alkali = lithium, sodium, potassium, or sodium and potassium (mole ratio 1:1). The fining agent was 0.2 wt% Sb_2O_3. These details should be kept in mind when the fundamental experiments are described.

3.8.1 The Reaction Mechanism

The reaction is basically an oxidation of oxide ions of the melt at the ZS surface by traces of redox impurities of the refractory, for example Fe_2O_3, frozen-in in a highly oxidized state at the high temperatures of ZS fabrication. The first step is the formation of a phase boundary redox equilibrium between melt and ZS,

$$2O^{2-}(s = 0) + 4Fe^{3+}(s = 0) \rightleftarrows O_2(s = 0) + 4Fe^{2+}(s = 0) , \qquad (3.173)$$

where s is the distance from the interface melt/ZS. This equilibrium is not maintained because the reaction continues into the ZS (and, *per se*, could not possibly lead to the formation of oxygen bubbles because of the small amounts involved in the equilibrium). The continuation of the reaction, however, necessitates (a) the transport of electrons from the ZS surface, where they are given off by the oxides, to the site in the refractory where they are consumed by threefold positive iron and (b) a balance of the electron charge, introduced into the ZS, by a transfer of alkali ions of the melt into the solid [3.283]. The experimental preconditions of both processes are met by the well-known finite electronic conductivity of ZS materials commonly used as refractories in glass melting, which have also been found to be sufficiently good alkali ion conductors at the high temperatures concerned [3.288].

The total process, which is more complicated than (3.173) suggests, consists of five reaction steps.

(1) Oxidation of oxide ions at the ZS surface,

$$2O^{2-}(s = 0) \rightarrow O_2(s = 0) + 4e^-(s = 0) , \tag{3.174}$$

(2) transport of electrons from the ZS surface to the time-dependent depth $s(t)$ in the ZS,

$$4e^-(s = 0) \rightarrow 4e^-(s(t)) , \tag{3.175}$$

(3) consumption of the electrons by the internal reduction of ferric ions,

$$4e^-(s(t)) + 4Fe^{3+}(s(t)) \rightarrow 4Fe^{2+}(s(t)) , \tag{3.176}$$

(4) balance of the charge difference of the phases, caused by the electron introduction, by a transfer of alkali ions from the melt into the solid,

$$M^+(melt, s = 0) \rightarrow M^+(s = 0) , \tag{3.177}$$

and, finally, (5) transport of the alkali ions from the ZS surface to the site of the internal reduction of ferric to ferrous ions,

$$M^+(s = 0) \rightarrow M^+(s(t)) . \tag{3.178}$$

Equation (3.179),

$$2O^{2-}(s = 0) + 4Fe^{3+}(s(t)) + 4M^+(melt, s = 0)$$
$$\rightarrow O_2(s = 0) + 4Fe^{2+}(s(t)) + 4M^+(s(t)) , \tag{3.179}$$

which is a summary of (3.174)–(3.178), describes the total process of the internal reduction of ferric ions in ZS by the oxidation of oxide at the melt/ZS interface. However, the transport of electrons, (3.175), and the single steps of alkali ion transport, (3.177) and (3.178), which cancel in (3.179), must additionally be taken into account when kinetic viewpoints are considered. It is clear by this analysis that the rate of the reaction decreases with time because of the increasing layer thickness which determines the overall diffusion rate $s(t)$. *Schmalzried* has investigated and thoroughly treated internal reactions in crystalline phases [3.289]. The internal reduction at hand, however, is different as it is a reaction of material traces and excludes phase changes.

The existence of a rather uniform reduced layer with a steep boundary between reduced and unchanged ZS, which is the basis of the internal reduction mechanism, was proved experimentally by concentration profiles of the charge-balancing alkali ions, which were found to be step profiles [3.283, 284]. Besides, the distinct sharp boundary between flesh-coloured highly oxidized zirconium silicate and light-grey reduced surface layers moved into the ZS during the reaction and slowed down with increasing distance from the interface. It was accompanied by an initially strong and slowly decreasing formation of oxygen bubbles at the ZS surface. These effects could often be repeated by removing the grey surface layers, which had formed during the preceding reactions, by grinding [3.283].

3.8.2 A Subsurface Electrochemical Cell

As sketched in Fig. 3.56, the reduced surface layer, together with the contacting oxidic melt and the unchanged ZS, form an internally short-circuited galvanic cell without transference [3.283, 284]. The melt is the (fictitious) negative electrode (anode) whose electrode reaction is given by oxide oxidation, (3.174), and alkali ion transfer, (3.177), the unchanged ZS forms the positive electrode (cathode) whose electrode reaction is the sum of internal reduction of ferric ions, (3.176), and the acceptance of alkali ions according to (3.178), and the reduced ZS layer constitutes both the electrolyte transporting alkali ions, (3.178), and the short-circuiting electron conductor, (3.175). The cell reaction is the sum of these electrode reactions and is given by (3.179).

Against appearances, the fact that the reaction layer forms a galvanic cell is not merely formal. This is obvious from a comparison with the well-known electrochemical cell in aqueous solution, see for instance [3.290]

$$\left|\mathrm{Pt,H_2}\left|\mathrm{HCl(solution,Pt)}\right|\!\!\left|\mathrm{HCl(solution,AgCl)}\right|\mathrm{AgCl}(s)\left|\mathrm{Ag}\right| \ , \right. \tag{3.XX}$$

which, assuming an appropriate hydrogen pressure and hydrochloric acid concentration, is subject to the following reactions:

(1) anodic hydrogen oxidation,

$$\mathrm{H_2(Pt)} \rightarrow 2\mathrm{H}^+(\mathrm{solution, Pt}) + 2\mathrm{e}^-(\mathrm{Pt}) \ , \tag{3.180}$$

(2) electron transport from platinum to silver,

$$2\mathrm{e}^-(\mathrm{Pt}) \rightarrow 2\mathrm{e}^-(\mathrm{Ag}) \ , \tag{3.181}$$

(3) reduction of silver ions of the silver chloride,

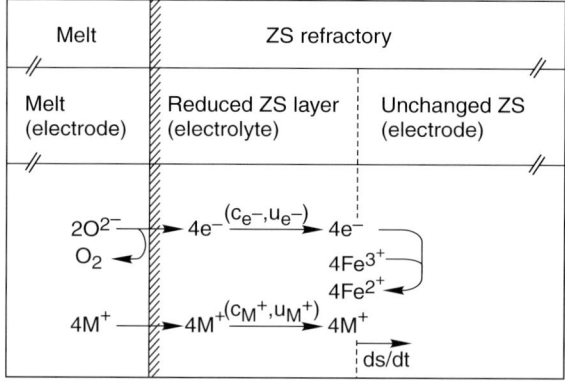

Fig. 3.56. Schematic presentation of the mechanism of the redox reaction between oxidic glass melts and zirconium silicate refractories containing a redox impurity in a highly oxidized state; c = concentration, u = mobility of the species indicated

$$2e^-(Ag) + 2AgCl(s) \rightarrow 2Ag + 2Cl^-(solution, AgCl) \,, \tag{3.182}$$

(4) shift of hydrogen ions from the HCl solution in the platinum/hydrogen to that in the siver/silver chloride electrode half-cell according to their transport number t_+,

$$2t_+H^+(solution, Pt) \rightarrow 2t_+H^+(solution, AgCl) \,, \tag{3.183}$$

(5) and shift of chloride ions in the reverse direction according to the transport number $t_- = 1 - t_+$ of chloride ions,

$$2(1 - t_+)Cl^-(solution, AgCl) \rightarrow 2(1 - t_+)Cl^-(solution, Pt) \,. \tag{3.184}$$

The overall cell reaction

$$\begin{aligned} H_2(Pt) + 2AgCl(s) \rightarrow\ & 2(1 - t_+)HCl(solution, Pt) \\ & +2t_+HCl(solution, AgCl) + 2Ag \end{aligned} \tag{3.185}$$

reduces to the simpler form

$$H_2(Pt) + 2AgCl(s) \rightarrow 2HCl(solution, AgCl) + 2Ag \tag{3.186}$$

if unit transport number, $t_+ = 1$, of hydrogen ions, for instance by a proton-conducting diaphragm, is assumed. The processes (3.180)–(3.184) correspond exactly to those of the ZS layer cell, (3.174)–(3.178), and also the cell reactions, (3.179) and (3.186), correspond to each other. The main difference, besides the physical state of the electrolytes, is the kind of short circuit, which in the Pt|Ag cell is via an external electron conductor and therefore controllable, whereas in the ZS layer cell it is through the reduced-ZS electrolyte and therefore uncontrollable. It will be seen below that the analogy is helpful even when the cell voltage of the ZS layer cell is formulated.

The redox reaction between oxidic melts and ZS refractories is an almost unparalleled case in electrochemistry. Generally, it is tried to subdivide a given reaction proposed as a cell reaction, for instance (3.186), into electrode reactions, for example (3.180) and (3.182), which can be verified experimentally, for example in cell (3.XX), so that, for instance, the energy or free energy of the reaction can be utilized as electric energy or emf, respectively, of a corresponding battery or galvanic cell, see for instance [3.291]. By contrast, the cell reaction between ZS and oxidic melts divides itself into two redox electrode reactions, which take place at different locations in the refractory, i.e., at the ZS surface and at a moving site in the bulk of the solid. The reason for this unique behaviour is the fortuitous combination of several properties, an electronic as well as ionic conductivity of the ZS, the presence of a redox ion couple in a non-equilibrium state in the ZS as well as of a redox couple at the melt/ZS interface, both in suitable energetic states, the presence of cations fitting the space in the solid, and conditions, for instance high temperatures, that allow the cell reaction to take place at reasonable rates.

EMF and Cell Voltage

The emf of the ZS layer cell defined as the difference of internal electrode potential minus melt potential according to (3.179) is given by

$$E_{ZS|m} = E^0_{ZS|m} + \frac{RT}{4F} \ln \frac{a^4_{Fe^{3+},ZS}\, a^2_{O^{2-},m}\, a^4_{M^+,m}}{a^4_{Fe^{2+},ZS}\, p_{O_2,m}\, a^4_{M^+,m}} \; , \tag{3.187}$$

where the indices ZS and m denote unchanged ZS and melt, respectively, and $p_{O_2,m}$ is the partial pressure in equilibrium with the activity of oxygen dissolved in the melt. $E_{ZS|m}$ is constant at a given temperature and melt composition because of the internal step profile of the ZS, which supplies a constant ferric/ferrous ion ratio to the layer cell. However, neither can the cell, Fig. 3.56, be subjected to direct measurements because of the fictitious electrode melt nor can the emf, (3.187), be measured indirectly (except by extrapolation of the cell voltage, see below) because of the unavoidable short circuit, which reduces the emf to the cell voltage.

The cell voltage is derived in the usual manner, see for instance [3.292]. Thus, in a galvanic cell, for instance cell (3.XX) above, which is under an external load iR_a, the emf E is the driving force of the uniform current i through the entire circuit,

$$E = i(R_a + R_i) \; , \tag{3.188}$$

if R_i is the internal cell resistance and R_a the external resistance, whereas the cell voltage E_c, which is available for work to be done by the cell, is equal merely to the voltage drop over the external resistance,

$$E_c = iR_a \; . \tag{3.189}$$

Combination of these expressions yields the cell voltage as a function of the internal resistance and the cell current,

$$E_c = E - iR_i \; , \tag{3.190}$$

where the second term on the right side could be called the internal voltage drop or internal iR drop.

Corresponding conditions are valid in the ZS layer cell, where the internal resistance is represented by the electrolytic resistance of the reduced-ZS layer, $R_i = R_{M^+}$, and the external resistance by the electronic resistance of the layer, $R_a = R_{e^-}$. The current is given by the ("internal") cation as well as the ("external") electronic current, whose (absolute) magnitudes are equal because of electroneutrality, $|i_{e^-}| = i_{M^+}$. The general equation (3.190) applied to the ZS layer cell is thus given by

$$E_c(t) = E_{ZS|m} - (i_{M^+}R_{M^+})(t) = E_{ZS|m} - (i_{e^-}R_{M^+})(t) \; , \tag{3.191}$$

where (t) indicates time-dependent terms. Differentiation, finally, yields the time dependence of the cell voltage,

$$\frac{\mathrm{d}E_c}{\mathrm{d}t} = -\frac{\mathrm{d}(i_{\mathrm{M}^+} R_{\mathrm{M}^+})}{\mathrm{d}t} = -\frac{\mathrm{d}(i_{\mathrm{e}^-} R_{\mathrm{M}^+})}{\mathrm{d}t} \ , \tag{3.192}$$

which is determined by the smallest time dependence of the iR drops of the ZS layer cell in (3.192).

Measurement of the Cell Voltage

The cell voltage of the ZS layer cell is measured by combining the melt/ZS system with a zirconia electrode [3.4],

$$\mathrm{Pt,O_2|ZrO_2|melt|ZS|Pt} \ , \tag{3.XXI}$$

or a platinum electrode,

$$\mathrm{Pt|melt|ZS|Pt} \ . \tag{3.XXII}$$

Figure 3.57 presents details and shows that these cells consist of the internal redox electrode and of two competing oxygen/oxide electrodes, melt/ZS and zirconia or platinum electrode. The complete cell reaction is thus given by

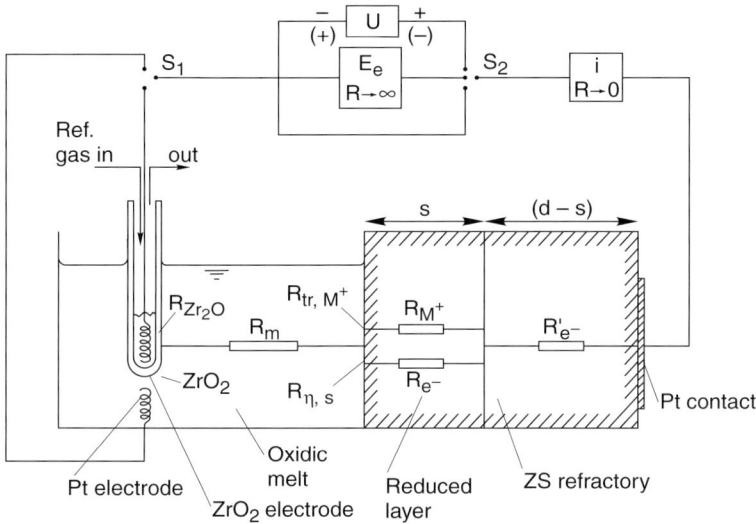

Fig. 3.57. Experimental arrangement for investigating the redox reaction between oxidic glass melts and redox impurity-containing ZS: measurement of time-dependent cell voltage $E_e(t)$, polarization of ZS by external voltage U, and short-circuiting of cell. S_1 and S_2 are switches for selecting external electrode and mode of measurement, respectively

$$4Fe^{3+}(ZS) + 2nO^{2-}(m, Pt) + 2(1-n)O^{2-}(m, ZS) + 4M^+(m)$$
$$\rightarrow 4Fe^{2+}(ZS) + nO_2(m,Pt) + (1-n)O_2(m, ZS) + 4M^+(ZS) , \qquad (3.193)$$

where n is the part of the oxygen/oxide reaction taking place at the zirconia or platinum electrode both indicated by (m, Pt), and $(1-n)$ is the part at the ZS surface indicated by (m, ZS). The emf of this cell,

$$E_{ZS|m|Pt} = E^0_{ZS|m|Pt} + \frac{RT}{4F} \ln \frac{a^4_{Fe^{3+},ZS}\, a^{2n}_{O^{2-},m,Pt}\, a^{2(1-n)}_{O^{2-},m,ZS}\, a^4_{M^+,m}}{a^4_{Fe^{2+},ZS}\, p^n_{O_2,m,Pt}\, p^{(1-n)}_{O_2,m,ZS}\, a^4_{M^+,ZS}}$$

$$(3.194)$$

becomes independent of n for uniform melt and equal oxygen partial pressures of melt and zirconia electrode (or uniform partial pressure of melt if the platinum electrode is considered):

$$E_{ZS|m|Pt} = E^0_{ZS|m|Pt} + \frac{RT}{4F} \ln \frac{a^4_{Fe^{3+},ZS}\, a^2_{O^{2-},m}\, a^4_{M^+,m}}{a^4_{Fe^{2+},ZS}\, p_{O_2,m}\, a^4_{M^+,ZS}} . \qquad (3.195)$$

The interesting result is thus that, under these conditions, which can easily be verified experimentally, the emf, (3.195), of the practical cells, (3.XXI) and (3.XXII), is equal to the emf of the internal ZS layer cell, (3.187), since also the standard potentials are equal, $E^0_{ZS|m|Pt} = E^0_{ZS|m}$ under these conditions. Consequently, the cell voltage E_e of cells (3.XXI) and (3.XXII) is also equal to the cell voltage E_c of the internal layer cell,

$$E_e(t) = E_c(t) , \qquad (3.196)$$

so that the time dependence of the cell voltage of cells (3.XXI) and (3.XXII) directly yields the time dependence of the iR drop of the ZS layer cell, (3.192). This allows meaningful quantitative measurements on cells (3.XXI) and (3.XXII) as indicated in Fig. 3.57. Thus, the time course of the reaction was measured under open-circuit and short-circuited conditions, and the reaction was accelerated and retarded by an external voltage source with its negative and positive pole connected to the platinum contact of the ZS, respectively. Table 3.10 gives an overview of the experiments, and it is helpful for the discussion to realize the resistances of the zirconium silicate specimen, Fig. 3.57:

(1) the electrolytic resistance of the ZS layer cell ,

$$R_{M^+}(t) = \frac{s(t)}{AFc_{M^+}u_{M^+}} , \qquad (3.197)$$

(2) the electronic resistance of the reduced-ZS layer ,

$$R_{e^-}(t) = \frac{s(t)}{AFc_{e^-}u_{e^-}} , \qquad (3.198)$$

and

(3) the electronic resistance of the unchanged zirconium silicate, which is much thicker than the ZS layer, $(d - s(t)) \gg s(t)$,

$$R'_{e^-}(t) = \frac{d - s(t)}{AFc'_{e^-} u'_{e^-}} . \tag{3.199}$$

c and u are concentrations and mobilities, respectively, of the species indicated, primed quantities refer to unchanged ZS, and A is the active cross section of the ZS specimen. The transfer resistance, R_{tr,M^+}, of alkali ions from the melt into ZS, Fig. 3.57, the polarization resistance $R_{\eta,s}$ of ZS, and the resistances of melt and zirconia, R_m and R_{ZrO_2}, respectively, have been found negligible.

Open-Circuit Measurements – Internal Cell Reaction. (Compare line 4 of Table 3.10.) The undisturbed, spontaneous cell reaction is characterized by an equal (absolute) cationic and electronic current density through the reduced-ZS layer of the internal cell. Figure 3.58a presents the time-dependent cell voltage of ZS 1300 in alkali calcium silicate melts with lithium, sodium, potassium, and sodium and potassium (mole ratio 1:1) and shows not only the expected increase of the iR drop of the internal cell with time but also a strong dependence of the change rate on the alkali species of the melt. The rate of the internal reaction is thus determined by the diffusion rate of the alkali ions in the reduced-ZS layer. The electrolytic resistance is much larger than the electronic resistance of the layer, $R_{M^+} \gg R_{e^-}$. The reaction shows no mixed-alkali effect.

Extrapolation of the monotonically decreasing parts of the plots, see Fig. 3.58a, to zero time yields approximate emfs of the ZS-layer cell, $E_e(t \to 0) = E_{ZS|m}$, which are between $+0.92\,V$ and $+0.98\,V$ for the four melt compositions. This narrow range can be understood because the oxygen partial pressures of the melts and in the zirconia electrode were equal (1 bar) and the optical basicities of the melts were between $\Lambda = 0.56$ (lithium-containing melt) and $\Lambda = 0.59$ (potassium-containing melt). The corresponding free energies under experimental conditions (not standard free energies) of the reactions are between $-355\,kJ\,mol^{-1}$ and $-378\,kJ\,mol^{-1}$.

Figure 3.58b shows the temperature-dependence of the cell voltage in the sodium-containing melt. As in Fig. 3.58a, a monotonic decrease of E_e with time is observed after several hours. The maximum slopes after the initial period yield an activation energy of $(204 \pm 10)\,kJ\,mol^{-1}$, which obviously implies solid-state diffusion of sodium ions because diffusion in melt penetrated into ZS pores would have smaller activation energies, for example $65\,kJ\,mol^{-1}$ at $1025\,°C$ and $56\,kJ\,mol^{-1}$ at $1350\,°C$ for the glass composition applied [3.1]. In addition, extrapolation of the monotonically decreasing parts to zero reaction time results in an approximate entropy of the layer cell reaction of $340\,J\,K^{-1}\,mol^{-1}$, which is a reasonable order of magnitude for solid-state diffusion.

Table 3.10. Schematic of cells (3.XXI) and (3.XXII) (see Fig. 3.57), demonstrating the effect of an internal reduction plane in ZS on electron and cation current densities caused by positive and negative polarization of the ZS in an oxidic glass melt. The internal reaction acts as a sink for electrons and cations, except at high positive polarization when it turns into an electron and cation source. Current densities within ZS (4th column) are defined to be positive for alkali ions and negative for electrons when they migrate from either ZS surface towards the reduction plane, and vice versa

Line	Electrochemical cell	Current densities to and from reduction plane	Characterization of circuit and reactions

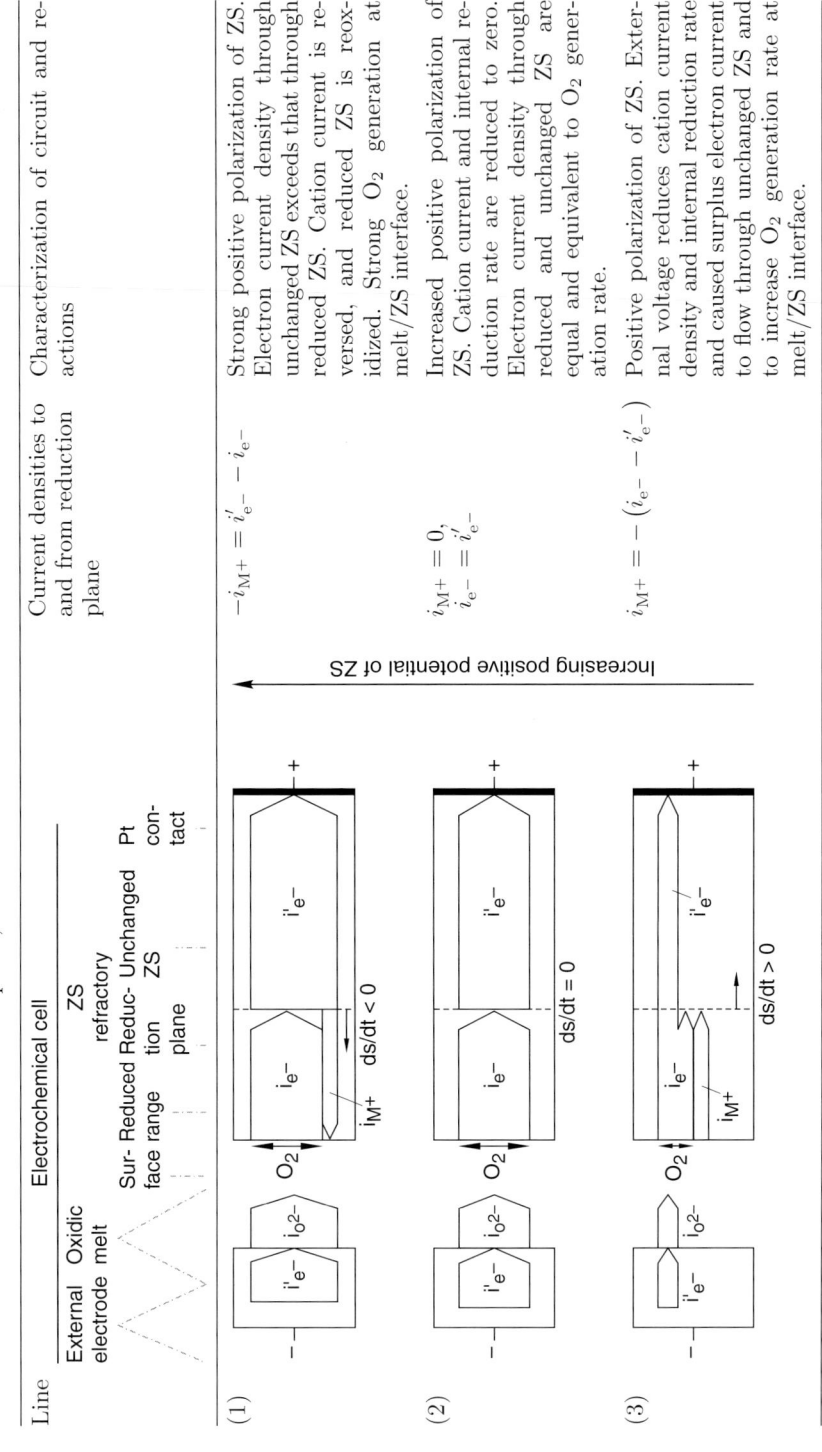

(1) $-i_{M+} = i'_{e-} - i_{e-}$ Strong positive polarization of ZS. Electron current density through unchanged ZS exceeds that through reduced ZS. Cation current is reversed, and reduced ZS is reoxidized. Strong O_2 generation at melt/ZS interface.

(2) $i_{M+} = 0$, $i_{e-} = i'_{e-}$ Increased positive polarization of ZS. Cation current and internal reduction rate are reduced to zero. Electron current density through reduced and unchanged ZS are equal and equivalent to O_2 generation rate.

(3) $i_{M+} = -(i_{e-} - i'_{e-})$ Positive polarization of ZS. External voltage reduces cation current density and internal reduction rate and caused surplus electron current to flow through unchanged ZS and to increase O_2 generation rate at melt/ZS interface.

Table 3.10 (continued)

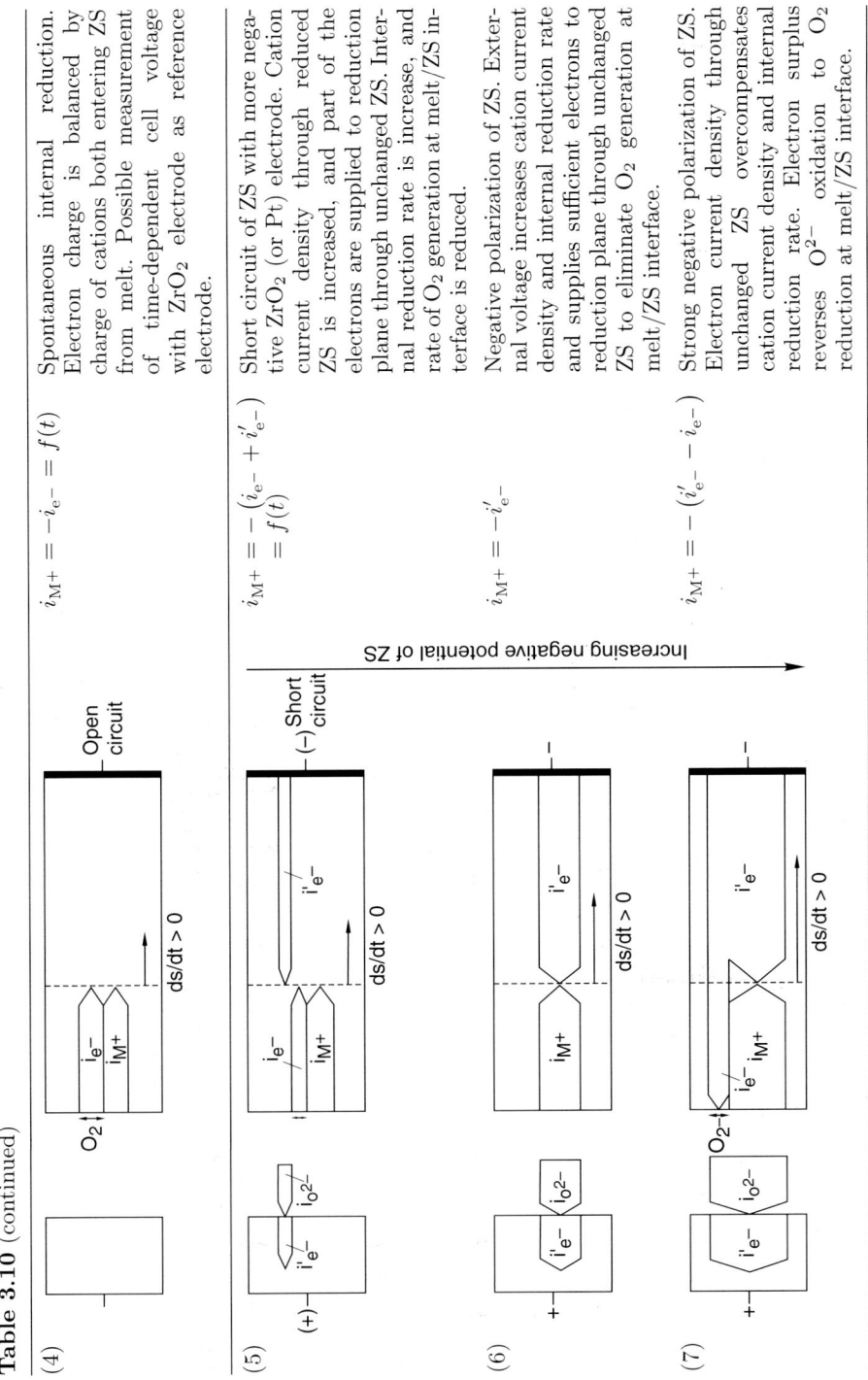

(4) $i_{M^+} = -i_{e^-} = f(t)$

Spontaneous internal reduction. Electron charge is balanced by charge of cations both entering ZS from melt. Possible measurement of time-dependent cell voltage with ZrO$_2$ electrode as reference electrode.

(5) $i_{M^+} = -(i_{e^-} + i'_{e^-}) = f(t)$

Short circuit of ZS with more negative ZrO$_2$ (or Pt) electrode. Cation current density through reduced ZS is increased, and part of the electrons are supplied to reduction plane through unchanged ZS. Internal reduction rate is increase, and rate of O$_2$ generation at melt/ZS interface is reduced.

(6) $i_{M^+} = -i'_{e^-}$

Negative polarization of ZS. External voltage increases cation current density and internal reduction rate and supplies sufficient electrons to reduction plane through unchanged ZS to eliminate O$_2$ generation at melt/ZS interface.

(7) $i_{M^+} = -(i'_{e^-} - i_{e^-})$

Strong negative polarization of ZS. Electron current density through unchanged ZS overcompensates cation current density and internal reduction rate. Electron surplus reverses O^{2-} oxidation to O$_2$ reduction at melt/ZS interface.

Increasing negative potential of ZS

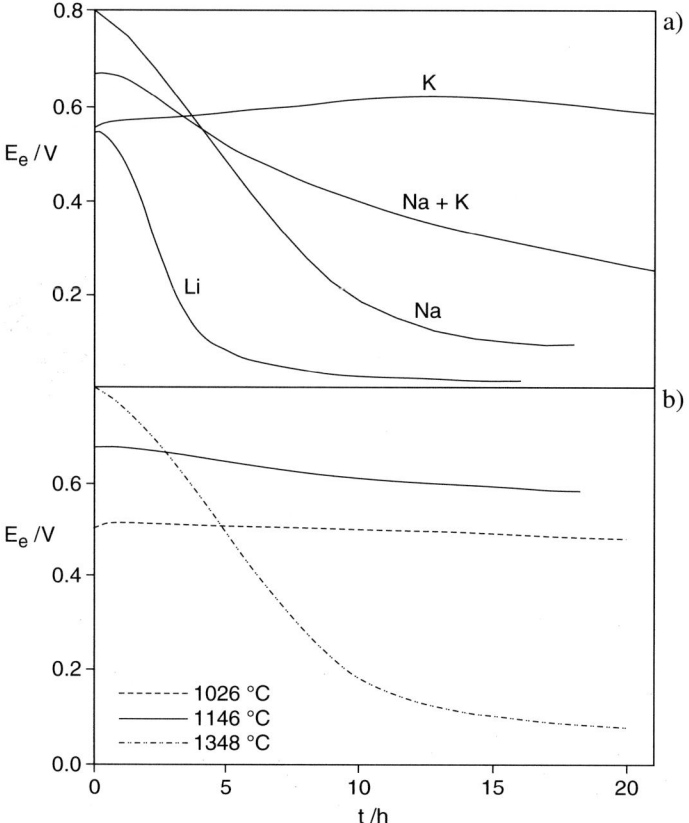

Fig. 3.58. Time-dependent cell voltage E_e of the cell shown in Fig. 3.57. ZS 1300 (**a**) in alkali calcium silicate melts with different alkali species at 1348 °C, (**b**) in sodium calcium silicate melt at various temperatures

Open-circuit measurements also yielded direct confirmation of the reaction mechanism. Thus, as expected, application of alkali-free (e.g. boron trioxide) melts in cell (3.XXI) excludes the reaction with ZS because the necessary charge balance, (3.178), is not possible. Neither formation of oxygen bubbles nor the typical time course of the open-circuit cell voltage, Fig. 3.59, are observed. Rather, the cell voltage indicates the formation of the phase boundary redox equilibrium, (3.173), and its shift to the left by an uptake of oxygen by the melt because, in this case, the oxygen partial pressure is below 1 bar due to the missing oxygen formation at the ZS surface. Addition of sodium oxide, however, releases the hindrance, immediately causes oxygen bubbles at the ZS surface, and subsequently brings about the time dependence of the cell voltage usually observed with ZS in alkali-containing melts. Extrapolation of the cell voltage to the time of Na_2O addition yielded the emf $E_{ZS|m} = 0.96$ V, which is of the order of those found with silicate melts.

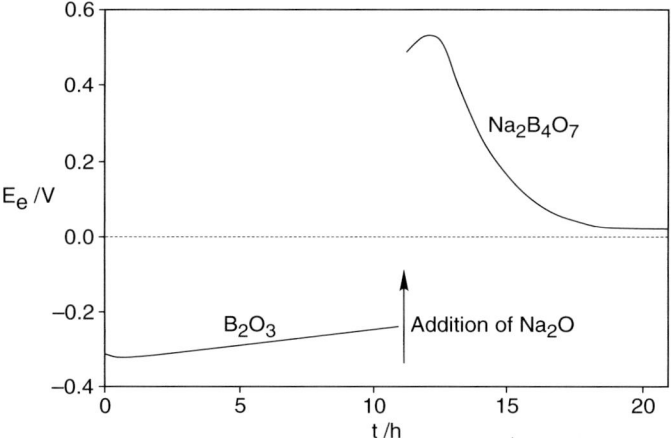

Fig. 3.59. Experiment verifying the mechanism of the redox reaction between oxidic glass melt and redox impurity-containing ZS as schematically presented in Fig. 3.56. In addition to the jump of the cell voltage, strong oxygen bubble formation is observed upon addition of sodium oxide

Negative Polarization of ZS. (Compare lines 5–7 of Table 3.10.) Negative polarization of the ZS by short-circuiting with the more negative zirconia or platinum electrode or by an external current source U, Fig. 3.57, has three effects.

- The charge-balancing electron current through the ZS layer cell is reduced, the difference being supplied to the internal redox electrode through the unchanged ZS.
- The cationic current from the interface melt|ZS to the internal redox electrode increases, and the internal reduction rate dn/dt therefore increases too.
- As n, the part of the oxygen/oxide redox reaction taking place at the competing zirconia or platinum electrode, (3.193), is no longer zero (as at open circuit), oxygen generation is observed also at these electrodes.

For example, the generation of oxygen bubbles at the ZS surface stops immediately after short-circuiting and is observed at the platinum electrode instead. The corresponding oxygen formation is measured in a zirconia electrode when it is short-circuited with the ZS. The electronic resistance of the unchanged ZS is thus smaller than that of the reduced layer, $R'_{e^-} < R_{e^-} < R_{M^+}$, and this is even more valid for the specific resistances $\rho'_{e^-} \ll \rho_{e^-}$, because of the vastly different thicknesses of unchanged ZS and ZS layer, $(d - s(t)) \gg s(t)$, (3.198) and (3.199). Because of the growing thickness of the ZS layer, the rate of the internal reduction decreases with time and the rate of oxygen formation at the competing zirconia (or platinum) electrode increases, $dn/dt > 0$. An increasing negative polarization of the

ZS increases these effects until the charge-balancing electronic current is entirely supplied to the internal reduction by the zirconia or platinum electrode ($n = 1$) through the unchanged ZS and the oxygen formation at the ZS|melt interface is completely eliminated, line 6 of Table 3.10. Further increasing the negative polarization even results in a reduction of oxygen at the melt|ZS interface and in a reversed electron conduction through the ZS layer cell, line 7 of Table 3.10.

Measurement of the time-dependent current through short-circuited cells (3.XXI) or (3.XXII) and of their cell voltage E_e under standardized conditions, Fig. 3.57, represents a useful method of testing ZS materials for their electrochemical properties because it is independent of the formation of other gas bubbles, including non-electrochemically generated oxygen bubbles. The ZS can be applied as an inverted electrode during this test [3.283]. On the other hand, the displacement of oxygen bubble formation to other, short-circuited, electrodes is not feasible in technical glass melting units because of the high ZS resistivity at the low temperatures of outer ZS refractory regions of melting tanks.

Positive Polarization of ZS. (Compare lines 1–3 of Table 3.10.) Positive polarization of the ZS reduces the cationic current through the reduced-ZS layer and the rate of the internal reduction and increases the electronic current through both reduced and unchanged ZS. Figure 3.60 presents the limiting case of zero internal reduction, ds/dt, corresponding to line 2 of Table 3.10, which can be retained for an unlimited time and was kept for 17 h in the example. After the halt, the reaction continues in the same way it would have done without the stop, as indicated by the identical time courses of the cell voltage. It could not be verified experimentally, however, that an additional

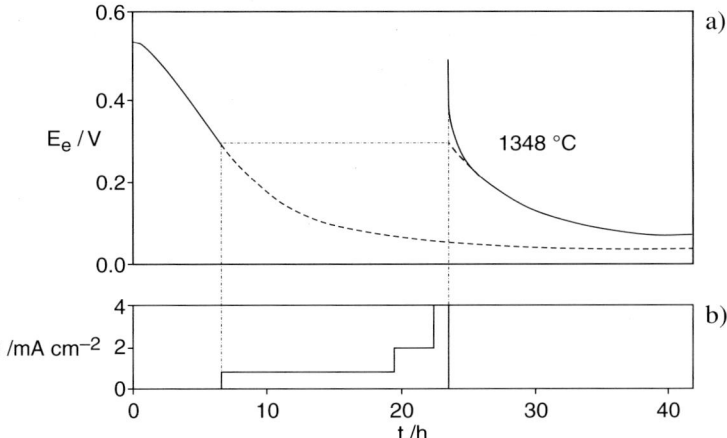

Fig. 3.60. Halt of the redox reaction between an oxidic glass melt and ZS 1300 by positive polarization of the ZS (see line 2 of Table 3.10) and release of the halt after 17 h. (**a**) Cell voltage E_e and (**b**) current density caused by the polarizing voltage

increase of the positive ZS polarization reverses the internal reduction to an internal oxidation, $ds/dt < 0$, according to line 1 of Table 3.10, because the connected large rate of oxygen bubble generation at the ZS surface caused strong foaming of the melt, which excluded further voltage increases.

3.8.3 General Reaction Mechanism, Technical Application

The reaction mechanism is independent not only of the cation species that balances the electron charge introduced into the solid by the internal reduction but also of the anion species of the melt if (a) the free energy of the cell reaction, (3.179), is negative, (b) alkali ions with a suitable radius are available in the melt for charge balance, and (c) the temperature allows a sufficiently fast rate-determining step, which is the cation transport in ZS [3.283, 284]. These conditions are met, for instance, by alkali halides:

$$2Fe^{3+}(ZS) + 2A^-(m) + 2M^+(m) \rightarrow 2Fe^{2+}(ZS) + A_2(gas) + 2M^+(ZS) \,, \tag{3.200}$$

where M means alkali and A means Cl, Br, and I. Figure 3.61 demonstrates this by the time-dependent cell voltage of cell (3.XXI) employing a sodium calcium silicate glass melt and ZS dense that had been contacted by sodium chloride for 24 h at 1200 °C (lower plot). After an initial period of 3 h, the cell voltage is constant and below +0.02 V, whereas that of the untreated sample still indicates strong oxygen bubble formation (upper plot). The general reaction, (3.200), for instance in sodium chloride melt, which is much less corroding than oxidic melts, has thus been proposed as a technical process for preconditioning ZS refractories [3.293], which would eliminate the often long

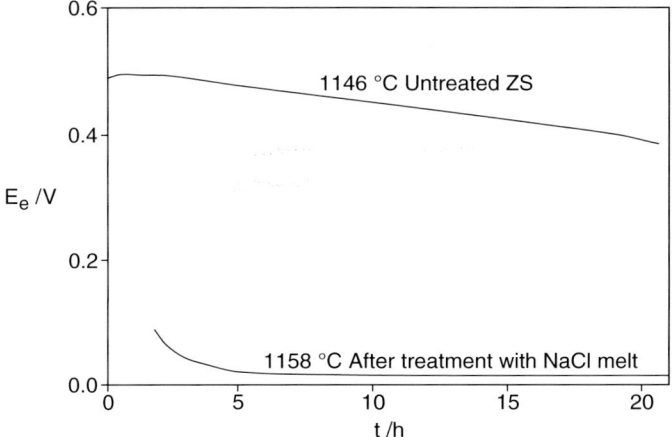

Fig. 3.61. Experiment demonstrating the validity of the reaction mechanism between ZS and oxidic glass melts as sketched in Fig. 3.56 also for the reaction between ZS and sodium chloride melt

lasting and disturbing initial bubble formation in newly constructed melting units. The reaction mechanism with oxygen was later proposed by *Krämer* [3.294] as the basis of a so-called pumping effect of gases observed during temperature changes of AZS (aluminium zirconium silicate) materials.

Electrochromic Character of the Reaction

The reaction between ZS and oxidic or halide melts with electrons and alkali ions moving in the same direction in the solid is not quite as unique as may appear at first sight. In fact, it belongs to a group of reactions which are called electrochromic [3.286, 295]. Figure 3.62a demonstrates this by comparing the reaction between ZS and an oxidic melt and the electrochromic reaction of tungsten trioxide. Both processes involve an internal reduction, ferric to ferrous ions in ZS and tungsten trioxide to its bronze HWO_3 with reduced, fivefold positive tungsten in WO_3, a corresponding oxidation at the surface,

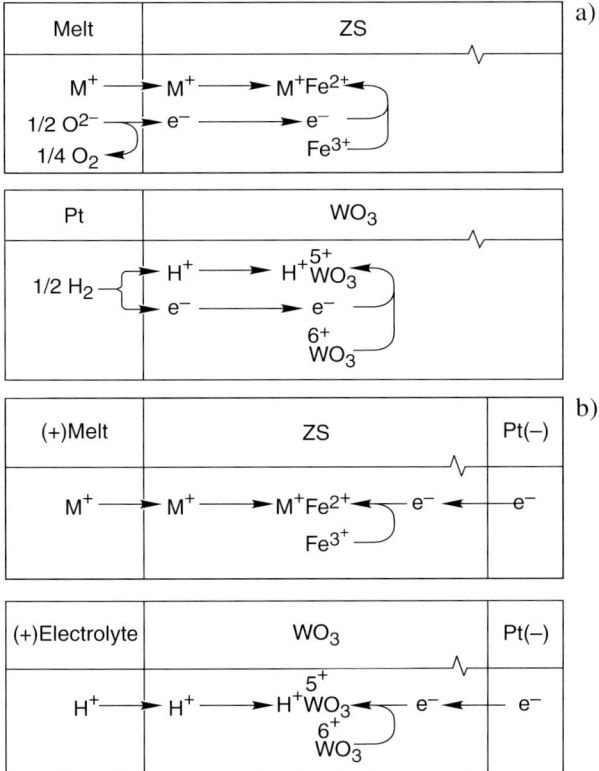

Fig. 3.62. Schematic demonstrating the electrochromic character of the redox reaction between oxidizable melts and redox impurity-containing ZS. (**a**) Spontaneous reactions, (**b**) field-driven reactions

oxide to oxygen at the ZS/melt interface and hydrogen to hydrogen ions at the tungsten trioxide/platinum interface, an electron transport between the locations of oxidation and internal reduction, and a charge balance by alkali ions in ZS and by hydrogen ions in tungsten trioxide. Also, the partly reduced materials are coloured. Reduced ZS containing ferric and ferrous ions has a flesh-like colour in contrast to the grey, more oxidized, refractory with mainly ferric ions, and the tungsten bronze containing five- and sixfold positive tungsten is deep-blue whereas tungsten trioxide is nearly colourless. Both reactions can also be conducted in the field-driven mode [3.286, 295]. For example, Fig. 3.62b compares the reaction of ZS with negatively polarized ZS (see line 6 of Table 3.10) with the field-assisted electrochromic reaction of tungsten trioxide, where the hydrogen ions, which, incidentally, can be exchanged for lithium ions, are supplied by an electrolyte. Also, a layer mechanism has been verified for the ZS/melt reaction [3.283, 284] and proposed for the field-driven electrochromic reaction [3.296, 297]. Because the reactions take place at vastly different temperatures, their similarity may not be of direct practical interest, but comparing them may yield valuable information for the understanding of both processes.

Technical Significance

Based on the knowledge of the reaction of ZS with oxidic melts, its technical significance can be quantified. It is thus concluded that $1\,m^2$ of a ZS refractory containing $0.1\,wt\%$ iron as an impurity, of which 10% are transferred from the oxidized to the reduced state within a layer thickness of 1 mm, produces 53 800 oxygen bubbles with 2 mm diameter or 430 600 oxygen bubbles with 1 mm diameter at $1200\,°C$ if the total oxygen generated appears as bubbles. This corresponds, respectively, to 5.3 and 43.6 bubbles per minute and square metre tank surface if seven days of duration of the reaction with a constant reaction rate are assumed. This latter assumption is certainly unrealistic, but the example demonstrates impressively that the effect of 100 ppm ferric ions, referred to the 1 mm thick refractory layer causing them, can well determine the economy and practical feasibility of a large-scale technical process, such as glass melting, for a certain time, which is normally assumed to be independent of impurity traces. Even more striking, in this case the impurity is part of the reactor material and does not enter the product.

   In many glass melts the bubble formation fortunately starts at a maximum rate, decreases, and eventually stops after some time. However, larger corrosion rates of ZS caused by different melts will cause a steady-state reaction layer and can result in a continuous, although less strong, oxygen bubble formation, which changes the time-limited initial effect into a permanent source of production difficulties. We therefore recommend obtaining as much information as possible about new charges of ZS (and other) refractories and their interactions with the melts to be produced so as to exclude unforeseen deleterious effects during later production periods.

3.9 The Conductivity of Glass-Forming Melts

Friedrich G.K. Baucke

For economical and ecological reasons, heating technical furnaces by electricity becomes increasingly exigent. Besides, this kind of energy application allows much better process control than heating by flames and thus strongly improves the chemical and physical qualities as well as the homogeneity of the glass products. For this kind of heating, the Joule effect is applied: the glass melt, which generally has a high electric conductance, is used as the heat resistor, and electric contacts to the melt are made by metal electrodes, for instance platinum or, if possible, more economical ones, for example molybdenum electrodes.

Electrical heating of glass melts at first sight may seem deceptively simple. Due to the high conductance of most melts, relatively low voltages and correspondingly high currents are applied to transfer the necessary power into melting units. Because in mass production the electrodes are usually molybdenum or related metal rods with relatively small volumes and surface areas, the current densities near the electrodes are much larger than the current densities in the bulk of the melt far from the electrodes. This effect is supported by the strong temperature dependence of the conductance so that the temperature distribution in the melt is not uniform but has maxima near the electrodes and particularly at the electrode surfaces, which can become critical if the melts contain oxides that undergo temperature-dependent reactions.

Apart from knowing the conductance of technically produced glass-forming melts at a particular temperature, it is thus mandatory to know the temperature dependence of the conductance if a specific melting process and all possibly connected side reactions are to be optimally controlled. Recognizing these facts, Schott Glas in the early 1970s asked the Electrochemical Laboratory to arrange for the measurement of temperature-dependent conductances, which meant nothing less than to set up an appropriate conductance cell. Much to our surprise, however, no description of a cell construction satisfying all requirements could be found in the literature. We therefore decided to follow an idea that had developed from the analysis of the reported cells [3.298], and to construct a new kind of conductance cell that was largely free from most of the drawbacks of conventional cells [3.298]. After gaining some experience, we optimized the design of the new cell to make its application easier and less time-consuming and at the same time improved its accuracy [3.299–302]. This arrangement is still in use for technical glass melts and for melts of glasses under development at Schott Glas today. Meanwhile, the new cell is also applied elsewhere, as we have learned from the literature [3.303, 304] and from frequent correspondence on the subject. The principles and limitations of this final cell design and its application to an especially critical case, a mixed-alkali glass system [3.1, 300, 302], will be reported in some detail in this section. For a better understanding of the experimental problems involved, which often seem deceptively trivial, the basic difficulties of conductance measurements of glass melts will first be discussed on the basis of the literature analysis [3.298]. It is noted that transport phenomena in glass-forming melts have recently been treated in a summarizing, fundamental publication [3.305].

3.9.1 A Critical Analysis of Reported Conductivity Cells

Basic Expressions

The conductance of melts, as of any other liquids, is measured by applying Ohm's law,

$$R(T) = \frac{U}{i} , \tag{3.201}$$

where U is the voltage applied and i is the current measured. $R(T)$, the resulting temperature-dependent resistivity of the cell containing the melt, is given by

$$R(T) = \rho(T)\frac{l}{A} = \frac{1}{\sigma(T)}\frac{l}{A} , \tag{3.202}$$

where $\rho(T)$ is the temperature-dependent specific resistance and $\sigma(T)$ the specific conductivity (or conductance) of the melt, and l and A are the cell length and cross section, respectively, of the measuring cell provided the electric field between the electrodes is homogeneous. The ratio of cell dimensions l and A is also called the cell constant,

$$C = \frac{l}{A} . \tag{3.203}$$

Thus, the principle of the measurement is simple. However, the experimental conditions, in particular the high temperature, the large temperature coefficient of the specific conductivity, and the aggressiveness of most glass melts complicate these measurements to a high degree. For instance, the temperature dependence of the cell dimensions and thus of the cell constant are not negligible in the wide temperature range applied, so that the cell constant must actually be expressed by

$$C(T) = \frac{l(T)}{A(T)} . \tag{3.204}$$

Two modifications of the determination of cell constants can be applied [3.298].

- The cell is calibrated at the calibrating temperature T_c by applying a standard electrolyte, a molten salt or an aqueous electrolyte solution, with a known specific conductivity $\sigma(T_c)$ at the calibrating temperature T_c,

$$C(T_c) = R(T_c)\sigma(T_c) \ . \tag{3.205}$$

For the standardization, the magnitude of the electric field need not be known, nor need the field be homogeneous.

- The cell constant is calculated from the cell dimensions measured at room temperature T_0,

$$C(T_0) = \frac{l(T_0)}{A(T_0)} \ , \tag{3.206}$$

which, however, requires that the electric field is either homogeneous and, for instance, is an axial-parallel cylindrical field, or can be analysed to yield $l(T_0)$ and $A(T_0)$, for instance, if it is a radial cylindrical field.

For exact results, both modifications necessitate temperature corrections because the calibration temperatures T_c and T_0 generally differ appreciably from the measuring temperature T_m.

Problems Caused by Electrodes and Crucibles

Reported cells employ either two platinum wires, rods or plates as electrodes [3.306], or a platinum [3.307] or a molybdenum [3.308] crucible serves as one and a centre rod of the respective metal as the other electrode. Also applied are two coaxial platinum cylinders, which are [3.309, 310] or are not closed by half-spheres [3.311] at their lower end. Similarly, a four-electrode arrangement consisting of two voltage-supplying and two voltage-measuring electrodes has been described [3.311]. Nearly all of these cells were used in an open state so that the glass melts investigated were not protected from evaporation and connected possible composition changes of their near-surface regions. Neither were the electrodes protected from being partly shortened by the hot vapour above the melt.

Electrode polarization is of serious concern but has not been mentioned in the appropriate publications. Two different kinds of polarization must be taken into account.

- Chemical polarization caused by different electrode materials adds a disturbing emf to the measurement. This effect is not restricted to metals as different as molybdenum and platinum, which are subject to potential differences of up to several hundred mV, see Sects. 3.4.5 and 3.7, but is also observed with different platinum alloys, for instance platinum-iridium and platinum-rhodium [3.4], see Sect. 3.4.2.

- Electrical polarization is caused by different current densities at electrodes with different shapes and sizes, for example at edges and tips of rod and wire electrodes. The different concentrations of, for instance, oxygen, which are thus generated at the differently shaped electrodes, cause a potential difference despite the identical electrode metals.

Cell Calibration by Means of Standard Melts and Solutions

The cell constant is obtained from measurements employing a standard melt with known conductance according to (3.205). However, the thermal expansion of cell length and cross section due to the usually large difference between calibrating and measuring temperature causes serious differences of the cell constant at these temperatures. The effect is particularly large when aqueous standard solutions are applied at ambient temperature. Molten salts used as standard melts require the measurement of the frequency dependence of the cell resistance and extrapolation to infinite frequency because of their strong tendency to polarize platinum electrodes [3.298, 309, 310, 312]. Besides, the specific conductivity of molten salts is only rarely known with sufficient accuracy. In addition, the cell constant depends on the electrode areas in contact with the electrolyte [3.308] and changes with varying meniscus of the melt. It thus depends on temperature changes [3.313, 314] and on the temperature-dependent surface tension, which generally differs strongly between standard electrolyte and glass melt. The application of aqueous standard electrolytes requires careful platinization of platinum electrodes [3.309, 310, 315, 316], which, however, has not always been carried out [3.306, 307]. Although the choice of the electrolyte concentration of aqueous standard solutions may be critical, dilute [3.307] and highly concentrated solutions [3.306, 309, 313] have been applied but were not further checked for their suitability.

Geometrical Determination of Cell Constant

The cell constant is calculated from the cell dimensions at room temperature according to (3.206) and must be corrected for thermal expansion. The effect of the surface tension of the glass melt on the cell constant is practically unknown and even more critical than with the use of standard electrolytes. U-shaped cell tubes of alumina and silica [3.311] eliminate the difficulties with evaporation and surface tension because the electrodes are submerged in the melt. But they introduce the problem of uniform temperature distribution over the entire cell length. Besides, the uniformity of the electric field in the bent part of the tubes is questionable and makes the calculation of the cell constant a problem. Surprisingly, application of standard electrolytes in U-shaped cell tubes has never been reported. Vertical straight tubes of alumina and silica [3.317, 318] also eliminate the problems caused by surface tension. But they must be sealed at their lower end, through which one of the electrodes is introduced, and this is a nearly insoluble problem, independent of whether mechanical seals or cements [3.317] are used, because of thermal expansion and the aggressiveness of the glass melts.

3.9.2 Conductance Cell Designed at Schott Glas

Cell Design

The analysis of the reported conductance cells given in the above paragraphs resulted in the construction of an amended cell, which is best characterized as an immersion-type cell yielding absolute conductances of glass melts [3.300–302]. Figure 3.63 presents a vertical cross section. The melt is contained in a Pt/10Rh crucible, and the actual cell volume is a volume within the melt, which is confined by the vertical alumina (or silica) cell tube introduced in the centre of the melt and an upper and a lower platinum (or Pt/10Rh) electrode. The lower electrode is formed by the upper flat Pt surface of a short cylinder protruding from the crucible bottom into the melt and snuggly surrounded by the lower end of the alumina cell tube. The lead to the upper electrode is surrounded by a four-bore alumina tube, which carries also the leads to a thermocouple situated directly above the upper electrode. The lower electrode is connected to an alumina-insulated lead via the platinum crucible. The crucible rests on four alumina support bars, which are positioned in a vertically movable supporting piston. The entire arrangement is contained in a platinum tube resistance furnace (Pt/20Rh). The crucible can be moved in a vertical direction into and out of the upper opening of the furnace by the piston, which is driven by an electric motor below the furnace. During measurements, the assembly is closed at its upper end by the cover with a cylindrical bore for the cell tube, which also carries the four-bore alumina tube and the insulated lead to the crucible.

The upper electrode is designed as a horizontal cross allowing the removal of accidentally captured bubbles, which usually stick rather firmly to horizontal lower surfaces. They are easily removed by slowly rotating the electrode because the highly viscous melts act as stagnant media so that the bubbles are sheared off and can rise through the bars of the cross-shaped electrode to the melt surface. The slight non-uniformity of the electric field introduced by the electrode shape is corrected for by a conductivity- and cell length-dependent correction factor, which is obtained by comparison with results obtained by means of a circular plate electrode and extremely degassed glass melts [3.300, 301], Fig. 3.64. For highest accuracy, the correction factor is applied iteratively. That the correction factor slightly exceeds unity at small conductances and large cell length, Fig. 3.64, is not an artifact but arises because the actual cell constant includes necessarily a small part of the electrode-supporting wire and melt volume above the upper electrode. At low inhomogeneities, that is, at small conductances and large cell lengths,

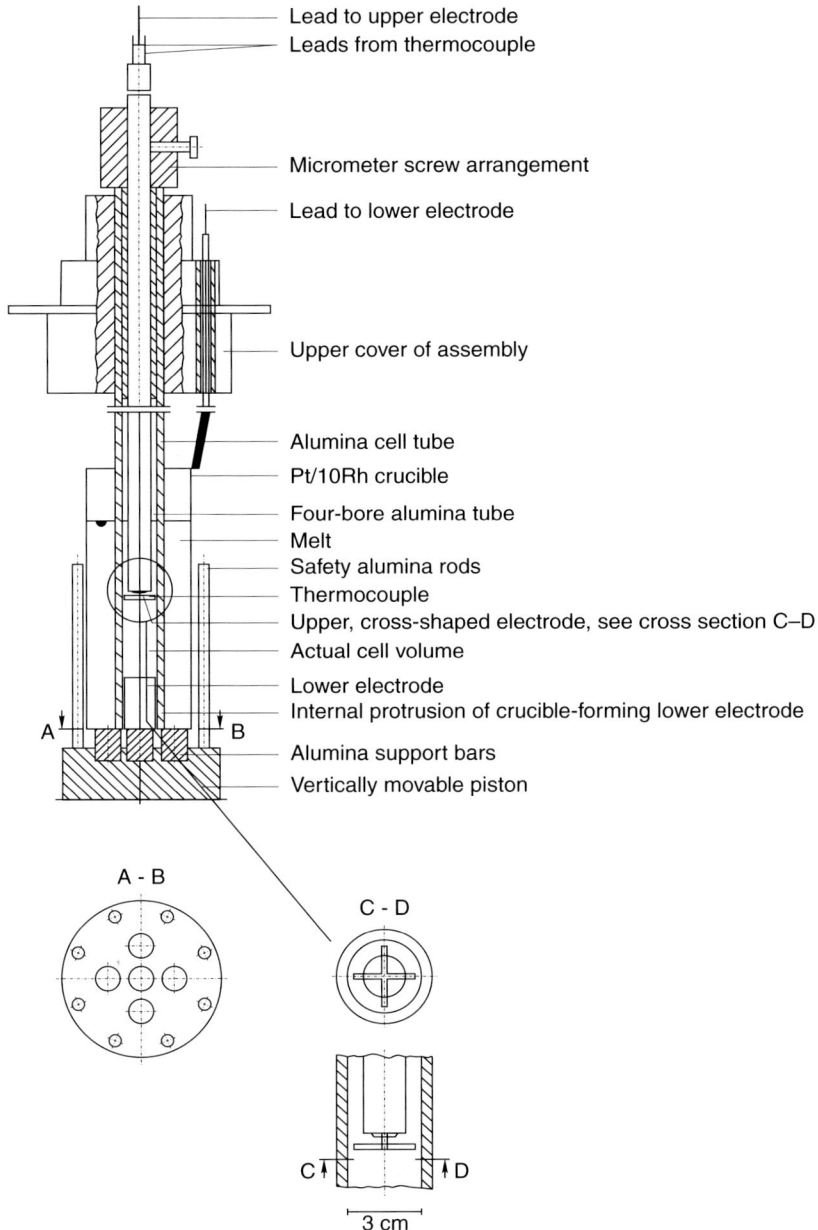

Lead to upper electrode
Leads from thermocouple

Micrometer screw arrangement

Lead to lower electrode

Upper cover of assembly

Alumina cell tube
Pt/10Rh crucible
Four-bore alumina tube
Melt
Safety alumina rods
Thermocouple
Upper, cross-shaped electrode, see cross section C–D
Actual cell volume

Lower electrode
Internal protrusion of crucible-forming lower electrode

Alumina support bars
Vertically movable piston

A - B

C - D

3 cm

Fig. 3.63. Vertical cross section of the conductance cell for glass and salt melts developed at Schott Glas

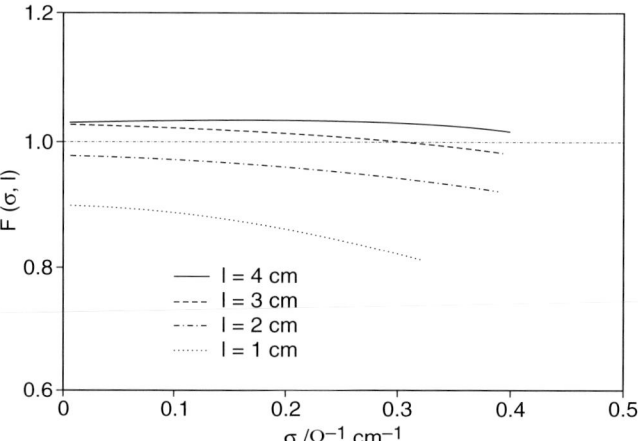

Fig. 3.64. Correction factor for the inhomogeneity of electric field caused by cross-shaped upper electrode of the cell as a function of conductance and cell length l

the corresponding "positive" effect on the electric field overrides the "negative" effect caused by the inhomogeneity due to the electrode shape.

In preparing for a measurement, the internal cross section of the cell tube and the distance between the upper and lower electrode are measured at room temperature using the micrometer screw arrangement above the cover. The crucible is then filled with glass grit above the furnace and lowered into the furnace. The glass is melted at a temperature sufficiently high to allow the subsequent slow introduction of cell and four-bore tube into the melt. The exact cell length is measured at each temperature by means of the micrometer screw, whereas the internal cross section of the cell tube is corrected for temperature changes by means of the coefficient of the thermal expansion of alumina.

Cell Characteristics

The advantages of the cell can be summarized in the following way.

- The cell yields absolute conductances according to (3.206) because the cell constant is determined from the cell tube dimensions at each measuring temperature T_m,

$$C(T_m) = \frac{l(T_m)}{A(T_m)} \ , \tag{3.207}$$

  thus excluding the application of standard melts.
- The cell arrangement has high mechanical stability because of the fixed positioning of the cell tube in crucible and cell cover.

- Due to the location of the actual cell volume within the bulk of the melt, the cell constant is independent of the temperature-dependent density and connected volume changes of the melt.
- For the same reason, the cell constant is independent of the (temperature-dependent) surface tension of the melt.
- The effect of evaporation of glass components is eliminated by the position of the cell volume below the melt surface.
- Stray currents through the hot atmosphere above the melt are excluded by two insulating alumina tubes, the cell tube and the four-bore tube around the lead to the upper electrode.
- The cross shape of the upper electrode allows the removal of bubbles sticking to its lower surface by rotating the electrode. A slight non-uniformity of the electric field introduced by the electrode shape is corrected by a correction factor determined by using a plate electrode in highly degassed melt.
- A variation of the cell length during measurements allows one to choose the optimal cell constant at each temperature and to check the cell tube for intactness and corrosion, the upper electrode for sticking bubbles, and the cell for impairing potential differences and polarization.

However, a successful application of the conductance cell depends on two conditions.

- The alumina of cell and four-bore tubes must have a high corrosion resistance. This is ensured by separate experiments prior to the measurements and is additionally confirmed by a "zipper-like" sequence of the measurements during increasing and decreasing the measuring temperature. Unfortunately, conductance measurements of glasses with high lead contents present a problem according to our experience, although a successful application of the cell to 24% PbO lead crystal has been reported [3.303, 304]. Silica tubes are sometimes an alternative when alumina tubes cannot be used because of corrosion.
- Moreover, the alumina must be sufficiently insulating to confine the electric field to the cell volume. Because at a maximum cell length of about 5 cm the ratio of ceramic tube to melt resistivity must be above 10^3 at maximum measuring temperature [3.300, 301], the alumina specific resistance is to be $\rho \geq 5 \times 10^{10}\,\Omega$ cm, which is a high requirement for a ceramic. The number of suppliers of alumina tubes with these characteristics is correspondingly small (Purox® tubes, formerly supplied by Morgan Refractories, Neston, Cheshire, UK; now sapphire tubes, supplied by Rostox - N, Naginsky District, 142432 Chomgolovka, Russia). New charges of tubes are always checked prior to application.

Instrumentation

The small resistivities of the cell to be measured require a careful choice of instruments. For example, an impedance bridge with 0.003% metre resolution and 0.001% analogue-voltage output (300 mV amplitude) and frequencies 100 Hz, 1 kHz, 10 kHz, and 100 kHz (General Radio, West Concord, USA, 1654 Impedance Comparator) is applied in connection with a standard resistor (General Radio, Decade Resistor 1412 BC) and a standard capacitance (General Radio, Decade Capacitor 1434 G). This combination has also successfully been replaced by an HP 4284 Precision LCR Meter, 80 Hz to 1 MHz. In nearly all cases, an additional constant condensor with approximately 50 nF used in parallel to the conductance cell was required. The resistivity of all leads of the assembly is smaller than one per thousand of the total resistivity measured. Measuring frequencies of 1 kHz and 10 kHz yielded identical results in all cases.

Reproducibility and Uncertainties

The conductance is obtained by using the equation

$$\sigma = \frac{4l_0}{\pi d_0^2 (1 + \alpha \Delta T) R_m F(\sigma, l)} \ , \tag{3.208}$$

where l_0 and d_0 are the cell length and diameter at room temperature, respectively, α is the linear coefficient of thermal expansion of alumina ($\alpha = 2.479 \times 10^{-7} + 1.134 \times 10^{-6} \ln T_m$ [3.319]), $\Delta T = (T_m - T_0)$ is the difference between measuring and room temperature, R_m is the measured resistance, and $F(\sigma, l)$ is the conductance- and cell length-dependent correction factor $F(\sigma, l) = R/R_m = C/C_{act}$ (C_{act} = actual cell constant). The overall reproducibility is $\Delta \sigma = \pm 0.015\sigma$, and the overall uncertainty amounts to approximately 4% of the conductance measured [3.300]. It contains the uncertainty of the temperature, which is the highest uncertainty of the measurements, but is not specific to the conductance cell. Besides, the uncertainty of the absolute conductance is rarely, if at all, of interest because the temperature dependence of the conductance is of much higher significance for research and development and glass production than the conductance.

3.9.3 Application of the Schott Cell to Mixed-Alkali Glass Melts

The cell arrangement described has also been applied to a system of so-called mixed-alkali glasses. The results are reported in the last part of this section because they yielded some interesting, yet unexplained results. Mixed-alkali glasses are glasses containing a constant sum and various ratios of mole fractions of two alkali ions. They are of particular theoretical interest because the dependence of many of their properties on the ratio of their alkali mole fraction exhibits extreme values [3.320, 321]. The mixed-alkali effect has been known for a long time and has been a challenging topic for continuing discussions and theoretical treatments without ever being completely explained; see, for example, [3.322–339]. The conductance of solid glasses, which is the property most obviously exhibiting this effect, has been measured with many glasses with different glass formers and for any combination of any two alkali ions.

It is of special interest that the mixed-alkali effect has also been found in the molten state of glasses although the number of such publications on glass melts is rather small. Thus, it has been reported for lithium-sodium, sodium-potassium [3.307, 340–342] and lithium-potassium silicate [3.307, 341, 342], and for borate melts [3.338] with one [3.307] and three [3.340–342] different total alkali concentrations. The reported data, however, are restricted to temperatures below $1400\,°C$, relatively large temperature intervals, and only three mole fraction ratios of each alkali couple besides the respective single alkalis. The reason for the small number of investigations is probably the lack of conductivity cells yielding conductances that are sufficiently reproducible for comparing different glasses and for a reliable evaluation of activation energies of the conductance.

Accordingly, the glass system $(1-x)Na_2O \cdot x K_2O \cdot 0.7CaO \cdot 4.8SiO_2$, consisting of nine glasses with different mole fraction $x (x = 0, 0.125, 0.250 \ldots 1.0)$ and containing $0.04\,mol\%$ Sb_2O_3 as a fining agent, was investigated [3.1, 300, 302]. Conductance measurements were carried out at $50\,K$ temperature intervals between $900\,°C$ and $1550\,°C$. All precautions were observed, for instance four changes in cell length per temperature and a "zipper-like" measuring sequence with increasing and subsequently decreasing temperature, were applied. During the total measuring time of $32\,h$ per melt, the conductances of all glasses of the system remained unchanged, which implies practically zero corrosion of the alumina tubes by this glass system. The results were evaluated by means of the computer program ASYST (Hewlett Packard, Palo Alto, California, USA). Due to the consistency of the resulting conductances, it was dared to extrapolate the data from $1550\,°C$ to $2000\,°C$ for qualitative reasons.

Figures 3.65 and 3.66 present the logarithm of the conductances and the activation energies of conductance, respectively, as a function of the potassium mole fraction. As in the solid state, both sets of curves show extreme values, which, below approximately $1300\,°C$, are positioned at a potassium mole fraction $x = 0.7 \pm 0.02$. However, the minimum of the conductance curves starts to move to higher potassium mole fractions at about $1300\,°C$, which is still in the temperature range of the measurements, Fig. 3.67, and disappears at approximately $1900\,°C$, whereas the maximum of the activation energy curve is retained up to temperatures well above $2000\,°C$, Figs. 3.66 and 3.67.

These results lead to the following conclusions:

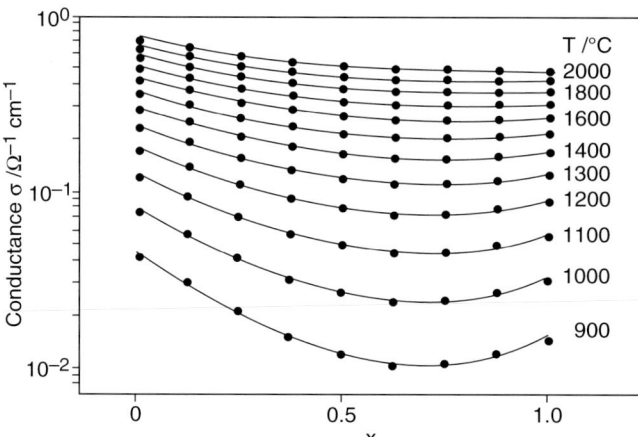

Fig. 3.65. Conductance as a function of potassium mole fraction x of the mixed-alkali glass system $(1 - x)\mathrm{Na_2O \cdot xK_2O \cdot 0.7CaO \cdot 4.8SiO_2}$ at various temperatures

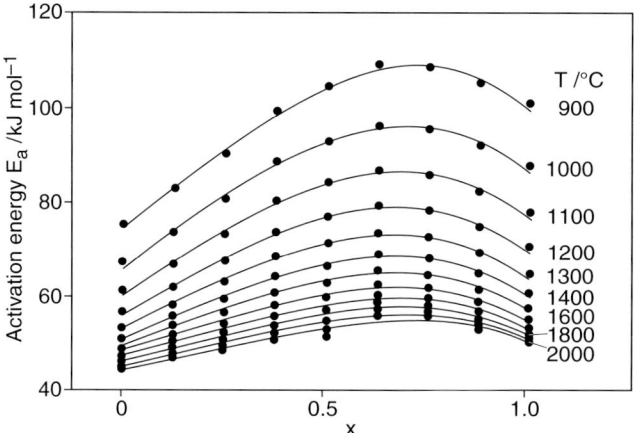

Fig. 3.66. Activation energy of conductance as a function of potassium mole fraction x of the mixed-alkali glass system $(1-x)\mathrm{Na_2O \cdot xK_2O \cdot 0.7CaO \cdot 4.8SiO_2}$ at various temperatures

- The existence of extreme values for both conductance and activation energy of the investigated mixed-alkali system confirms that the mixed-alkali effect of glasses exists also in the molten state. Obviously, the melt retains part of the glass structure at these relatively low temperatures, and it would be highly interesting to repeat the measurements after the melts have been heated to extremely high temperatures. At any rate, the results should be taken into account for future theoretical work on the mixed-alkali effect as well as for investigations on the structure of glass melts.

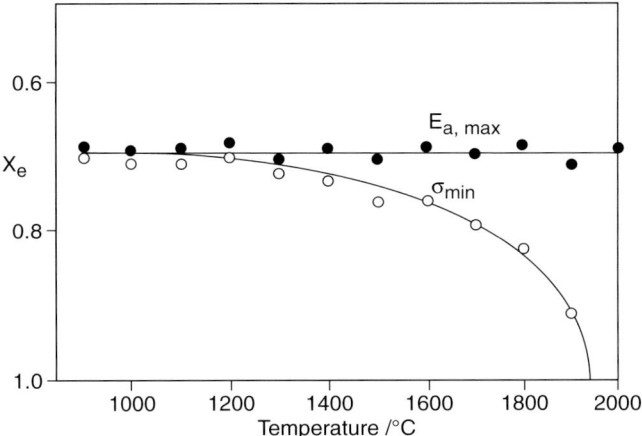

Fig. 3.67. Potassium mole fractions x_e with extreme values of conductance and activation energy of conductance of the mixed-alkali system investigated as functions of temperature

- The finding that only the mixed-alkali effect of the activation energy "survives" at temperatures above approximately 1900 °C raises the question of whether the physical phenomenon underlying the mixed-alkali effect is sufficiently and correctly characterized by extreme values, which are always seen as the typical property of this effect. The main cause of the phenomenon could, for instance, also be the mere deviation of the melt from "ideal" behaviour, which would probably be characterized by the deviation of the experimental values from the straight line connecting the data for the single alkalis.

Figure 3.68 shows this "non-ideality" of the conductance as defined by

$$\Delta\sigma(x_e) = 100\frac{\sigma(x_e) - \sigma_{\text{ideal}}(x_e)}{\sigma_{\text{ideal}}(x_e)} \tag{3.209}$$

and of the activation energy as defined by

$$\Delta E_a(x_e) = 100\frac{E_a(x_e) - E_{a,\text{ideal}}(x_e)}{E_{a,\text{ideal}}(x_e)} \tag{3.210}$$

as functions of the reciprocal absolute temperature. $\sigma(x_e)$ and $E_a(x_e)$ are the measured conductance and activation energy, respectively, at the mole fraction x_e at which extreme values are observed, and $\sigma_{\text{ideal}}(x_e)$ and $E_{a,\text{ideal}}(x_e)$ are the "ideal" conductance and activation energy, respectively, obtained by linearly connecting the data of the single alkali melts. (Note, however, that the mole fraction at which the maximum difference between measured and "ideal" conductance and the activation energy occurs does not completely coincide with the mole fraction of the extreme values.) Surprisingly, both plots

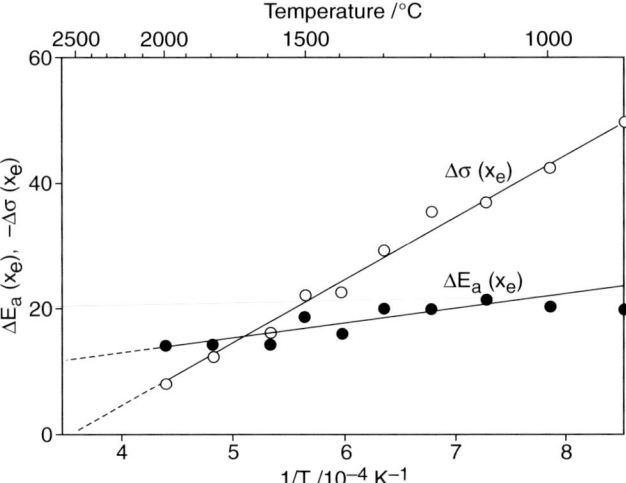

Fig. 3.68. Non-idealities of conductance and of activation energy of conductance as defined by (3.209) and (3.210), respectively, as functions of reciprocal absolute temperature

are linear, with values decreasing with increasing temperature, within the entire temperature range between $900\,°\mathrm{C}$ and $2000\,°\mathrm{C}$. However, the slope of the non-ideality of the conductance is much larger than that of the non-ideality of the activation energy. An extrapolation to high temperatures, if allowed, shows that the non-ideality of the conductance disappears at approximately $2500\,°\mathrm{C}$, whereas the non-ideality of the activation energy is retained and becomes approximately zero only at zero reciprocal temperature. Consequently, the treatment on the basis of non-ideality, in principle, yields qualitatively the same result as the treatment based on extreme values: at sufficiently high temperatures, above $2500\,°\mathrm{C}$, the conductance exhibits ideal behaviour, whereas the activation energy shows non-ideality up to the highest temperatures. It thus remains to explain the results obtained by the reported conductance measurements on this mixed-alkali glass system, and, for instance, to explore whether other mixed-alkali glass systems exhibit the same or a similar effect.

References

3.1 F.G.K. Baucke, R.-D. Werner: "Mixed alkali effect of electrical conductivity in glass-forming silicate melts", Glastechn. Ber. **62**, 182–186 (1989)

3.2 R.A. Robinson, R.H. Stokes: *Electrolyte Solutions*, 2nd ed. (Butterworths, London 1968) p. 462

3.3 J. Kieffer, G. Borchardt: "Kinetic model of silicate melts – equilibrium case", Glastechn. Ber. **62**, 337–344 (1989)

3.4 F.G.K. Baucke: "High-temperature sensors for oxidic glass-forming melts", in *Sensors. A Comprehensive Survey*, Chemical and Biochemical Sensors, Vol. 3, ed. by W. Göpel et al. (VCH, Weinheim 1992) pp. 1155–1180

3.5 B. Douglas, D. McDaniel, J. Alexander: *Concepts and Models of Inorganic Chemistry*, 3rd ed. (Wiley, New York 1993) Chap. 7

3.6 J.E. Huheey, E.A. Keiter, R.L. Keiter: *Inorganic Chemistry*, 4th ed. (Harper Collins, New York 1993) Chap. 9

3.7 I.S. Butler, J.F. Harrod: *Inorganic Chemistry* (Cummings, Redwood City 1989) Chap. 20

3.8 T. Moeller: *Inorganic Chemistry* (Wiley, New York 1952) Chap. 9

3.9 L.P. Hammett: *Physical Organic Chemistry*, 2nd ed. (McGraw-Hill, New York 1970) pp. 267–272

3.10 H. Lux: "'Säuren' und 'Basen' im Schmelzfluß: Die Bestimmung der Sauerstoffionen-Konzentration", Z. Elektrochem. **45**, 303–309 (1939)

3.11 H. Flood, T. Förland: "The acidic and basic properties of oxides", Acta Chem. Scand. **1**, 592–604 (1947)

3.12 E.A. Guggenheim: "The conceptions of electrical potential difference between two phases and the individual activities of ions", J. Phys. Chem. **33**, 842–849 (1929)

3.13 K.H. Sun: "A scale of acidity and basicity in glass", Glass Ind. **29**, 73–74, 98 (1948)

3.14 R.A. Cameron: "Kinetics of arsenic-antimony fining", in *67th Ann. Meeting* (Am. Ceram. Soc., Philadelphia 1965)

3.15 S. Holmquist: "Oxygen ion activity and the solubility of sulfur trioxide in sodium silicate melts", J. Am. Ceram. Soc. **49**, 467–473 (1966)

3.16 S. Holmquist: "Sodium oxide activities in molten sodium sulfate and sodium silicates", Phys. Chem. Glasses **9**, 32–34 (1968)

3.17 H. Franz, H. Scholze: "Die Löslichkeit von H_2O-Dampf in Glasschmelzen verschiedener Basizität", Glastechn. Ber. **36**, 347–356 (1963)

3.18 M.L. Pearce: "Solubility of carbon dioxide and variation of oxygen ion activity in soda-silica melts", J. Am. Ceram. Soc. **47**, 342–347 (1964)

3.19 M.L. Pearce: "Solubility of carbon dioxide and variation of oxygen ion activity in sodium borate melts", J. Am. Ceram. Soc. **48**, 175–178 (1965)

3.20 H. Franz: "Oxygen ion activity and reaction equilibria in glass melts", J. Can. Ceram. Soc. **38**, 89–93 (1969)

3.21 F.W. Krämer: "Contribution to basicity of technical glass melts in relation to redox equilibria and gas solubilities", Glastechn. Ber. **64**, 71–80 (1991)

3.22 C. Bodsworth, H.B. Bell: *Physical Chemistry of Iron and Steel Manufacture*, 2nd ed. (Longman, Harlow, Essex, England 1972) pp. 177 and 445

3.23 A. Paul: "Acid-base concepts in relation to the structure of borate and silicate glasses", Trans. Indian. Ceram. Soc. **28**, 63–81 (1969)

3.24 J.A. Duffy, M.D. Ingram: "Establishment of an optical scale for Lewis basicity in inorganic oxyacids, molten salts and glasses", J. Am. Chem. Soc. **93**, 6448–6454 (1971)

3.25 J.A. Duffy, M.D. Ingram: "An interpretation of glass chemistry in terms of the optical basicity concept", J. Non-Cryst. Solids **21**, 373–410 (1976)

3.26 J.A. Duffy: *Bonding, Energy Levels and Bands in Inorganic Solids* (Longman, Harlow, Essex, England 1990) Chaps. 6 and 8

3.27 J.A. Duffy: "A review of optical basicity and its applications to oxidic systems", Geochim. Cosmochim. Acta **57**, 3961–3970 (1993)

3.28 W.W. Porterfield: *Inorganic Chemistry – A Unified Approach* (Addison-Wesley, Reading, MA 1984) Chap. 6

3.29 C.K. Jørgensen: *Orbitals in Atoms and Molecules* (Academic Press, New York 1962) Chap. 4

3.30 J.A. Duffy, M.D. Ingram: "A new correlation between s–p spectra and the nephelauxetic ratio: applications in molten salt and glass chemistry", J. Chem. Phys. **54**, 443–444 (1971)

3.31 J.A. Duffy, E.I. Kamitsos, G.D. Chryssikos, A.P. Patsis: "Trends in local optical basicity in sodium borate glasses and relation to ionic mobility", Phys. Chem. Glasses **34**, 153–157 (1993)

3.32 J.H. Binks, J.A. Duffy: "A molecular orbital treatment of basicity of oxyanion units", J. Non-Cryst. Solids **37**, 387–400 (1980)

3.33 A. Dietzel: "Glasstruktur und Glaseigenschaften", Glastechn. Ber. **22**, 41–50, 81–86, 212–224 (1948–49)

3.34 W.A Weyl, E.G. Marboe: *The Constitution of Glasses* (Interscience, New York 1962)

3.35 K. Fajans, G. Joos: "Molrefraktion von Ionen und Molekülen im Lichte der Atomstruktur", Z. Physik **23**, 1–46 (1924)

3.36 P. Baltã, C. Spurcaciu: "Some new ideas concerning the basicity of glasses", 9-IBAUSIL, Weimar 1985. Sektion 4, pp. 21–26

3.37 P. Baltã: "The basicity of glasses", in *Proc. Fifth ESG Conf. on Glass Science and Technology for the 21st Century*, Prague, 1999 (Czech Glass Soc., Prague 1999) B 3, pp. 125–130

3.38 J.R. Tessman, A.H. Kahn, W. Shockley: "Electronic polarizabilities of ions in crystals", Phys. Rev. **92**, 890–895

3.39 J.A. Duffy: "Optical basicity and glass chemistry", in *Fundamentals of the Glass Manufacturing Process 1991*, Proc. First Conf. of the European Society of Glass Science and Technology, Sheffield (Society of Glass Technology, Sheffield 1991) pp. 42–44

3.40 A. Klonkowski: "Changes of optical basicity in the glass system M(II)-P$_2$O$_5$", Phys. Chem. Glasses **22**, 163–167 (1981)

3.41 N. Iwamoto, Y. Makino, S. Kasahara: "Correlation between refraction basicity and theoretical optical basicity", J. Non-Cryst. Solids **68**, 379–388 (1984)

3.42 F.G.K. Baucke, J.A. Duffy: "Use of thallium(I) probe for identifying sites of mobile cations in glass during electrolysis", J. Chem. Soc. Faraday Trans. I **79**, 661–667 (1983)

3.43 J.A. Duffy, E.I. Kamitsos, G.D. Chryssikos: "Chemical bonding analysis of alkali oxidic glass systems: charges on metal ions and network sites", Phys. Chem. Glasses **36**, 53–58 (1995)

3.44 M.D. Ingram: "Optical basicities and structural dynamics in glassy materials", J. Non-Cryst. Solids **222**, 42–49 (1997)

3.45 A. Matthai, D. Ehrt, C. Rüssel: "Redox behaviour of polyvalent ions in phosphate glass melts and phosphate glasses", Glastechn. Ber. **71**, 187–192 (1998)

3.46 O. Claussen, C. Rüssel: "Voltammetric study of the thermodynamics of the Fe^{3+}/Fe^{2+} equilibrium and the self diffusivity of iron in glasses with the basic composition 74SiO$_2$·(26−x)Na$_2$O·xCaO", Phys. Chem. Glasses **39**, 200–205 (1998)

3.47 H.A. Schaeffer: "Scientific and technological challenges of industrial glass melting", Solid State Ionics **105**, 265–270 (1998)

3.48 H.D. Schreiber, L.J. Peters, J.W. Beckman, C.W. Schreiber: "Redox chemistry of iron-manganese and iron-chromium interactions in soda lime silicate glass melts", Glastechn. Ber. **69**, 269–277 (1996)

3.49 R. Brückner, H. Hessenkemper: "Influence of water content and basicity on redox ratio. – Consequences on radiation heat absorption and emission of glass melts during fusion and procession", Glastechn. Ber. **66**, 245–253 (1993)

3.50 H. Müller-Simon: "On the interaction between oxygen, iron and sulfur in industrial glass melts", Glastechn. Ber. **67**, 297–303 (1994)

3.51 K.H. Karlsson: *Chemistry of Glass Forming Silicate Melts* (Åbo Akademi University Press, Turku, Finland 1995) pp. 1–45

3.52 K. Takahashi, Y. Miura: "Electrochemical behaviour of glass melts", J. Non-Cryst. Solids **95/96**, 119–130 (1987)

3.53 R.J. Araujo, N.F. Borrelli: SPIE Submolecular Glass Chem. Phys. **1590**, 138 (1991)

3.54 M. Cable, Z.D. Xiang: "The optical spectra of copper ions in alkali-lime-silica glasses", Phys. Chem. Glasses **33**, 154–160 (1992)

3.55 C. Rüssel, E. Freude: "Voltammetric studies of the redox behaviour of various multivalent ions in soda-lime-silica glass melts", Phys. Chem. Glasses **30**, 62–68 (1989)

3.56 R. Pyare, P. Nath: "Stannous-stannic equilibrium in molten binary alkali silicate and ternary silicate glasses", J. Am. Ceram. Soc. **65**, 549–554 (1982)

3.57 R. Pyare, S.P. Singh, A. Singh, P. Nath: "The As^{3+}–As^{5+} equilibrium in borate and silicate glasses", Phys. Chem. Glasses **23**, 158–168 (1982)

3.58 A. Paul, R.W. Douglas: "Cr^{3+}–Cr^{6+} equilibrium in binary alkali silicate glasses", Phys. Chem. Glasses **6**, 197–202 (1965)

3.59 A. Paul, R.W. Douglas: "Ferrous-ferric equilibrium in binary alkali silicate glasses", Phys. Chem. Glasses **6**, 207–211 (1965)

3.60 A. Paul, R.W. Douglas: "Cerous-ceric equilibrium in binary alkali silicate glasses", Phys. Chem. Glasses **6**, 212–215 (1965)

3.61 A. Paul, R.W. Douglas: "Mutual interaction of different redox pairs in glass", Phys. Chem. Glasses **7**, 1–13 (1966)

3.62 J.A. Duffy, M.D. Ingram, I.D. Sommerville: "Acid-base properties of molten oxides and metallurgical slags", J. Chem. Soc. Faraday Trans. (1) **74**, 1410–1419 (1978)

3.63 D.R. Gaskell: "On the correlation between the distribution of phosphorus between slag and metal and the theoretical optical basicity of the slag", Trans. Iron Steel Inst. Jpn. **22**, 997–1000 (1982)

3.64 D.R. Gaskell: "Optical basicity and the thermodynamic properties of slags", Met. Trans. **20 B**, 113–118 (1989)

3.65 T. Mori: "On the phosphorus distribution between slag and metal", Bull. Jpn. Inst. Metals **23**, 354–361 (1984)

3.66 D.J. Sosinsky, I.D. Sommerville: "The composition and temperature dependence of the sulfide capacity of metallurgical slags", Met. Trans. **17 B**, 331–337 (1985)

3.67 S. Sumita, Y. Matsumoto, K. Morinaga, T. Yanagase: "The optical basicity and Fe^{2+}–Fe^{3+} redox in oxyacid salt systems", Trans. Jpn. Inst. Metals **23**, 360–367 (1982)

3.68 N.N: *3rd Int. Conf. Molten Slags and Fluxes*, Glasgow, 1988 (The Institute of Metals, London 1989) pp. 29, 60, 86, 91–94, 107, 146–149, 150–153, 154–156, 157–162, 166–168, 241–245, 277–282, 313–316

3.69 R.W. Young, J.A. Duffy, G.J. Hassall, Z.Xu: "Use of optical basicity concept for determining phosphorus and sulfur slag-metal partitions", Ironmaking and Steelmaking **19**, 201–219 (1992)

3.70 F. Mitchell, D.H. Sleeman, J.A. Duffy, M.D. Ingram, R.W. Young: "Optical basicity of metallurgical slags: a new computer based system for data visualisation and analysis", Ironmaking and Steelmaking **24**, 306–320 (1997)

3.71 L. Pauling: "The modern theory of valency", J. Chem. Soc., 1461–1467 (1948)

3.72 C.K. Jørgensen: *Oxidation Numbers and Oxidation States* (Springer, Berlin, Heidelberg 1969) Chap. 5

3.73 G. Jeddeloh: "The redox equilibrium in silicate melts", Phys. Chem. Glasses **25**, 163–164 (1984)

3.74 F.G.K. Baucke, J.A. Duffy: "The effect of basicity on redox equilibria in molten glasses", Phys. Chem. Glasses **32**, 211–218 (1991)

3.75 J.A. Duffy, M.D. Ingram: "Optical basicity. Part V: A correlation between the Lewis (optical) basicity of oxyanions and the strengths of Bronsted acids in aqueous solution", J. Inorg. Nucl. Chem. **38**, 1831–1833 (1976)

3.76 F.G.K. Baucke, J.A. Duffy: "Oxidation states of metal ions in glass melts", Phys. Chem. Glasses **35**, 17–21 (1994)

3.77 J.A. Duffy, F.G.K. Baucke: "Corrosion of metals in molten silicates: Relationship with electrode potentials in aqueous solution", J. Phys. Chem. **99**, 9189–9193 (1995)

3.78 J.A. Duffy, F.G.K. Baucke: "Effect of basicity on reduction of metal ions to the metallic state in glass melts", Phys. Chem. Glasses **38**, 25–26 (1997)

3.79 J.A. Duffy: "Effect of glass basicity on the ultraviolet spectra of thallium(I) and thallium(III)", Phys. Chem. Glasses **32**, 55–57 (1991)

3.80 J.A. Blair, J.A. Duffy: "Effect of temperature on redox equilibria in phosphate glasses and melts", Phys. Chem. Glasses **36**, 73–76 (1995)

3.81 F.G.K. Baucke, J.A. Duffy: "Redox reactions between cations of different polyvalent elements in glass melts: an optical basicity study", Phys. Chem. Glasses **34**, 158–163 (1993)

3.82 R.C. Weast (Ed.): *Handbook of Chemistry and Physics*, 55th ed. (CRC, Cleveland, OH 1974–75)

3.83 J.H. Campbell, E.P. Wallterstein, J.S. Hayden, D.L. Sapak, D. Warrington, A.J. Marker, H. Toratani, H. Meissner, S. Nakajima, T. Izumitani: "Elimination of platinum inclusions in phosphate laser glasses", LLNL Report **UCRL 53932** (Lawrence Livermore National Laboratory, Livermore, CA 1989)

3.84 G. Gliemeroth, U. Eichhorn, E. Hölzel: "Zur Beeinflussung der Eigenschaften silberhalogenidhaltiger fototroper Gläser", Glastechn. Ber. **54**, 162–174 (1981)

3.85 J.S. Strout: "Optical absorption and color caused by selected cations in high-density, lead silicate glass", J. Am. Ceram. Soc. **54**, 401–406 (1971)

3.86 C.R. Bamford: "Colour generation and control in glass", in *Glass Science and Technology*, Vol. 2, ed. by D.R. Uhlman, N.J. Kreidl (Elsevier, Amsterdam 1977) pp. 35–38

3.87 M. Cable: "Principles of glass melting", in *Glass Science and Technology*, Vol. 2, ed. by D.R. Uhlman, N.J. Kreidl (Academic Press, Orlando, 1984) pp. 1–44

3.88 B. Stahlberg: *Bestimmung thermodynamischer Größen des Sb^{3+}/Sb^{5+}-Gleichgewichts in Silikatglas-Schmelzen unter Verwendung der Mößbauer-Spektroskopie und elektro-chemischer Messungen*, PhD Thesis (Münster 1987)

3.89 B. Stahlberg, B.D. Mosel, W. Müller-Warmuth, F.G.K. Baucke: "Combined electrochemical and Mössbauer studies of the Sb^{3+}/Sb^{5+} equilibrium in a silicate glass-forming melt", Glastechn. Ber. **61**, 335–340 (1988)

3.90 E. Freude, C. Rüssel: "Voltammetric methods for determining polyvalent ions in glass melts", Glastechn. Ber. **60**, 202–204 (1987)

3.91 C. Montel, C. Rüssel, E. Freude: "Square-wave-voltammetry as a method for the quantitative in-situ determination of polyvalent elements in molten glass", Glastechn. Ber. **61**, 59–63 (1988)

3.92 C. Rüssel: "Polyvalent ions in glass melts", Glastechn. Ber. **63K**, 197–201 (1990)

3.93 A.W.M. Wondergrem-de Best: *Redox Behaviour and Fining of Molten Glass*, PhD Thesis (Eindhoven 1994)

3.94 T. Pfeiffer: "Square wave voltammetry", Parts 1 and 2, Labornotiz (Schott Glas, Mainz 1994)

3.95 M. Zink, C. Rüssel, H. Müller-Simon, K.W. Mergler: "Voltammetric sensor for glass tanks", Glastechn. Ber. **65**, 25–31 (1992)

3.96 K. Kiukkola, C. Wagner: "Galvanic cells for the determination of the standard molar free energy of formation of metal halides, oxides, and sulfides at elevated temperatures", J. Electrochem. Soc. **104**, 308–316 (1957)

3.97 K. Kiukkola, C. Wagner: "Measurements on galvanic cells involving solid electrolytes", J. Electrochem. Soc. **104**, 379–387 (1957)

3.98 J. Besson, C. Deportes, M. Darcy: "Sur un electrode de comparaison utilisable en bains de sels oxygenes a haute temperature", Compt. Rend. Acad. Sci. **251**, 1630–1632 (1960)

3.99 E. Plumat, F. Toussaint, M. Boffe: "Formation of bubbles by electrochemical processes in glass", J. Am. Ceram. Soc. **49**, 551–558 (1966)

3.100 W.A. Fischer, D. Janke: *Metallurgische Elektrochemie* (Springer, Berlin, Heidelberg 1975)

3.101 V.S. Stubican, R.C. Hink, S.P. Ray: "Phase equilibria and ordering in the system ZrO_2–Y_2O_3", J. Am. Ceram. Soc. **61**, 17–21 (1978)

3.102 K.S. Goto: *Solid State Electrochemistry and Its Applications to Sensors and Electronic Devices*, Materials Science Monographs, Vol. 45 (Elsevier, Amsterdam 1988) pp. 283–288

3.103 F.G.K. Baucke: "Sauerstoffsensoren für Metall- und Glasschmelzen", Dechema-Monographien, Vol. 126 (VCH, Weinheim 1992) pp. 345–361

3.104 F.G.K. Baucke: "High-temperature oxygen sensors for glass-forming melts", Fresenius' J. Anal. Chem. **356**, 209–214 (1996)

3.105 F.G.K. Baucke: "Development of electrochemical cells employing oxide ceramics for measuring oxygen partial pressures in laboratory and technical glass melts", Glastechn. Ber. **56K**, 307–312 (1983)

3.106 Th. Frey, H.A. Schaeffer, F.G.K. Baucke: "Entwicklung einer Sonde zur Messung des Sauerstoffpartialdrucks in Glasschmelzen", Glastechn. Ber. **53**, 116–123 (1980)

3.107 H.A. Schaeffer, Th. Frey, I. Löh, F.G.K. Baucke: "Oxidation state of equilibrated and non-equilibrated glass melts", J. Non-Cryst. Solids **49**, 179–189 (1982)

3.108 F.G.K. Baucke: "Electrochemical cells for on-line measurements of oxygen fugacities in glass-forming melts", Glastechn. Ber. **61**, 87–90 (1988)

3.109 F.G.K. Baucke, W. Frank, G. Röth: "Meßanordnung zur Messung von Sauerstoff-Partialdrücken", German Patent P 30 28 270 (1986)

3.110 F.G.K. Baucke, Th. Frey, H.A.Schaeffer: "Meßsonde zur Bestimmung des Sauerstoffpartialdruckes in heißen Medien", German Patent P 29 08 368 (1979)

3.111 F.G.K. Baucke, G. Röth: "Sonde zur Messung von Sauerstoffpartialdrücken in hochaggressiven Medien", German Patent P 31 09 454 (1987)

3.112 F.G.K. Baucke, G. Röth: "Sauerstoffsonde unter Verwendung eines nichtleitenden Keramikrohres", German Patent GBM 85 13 976 (1985)

3.113 S.L. Fridman, S.F. Pal'guev, V.N. Chebotin: "Thermoelectromotive force in solid ZrO_2 + Y_2O_3 electrolytes", Éktrokhimiya **5**, 357–358 (1969)

3.114 W. Fischer: "Die Thermokraft von kubisch stabilisiertem Zirkondioxid zwischen Sauerstoffelektroden", Z. Naturforsch. **22a**, 1575–1581 (1967)

3.115 S. Pizzini, C. Riccardi, V. Wagner, C. Sinistri: "On the thermoelectric power of stabilized zirconia", Z. Naturforsch. **25a**, 559–565 (1970)

3.116 J.A. Veith: *Ermittlung von Standard-Seebeck-Koeffizienten von Yttrium-dotierten Zirkondioxid-Keramiken zwischen 700° C und 1500° C*, Diploma Thesis (FH Rheinland-Pfalz, Bingen 1983)

3.117 F.G.K. Baucke, G. Röth, R.-D. Werner: "Meßvorrichtung zum Messen des Sauerstoffpartialdruckes in aggressiven Flüssigkeiten hoher Temperatur", German Patent 38 11 865 (1989)

3.118 F.G.K. Baucke, G. Röth, R.-D. Werner: "Meßvorrichtung zum Messen des Sauerstoffpartialdruckes in aggressiven Flüssigkeiten hoher Temperatur", German Patent 38 11 864 (1990)

3.119 F.G.K. Baucke, G. Röth, R.-D. Werner: "Meßvorrichtung zum Messen des Sauerstoffpartialdruckes in aggressiven Flüssigkeiten hoher Temperatur", German Patent 38 11 915 (1990)

3.120 F.G.K. Baucke, G. Röth: "Referenzelektrodenanordnung einer Meßkette zur Messung des Sauerstoffpartialdrucks in aggressiven Medien von hoher Temperatur", German Patent 41 38 409 (1993)

3.121 J.P. Coughlin: *Contributions to the Data on Theoretical Metallurgy. XII. Heats and Free Energies of Formation of Inorganic Oxides, Bureau of Mines*, Bulletin 542 (US Government Printing Office, Washington, DC 1954) pp. 7–10

3.122 G.N. Lewis, M. Randall: *Thermodynamics*, 2nd ed., revised by K.S. Pitzer, L. Brewer (McGraw-Hill, New York 1961) p. 672

3.123 K.S. Goto: *Solid State Electrochemistry and Its Applications to Sensors and Electronic Devices*, Materials Science Monographs, Vol. 45 (Elsevier, Amsterdam 1988) pp. 231–265

3.124 D.J.G. Ives: "Oxide, oxygen, and sulfide electrodes", in *Reference Electrodes. Theory and Practice*, ed. by D.J.G. Ives, G.J. Janz (Academic Press, New York 1961) pp. 322–392

3.125 F.G.K. Baucke, Th. Pfeiffer, S. Biedenbender, G. Röth, R.-D. Werner: "Verwendung einer Metall/Metalloxid-Elektrode", German Patent 4 324 922 (1995)

3.126 G.V. Samsonov: *The Oxide Handbook*, 2nd ed. (IFI/Plenum, New York 1982) pp. 44–48

3.127 W.A. Fischer, D. Janke: *Metallurgische Elektrochemie* (Springer, Berlin, Heidelberg 1975) pp. 192 ff., 244 ff., 318 ff.

3.128 K.S. Goto: *Solid State Electrochemistry and Its Applications to Sensors and Electronic Devices*, Materials Science Monographs, Vol. 45 (Elsevier, Amsterdam 1988) pp. 299–332

3.129 L. Nemec, M. Muhlbauer: "Verhalten von Gasblasen in der Glasschmelze bei konstanter Temperatur", Glastechn. Ber. **54**, 99–108 (1981)

3.130 M. Cable: "Principles of glass melting", in *Glass Science and Technology*, Vol. 2. *Processing*, ed. by D.R. Uhlmann, N.J. Kreidl (Academic Press, Orlando, FL 1984) Chap. 1, pp. 16–28

3.131 F. Krämer: "Mathematisches Modell der Veränderung von Gasblasen in Glasschmelzen", Glastechn. Ber. **52**, 43–50 (1979)

3.132 L. Nemec: "The behaviour of bubbles in glass melts, Part 1. Bubble size controlled by diffusion", Glass Technology **21**, 134–138 (1980)

3.133 L. Nemec: "The behaviour of bubbles in glass melts, Part 2. Bubble size controlled by diffusion and chemical reaction", Glass Technology **21**, 139–143 (1980)

3.134 M.C. Weinberg, P.I.K. Onorato, D.R. Uhlmann: "Behavior of bubbles in glass melts. I. Dissolution of a stationary bubble containing a single gas", J. Am. Ceram. Soc. **63**, 175–180 (1980)

3.135 M.C. Weinberg, P.I.K. Onorato, D.R. Uhlmann: "Behavior of bubbles in glass melts. II. Dissolution of a stationary bubble containing a diffusing and a nondiffusing gas", J. Am. Ceram. Soc. **63**, 435–438 (1980)

3.136 P.I.C. Onorato, M.C. Weinberg, D.R. Uhlmann: "Behavior of bubbles in glass melts. III. Dissolution and growth of a rising bubble containing a single gas", J. Am. Ceram. Soc. **64**, 676–682 (1981)

3.137 M.C. Weinberg, R.S. Subramanian: "Dissolution of multicomponent bubbles", J. Am. Ceram. Soc. **63**, 527–531 (1980)

3.138 J.I. Ramos: "Behavior of multicomponent gas bubbles in glass melts", J. Am. Ceram. Soc. **69**, 49–54 (1986)

3.139 M. Cable, J.R. Frade: "The diffusion-controlled dissolution of spheres", J. Mater. Sci. **22**, 1894–1900 (1987)

3.140 M. Cable, J.R. Frade: "Theoretical analysis of the dissolution of multicomponent gas bubbles", Glastechn. Ber. **60**, 355–362 (1987)

3.141 M.C. Weinberg: "Dissolution of a stationary bubble in a glass melt with a reversible chemical reaction: rapid forward reaction rate constant", J. Am. Ceram. Soc. **65**, 479–485 (1982)

3.142 H. Hübenthal, G.H. Frischat: "Formation and behaviour of nitrogen bubbles in glass melts", Glastechn. Ber. **60**, 1–10 (1987)

3.143 R.G.C. Beerkens: "Chemical equilibrium reactions as driving forces for growth of gas bubbles during refining", Glastechn. Ber. **63K**, 222–242 (1990)

3.144 H. Yoshikawa, Y. Kawase: "Significance of redox reactions in glass refining processes", Glastechn. Ber. Glass Sci. Technol. **70**, 31–40 (1997)

3.145 F. Krämer: "Gasprofilmessungen zur Bestimmung der Gasabgabe beim Glasschmelzprozeß", Glastechn. Ber. **53**, 177–188 (1980)

3.146 S. Takeshita et al.: "Refining of glasses under sub-atmospheric pressures III", in *Proc. XVI International Congress on Glass*, Madrid, 1992. Vol. 6 (S.E. de Ceramica y Vidrio, Madrid 1992) pp. 173–178

3.147 V. Klein: "Die Entgasung von Glasschmelzen durch Schallwellen", Glastechn. Ber. **16**, 232–233 (1938)

3.148 F. Krüger: "Über die Entgasung von Glasschmelzen durch Schallwellen", Glastechn. Ber. **16**, 233–236 (1938)

3.149 E.D. Spinosa, D.E. Ensminger: "Sonic energy as a means to reduce energy consumption during glass melting", Ceram. Eng. Sci. Proc. **7**, 410–425 (1986)

3.150 A. Eller: "Force on a bubble in a standing acoustic wave", J. Acoustic Soc. Am. **43**, 170–171 (1968)

3.151 C. Eden: "Ultraschall-Entgasung von Glasschmelzen im Hochfrequenzinduktionsofen", Glastechn. Ber. **25**, 83–86 (1952)

3.152 K. Högerl, G.H. Frischat: "Homogenization of glass melts by bubbling", in *Proc. International Congress on Glass*, Madrid, 1992. Vol. 6 (S.E. de Ceramica y Vidrio, Madrid 1992) pp. 179–184

3.153 L. Nemec: "Refining in the glass melting", J. Am. Ceram. Soc. **60**, 436–440 (1977)

3.154 H.D. Schreiber, S.J. Kozak, P.G. Leonhard, K.K. McManus: "Sulfur chemistry in a borosilicate melt, Part I: Redox equilibria and solubility", Glastechn. Ber. **60**, 389–398 (1987)

3.155 H. Müller-Simon: "Oxygen balance in sulfur-containing melts", Glastechn. Ber. **71**, 157–165 (1998)

3.156 S.M. Budd: Letter to the Editor, without title, Phys. Chem. Glasses **7**, 210–213 (1966)

3.157 H.O. Mulfinger: "Gase (Blasen) in der Schmelze", in *Glastechnische Fabrikationsfehler*, ed. by H. Jebsen-Marwedel, R. Brückner (Springer, Berlin, Heidelberg 1980) pp. 193 ff.

3.158 R.G.C. Beerkens, L. Zamann, P. Laimböck, S. Kobayashi: "Impact of furnace atmosphere and organic contamination of recycled cullet on redox state fining of glass melts", Glastechn. Ber. Glass Sci. Technol. **72**, 127–144 (1999)

3.159 K. Takahashi, Y. Miura: "Electrochemical studies on diffusion and redox behavior of various metal ions in some molten glasses", J. Non-Cryst. Solids **39**, 527–532 (1980)

3.160 K. Takahashi, Y. Miura: "Electrochemical studies on redox behavior of metallic ions in molten oxide glasses", Glastechn. Ber. **56K**, 928–933 (1983)

3.161 B. Strzelbicka, A. Bogacz: "Chronopotentiometric investigation of Pb(II)/Pt electrode processes in molten $Na_2Si_2O_5$", Electrochim. Acta **30**, 865–870 (1985)

3.162 G.C. Barker: "Square-wave voltammetry and some related techniques", Anal. Chim. Acta **18**, 118–131 (1958)

3.163 L. Ramaley, M.S. Krause, Jr.: "Theory of square wave voltammetry", Anal. Chem. **41**, 1362–1365 (1969)

3.164 M.S. Krause, L. Ramaley: "Analytical application of square wave voltammetry", Anal. Chem. **41**, 1365–1369 (1969)

3.165 J.G. Osteryoung, R.A. Osteryoung: "Square wave voltammetry", Anal. Chem. **57**, 101A–110A (1985)

3.166 J.G. Osteryoung, J.J. O'Dea: "Square wave voltammetry", in *Electroanalytical Chemistry*, Vol. 14, ed. by A.J. Bard (Dekker, New York 1986) pp. 209–308

3.167 G. Bouquet, S. Dobos, Z. Boksay: "Untersuchung der Oberflächenschicht des Glases", Ann. Univ. Sci. Budapest (Rolando Eötvös Nominatae), Sect. Chim. **6**, 5–13 (1964)

3.168 H. Bach, F.G.K. Baucke: "Measurement of ion concentration profiles in surface layers of leached ('swollen') glass electrode membranes by means of luminescence excited by ion sputtering", Electrochim. Acta **16**, 1311–1319 (1971)

3.169 T. Hayashi, W.G. Dorfeld: "Electrochemical study of As^{3+}/As^{5+} equilibrium in a barium borosilicate melt", J. Non-Cryst. Solids **177**, 331–339 (1994)

3.170 R. Akiyama, A. Takenaka, M. Sugazaki: "Determination of antimonic(III) and antimonic(V) in glasses by ion chromatography/inductively coupled plasma atomic emission spectroscopy", Rep. Res. Lab. Asahi Glass **44**, 13–18 (1994)

3.171 H.D. Schreiber: "An electrochemical series of redox couples in silicate melts: A review and applications to geochemistry", J. Geophys. Res. **92**, 9223–9232 (1987)

3.172 H.D. Schreiber: "Redox chemistry in glass-forming melts. – Electron exchanges", Glass Res. Bull. Glass Sci. Eng. **3**, 6–7 (1983)

3.173 H.D. Schreiber, B.K. Kochanowski, C.W. Schreiber, A.B. Morgan, M.T. Coolbaugh, T.G. Dunlap: "Compositional dependence of redox equilibria in sodium silicate glasses", J. Non-Cryst. Solids **177**, 340–346 (1994)

3.174 H.D. Schreiber, R.W. Fowler, C.C. Ward: "Sulphate as a selective redox buffer for borosilicate melts", Phys. Chem. Glasses **34**, 66–70 (1993)

3.175 A. Kumar, S.P. Singh: "Oxygen ion activity and its influence on redox equilibria in a ternary soda-lime-silica glass system", Glastechn. Ber. **65**, 69–72 (1992)

3.176 L. Ortmann, D. Höhne, G. Nölle: "Equilibrium constant-determination and influence on redox reactions in soda-lime-silica glass melts", Glastechn. Ber. **69**, 235–241 (1996)

3.177 R.L. Mössbauer: "Kernresonanzfluoreszenz von Gammastrahlung in Ir[191]", Z. Phys. **151**, 124–143 (1958)

3.178 H. Frauenfelder: *The Mössbauer Effect* (Benjamin, New York 1963)

3.179 G.K. Wertheim: *Mössbauer Effect: Principles and Applications* (Academic Press, New York 1964)

3.180 H. Wegener: *Der Mößbauer Effekt* (Bibliographisches Institut, Mannheim 1965)

3.181 V.I. Goldanskii: "Zur Gamma-Resonanzspektroskopie (Mössbauerspektroskopie) in der Chemie", Angew. Chem. **79**, 844–858 (1967)

3.182 V.I. Goldanskii, R.H. Herber: *Chemical Application of Mössbauer Spectroscopy* (Academic Press, New York 1968)

3.183 P. Gütlich: "Physikalische Methoden in der Chemie: Mößbauer-Spektroskopie I", Chemie in unserer Zeit **4**, 133–144 (1970)

3.184 P. Gütlich: "Physikalische Methoden in der Chemie: Mößbauer-Spektroskopie II", Chemie in unserer Zeit **5**, 131–141 (1971)

3.185 R.L. Mößbauer: "Gammastrahlen-Resonanzspektroskopie und chemische Bindung", Angew. Chem. **83**, 524–534 (1971)

3.186 N.N. Greenwood, T.C. Gibb: *Mössbauer Spectroscopy* (Chapman and Hall, London 1971)

3.187 T.C. Gibb: *Principles of Mössbauer Spectroscopy* (Chapman and Hall, London 1976)

3.188 G.K. Shenoy, F.E. Wagner: *Mössbauer Isomer Shifts* (North Holland, Amsterdam 1978)

3.189 D. Barb: *Grundlagen und Anwendungen der Mössbauer-Spektroskopie* (Ed. Acad. Rep. Soc. Romania, Bucharest 1980)

3.190 W. Müller-Warmuth, H. Eckert: "Nuclear magnetic resonance and Mössbauer Spectroscopy of Glasses", Physics Reports (Rev. Sect. of Phys. Lett.) **88**, 91–149 (1982)

3.191 H. Cremers, B.D. Mosel, W. Müller-Warmuth, G.H. Frischat, V. Braetsch: "^{121}Sb Mössbauer studies of glasses in the system Ge–Sb–Se", Phys. Chem. Glasses **30**, 79–82 (1989)

3.192 H.O. Mulfinger, A. Dietzel, O. v. d. Rhön, J.M.F. Navarro: "Physikalische Löslichkeit von Helium, Neon und Stickstoff in Glasschmelzen", Glastechn. Ber. **45**, 389–396 (1972)

3.193 A.W.M. Wondergem-de Best: *Redox Behaviour and Fining of Molten Glass*, PhD Thesis (Eindhoven 1994) p. 266

3.194 M. Nakashima, H Yamashita, T. Maekawa: "Electrochemical study of Fe ions in alkali borate melts", J. Non-Cryst. Solids **223**, 133–140 (1998)

3.195 B. LaFage, P. Taxil: "Titration of molten soda lime silicate glasses by square wave voltammetry", J. Electrochem. Soc. **140**, 3089–3093 (1993)

3.196 K. Takahashi, Y. Miura: "Application of chronopotentiometry to electrode reaction of metal ions in molten sodium borate", J. Ceram. Soc. Jpn. 87, 95–104 (1979)

3.197 K. Takahashi, Y. Miura: "Chronopotentiometric analysis of various electrode reactions of metal ions in molten sodium borate", J. Ceram. Soc. Jpn. 87, 189–197 (1979)

3.198 T. Berzins, P. Delahay: "Oscillographic polarographic waves for the reversible deposition of metals on solid electrodes", J. Am. Chem. Soc. **75**, 555–559 (1953)

3.199 H. Matsuda, Y. Ayabe: "Zur Theorie der Randles-Sevcikschen Kathodenstrahl-Polarographie", Z. Elektrochem. **59**, 494–503 (1955)

3.200 A. Sasahira, T. Yokogawa: "Ce^{4+}/Ce^{3+} redox equilibrium in Na_2O–B_2O_3 melts by linear sweep voltammetry", Electrochim. Acta **29**, 533–540 (1984)

3.201 A. Sasahira, T. Yokogawa: "Fe^{3+}/Fe^{2+} redox equilibrium in the molten Na_2O–B_2O_3 system by linear sweep voltammetry", Electrochim. Acta **30**, 441–448 (1985)

3.202 A. Lenhart, H.A. Schaeffer: "The determination of oxidation state and redox behavior of glass melts using electrochemical sensors", in *XIV. Int. Congr. on Glass*, Vol. 1 (Indian Ceram. Soc., New Delhi 1986) pp. 147–154

3.203 K. Takahashi, Y. Miura: "Electrochemical studies on ionic behavior in molten glasses", J. Non-Cryst. Solids **80**, 11–19 (1986)

3.204 J.-Y. Tilquin, P. Duveiller, J. Glibert, P. Claes: "High-temperature study of multivalent elements in glass-forming melts: the particular case of iron", Ber. Bunsenges. Phys. Chem. **100**, 1489–1492 (1996)

3.205 J.-Y. Tilquin, P. Duveiller, J. Glibert, P. Claes: "Effect of basicity on redox equilibria in soda silicate melts: An in situ electrochemical investigation", J. Non-Cryst. Solids **211**, 95–104 (1997)

3.206 J.-Y. Tilquin, P. Duveiller, J. Glibert, P. Claes: "Electrochemical behaviour of sulfate in sodium silicates at 1000 °C", Electrochim. Acta **42**, 2339–2346 (1997)

3.207 O. Claußen, C. Rüssel: "Diffusivities of polyvalent elements in glass melts", Solid State Ionics **105**, 289–296 (1998)

3.208 C. Rüssel, G. Sprachmann: "Electrochemical methods for investigations in molten glass, illustrated at iron- and arsenic-doped soda-lime-silica glass melts", J. Non-Cryst. Solids **127**, 197–206 (1991)

3.209 M. Leister, D. Ehrt: "Redox behaviour of iron and vanadium ions in silicate melts at temperatures up to 2000 °C", Glastechn. Ber. Glass Sci. Technol. **72**, 153–160 (1999)

3.210 T.Y. Tilquin, E. Herman, J. Glibert, P. Claes: "*In situ* electrochemical investigation of copper in binary sodium silicate melts at 1000 °C", Electrochim. Acta **40**, 1933–1938 (1995)

3.211 M. Yokozeki, T. Moriyasu, H. Yamashita, T. Maekawa: "Electrochemical studies of the redox behavior of antimony ions in sodium borate and silicate melts", J. Non-Cryst. Solids **202**, 241–247 (1996)

3.212 P. Claes, P. Duveiller, J.Y. Tilquin, J. Glibert: "In situ electrochemical and spectrophotometric investigation of the oxygen pressure dependence of the [Cr(VI)]/[Cr(III)] ratio in a borosilicate melt", Ber. Bunsenges. Phys. Chem. **100**, 1479–1483 (1996)

3.213 J.-Y. Tilquin, J. Glibert, P. Claes: "Anodic polarization in molten silicates", J. Non-Cryst. Solids **188**, 266–274 (1995)

3.214 P. Claes, Ch. Dauby, C. Dupont, L. van Cangh: "Method of and apparatus for monitoring the redox state of elements in glass", US Patent 4 557 743 (Dec. 1985)

3.215 C. Rüssel: "The electrochemical behavior of some polyvalent elements in a soda-lime-silica glass melt", J. Non-Cryst. Solids **119**, 303–309 (1990)

3.216 C. Rüssel: "Polyvalent ions in glass melts", Glastechn. Ber. **63K**, 197–211 (1990)

3.217 C. Rüssel: "Voltammetric studies of the redox behaviour of chalcogenides in a soda-lime-silica glass melt", Phys. Chem. Glasses **32**, 138–141 (1991)

3.218 C. Rüssel: "On-line measurements of redox properties in glass forming melts", Ceram. Trans. **29**, 259–266 (1993)

3.219 C. Rüssel: "Voltammetry in molten glasses", in *Proc. Int. Congr. on Glass*, Vol. 1 (Chinese Ceram. Soc., Beijing 1995) pp. 321–330

3.220 O. Claußen, C. Rüssel: "Quantitative in-situ determination of iron in a soda-lime-silica glass melt with the aid of square-wave voltammetry", Glastechn. Ber. **69**, 95–100 (1996)

3.221 O. Claußen, C. Rüssel: "Voltammetry in silicate and borate glass melts", Ber. Bunsenges. Phys. Chem. **100**, 1475–1478 (1996)

3.222 D. Köpsel: *Modellierung der Läuterung mit Na₂SO₄ unter oxidierenden Be-*
      dingungen, PhD Thesis (TU Bergakademie, Freiberg 1991)
3.223 W.L. Konijnendijk, J.H.J.M Buster: "Raman scattering measurements of
      silicate glasses containing sulphate", J. Non-Cryst. Solids **23**, 401–418 (1977)
3.224 T. Kordon, C. Rüssel: "Voltammetric investigations in Na₂SO₄-refined soda-
      lime-silica glass melts", Glastechn. Ber. **63**, 213–218 (1990)
3.225 A.A. Ahmed, N.A. Sharaf, R.A. Condrate, Sr.: "Raman microprobe inves-
      tigation of sulphur-doped alkali borate glasses", J. Non-Cryst. Solids **210**,
      59–69 (1997)
3.226 J.E. Shelby: *Handbook of Gas Diffusion in Solids and Melts* (ASM Interna-
      tional, Materials Park, OH 1996)
3.227 K. Papadopoulos: "The solubility of SO₃ in soda-lime-silica melts", Phys.
      Chem. Glasses **14**, 60–65 (1973)
3.228 E. Kordes, B. Zöfelt, H. Pröger: "Die Mischungslücke im flüssigen Zustand
      zwischen Na-Ca-Silicaten und Na₂SO₄", Z. Anorg. Allg. Chemie **264**, 255–
      271 (1973)
3.229 M.L. Pearce, J.F. Beisler: "Miscibility gap in the system sodium oxide-silica-
      sodium sulfate at 1200 °C", J. Am. Ceram. Soc. **48**, 40–42 (1965)
3.230 E. Raask, R. Jessop: "Miscibility gap in the potassium sulphate – potassium
      silicate system at 1300 °C", Phys. Chem. Glasses **7**, 200–201 (1966)
3.231 Z. Karch: "Läuterung der Glasmasse durch Anhydrit (CaSO₄) – Theoreti-
      sche Grundlagen des Auftretens von Sulfatgalle", Sprechsaal **118**, 767–773
      (1985)
3.232 I. Barin, O. Knacke: *Thermochemical Properties of Inorganic Substances*
      (Springer, Berlin, Heidelberg 1973)
3.233 R.J. Charles: "Activities in Li₂O–, Na₂O– and K₂O–SiO₂ solutions", J. Am.
      Ceram. Soc. **50**, 631–641 (1967)
3.234 B.A. Shakhmatkin, M.M. Shul'ts: "Thermodynamic properties of glass-
      forming melts in the system Na₂O–SiO₂ between 800–1200 °C", Sov. J. Glass
      Phys. Chem. **6**, 89–94 (1980)
3.235 D.A. Neudorf, J.F. Elliott: "Thermodynamic properties of Na₂O–SiO₂–CaO
      melts at 1000 to 1100 °C", Metall Trans. **11B**, 607–614 (1980)
3.236 S. Yamaguchi, A. Imai, K.S. Goto: "Activity measurement of Na₂O in Na₂O–
      SiO₂ melts using beta-alumina as the solid electrolyte", Scand. J. Metal-
      lurgy **11**, 263–264 (1982)
3.237 C.J.B. Fincham, F.D. Richardson: "The behaviour of sulphur in silicate and
      aluminate melts", Proc. Royal Soc. A **233**, 40–62 (1954)
3.238 S. Nagashima, T. Katsura: "The solubility of sulfur in Na₂O–SiO₂ melts
      under various oxygen partial pressures at 1100 °C, 1250 °C, 1300 °C", Bull.
      Chem. Soc. Jpn. **46**, 3099–3103 (1973)
3.239 F.J. Kohl, C.A. Stearns, G.C. Fryburg: "Sodium sulfate: vaporization ther-
      modynamics and role in corrosive flames", NASA Techn. Rep. TM X-71641
3.240 K.H. Lau, D. Cubicciotti, D.L. Hildenbrand: "Effusion studies of the va-
      porization/decomposition of potassium sulfate", J. Electrochem. Soc. **126**,
      490–495 (1979)
3.241 J.C. Halle, K.H. Stern: "Vaporization and decomposition of Na₂SO₄, ther-
      modynamics and kinetics", J. Phys. Chem. **84**, 1699–1704 (1980)
3.242 K.H. Lau, R.D. Brittain, R.H. Lamoreaux, D.L. Hildenbrand: "Studies of
      the vaporization/decomposition of alkali sulfates", J. Electrochem. Soc. **132**,
      3041–3048 (1985)
3.243 F.W. Krämer: "Ersatz der Antimonläuterung bei Alkali-Erdalkali-Silikatglä-
      sern", Laboratory Report **14** (Schott Glas, Mainz 1994)

3.244 Y. Kokubu, J. Chiba, T. Okamura: "The behavior of sodium sulfate during glass melting process", *Proc. 11. ICG Congress*, Vol. 4, (Prague 1977) pp. 147–154

3.245 S. Manabe, K. Kitamura: "Effect of sodium sulfate and temperature on the fining of float glass", J. Non-Cryst. Solids **80**, 630–636 (1986)

3.246 A.R. Conroy, W.H. Manring, W.C. Bauer: "The role of sulfate in the melting and fining of glass batch", Glass Ind. **47**, 84–89, 110 (1966)

3.247 C. Tanaka, Y. Nakao, Y. Kokubu, T. Mori: "Decomposition behavior of sodium sulfate in two stage melting process (rough melting and fining) for manufacturing of sheet glass", J. Ceram. Soc. Jpn. **94**, 615–620 (1986)

3.248 H.P. Williams: "Einflußdes Oxidationszustandes des Gemenges auf die Glasläuterung mit schwefelhaltigen Läutermitteln", Glastechn. Ber. **53**, 189–194 (1980)

3.249 G. Nölle, M.A. Al Hamdam: "Kohlenstoff in Glasrohstoffgemengen", Silikattechnik **41**, 192–193 (1990)

3.250 W. Simpson, D.D. Myers: "The redox number concept and ist use by the glass technologist", Glass Technology **19**, 82–85 (1978)

3.251 W.H. Manring, R.E. Davis: "Controlling redox conditions in glass melting", Glass Ind. **59**, 13–30 (1978)

3.252 W.H. Manring, G.M. Diken: "A practical approach to evaluating redox phenomena involved in the melting and fining of soda–lime glasses", J. Non-Cryst. Solids **38/39**, 813–818 (1980)

3.253 H. Müller-Simon, K.W. Mergler: "Electrochemical measurements of oxygen activity of glass melts in glass melting furnaces", Glastechn. Ber. **61**, 293–299 (1988)

3.254 H. Müller-Simon, K.W. Mergler, H.A. Schäffer: "Oxygen activity measurements of melts in glass tanks using electrochemical sensors", in *Glass 89, Proc. XV. Int. Congr. on Glass*, Leningrad 1989, Vol. 1a, ed. by O.V. Mazurin (Nauka, Leningrad 1989) pp. 150–155

3.255 Kühnreich & Meixner: "Redox control in glass melts. ZrO$_2$ sensor for continuous measurement in glass production", Advertisement in Glastechn. Ber. Glass Sci. Technol., for instance in **71**(7), V (1998)

3.256 F.G.K. Baucke, R.-D. Werner, H. Müller-Simon, K.W. Mergler: "Application of oxygen sensors in industrial glass melting tanks", Glastechn. Ber. Glass Sci. Technol. **69**, 57–63 (1996)

3.257 A. Lenhart, H.A. Schäffer: "Elektrochemische Messung der Sauerstoffaktivität in Glasschmelzen", Glastechn. Ber. **58**, 139–147 (1985)

3.258 H. Müller-Simon, K.W. Mergler: "Sensor for oxygen activity measurements in glass melts", Glastechn. Ber. **64**, 49–51 (1991)

3.259 F.G.K. Baucke: "Zur elektrolytischen Läuterung", Laboratory Report **60/92** (Schott Glas, Mainz 1992)

3.260 F.G.K. Baucke, T. Pfeiffer: "Verfahren zur Läuterung oxidischer Schmelzen", German Patent 42 07 059 (Oct. 1993)

3.261 C. Schwand, W. Weppner: "Variation of the oxygen exchange rate of zirconia-based electrodes by electrochemical pretreatment", Solid State Ionics **112**, 229–236 (1998)

3.262 K.S. Goto: *Solid State Electrochemistry and Its Application to Sensors and Electronic Devices*, Materials Science Monographs, Vol. 45 (Elsevier, Amsterdam 1988) Chap. 5, pp. 90–124

3.263 J. Richter: "Thermal diffusion in ionic melts", Electrochim. Acta **22**, 1035–1042 (1972)

3.264 W. Jost, K. Hauffe: *Diffusion, Methoden der Messung und Auswertung*, 2nd ed., Fortschritte der physikalischen Chemie (Steinkopff, Darmstadt 1972) Chap. 7, pp. 247–254

3.265 Y. Ukyo, K.S. Goto: "Coupling phenomena in molten iron alloys and slags at high temperature", Tetsu to Hagane **68**, 1971–1980 (1982)

3.266 H. Reuther, J. Wiegmann, W. Hinz: "Thermotransport in Silicatgläsern", Part 1: "Alkalisilicatgläser", Glastechn. Ber. **56**, 19–25 (1983); Part 2: "Untersuchungen an Kieselglas und allgemeine Schlußfolgerungen", Glastechn. Ber. **56**, 47–50 (1983)

3.267 K. Mücke: *Thermotransport in Glasschmelzen*, Diploma Thesis (FH Fresenius, Wiesbaden 1984)

3.268 F.G.K. Baucke, K. Mücke: "Measurement of standard Seebeck coefficients in nonisothermal glass melts by means of ZrO_2 electrodes", J. Non-Cryst. Solids **84**, 174–182 (1986)

3.269 F.G.K. Baucke: "Measurement and significance of standard Seebeck coefficients in oxidic glass-forming melts", in *Proc. XV International Congress on Glass 1989*, ed. by O.V. Mazurin (Leningrad NAUKA, Leningrad 1989) Vol. 2b, pp. 263–266

3.270 W. Oldekop: "Über thermoelektrische Erscheinungen an Gläsern", Glastechn. Ber. **29**, 73–78 (1956)

3.271 D.E. Carlson, C.E. Trzeciak: "Thermoelectric effects in ion conducting glasses", Phys. Chem. Glasses **14**, 10–15 (1973)

3.272 N. Cusack, P. Kendall: "The absolute scale of thermoelectric power at high temperature", Proc. Phys. Soc. (London) **72**, 898–901 (1958)

3.273 C.D. Scholz: *Entwicklung eines Thermogradientenofens*, Diploma Thesis (FH Rheinland-Pfalz, Bingen 1991)

3.274 E. Plumat: "Étude des phénomènes de contact entre verre et oxyde à haute température par les mesures de potentiel électrique", Silicates Industriels **19**, 141–154 (1954)

3.275 J.H. Cowan, W.M. Buehl, J.R. Hutchins, III: "An electrochemical theory for oxygen reboil", J. Am. Ceram. Soc. **49**, 559–562 (1966)

3.276 E.J. Horniak, Jr., P.D. Perry: "Electric forehearth and method of melting therein", US Patent 4 227 909 (1979)

3.277 E.J. Horniak, Jr., P.D. Perry: "Verfahren und Vorrichtung zur Herstellung von bläschenfreiem erschmolzenen Glas", DOS 30 22 091, January 1981

3.278 V.M. Shostak, et al.: British Patent GB 2 175 985, July 1984

3.279 P. Bedroš, J. Štverák: "Studium der Ursachen der Entstehung von Blasen auf einem Platinüberzug in einer Speiserrinne und Maßnahmen dagegen", Sklár Keram. **35**, 142–143 (1985)

3.280 G. Brooks: "Electrolysis under control", Glass **74**, 393–395 (1997)

3.281 A.G. Bossard, E.R. Begley: "Refractory blistering in glass", in *Symposium on Defects in Glass*, Ann. Meeting ICG (Tokyo, Kyoto 1966) pp. 69–81

3.282 P. Bedros, M. Fojtková: "Bubble formation at zircon refractories in Simax glass" (Orig. Czech.), Sklár Keram. **34**, 349–354 (1984)

3.283 F.G.K. Baucke, G. Röth: "Electrochemical mechanism of the oxygen bubble formation at the interface between oxidic melts and zirconium silicate refractories", Glastechn. Ber. **61**, 109–118 (1988)

3.284 F.G.K. Baucke, G. Röth: "Electrochemical mechanism of the oxygen bubble formation at the interface between oxidic melts and zirconium silicate refractories", in *Advances in the Fusion of Glass*, Proc. 1st Int. Conf. on Adv. in the Fusion of Glass, Alfred Univ., Alfred, New York, June 14-17, 1988, ed. by D.F. Bickford et al. (Am. Ceram. Soc., Westerville, OH 1988) pp. 54.1–54.16

3.285 F.G.K. Baucke: "Reaktionsmechanismus aufgeklärt" (Engl. Version: "Reaction mechanism elucidated"), Schott Information **50**, 10–11 (1989)

3.286 F.G.K. Baucke: "Electrochromic applications", Mater. Sci. Eng. **B10**, 285–292 (1991)

3.287 Corhart Refractories, Ceramic Products Div., Corning Glass Works (USA): "Corhart® ZS dense zircon refractory", in *Fiberglass and Specialty Refractories*, 1985, pp. 1.00 ff., 1.01 ff.

3.288 R. Hammerschmidt, H. Hausner: "Elektrische Leitfähigkeit von Wannensteinen vor und nach dem Einsatz in Glasschmelzwannen", Glastechn. Ber. **55**, 30–36 (1982)

3.289 H. Schmalzried: *Chemical Kinetics of Solids* (VCH, Weinheim 1995) pp. 209–233

3.290 D.J.G. Ives, G.J. Janz: "The concept of electrode potential", in *Reference Electrodes. Theory and Practice*, ed. by D.J.G. Ives, G.J. Janz (Academic Press, New York 1961) pp. 3–14

3.291 G. Kortüm: *Lehrbuch der Elektrochemie* (VCH, Weinheim 1957) pp. 241–271

3.292 R. Brdička: *Grundlagen der physikalischen Chemie*, 7th ed. (VEB Deutscher Verlag der Wissenschaften, Berlin 1968) pp. 672–673

3.293 F.G.K. Baucke, G. Röth: "Verfahren und Vorrichtung zum Konditionieren von Schmelzwannenauskleidungselementen aus Zirkonsilikat", German Patent 41 09 652 (1992)

3.294 F.W. Krämer: "Analysis of gases evolved by AZS refractories and by refractory/glass melt reactions. Techniques and results. Contribution to the bubble formation mechanism of AZS material", Glastechn. Ber. **65**, 93–98 (1992)

3.295 F.G.K. Baucke: "Electrochromic mirrors with variable reflectance", Solar Energy Mat. **16**, 67–77 (1987)

3.296 R.S. Crandall, B.W. Faughnan: "Dynamics of coloration of amorphous electrochromic films of WO_3 at low voltages", Appl. Phys. Lett. **28**, 95–97 (1976)

3.297 B.W. Faughnan, R.S. Crandall, M.A. Lampert: "Model for the bleaching of WO_3 electrochromic films by an electric field", Appl. Phys. Lett. **27**, 275–277 (1975)

3.298 F.G.K. Baucke, W.A. Frank: "Conductivity cell for molten glasses and salts", Glastechn. Ber. **49**, 157–161 (1976)

3.299 S. Kropp: *Vereinfachung einer Methode zur Messung der elektrischen Leitfähigkeit von Salz- und Glasschmelzen*, Diploma Thesis (FH Fresenius, Wiesbaden 1980)

3.300 R.-D. Werner: *Elektrochemische Untersuchungen in oxidischen Glasschmelzen. Der Mischalkali-Effekt im flüssigen Zustand*, Diploma Thesis (FH Rheinland-Pfalz, Bingen 1987)

3.301 F.G.K. Baucke, J. Braun, G. Röth, R.-D. Werner: "Accurate conductivity cell for molten glasses and salts", Glastechn. Ber. **62**, 122–126 (1989)

3.302 F.G.K. Baucke, R.-D. Werner: "Temperature-dependent mixed alkali effect in silicate melts", in *Glass 89*, Proc. of the XV. Int. Congress on Glass 1989, Vol. 2a, *Properties of Glass. New Methods of Glass Formation. Techn. Sessions*, ed. by O.V. Mazurin (NAUKA, Leningrad 1989) pp. 242–246

3.303 O. Svensson: "Electrical conductivity of glasses in the composition range of 24% PbO lead crystal", Glasteknisk Tidskrift **35** (1), 5–11 (1980)

3.304 O. Svensson: "Electrical conductivity of glasses in the composition range of 24% PbO lead crystal – complementary measurements, Part II", Glasteknisk Tidskrift **35** (2), 37–40 (1980)

3.305 Th. Pfeiffer, R. Müller, R.D. Werner: "Transport phenomena in oxidic glass-forming melts", Ber. Bunsenges. Phys. Chem. **100**, 1503–1507 (1996)

3.306 K.-P. Müller: "Struktur und Eigenschaften von Gläsern und glasbildenden Schmelzen, Teil I. Elektrische Leitfähigkeit geschmolzener Alkaliborate und -phosphate", Glastechn. Ber. **42**, 1–9 (1969)

3.307 K. Endell, J. Hellbrügge: "Über den Einfluß des Ionenradius und der Wertigkeit der Kationen auf die elektrische Leitfähigkeit von Silikatschmelzen zwischen 1250 und 1450 °C", Glastechn. Ber. **20**, 277–287 (1942)

3.308 J. O'M. Bockris, J.A. Kitchener, S. Ignatowicz, et al.: "Electrical conductance in liquid silicates", Trans. Faraday Soc. **48**, 75–91 (1952)

3.309 C. Kröger, P. Weisgerber: "Zur Bestimmung der elektrischen Leitfähigkeit von Natrium-Silikatschmelzen", Z. Phys. Chem. N.F. **18**, 90–109 (1958)

3.310 C. Kröger: *Das elektrische und Wärme-Leitvermögen von Glasgemengen und Glasschmelzen*, Forsch. Ber. des Landes Nordrhein-Westfalen, No. 863 (Westdeutscher Verlag, Köln 1960)

3.311 K.A. Kostanyan, K.S. Saakyan: "The electrical conductivity of industrial glasses and their tendency towards 'automisregulation' in the melt", Glass and Ceramics **25**, 159–161 (1969)

3.312 K. Matiasovský, V. Danek, B. Lillebuen: "On the frequency- and temperature-dependence of the conductivity in molten salts", Electrochim. Acta **17**, 463–469 (1972)

3.313 J. Stanek: "Probleme der modernen Glasschmelz- und Verarbeitungstechnologie, T. 1", Silikattechnik **25**, 336–339 (1974)

3.314 E.N. Boulos, J.W. Smith, C.T. Moynihan: "Rapid and accurate measurements of electrical resistivity on glass melts", Glastechn. Ber. **56K**, 509–514 (1983)

3.315 R.A. Robinson, R.H. Stokes: *Electrolyte Solutions*, 2nd ed. (Butterworths, London 1968) pp. 87, 91

3.316 A.M. Feltham, M. Spiro: "Platinized platinum electrodes", Chem. Rev. **71**, 177–193 (1971)

3.317 A. Piechurowski: "Method of measuring glass resistance at high temperatures" (Orig. Pol.), Szklo. Ceram. **26**, 2–5 (1975)

3.318 H. Wakayabashi, A. Terai: "Measurement of electrical conductivity for molten salts", Bull. Govt. Ind. Res. Inst. Osaka **35**, 58–61 (1984)

3.319 Landolt-Börnstein: *Zahlenwerte und Funktionen aus Physik, Chemie, Astronomie, Geophysik und Technik*, 6th ed., Vol. 2, Part 1 (Springer, Berlin, Heidelberg 1971) p. 587

3.320 J.O. Isard: "The mixed alkali effect in glass", J. Non-Cryst. Solids **1**, 235–261 (1969)

3.321 D.E. Day: "Mixed alkali glasses – their properties and uses", J. Non-Cryst. Solids **21**, 343–372 (1976)

3.322 R.M. Hakim, D.R. Uhlmann: "On the mixed alkali effect in glasses", Phys. Chem. Glasses **8**, 174–177 (1967)

3.323 J.R. Hendricksen, P.J. Bray: "A theory for the mixed alkali effect in glass. Part 1", Phys. Chem. Glasses **13**, 43–49 (1972)

3.324 J.R. Hendricksen, P.J. Bray: "A theory for the mixed alkali effect in glass. Part 2", Phys. Chem. Glasses **13**, 107–115 (1972)

3.325 R.H. Doremus: "Mixed-alkali effect and interdiffusion of Na and K ions in glass", J. Am. Ceram. Soc. **57**, 478–480 (1974)

3.326 V.N. Filipovich: "Theory of the electrical conductivity of two-alkali silicate glasses and the mixed-alkali effect", Fiz. Khim. Stekla **6**, 369–382 (1980) (Transl. pp. 245–257)

3.327 C.T. Moynihan, N.S. Saad, D.C. Tran, A.V. Lesikar: "Mixed-alkali effect in the dilute foreign-alkali region. Failure of the strong electrolyte/cationic interaction model", J. Am. Ceram. Soc. **63**, 458–464 (1980)

3.328 H. Jain, N.L. Peterson, H.L. Downing: "Tracer diffusion and electrical conductivity in sodium-cesium silicate glasses", J. Non-Cryst. Solids **55**, 283–300 (1983)

3.329 G. Tomandl, H.A. Schaeffer: "The mixed-alkali effect – a permanent challenge", J. Non-Cryst. Solids **73**, 179–196 (1985)

3.330 R. Wäsche, R. Brückner: "The structure of mixed alkali phosphate melts as indicated by their non-Newtonian flow behaviour and optical birefringence", Phys. Chem. Glasses **27**, 87–94 (1986)

3.331 J.M. Hyde, M. Tomozawa, M. Yoshiyagawa: "A comparison of the dielectric characteristics of single alkali and mixed alkali glasses", Phys. Chem. Glasses **28**, 174–176 (1987)

3.332 P. Mazzoldi, A. Miotello: "Mixed alkali effect in glasses: a new model using the thermodynamics of irreversible processes", J. Non-Cryst. Solids **95/96**, 897–904 (1987)

3.333 W.C. LaCourse: "A defect model for the mixed alkali effect", J. Non-Cryst. Solids **95/96**, 905–912 (1987)

3.334 R. Terai, H. Wakayabashi, H. Yamanaka: "Haven ratio in mixed alkali glasses", J. Non-Cryst. Solids **103**, 137–142 (1988)

3.335 G. De Marchi, P. Mazzoldi, A. Miotello: "Ionic conductivity in glass network", J. Non-Cryst. Solids **123**, 321–323 (1990)

3.336 Z. Boksay: "Effect of mixing mobile ions in glasses on transport processes", J. Non-Cryst. Solids **123**, 324–327 (1990)

3.337 M.D. Ingram, P. Maas, A. Bunde: "Ionic conductivity and memory effects in glassy electrolytes", Ber. Bunsenges. Phys. Chem. **95**, 1002–1006 (1991)

3.338 M. Tomozawa: "Alkali ion transport in mixed alkali glasses", J. Non-Cryst. Solids **152**, 59–69 (1993)

3.339 R.M. Wenslow, K.T. Mueller: "Cation sites in mixed-alkali phosphate glasses", J. Non-Cryst. Solids **231**, 78–88 (1998)

3.340 K.A. Kostanyan: "Investigation of the conductivity neutralization effect in fused borate glasses", in *The Structure of Glass*, Vol. 2, Proc. Third All-Union Conf. on the Glassy State, Leningrad 1959 (Consultants Bureau, New York 1960) pp. 234–236

3.341 R.E. Tickle: "The electrical conductance of molten alkali silicates. Part 1. Experiments and results", Phys. Chem. Glasses **8**, 101–112 (1967)

3.342 R.E. Tickle: "The electrical conductance of molten alkali silicates. Part 2. Theoretical discussion", Phys. Chem. Glasses **8**, 113–124 (1967)

List of Contributors

Friedrich G.K. Baucke
Schott Glas, Mainz [1]
retired
(home address:
Kaiserstraße 36,
55116 Mainz
Germany)

John Duffy
University of Aberdeen
Department of Chemistry
Meston Walk,
Old Aberdeen AB9 2UE
Scotland

Dieter Krause
Schott Glas, Mainz [1]
retired

Detlef Köpsel
Schott Glas, Mainz [1]

[1] Hattenbergstraße 10, 55122 Mainz, Germany

Sources of Figures and Tables

We are indebted to the following editors and authors, respectively, for the kind permission to reproduce copyrighted materials.

Material	Source	Original Publisher
Fig. 1.3	[1.87]	Deutsche Glastechnische Gesellschaft, Mendelssohnstr. 75–77, 60325 Frankfurt, Germany
Fig. 2.5	[2.22]	The Royal Society of Chemistry, Thomas Graham House, Science Park, Milton Road, Cambridge CB4 4W, UK
Fig. 2.7 Fig. 2.8	[2.88]	American Ceramic Society, 735 Ceramic Place, Westerville, OH 43081, USA
Fig. 2.12	[2.87]	Springer-Verlag Heidelberg, Berlin, P.O. Box 10 52 80, 69042 Heidelberg, Germany
Fig. 2.17	[2.190]	Elsevier Science Publishers, P.O. Box 103, 1000 AC Amsterdam, The Netherlands
Fig. 2.19	[2.97]	Trans Tech Publications Ltd., Bandrain 6, 8707 Uetikon a.S., Switzerland
Fig. 2.21	[2.55]	Elsevier Science Publishers, P.O. Box 103, 1000 AC Amsterdam, The Netherlands
Fig. 2.22	[2.152]	Trans Tech Publications Ltd., Bandrain 6, 8707 Uetikon a.S., Switzerland
Fig. 2.23	[2.55]	Elsevier Science Publishers, P.O. Box 103, 1000 AC Amsterdam, The Netherlands
Fig. 2.31	[2.54]	American Ceramic Society, 735 Ceramic Place, Westerville, OH 43081, USA
Fig. 2.38	[2.56]	Elsevier Science Publishers, P.O. Box 103, 1000 AC Amsterdam, The Netherlands
Fig. 2.39	[2.55]	Elsevier Science Publishers, P.O. Box 103, 1000 AC Amsterdam, The Netherlands
Fig. 2.41 Fig. 2.42	[2.56]	Elsevier Science Publishers, P.O. Box 103, 1000 AC Amsterdam, The Netherlands
Fig. 2.43	[2.55]	Elsevier Science Publishers, P.O. Box 103, 1000 AC Amsterdam, The Netherlands

Material	Source	Original Publisher
Fig. 2.44 Fig. 2.45	[2.177]	Deutsche Glastechnische Gesellschaft, Mendelssohnstr. 75–77, 60325 Frankfurt, Germany
Fig. 2.46	[2.207]	Society of Glass Technology, 20 Hallam Gate Road, Sheffield S10 5BT, UK
Fig. 2.48	[2.75]	Society of Glass Technology 20 Hallam Gate Road, Sheffield S10 5BT, UK
Fig. 2.50– Fig. 2.53	[2.192]	Elsevier Science Publishers, P.O. Box 103, 1000 AC Amsterdam, The Netherlands
Fig. 2.54 Fig. 2.55	[2.207]	Society of Glass Technology, 20 Hallam Gate Road, Sheffield S10 5BT, UK
Fig. 2.56 Fig. 2.57 Fig. 2.61	[2.188]	Plenum Publishing Corporation 233 Spring Street, New York 10013-1578, USA
Fig. 2.62	[2.87]	Springer-Verlag Heidelberg, Berlin, P.O. Box 10 52 80, 69042 Heidelberg, Germany
Fig. 2.63– Fig. 2.65	[2.219]	Elsevier B.V., P.O. Box 211, 1000 AE Amsterdam, The Netherlands
Fig. 2.67	[2.72]	VCH Verlagsgesellschaft, Pappelallee 3, 69469 Weinheim
Fig. 2.69– Fig. 2.71	[2.106]	American Chemical Society, 1155 Sixteenth Street, NW, Washington DC 20036, USA
Fig. 2.72	[2.294]	Urban & Fischer Verlag P.O. Box 20 19 30, München, Germany
Fig. 2.73 Fig. 2.74	[2.108]	American Chemical Society, 1155 Sixteenth Street, NW, Washington DC 20036, USA
Table 2.1	[2.95]	American Chemical Society, 1155 Sixteenth Street, NW, Washington, DC 20036, USA
Table 2.2	[2.87]	Springer-Verlag Heidelberg, Berlin P.O. Box 10 52 80, 69042 Heidelberg, Germany
Fig. 3.3	[3.26]	Longman Scientific & Technical Edinburgh Gate, Harlow, Essex, UK
Fig. 3.7	[3.31]	Society of Glass Technology, 20 Hallam Gate Road, Sheffield S10 5BT, UK
Fig. 3.8	[3.42]	The Royal Society of Chemistry, Thomas Graham House, Science Park, Milton Road, Cambridge CB4 4W, UK
Fig. 3.10– Fig. 3.12	[3.76]	Society of Glass Technology, 20 Hallam Gate Road, Sheffield S10 5BT, UK

Material	Source	Original Publisher
Fig. 3.14	[3.4]	VCH Verlagsgesellschaft, Pappelallee 3, 69469 Weinheim
Fig. 3.15	[3.103]	VCH Verlagsgesellschaft, Pappelallee 3, 69469 Weinheim
Fig. 3.18 Fig. 3.21	[3.4]	VCH Verlagsgesellschaft, Pappelallee 3, 69469 Weinheim
Fig. 3.23 Fig. 3.24	[3.89]	Deutsche Glastechnische Gesellschaft, Mendelssohnstr. 75–77, 60325 Frankfurt, Germany
Fig. 3.27	[3.55]	Society of Glass Technology, 20 Hallam Gate Road, Sheffield S10 5BT, UK
Fig. 3.29	[3.165]	American Chemical Society, 1155 Sixteenth Street, NW, Washington, DC 20036, USA
Fig. 3.31	[3.229]	American Ceramic Society, 735 Ceramic Place, Westerville, OH 43081, USA
Fig. 3.35	[3.248]	Deutsche Glastechnische Gesellschaft, Mendelssohnstr. 75–77, 60325 Frankfurt, Germany
Fig. 3.36– Fig. 3.38	[3.256]	Deutsche Glastechnische Gesellschaft, Mendelssohnstr. 75–77, 60325 Frankfurt, Germany
Fig. 3.56– Fig. 3.62	[3.283]	Deutsche Glastechnische Gesellschaft, Mendelssohnstr. 75–77, 60325 Frankfurt, Germany
Fig. 3.63 Fig. 3.64	[3.301]	Deutsche Glastechnische Gesellschaft, Mendelssohnstr. 75–77, 60325 Frankfurt, Germany
Fig. 3.65– Fig. 3.68	[3.1]	Deutsche Glastechnische Gesellschaft, Mendelssohnstr. 75–77, 60325 Frankfurt, Germany
Table 3.4 Table 3.6	[3.77]	American Chemical Society, 1155 Sixteenth Street, NW, Washington, DC 20036, USA
Table 3.9	[3.256]	Deutsche Glastechnische Gesellschaft, Mendelssohnstr. 75–77, 60325 Frankfurt, Germany

Index

Printing: Mercedes-Druck, Berlin
Binding: Buchbinderei Lüderitz & Bauer, Berlin